Signals in the Soil

Abdul Salam • Usman Raza

Signals in the Soil

Developments in Internet of Underground Things

Springer

Abdul Salam
Department of Computer and Information Technology
Purdue University
West Lafayette, IN, USA

Usman Raza
Department of Computer and Information Technology
Purdue University
West Lafayette, IN, USA

ISBN 978-3-030-50860-9 ISBN 978-3-030-50861-6 (eBook)
https://doi.org/10.1007/978-3-030-50861-6

This Springer imprint is published by the registered company Springer Nature Switzerland AG
The registered company address is: Gewerbestrasse 11, 6330 Cham, Switzerland

This book is dedicated to our families.

Preface

The farming sector is mainly responsible for the world's food needs; however, land degradation, contamination, poor agricultural techniques and practices, and urbanization are diminishing this resource. Therefore, given the rapidly increasing world's population, soil must be considered to be just as important as other natural resources like water and air. Soil has billions of organisms which initiate chemical, biological, and physical processes required for fiber and food production, growth of the plant, and removal of contaminants from water, hence, an integral component of an efficient farming ecosystem. Soil has many applications in human life: as a foundation of structures that cannot be supported by rocks, as a widely used material in the construction section, as a way to supply antibiotics to fight diseases, and as storage media for important gases (e.g. O_2, CO_2, and CH_3). Moreover, measures should be taken to prevent corrosion, soil movement, groundwater seepage, and other environmental effects to ensure that underground infrastructure (pipelines, tunnels, and basements) can function properly. The U.S. government spends billions of dollars to find and solve groundwater and soil contamination due to hazardous material, e.g., deliberate and accidental chemical spills, nuclear weapons production, pipeline ruptures, etc. Therefore, it is imperative to understand the soil ecosystems so that it can efficiently be used in meeting world's food demands. Moreover, this quote by Franklin D. Roosevelt *A nation that destroys its soils destroys itself* still holds true today.

Many credible national and international organizations like National Science Foundation (NSF), The Directorates for Engineering (ENG) and Geosciences (GEO), the Science and Technology Facilities Council (STFC) of United Kingdom Research and Innovation (UKRI), the Division of Integrative Organismal Systems in the Directorate for Biological Sciences (BIO/IOS), the Division of Computer and Network Systems in the Directorate for Computer and Information Science and Engineering (CISE/CNS) in collaboration with the US Department of Agriculture National Institute of Food and Agriculture (USDA NIFA), the Engineering and Physical Sciences Research Council (EPSRC), the Natural Environment Research Council (NERC), and the Biotechnology and Biological Sciences Research Council (BBSRC) encourage countries to do convergence and utilize underlying soil capabilities to their full extent using sensor and modeling systems. To that end, a collaborative effort from interdisciplinary researchers, scientific community, and funding agencies is required to develop advanced sensors, sensing systems, wireless communication systems, soil models, and cyber systems for complex problem-solving through education and outreach.

Lack of in-situ site-specific measurements of biological, chemical, and physical properties hinders the knowledge of dynamic soil changes. Currently, these properties are being measured by either laboratory methods using soil samples from the sites or soil models to predict soil states. However, these methods are based on limited data and unreliable assumptions. Therefore, it is important to develop sensing systems which provide advancement in detecting spatio-temporal dynamic soil changes. The sensor systems must have the capabilities to communicate wireless data being generated from the sensor, integrate with other available information, and generate analytics to observe spatial and temporal properties of managed and unmanaged soils. It is not possible to detect the soil health through signals, the term "healthy soil" has a proper definition. Therefore, it is important to integrate fundamental science and engineering knowledge to design a capable sensing system. To that end, skills from different departments, e.g., biological, atmospheric, hydrological, biogeochemical, geological sciences, and engineering are combined for assessment and monitoring of functional and sustainable soil. Furthermore, the sensor system will need advanced ground penetration, data transmission, dynamic modeling, data analytics, and visualization tools. The purpose of the research is to develop awareness and understanding about the soil so that it can be managed using new and innovative ways. It will also help in understanding the interaction between soil and the life (e.g., plants, microorganisms, etc.) supported by it.

Signals in the Soil adds new empirical and analytical results to the body of knowledge which enables researchers and industry professionals from the public and private sectors across the world to accelerate research on underground wireless communication, sensing, and networking technologies. It presents diverse and unique next-generation IoT applications in urban and rural areas. It is an excellent book for graduate students, academic researchers, and industry professionals, involved in communication, sensing, agriculture, and bio-systems research.

This book provides an understanding of the most recent developments in Signals in the Soil (SitS), from both the theoretical and practical perspectives. It identifies and discusses technical challenges and recent results related to improving wireless underground communications and sensing in Internet of Underground Things (IOUT). It covers both existing network technologies and those currently in development in three major areas of SitS: wireless underground communications, subsurface sensing, and antennas in the soil medium. It explores new applications of Internet of Underground Things in digital agriculture and urban underground infrastructure monitoring.

Signals in the Soil is an essential reference book for advanced students on courses in wireless underground communications and Internet of Things. It will also be of interest to researchers, communication engineers, system and network planners, technical managers, and other professionals in these fields. There is no book on the market that discusses wireless underground communications and Internet of Underground Things. Currently, this audience gets information about this topic from different sources (e.g., IEEE conferences, COMSOC tutorials, IEEE journals and transactions articles, and web forums). The purpose of this book is to transform information from these scattered sources into a comprehensive and easily accessible knowledge body.

Signals in the Soil is split into four parts: Physical Layer for Wireless Underground Communications, Underground Antenna, and Radio Interface Technologies; Novel Soil Sensing Techniques; Internet of Underground Things Advancements; Applications in Urban Underground Infrastructure Monitoring and Real-time Soil Monitoring and Irrigation Automation. It starts by introducing emerging technologies in wireless communications in the soil medium before moving on to cover propagation models, soil properties, and beyond; beamforming, antenna arrays, and MIMO; capacity and path loss analysis; empirical and statistical channel models in IOUT; underground antenna design; cross-layer and environment-aware protocol design and energy harvesting and power transfer, moving toward the applications in digital agriculture and urban underground infrastructure monitoring; and more. This valuable resource:

- Provides a comprehensive reference for all aspects of Signals in the Soil,
- Focuses on fundamental issues of wireless underground communication and subsurface sensing in an easy language that is understandable by a wide audience,
- Includes advanced treatment of IOUT custom applications of variable technologies in the field of digital agriculture and covers protocol design and energy harvesting,
- Features research developments and open research challenges in subsurface sensing,
- Provides a detailed set of path loss, antenna, and wireless underground channel measurements in novel indoor and field testbeds.

The intended audience for *Signals in the Soil: Developments in Internet of Underground Things* are graduate students, precision equipment technicians, academic researchers, field applicators, industry professionals and managers, dealerships, and digital agriculture sales specialists. There are many such individuals in academia and digital agriculture technology business, which is mostly targeted to the applications of technology in big and small agricultural fields. Moreover, the use of digital agriculture technology in diverse topographic and soil texture fields is also expanding. To effectively utilize enormous data being generated from the agricultural farms, this book has introduced technologies which are useful for data collection, design of decision tools, interpretation, and real-time decision-making in the field.

West Lafayette, IN, USA
August 2020

Abdul Salam
Usman Raza

Contents

About the Authors

Dr. Abdul Salam is an Assistant Professor in the Department of Computer and Information Technology, Purdue University. He holds a Ph.D. from University of Nebraska–Lincoln. He is the Director of the Environmental Networking Technology Laboratory at Purdue University. He has also taught at the Bahauddin Zakariya University, Multan and Islamia University, Bahawalpur. He served in Pakistan Army for 9 years in a number of command, staff, and field roles. He held the principal position at the Army Public School and College, Thal Cantonment. He has won several awards including the ICCCN 2016 Best Student Paper Award, the Robert B. Daugherty Water for Food Institute Fellowship, the Gold Medal MS (CS) on securing first position in order of merit, and the 2016–2017 Outstanding Graduate Student Research Award from the University of Nebraska–Lincoln. He is the author of two books: Internet of Things for Sustainable Community Development and Signals in the Soil. He has published over 50 research articles in major journals and conferences. His recent research focuses on underground soil sensing, wireless communications, Internet of Underground Things in digital agriculture, sensor-guided irrigation systems, and vehicular communications. He has served as an Associate Editor for the IEEE GRSS Remote Sensing Code Library from 2016 to 2018. He serves as the Associate Editor of the Advanced Electromagnetics Journal and Array (Elsevier). Dr. Salam received his B.Sc. and M.S. degrees in computer science from Bahauddin Zakariya University, Multan, Pakistan, in 2001 and 2004, respectively, the M.S. degree in computer engineering from UET, Taxila, Pakistan, in 2012, and the Ph.D. degree in computer engineering from the Cyber-Physical Networking Laboratory, Department of Computer Science and Engineering, University of Nebraska–Lincoln, Lincoln, NE, USA.

Usman Raza is a Ph.D. student in computer and information technology at the Purdue University and a research assistant at Environmental Networking Technology Laboratory under the supervision of Dr. Abdul Salam. His research involves Internet of Underground Things, including the channel model, connectivity, and network protocols.

Part I
Physical Layer for Wireless Underground Communications

Chapter 1
Signals in the Soil: An Introduction to Wireless Underground Communications

1.1 Introduction

Wireless underground communication (WUC) is becoming popular because of its secured deployment methodology, i.e., concealed far below the ground. Underground communication was first observed in World War; however, its use was limited to radio propagation techniques only. V. Fritsch and R. Wundt conducted the experiments, in 1938–1940, to study the propagation of radio waves in underground coal mines using small transceivers deployed below the ground. Although the communication range varied depending upon the nature of the coal; however, they were successfully able to achieve an overall range of upto 1000 feet. In 1942, they conducted another experiment at the depth of 2000 feet; however, the experiments were conducted in 100 feet thick salt mine instead of coal mine. For the salt mine experiment, a battery operated horizontal dipole antenna was used as transmitter and receiver. They performed voice communication using the amplitude modulation. The experiment was performed with extreme care and intelligence to avoid extraneous noise or any other added radio signals at transmitter. It was made sure that no measurable wave existed on the earth surface so that true underground propagation can be studied. Moreover, transmitter and receiver were separated by a carefully planned distance. A range of 15 km, i.e., 9-1/2 miles, was successfully achieved for voice communication. Since then, underground communication has come long way with improvement in methodologies and equipment. This Chapter discusses the potential and challenges of underground communication. The organization the Chapter 1 is shown in Fig. 1.1.

Smart farming [6, 23, 36, 54, 114, 148, 179, 180, 182, 183, 228, 244] is an agricultural management process which exploits the spatio-temporal changes in crop, soil, management, and production with new technologies to improve the farming experience. Smart farming employs large number of wireless devices to sense crop-related data and send this data to a central control room or server center [158, 243]. In recent years, sensing technologies have evolved a lot. These advanced

A. Salam, U. Raza, *Signals in the Soil*,
https://doi.org/10.1007/978-3-030-50861-6_1

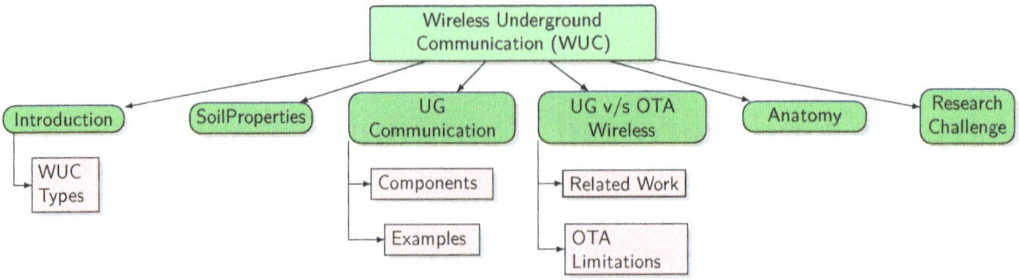

Fig. 1.1 Organization of the chapter

sensing methods are then combined with adaptive input applications (e.g., adaptive application of fertilizers) and soil mapping methods for efficient operation.

In recent years, evolution and advancement in sensing technologies have risen the demand of high data rates and increased communication range. As per the reports of Cisco's visual networking index [28], 11.6 billion devices are predicted to be connected via Internet by 2020. The vastness of this number can be realized by the fact that population of the world is predicted to be 10 billion by 2050, i.e., even less than the predicted number of connected devices by 2020. To fulfill food requirement of such a huge population of the world, it is imperative to utilize smart farming methodologies for a better and cost-efficient crop production through timely decision making and conserving natural resources. To that end, it is important to achieve an ubiquitous connectivity on the farms by using underground wireless communications channel [184, 243, 244].

Wireless underground communications (WUC) applications can be classified into various categories [154, 164, 172]. Some of them, for example, include: environment monitoring, e.g., precision agriculture and landslide monitoring, infrastructure monitoring, e.g., preventing leakage and urban infrastructure monitoring, application for determining location can be helpful in locating people stuck in disaster, and security monitoring applications, e.g., to detect infiltration at border through concealed underground devices. Figure 1.2 shows some of these applications [153].

WUC and conventional wireless networks differ mainly in the communication medium they use. WUC sensor nodes communicate through soil, whereas over-the-air (OTA) terrestrial wireless network uses air as a medium to communicate. The signal propagation in soil is never investigated properly before, in fact, electromagnetic (EM) wave propagation was not even considered a viable option for underground (UG) communication. Therefore, feasible options and solutions are explored to develop a power-efficient UG communications.

There is a lack of detailed wireless channel model because of the challenges experienced in developing a power-efficient UG communication system which also hinders the protocol development in WUC. To that end, existing literature was studied in detail along with a very detailed and time-intensive experiments [195–199]. The results from these experiments were analyzed over a period of 18 months to generalize performance of an UG communication channel. A summary of those results can be found in [191]. It was observed that many soil parameters (e.g.,

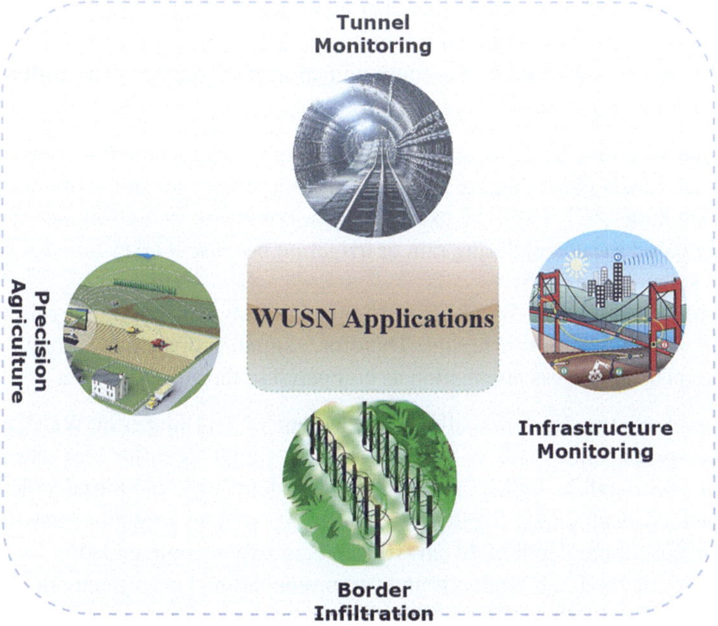

Fig. 1.2 Use of wireless underground communications (WUC) in different areas

soil texture and moisture and irregular soil surface) and antenna parameters (burial depth, antenna design, and operating frequency) have effect on UG communication. It substantiates the fact that performance of an underground channel can highly be affected by the spatio-temporal environmental factors leading to a unique correlation of communication systems, i.e., both information data and communication medium, with environment. Hence, in addition to operational and deployment factors, these parameters should also not be overlooked while analyzing an underground channel.

A wireless underground communications (WUC) model has been developed and presented in [191]. The model focuses on propagation model rather than antenna problem. This WUC model determines the total signal attenuation and the BER (bit error rate) using three-wave components (direct wave (DW), reflected wave (RW), and lateral wave (LW)), dielectric soil properties prediction model, and the signal superposition model. In contrast to existing literature, WUC model captures the gain from the directivity of special antennas instead of simple insulated dipole [182, 183]. However, to avoid over-complication of the model, antennas problem is not considered in this model because of a large number of antennas schemes.

Silva [191] conducted in-situ experiments without considering lateral wave component. However, if lateral waves are also considered along with special antennas, communication range can be increased with same transmitting power. The results obtained from the study helped in designing WUC systems. A strong multi-hop networking solution among the buried nodes can be achieved with long-range (distance >10 m) eliminating the topology dependency of aboveground devices.

In [191], authors have also shown that depth has high effect on communication performance. Through empirical evaluations, they observed that even a small change in depth can degrade the communication performance. The difference in communication performance between topsoil and subsoil is because of:

- **Soil parameters**. Both, topsoil and subsoil, have different soil texture and soil moisture levels [44] which is the reason for the difference in communication in both mediums [6, 7, 195]. For example, topsoil will have more soil moisture level as compared to subsoil during rain or irrigation because it takes time for water to reach subsoil area [225].
- **Soil surface effects**. LW and RW component plays an important role in high signal strength in topsoil region. Therefore, signal propagating through topsoil experiences much less attenuation as compared to the subsoil region.

It is preferred to have a shallow deployment of UG nodes in WUC because of shorter propagation path in the soil causing signal to suffer less attenuation. However, the depth is highly application-dependent, e.g., for intruder detection, recommended deployment depth is 10 cm and sport field irrigation; however, for precision agriculture depth of 40 cm–100 cm is mostly recommended.

Another method of underground communication, not given in Fig. 1.3, is Through-The-Earth (TTE). TTE is applied in areas like military UG communication, geophysical exploration, and mining. It is mainly used to communicate in emergency situations where people stuck in disasters, e.g., miners stuck in mines [16]. WUC & TTE, with all their similarities, face completely different set of challenges (see Table 1.1). For example, a typical depth considered for TTE deployment is very deep (hundreds of meters) as compared to WUC (few centimeters). Therefore, they are considered two different technologies in the literature [98, 178, 179].

It can be seen in Table 1.1 that most of the challenges are related to the physical layer. TTE struggles in traversing rocks with long-range communication, and WUSN struggles in long-range communication through soil. Soil moisture highly affects the subsoil communication [7, 195, 197], therefore, it requires cross-layer approach [7]. Moreover, WUC needs power-efficient nodes buried for long lasting operations.

Relative permittivity of a soil depends upon the signal frequency and volumetric water content (VWC), therefore, signal frequency indirectly affects the strength

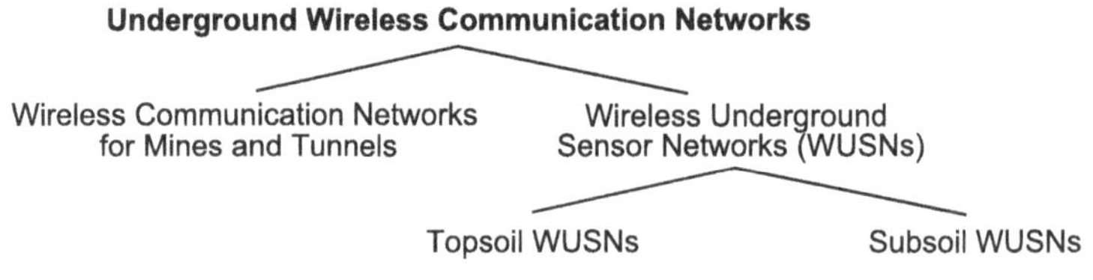

Fig. 1.3 The different types of networking in (WUC) [242]

Table 1.1 Typical aspects for Through-The-Earth (TTE) and WUC scenarios [191]

Aspect	TTE-based communication	WUSN
Frequency range	VLF / LF	VHF / UHF
Maximum range (soil path)	Up to hundred meters	5 cm to dozen meters
Bandwidth	Very small: bps	Small: Kbps
Network topology	One-hop	One-hop and multi-hop
Network density	Sender-receiver or few nodes	Hundred to thousand nodes
Underground channel noise	Very critical aspect	Small impact
Rock penetration	Feasible	Usually not feasible
Soil moisture	Small impact	Very critical aspect
Energy criticality	Relatively small impact	Very critical aspect
Node cost	Relatively high	Small
Communication protocol design	Emphasis on the physical layer	Cross-layer approach

of the signal [18, 29]. In addition to the frequency, soil conductivity also has an effect on signal attenuation. This is contrary to the popular belief that signal is less attenuated under smaller frequencies. Hence, signal attenuation cannot be estimated from soil permittivity only, other soil parameters also contribute to the attenuation [18, 135].

Soil permittivity estimation has been investigated for a specific range of frequencies. All such studies conclude that frequencies around 1 GHz produce reasonable soil permittivity values and are suitable for practical wireless systems under 300 MHz frequencies. However, as the frequency decreases, wavelength of the signal is increased, consequently, increasing the antenna size. Hence, very low frequencies, e.g., less than 300 MHz, are not feasible for WUC. In military WUC, the major requirement is to get longer communication range, e.g., less than 10 km. To that end, military WUC uses HF to LF frequency band filter with huge antennas consuming more power. It is shown that the signal suffers with much less attenuation under UHF bands (300 MHz–3 GHz), and frequencies ranging from 300 MHz–1 GHz [195] which makes them optimal to be used in practical WUC [158, 185].

1.2 Types of Wireless Underground Channel

WUC does not contain only UG nodes. Hybrid WUC is a combination of underground (UG) and aboveground (AG) nodes [6, 173, 180]. As hybrid WUC contains multiple types of devices, it also utilizes multiple type of links for communication between them, i.e., aboveground-to-underground (AG2UG), underground-to-underground (UG2UG), and underground-to-aboveground (UG2AG). Figure 1.4 shows one such hybrid WUC in an agricultural setup where various UG sensors nodes are communicating with each other from soil medium, through UG2UG

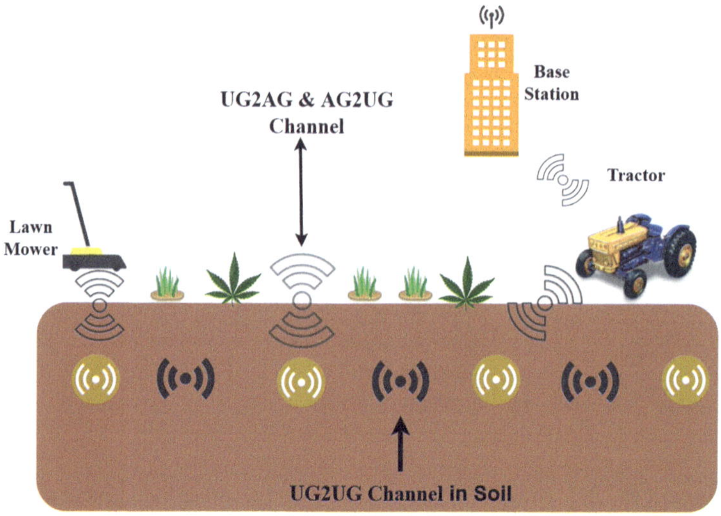

Fig. 1.4 Hybrid WUC architecture

channel, with different AG nodes and vice versa. AG nodes, i.e., agricultural equipment and base station, send data to UG nodes through AG2UG channel. Similarly, UG nodes send data to AG nodes through UG2AG channel. Here, this book focuses on the characterization of WUC UG2UG channel. Moreover, other WUC channels, i.e., AG2UG & UG2AG channels, can be characterized using WUC channel model.

Lateral waves have been extensively used in UG communications [25, 87, 151, 153] and empirically evaluated by Huang [70], King et al. [87], and Vaziri et al. [238]. Special antennas (eccentrically insulated traveling-wave (EITW) antenna) are used for empirical evaluation. Underground lateral wave communication is empirically evaluated through following UG2AG experiment setup: burial depth is 40 cm, aboveground antenna and soil surface were separated by the distance of 55 cm, transmit power level is maintained at 30 dBm, and frequency of 144 MHz is used. The study [87] was successful to achieve longer communication range of 50 m [170].

In [156, 171, 209], authors performed experiment to empirically evaluate UG2AG communication. The experiment setup for this study is given as follows: Terrestrial commodity sensors MicaZ [30] motes were used as UG node, operational frequency of 2.4 GHz, burial depths of 0 cm, 6 cm, and 13 cm, and transmit power level of 0 dBm was used. The experiments were performed in two sets of sender-receiver scenario. Both sets differed in distance between soil and receiver. For the first set, the receiver was kept on the soil (distance = 0 m) and the second set was performed with a distance of 1 m between soil and the receiver. The UG2AG communication was evaluated for two metrics: packet error rate (PER) and received signal strength (RSS). First experiment, with a distance of 0 m, was used as baseline experiment to compare it with the second experiment. It was observed that node buried at 13 m depth was able to communicate at maximum horizontal distance of 2.5 m and node at 6 cm depth achieved a maximum of 7 m horizontal communication range. Hence, it shows that attenuation is inversely proportional to the path covered by the signal in the soil. The study achieved the PER of 10 % [155, 157, 224].

In [22, 152, 161], a uni-directional UG2AG communication model was studied with an effect of reflection dielectric on the signal attenuation. The model is validated through laboratory experiments. The experiment setup was as follows: SoilNet was used as sensors node, operational frequency was 2.44 GHz, transmit power of 19 dBm was used, and sensor was buried at different depth ranging from 5 cm to 9 cm. The strength of the received signal was measured by a soil probe. It was observed that for soil width of 1 cm–7 cm, signal attenuation was increased up to 25 dB. However, 10 dBm of attenuation was observed with 0 %–35 % increase in VWC of the soil [162, 184]. Moreover, bulk density and bulk electrical conductivity had a negligible effect on signal attenuation. The results confirm the empirical results presented by Silva [191].

Tiusanen [227] and Salam [154] proposed a UG2AG communication model using a customized sensor node: *Soil Scout*. Following parameters were used for the experiments: operation frequency of 869 MHz, transmit power of +10 dBm, and an ultra-wide band elliptical antenna [139, 162, 167] was used for underground communication [226] and model validation. The model predicts signal attenuation on the basis of (a) reflection effects of a soil surface, (b) dielectric loss of the soil, and (c) refraction effect of EM waves at soil surface (angular defocusing). It was shown that wideband antenna radiation pattern is independent of soil texture and soil moisture and showed efficient radiation in different soil types with varying soil moisture levels. The study [159, 166, 227] was successful to achieve long communication range of 30 m and 150 m at the burial depths of 40 cm and 25 cm, respectively.

In [146, 163, 168], experiments are performed using customized sensor nodes. The experimental setup was as follows: burial depth was 10 cm, operational frequency of 869 MHz is used, and transmit signal power was +10 dBm. Silva and Vuran [197] performs experiments for evaluating AG2UG and UG2AG communication links. It uses Mica2 motes as sensor nodes, operational frequency and transmitting power are 433 MHz and +10 dBm, respectively. Moreover, they used an ultra-wide band antenna [139] which resulted in significantly improved communication range of 22 m and 37 m at the depths of 35 cm and 15 cm, respectively. Similarly, [199] performs AG2UG and UG2AG experiments with Mica2 motes for precision irrigation application [160, 169].

Although UG2UG communication has been investigated a lot in the existing literature; however, there is still a gap in literature for detailed UG2UG communication channel characterization in subsurface soil region. Only few studies [7, 164, 165, 195] have performed theoretical and empirical analyses of UG2UG communication link. Therefore, a detailed characterization of UG2UG communication channel is presented in the coming sections.

1.3 Underground Communications Overview

Most commercially available solutions use over-the-air (OTA) communication solutions. One of the major challenges in implementing OTA solutions is their unknown environmental impact. WUC uses soil as a medium for wireless underground communications. There are many license-free solutions (e.g., ZigBee, Bluetooth, and DASH7) available for short-range communication. These are used in Industrial, Scientific and Medical (ISM) bands. Recently, FCC has loosened the restriction on using the TV white space frequencies for farms [42] (Order No. DA 16- 307 Dated: Mar 24, 2016). Interference with other licensed band is not expected in this space.

1.3.1 Components of UG Communications

In UG communications, UG nodes are completely concealed. It reduces the operational cost and external impact from the environmental and weather changes [36]. UG nodes can communicate in any one of the two scenarios: (1) communication with devices above the ground termed as aboveground (AG) communication, (2) communication between the UG nodes is termed as underground (UG) communication. Furthermore, soil–air interface affects the AG communication links. Due to interface, these links are not symmetric and must be analyzed for signals propagating in both directions, i.e., UG-to-AG and AG-to-UG. It shows that in order to achieve multi-hop connectivity, a practical distance for UG communication is limited to 12 m. For AG communication, a communication range up to 200 m is possible. If the UG communication medium is soil, it can affect the communication in the following ways:

- **Changes in Soil Bulk Density and Soil Texture:** EM waves attenuate in the soil. Soil is composed of various components such as pore spaces, clay, soil, and silt particles. There can be 12 soil textures depending upon relative concentration of these components [45]. Bound water is the major component responsible for EM waves attenuation in the soil. The amount of bound water varies from one soil type to other. For example, sandy soil has less bound water from silt loam and silty loam, hence, it suffers from lower attenuation. Similarly, medium textured soils hold more water than coarse soils because of lower pore size.
- **Volumetric Water Content (VWC) of Soil:** The effective permittivity of a soil is a complex number. Therefore, in addition to diffusion attenuation, EM waves suffer attenuation due to absorption of water content by the soil [34, 36, 193]. Dielectric spectra conductivity of the soil is dependent on VWC or soil moisture. For a dry soil, dielectric constant is in the range of 2 and 6 and conductivity is in the range of 10^{-4} S/m–10^{-5} S/m. For a near-saturation level soil, range of dielectric constant is 5–15 and that of conductivity is in the 10^{-4} S/m to

10^{-5} S/m [235]. Coherence bandwidth of UG channel is a few hundred kHz [173, 175, 176]. Coherence bandwidth changes with the change in SM which makes the designing process more challenging.

- **Distance and Depth Variations:** EM waves attenuation also depends upon travel distance of the signals. WUC sensors are normally buried in the top sub-meter layer. Therefore, received strength of the signal varies with the distance and depth of antennas. In WUC, sensors are buried in both, subsoil and topsoil layers [197, 199]. Burial at higher depth results in higher attenuation [173].
- **Antenna in Soil:** Return loss of a buried antenna varies due to high permittivity of soil [35]. Change in soil moisture levels changes soil permittivity which in turn causes variations in return loss. Resonant frequency is shifted to lower frequency spectrum due to change in return loss. Moreover, achieving high overall system bandwidth also becomes challenging for UG communications.
- **Change in Frequency:** The path loss due to attenuation is frequency dependent [33]. High frequencies suffer high attenuation because of increased water absorption. The EM waves in soil have shorter wavelength as compared to EM waves in the air because of higher permittivity of the soil. Channel capacity in soil is also determined by operation frequency [35].
- **Lateral Waves:** Underground nodes communicate with each other using any one of the three major paths: direct, lateral, and reflected (LDR) waves [37, 175, 176, 181]. Direct and reflected waves are most effected by above-mentioned challenges because their complete travel path is through the soil. On contrary, lateral waves can travel along soil–air interface in air, hence, they experience lowest attenuation among all. Therefore, lateral waves are the most important component to consider while extending the UG communication range.
- **Developments in WUC:** UG communications have evolved a lot since its inception. A lot of work has been done in characterization of UG channel and cross-layer communication solutions are proposed to get long communication range and achieve high data rate. In [181], authors capture and analyze impulse response of UG channel through detailed experimentation.

A total of 1500 UG greenhouse testbeds has been developed to analyze the effect of soil moisture and soil texture on wireless UG communication channel. These experiments helped in developing main characteristic of wireless UG channel impulse response such as root mean square delay spread, coherence bandwidth, and power of multi-path components. It also validates main components of UG channel, i.e., direct, lateral, and reflected waves. The coherence bandwidth decreases with the increase in distance in soil, e.g., it is shown in [181] that a coherence bandwidth of less than 1.15 MHz can decrease further upto 418 kHz, if distance is increased for more than 12 m in soil [181]. Root mean square delay spread is affected by the soil moisture and it should adapt to change in soil moisture values. In [183], an important statistical model for UG multi-carrier communication and soil moisture adaptive beamforming is given for WUC solutions.

1.3.2 Examples of UG Wireless Communication Systems

WUC is being used in many applications: border patrol, precision agriculture, and environment monitoring. WUC mainly consists of two components: sensors and communication devices. These components are either completely or partially buried in the soil. WUC aims to provide real-time soil monitoring and sensing. In precision agriculture, WUC is mainly used for sensing and monitoring of soil and other related physical properties [6, 23, 36, 54, 57, 68, 84, 114, 173, 177, 182, 214, 226, 228]. The WUC is also being used to implement border monitoring for stop border infiltration [7, 218]. Other monitoring applications of WUC include pipeline monitoring and landslide monitoring [54, 214, 217].

Another important component of WUC is the wireless communication. There exist few models in the literature which represents underground communication. Underwater communication [19, 138] has same challenging medium as of underground communication. However, for underwater communication, acoustic waves [19] are used instead of EM waves due to very high attenuation of EM waves in the water. Acoustic propagation has its own disadvantages such as low quality of physical link and higher delays because of low speed of sound, extremely low bandwidth, challenging deployment, and size and cost of equipment. These disadvantages restrict the use of acoustic methods for WUC.

1.4 Why UG Wireless is Different from Over-the-Air Wireless?

Wireless underground communications with magnetic induction (MI) have also been studied in [93, 105, 114, 213, 219, 222]. However, signal strength of MI-based solutions attenuates with the inverse cube factor and suffers from very low data rates. MI communication is also dependent on relative position of receiver and sender as it cannot communicate if both receiver and sender are perpendicular to each other. Furthermore, long wavelength of the magnetic channel does not allow network to scale. These disadvantages and inability of communicating with the aboveground devices do not make MI solutions a feasible option for WUC.

Some literature [34, 241] has given UG channel models without empirical validation. Integration of WUC with precision agriculture cyber-physical systems and center-pivot systems is given in [36]. Underground channel is empirically evaluated in [193, 197]; however, they did not consider the antenna bandwidth for evaluation. A 2-wave path loss model is developed in [241], without considering the lateral wave component. Path loss prediction model has been proposed in [22]; however, they did not considered underground communication. In [214], authors present an underground communication model for mines and road tunnels. However, it cannot be applied to WUC due to difference in wave propagation mechanism in tunnel and soil. A model is proposed by Dong and Vuran [34] for closed-form path loss with lateral waves but this simple model cannot capture statistics and impulse response of the channel. Salam [182] presents the detailed characteristics of coherence bandwidth of the underground channel.

There is no detailed discussion about the channel capacity in the literature. Capacity of single-carrier underground communication channel has been discussed in [35]. This discussion, however, does not consider a practical modulation scheme and does not perform the empirical validation. In [173], the authors analyze the capacity of multi-carrier modulation in underground channel using empirical values of coherence bandwidth, channel transfer function, and return loss of antenna. They used three different types of soil and under varying levels of soil moisture conditions.

WUC antennas are different from traditional antennas used for OTA communication because of deployment in soil. In 1909, Sommerfeld's seminal work [208] laid the foundation of study of EM waves propagation. For the complete twentieth century, EM wave propagation in subsurface stratified media and effect of medium on EM waves have been investigated thoroughly in many works [12, 17, 20, 38, 62, 120, 202, 221, 245, 250]. These studies use infinitesimal dipole of unit electric moment for analysis of electromagnetic fields. However, it is desirable to use finite size antenna with already known field pattern, current distribution, and impedance for practical purposes. Field calculations and dipole numerical evaluations for lossy half space were first studied in [128]. In [245], authors extensively analyze the propagation of EM wave along the interface. However, this work does not apply to underground buried antenna. Buried dipole was analyzed in lossy half space in [120]. The authors presented the ground wave attenuation factor of far-field radiation from UG dipole and depth attenuation factor using two vector potentials. However, it does not consider the current reflected from the soil–air interface. In [17], authors calculate the field component per unit dipole using Hertz potential. The difference between the study in [120] and [17] is that the former ignores the displacement current in lossy half space. Authors in [221] give the Hertzian dipole in an infinite isotropic lossy medium. EM fields are improved by considering lateral waves and half space interface in [88, 253].

Studies in [47, 86] analyze antennas in a matter where antennas EM fields have been theoretically derived in half space and infinite dissipative medium. These analyses assume perfectly matched dipole antennas, hence do not consider the return loss. Relative gain expressions of underground antennas are developed in [62, 250] without empirical results. The impedance of dipole antenna inside the solutions is evaluated in [75]. It discusses the effect of antenna depth, dipole length, and solution's permittivity. However, this work cannot be used in WUC because of difference between soil and solutions permittivity. Moreover, it does not consider change in permittivity occurring because of soil moisture. Kesar and Weiss [82] studies the communication between the buried underground antenna without considering orientation and impedance of antenna. Another work [43] conducts the performance analysis of four antenna buried in refractory concrete. In this work, the transmitter is buried at 1 m depth and author does not consider the concrete–

air interface. Castorina [26] analyzes circularly polarized patch antenna. It does not consider the interface effect and antenna is buried at 3 cm depth in concrete.

Current WUC applications and experiments calculate the soil permittivity by using soil dielectric model [7, 135] which evaluates to actual wavelength used for the antenna design [227]. In [227], an WUC-based elliptical planar antenna is designed. It, using the same frequency, compares the antenna wavelength in soil and air to determine the size of the antenna. However, this methodology lacks in providing impedance match. Zemmour et al. [258] present results from the experiments on Impulse Radio Ultra-Wide Band (IR-UWB) WUC without considering the effect of soil–air interface. Tokan et al. [231] design surface-based lateral wave antenna and does not consider the underground scenarios.

Impedance change in soil causes disturbance. This is similar to the disturbance caused by impedance change of hand-held device in close proximity with human body [24, 229] or that by devices which are implanted in the human bodies [32, 52]. Experimental results obtained from these applications show that the human body contributes to performance degradation of antenna. Even these studies are similar, they still cannot be used in WUC because of the difference between the permittivity of soil and human bodies. Permittivity of human body is greater than the soil. Moreover, permittivity of human body is static, whereas soil has varying permittivity mainly dependent upon the moisture. For example, at frequency of 900 MHz, human body has permittivity of 50 [229] and that of soil with 5 % moisture is 5 [135].

Beamforming has been investigated for over-the-air wireless channel [11, 13, 39, 101, 127, 143, 251] and MI power transfer [92]. However, there exists no work in the literature on UG beamforming. Using beamforming, lateral components [86] in UG communications can go to the longer distance which is normally limited to 8 m–12 m owing to high level of attenuation suffered because of soil [182].

There has been discussion on soil permittivity and soil moisture in the literature. Here some of those techniques are discussed for comparison purpose. This comparison will highlight the difference and similarities between different techniques. Some of the method used for quantifying water content in the soil includes: gravimetric method, GPR, TDR, capacitance probes, hygrometric techniques, tensionmetry, nuclear magnetic resonance, resistive sensors, gamma ray attenuation, electromagnetic induction, remote sensing, neutron thermalization, and optical methods.

Firstly, techniques which are used in laboratory for the soil properties estimations are discussed laboratory based. Authors in [66] uses soil density, soil moisture and frequency to derive EM parameters of the soil. The model restricted soil moisture weight to 20 % and it need rigorous methods of sample preparation. Authors in [31] develop a probe-based lab equipment which uses vector network analyzer (VNA) in the frequency range of 45 MHz–265 MHz. In [247], a model for estimating a dielectric permittivity of soil is developed on the basis of empirical evaluation. Authors in [33] develops dielectric permittivity model for frequencies greater than 1.4 MHz. Peplinski in [135] modifies this model to work in the frequency range of 300 MHz–1.3 GHz. A detailed survey for soil permittivity estimations is given in [31]. All of these methods are performed in laboratories and required soil sample from the site. Collecting soil sample from the soil is very labor intensive and do not represent the in-situ soil conditions. Therefore, it is required to develop automated techniques for monitoring the soil moisture.

Another approach of measuring soil properties is given in [125]. It is based on TDR and requires refractive index and impedance of soil. Toro-Vazquez et al. [232] propose a technique to estimate EM properties of soils for detecting Dense Non-Aqueous Phase Liquids (DNAPLs) hazardous materials using Cross-Well Radar (CWR). This technique transmits wideband pulse waveform in the range of 0.5 GHz–1.5 GHz. It also calculates soil permittivity with transmission and reflection simulations in dry sand. The well-explained survey on measurement of time-domain permittivity in soils is presented in [239]. For TDR-based approaches, it is required to install sensors at each experiment location. However, in order to make effective decisions in agriculture, real-time soil moisture sensing is the primary requirement.

Many studies have been proposed to investigate antenna related soil properties. An attempt to measure electrical properties of earth using buried antenna has been proposed in [205, 206]. However, this method requires measuring the input reactance for obtaining electrical parameters of the material, and length of antenna is also required to be adjusted to get zero input reactance. Solimene et al. [207] use Fresnel reflection coefficients to estimate GPR-based soil permittivity with soil antenna. However, they do not provide empirical validation and also require a complex time-domain analysis. In [21], dielectric properties of soil are presented using wideband frequency domain and frequency range of 0.1 GHz–1 GHz. It uses impedance measurement equipment (LCR meter) and VNA. In [122, 249], complex dielectric properties of soil are measured using frequency domain method which requires placing soil in the probe.

Soil moisture and permittivity can also be measured by using GPR method. Hislop [67] estimates ground permittivity by correlating ground dielectric properties with cross-talk of early-time GPR signal. However, GPR method requires calibration and works only for shallow depths (0–20 cm). Furthermore, soil moisture measuring technique cannot be limited to a certain burial depth.

Another important method of measuring soil moisture is remote sensing. Remote sensing has a high range of measurement [236] and is sensitive to soil water content [78]. There are two major types of remote sensing: active and passive. Passive remote sensing [85] has low spatial resolutions which can be improved by active remote sensing technologies; however, active methods limit the soil moisture readings to few centimeters of the topsoil which highly affect the readings taken [204]. Table 1.2 summarizes the existing work done in WUC.

1.4.1 Limitations of Over-the-Air Wireless in Soil

There are many research challenges facing by the development and widespread of WUC. These challenges must properly be investigated. A centralized networking solution for WUC can be classified into two architectures: (1) One with only buried

Table 1.2 Summary on WUC systems

WUC systems		
Technology specific	MI-based	[89, 90, 93, 93, 100, 105, 114, 213, 215, 219, 220, 222] [55, 104, 106, 233]
	EM-based	[64, 109, 110, 187, 203, 246, 261] [5, 51, 194, 200, 241, 256, 256] [40, 53, 69, 76, 97, 157, 174, 176, 186, 211, 257, 259]
	Acoustic based	[3, 49, 83, 124, 129, 189] [4, 46, 58, 79, 134, 248] [9, 113, 201, 210, 216, 237, 255]
	Channel modeling	[111, 112, 192, 212, 213, 222, 254]
	Wired	[10, 15, 65, 81, 99, 118, 140, 188, 260, 262]
Application specific	Agriculture	[46, 53, 69, 129, 189, 194, 201, 246, 255, 257, 261] [61, 154, 171, 174, 176, 181, 259, 263]
	Drilling and telemetry	[4, 9, 49, 58, 59, 63, 73, 74, 77, 113, 113, 124, 134, 142, 145, 248]
	Oil and gas	[5, 63, 96, 103, 107, 149, 187, 188, 260, 262]
	Irrigation	[37, 117, 123, 126, 130–132, 263]
	Mining, monitoring and tracking	[1, 2, 15, 71, 72, 81, 94, 99, 114–116, 188] [83, 119, 121, 144, 170, 210, 216, 223, 237, 264]

UG nodes communicating with the AG node using UG links and (2) Hybrid WUC employing both UG and AG nodes (static and mobile) to communicate through UG and OTA links [6, 199]. Apart from OTA links, UG2AG and AG2UG links are also being used extensively. Therefore, multi-hop networking involving UG2UG links must be investigated in detail.

A detailed analysis of UG2UG communication must be performed to address the WUC challenges. Although all challenges cannot be solved owing to the challenging environment of WUC; however, identifying and proposing solutions for the major challenges is also an important contribution to its development. To that end, the WUC research challenges are discussed below:

A. Antenna problem. A radio communication can be analyzed theoretically in two phases: (1) the antenna problem and (2) the propagation problem. WUC model is an underground propagation model. A dipole antenna with an ideal isotropic radiation pattern can guarantee high accuracy with combination of generic antenna gains and initial decays. However, with unavailability of ideal antennas, more practical approach would be to introduce specialized antenna factor for DW, RW, and LW to achieve more accuracy. Furthermore, conducting empirical investigations using large number of different values for burial depths, transmit power levels, and VWC, can precisely address the antenna problem.
Underground channel modeling with antenna problem is a very complex task. The complexity level increases manifolds even if one component of antenna problem is considered for modeling. To understand this issue, consider an

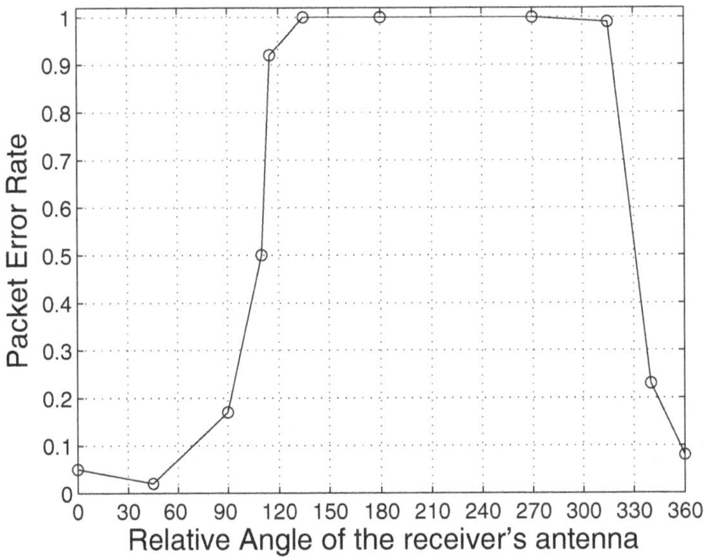

Fig. 1.5 Effects of the VWC on the ratio between antenna's length and wavelength of the signal [198]

example of the radiation pattern of the antenna and its implied directivity gain. Figures 1.5 and 1.6 show how VWC impacts the radiation pattern of an antenna. First, change in VWC changes the signal wavelength in soil which will also change the ratio between the signal's wavelength and antenna's fixed length (17.3 cm). The given values are from Mica2 mote (a 1/4 monopole antenna) antenna operating at 433 MHz. The ratio $\frac{\text{length}}{\text{wavelength}}$ considers two times of Mica2 antenna length, e.g., 34.6 cm, and half the wavelength of signal in soil or air. The two-fold increase in length is mandatory because a 1/4 monopole antenna is same as 1/2 dipole antenna with ground structure representing half of the antenna. Therefore, 1/2 ratio for a half-wave dipole is shown for the comparison. VWC causes decrease in wavelength which in turn increase the length-wavelength ratio.

Figure 1.6 plots the elevation pattern of a linear dipole antenna (oriented vertically) with length measured in terms of wavelength [234]. The change in ratio ($\frac{\text{length}}{\text{wavelength}}$) (Fig. 1.5) is represented using different radiation pattern (Fig. 1.6). VWC causes increase in ratio making the radiation pattern behavior monotonous.

The antenna problem differs with type of antenna and orientation of antenna and should be addressed accordingly for each antenna scheme. However, all antenna schemes are not suitable for underground communications. Therefore, it is recommended to identify antenna schemes which can improve the performance of UG2UG, UG2AG, and AG2UG links to support WUC channel model with adding more antenna models. A possible solution is suggested along with the

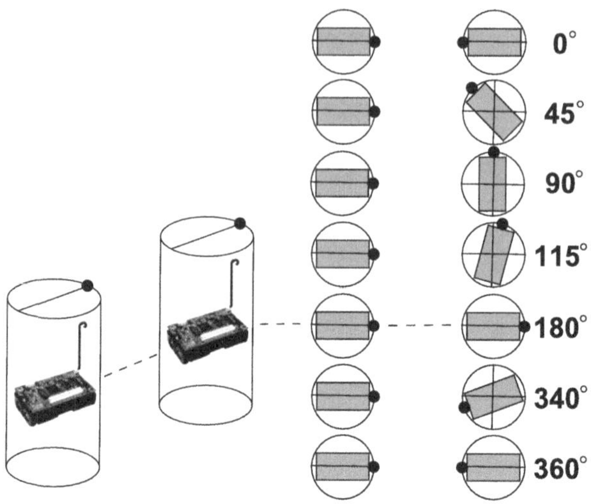

Fig. 1.6 Volumetric water content effecting radiation pattern monopole antenna [198]

results in [70, 87, 197, 199, 227, 238]. It uses an ultra-wide band antenna for UG2AG and AG2UG links and traveling-wave antenna to study lateral wave propagation in UG2UG links. An empirical investigation must be done to evaluate the solutions for different depth and transmit power level.

B. **Burial depth.** In WUC model, burial depth can be defined as the distance between antenna center and soil surface. The existing results show a strong correlation between depth and communication performance. Hence, burial depth of sensors and radio modules has no effect on the model but antenna's depth does. Adjusting to optimal depth can significantly extend the communication range along with a high-power transceiver. There are also some design constraints in WUC which cannot be violated, e.g., in crop irrigation, nodes must be below the topsoil region where plowing happens.
The challenge is to deploy antenna in topsoil such that they are not affected by the mechanical activities in their vicinity. One solution is installing and removing nodes during such activities; however, it will increase the deployment cost. Apart from the increased cost, installation and calibration of soil sensors are also a time taking process. In some scenarios, where sensor(s) and processors are permanently fixed in subsoil, easy installation/removal is only possible for communication module near to soil surface (see Fig. 1.7). In such cases, sensors are fixed and only removable component is the long-range communication module. This module requires a short-range transceiver (with deeply buried sensor nodes) and a transceiver which enables communication between aboveground devices and other long-range modules. There is a need to investigate optimal values of burial depth (including $d_{bg} = 0$) for such long-range module.

C. **Housing for the sensor nodes.** In some WUC, concealment of sensor nodes is more important than the high depths. One solution is to use plastic boxes which

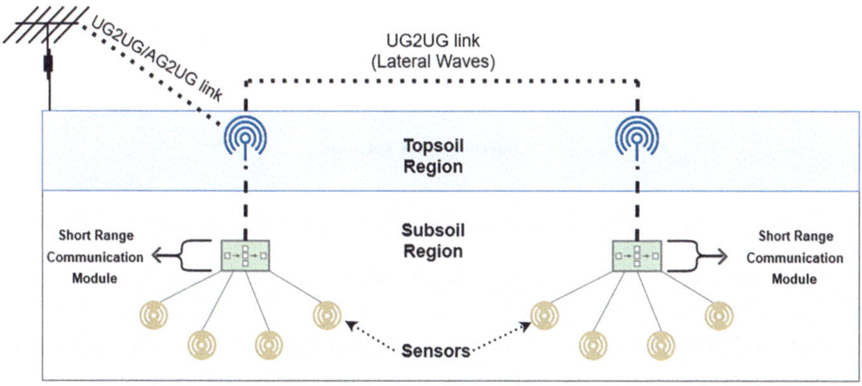

Fig. 1.7 The WUC deployment

can conceal processor, communication module, and antennas. However, it has never been investigated in detail for UG2UG communication and preliminary experiments show completely different effects on communication performance. A scenario using stratified media (air/soil) must also be analyzed for UG2UG links in WUC.

D. Direct and reflected waves. So far, communication through lateral waves has been presented as a power-efficient solution to achieve a long-range UG2UG communication. WUC model can be converted into a simple LW model. However, it is not recommended to do so, because the short-range communication is mainly based on DW (Fig.1.7). Some components of WUC model can also be used in development of UG2AG/AG2UG channel models. Inter-node distance can be increased using directional antennas and high-power transceiver.

E. Lateral waves. There is a need of detailed empirical and theoretical evaluation of lateral wave propagation for UG2UG links in WUC. The results discussed are highly limited by the power-efficient transceiver and antennas. Special antennas and high-power transceivers must be used to achieve long-range communication. It will contribute towards complete validation of WUC model. Effect of using terminated traveling-wave antennas needs to be studied. These antennas were used for underground communication previously [70, 87, 238]. Therefore, these studies can be re-investigated for a typical WUC scenario with modified deployment parameters. The power requirements of multi-hop LW/UG2UG technique and centralized one-hop UG2AG/AG2UG must be studied in detail to give extremely important power related guideline for developing WUC.

It is also important to study the impact of snow, water, and obstacles in surface on UG2UG links communicating using lateral waves. The results from such studies can further complement WUC model. These studies can be used for security purposes, e.g., detecting intruder in border patrol application. The detection process uses the disturbance of wireless channel (Fig. 1.8).

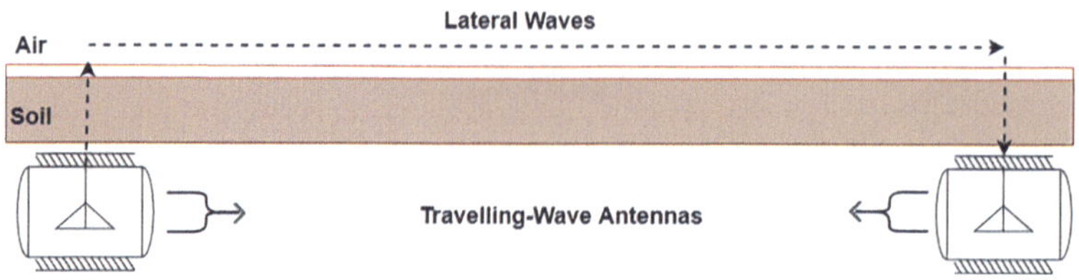

Fig. 1.8 Lateral waves can potentially be applied in security applications for WUC

F. **UG2AG and AG2UG channel models.** A detailed channel model for UG2AG and AG2UG links must be developed. There exists no generic model which can be applied to all WUC. There are some preliminary empirical investigation done by Silva and Vuran [197, 199]; however, an in-depth theoretical analysis is still needed. Overall energy consumption requirement for such solution also needs to be investigated. Lateral wave propagation already has its application in UG2UG links. However, a comparative study for the power budgets of multi-hop LW/UG2UG approaches and centralized one-hop UG2AG/AG2UG should be done.

1.5 Anatomy of a WUC Module

The underground nodes currently used in WUC testbeds suffer from several shortcomings. These shortcomings lead to reduced communications performance in WUC, reduced experimental effectiveness, and higher costs. To address these faults, there is need of nodes tailored to WUC. The following capabilities are desirable in these nodes [191, 252]:

Environmental Factors The current generation of WUC nodes is designed to support academic research being conducted primarily in a laboratory setting. Hence, the experiments do not consider many feature of uncontrolled outdoor environments. First, the WUC nodes cannot be reprogrammed without interfacing to a special hardware board. If the devices are to be reprogrammed in the field, they must either be dug up, or each mote should be deployed with an additional hardware programming board. Digging up the WUC nodes is a time consuming and a difficult process. Deploying additional underground hardware, to reprogram the WUC nodes, is expensive and can further complicate the deployment process [191, 252].

Secondly, remote charging of the nodes is not possible. If a node's battery ends during an experiment, a buried node must be dug up for the battery replacement. It is an extremely time-consuming operation, and the performance of an experiment may be suboptimal until the node is replaced [164, 169].

Propagation While the current experiments demonstrate the viability of WUC, the performance could be further enhanced by tailoring the radio of the mote as per the requirement of the underground networks. The radios of the current WUC nodes are designed to communicate over-the-air. The parameters of the radios are not well matched to the WUC environment in terms of transmit powers and frequencies. The existing WUC nodes can be modified to better match the desired parameters; however, it is not as effective as choosing a radio specifically matched to the needs of a WUC node [162].

Sensing The sensor packages that can be deployed with the current generation of WUC nodes do not collect all the information desired for an underground environment, or contain many extra sensors that are not useful for WUC. These added sensors increase the cost of deploying experimental testbeds.

All of these areas can be improved by using a node designed specifically for WUC. To address these challenges, a WUC node should be designed to operate on a limited power reserves, monitor the underground environment, and communicate the results to aboveground nodes. The design of the different desirable aspects of a WUC node is given below [191, 252].

(i) **Transmitter/Receiver.** A radio should have a high transmit power and be able to operate on a variety of sub-1 GHz frequencies that are suitable for WUC [160]. The radio implementation can be modified to meet the specific requirements of the antennas and RF environment of WUC-application. It will increase the transmission range and capabilities of a radio device.
(ii) **Microcontroller.** The microcontroller should be able to provide enough processing power [182]. One such example of a microcontroller is MSP430 which is extremely energy efficient and also extends the lifetime of the deployed sensors. The MSP430 can also interface to a variety of sensors, communication, and storage devices.
(iii) **Sensors.** The WUC node should contain a built-in accelerometer and temperature probe with an ability of interfacing with an external soil moisture sensor. The combination of multiple sensors enables a node to accurately measure the characteristics of the underground environment. These measurements can help the radio to adapt to its environment in real-time. Accordingly, the sensor readings can be used to assess the viability of energy harvesting through kinetic vibrations [173].
(iv) **Data Repository.** WUC nodes should have an on-board micro-SD card for storage. This large storage space can be used to store extensive sensor readings for a long-term monitoring of the underground environment. By adding a large storage capability, the system can sense at a much higher rate than it can transmit information. After an extended deployment, the information from nodes can be recovered, and a highly detailed model of an underground environment can be developed from the stored sensor readings [183].
(v) **Energy.** WUC node should support a variety of energy sources with energy harvesting and external power transfer support that enables the system to sense at higher rates and operate for longer periods of time than the current

generation of WUC nodes [191, 252]. Moreover, the nodes should also support recharging through a USB cable accessible from aboveground after the node has been deployed. Accordingly, the device can be recharged quickly in the field without removing and re-deploying a node in the testbed. The mote can also be enhanced with kinetic energy harvesting capabilities that will further increase the lifetime of the WUC nodes.

1.6 Research Challenges

The development in WUC has extended the research possibilities and brought some research challenges as well. Therefore, this section presents the research challenges in this area. Moreover, Table 1.3 shows the importance of these challenges in the different WUC applications.

Deployment

Deployment is a major issue in WUC applications because of the harsh underground environment [95]. The underground smart objects can easily be damaged by the aboveground activities, i.e., digging, plowing, harvesting. Therefore, node deployment is very difficult in WUC as compared to the terrestrial networks. The objects with high energy requirement should be deployed near to the surface so that frequent battery replacement can be done easily. High capacity batteries and power saving protocols can also be used to meet the requirement of high energy nodes. The deployment challenge become relatively severe in WUC applications such as seismic and Oil and gas exploration because of higher depth. Therefore, in [215], a MI-based WUC is used with managed and organized orientation of coils to minimize the power reflection. Kisseleff et al. [90] 6+reduce the complexity by using different deployment strategies (horizontal and vertical). One important

Table 1.3 Research challenges for IOUT applications [150]

Research challenge	Agriculture	Seismic exploration	Oil and gas
Deployment	Medium	High	High
Channel modeling	Medium	Medium	High
Transmission range	Low	High	Medium
Latency	Low	Low	Medium
Reliability	Low	Medium	High
Security	Medium	High	High
Scalability	Low	Medium	Medium
Robustness	Low	Medium	High
Networking	High	Medium	Medium
Cloud computing	High	Medium	Low
Fog computing	Low	Medium	High
Localization	Medium	High	Medium

issue to consider is the path loss occurring due to heterogeneous nature of soil. Unfortunately, there is a very limited work on efficient WUC deployment which aims to solve this challenge along with the consideration of different operational parameters [157, 184, 186].

Channel Modeling
The EM signal attenuation is much higher in soil as compared to the terrestrial networks [194]. The major factors contributing to high attenuation loss are the soil permittivity and conductivity which were also the reason for inception of the MI-based WUC. Each layer of the heterogeneous soil affects the magnetic field differently. Given this behavior, [246] assigns a scaling factor to different depths. In [91], the authors study propagation though the soil by calculating the skin depth of each layer. Sun and Akyildiz [213] characterizes the path loss for MI-based communication. Guo et al. [56] investigate the asymmetric transceiver to cope up with the case of coils misalignment in MI-based WUC. Path loss has been extensively studied for each type of the wireless channel; however, few efforts have been made for WUC systems. Therefore, this area of WUC needs special consideration.

Transmission Range
MI-based WUC with all its advantages, i.e., not effected by boundary effects and multipath fading [222], has a disadvantage of limited transmission range. This is because of high path loss in the soil. In [100, 215], authors proposed using relay coils to extend the transmission range. Similarly, [100, 133] proposed using super conductors and meta-materials for this purpose. Large coils were used in [107] with an aim of achieving high transmission range; however, it might not be a practical solution. Therefore, achieving the long communication range for buried nodes is an important research issue.

Latency and Reliable Communication
Latency and reliable communication is the primary requirement of all critical applications such as oil and gas exploration. Late or incorrect sensors reading can cause major disaster. WUC challenging environment is the major hindrance in achieving the reliable communication. Although the reduced latency and reliability is one of the major requirements of the conventional IoT as well [173, 182]; however, in WUC, this issue needs more deliberation due to tough operating parameters and regulations on subsurface environment. It is not possible to meet the WUC communication requirement with any single system. For example, wired communication provides reliability and low latency, whereas wireless solutions are scalable with low complex. Therefore, it is important to develop a WUC with low latency, lower transmission delays, and minimized sensor failures.

Security
Security is the least studied aspect of WUC systems. WUC security includes security of equipment and security of communication protocols. Node replication, jamming the signal, and worm hole are few potential security attacks that can occur in WUC systems. A security breach can be used to raise false alarms. Responding

to the frequent false alarms can exhaust network resources. In [147], authors discuss the security issues (e.g., forward and backward security) and malicious attacks (e.g., node compromise attack) on a cloud-based IoT. Authors in [41] use the data tagging technique for improved data security. They use information flow control (IFC) for this purpose. A secure IoT architecture using host identity protocol (HIP) and datagram transport layer security (DTLS) is presented in [48]. Sicari [190] provides an extensive security survey in IoT. These studies are targeted towards improving security in terrestrial networks; however, these can be modified to WUC environment by introducing underground operational constraints. For example, old oil and gas systems are being transformed to digital WUC systems. Therefore, it is required to update security of such globally connected systems which, otherwise, in an event of cyber attack, can lead to some disastrous situation. Blockchain technology can be also be used in WUC systems to deal with the cyber crimes [180, 185].

Scalability
Scalability issues can rise due to the factors such as higher network density, high energy-consumption of underground things, node failures, routing overhead, low memory of underground nodes, and vendor-specific nodes can cause interoperability issues. Gharbieh et al. [50] use spatio-temporal stochastic modeling to deal with the scalability in WUC. For tunnels, [102] proposed an adaptive structure-aware WUC system. Interoperability issue is discussed in [102] using middleware protocol. Heterogeneity of sensor nodes is studied in [240]. The mentioned work deals with the scalability of terrestrial IoT; however, these can be modified as per requirements of WUC systems. For example, high path loss in soil limits the deployment of large wireless network. This problem is studied in [152, 230] which uses the sink nodes to connect with the sparsely buried sensor nodes. It uses the energy harvesting to increase the lifetime of the nodes. Besides these solutions, it is important to efficiently develop a self-healing and self-organizing WUC systems which can overcome the scalability issues.

Robustness
An underground channel is very unpredictable facing the issues like energy constraints, dynamic topology, sparsity of nodes. Hence, achieving robustness is very critical in WUC systems. A small world model is proposed in [108] for the improvement of latency and robustness by considering the local importance of smart objects. Extensive literature exists for the improvement of robustness in terrestrial network [27]; however, work in robustness in WUC is limited to the mining application. For example, [80] improves the robustness of an underground mining by using a wireless mesh network. One of the major challenges in the WUC systems is to develop robust communication and data gathering techniques. Communication range of EM waves in soil is highly limited because of attenuation. However, magnetic induction is considered relative robust for communicating in the soil but requires perfect orientation of the coils. The research of MI-based WUC for robustness is still not mature and needs to be studied further.

Hybrid Sensing

Hybrid sensing systems include the usage of multiple sensor systems and integration of their signals, e.g., long-term underground fiber sensors can be combined with short-term ground penetrating radars for the purpose of detection and localization. SoilNet Systems [22] is a hybrid sensor system which combines ZigBee network with wired communication. ZigBee network is used for aboveground nodes and wired communication is used for the underground nodes. A combination of EM- and MI-based can be used for providing long-range downlink (EM-based) and short-range uplink (MI-based) communication [107]. Therefore, hybrid sensing systems can improve the efficiency of WUC systems.

Software-Defined Networking (SDN)

Software-Defined Networking (SDN) provides robustness, scalability, reliability, and secure networking solution for WUC systems. It is different form conventional networking solutions in that it separates the control logic from the networking hardware. These advantages make it suitable for the usage in underwater environment. A surface station can be any SDN controller which communicates with the underwater sensors through in/out-band control channels [8]. The SDN controller will separate the data plane and controller plane. Such technique can also be used for WUC systems [141]. SDN-based WUC will have lower network complexity, improved congestion control mechanism, increased network life, efficient utilization of network resources, and reduced latency. For example, SDN-based WUC for oil and gas can allow users to efficiently manage the system by providing the global view of buried sensors nodes. SDN-based WUC can also be used in agricultural applications for achieving a scalable network solution. Furthermore, data visualization can be used with SDN-controller for correlation of sensor data. These advantages of SDN paradigm force researcher to look into the possibilities of SDN-based WUC systems [155].

Big Data

Massive amount of data is generated by WUC applications (agriculture, seismic surveying, and oil/gas fields). This data should properly organized, correlated, and analyzed for making accurate decisions [60]. Integration of big data and traditional IoT is already being studied extensively, e.g., [14] presents the application of big data in IoT. In [136], authors study the application of context-aware computing in IoT. These works motivate and present an opportunity of integrating big data analytics with WUC system. For example, oil and gas WUC generates glut of data and managing that data is the major concern of respective industries [166, 167]. Similarly, geo-scientists spend major portion of their time (nearly 50 %) on managing data. Big data provides an opportunity to handle such big amount of data and perform analysis. Therefore, proper data analytics tools must be developed for the WUC systems.

Fog and Cloud Computing
Cloud/fog computing provides different feature (scalability, mobility, low delays, and location awareness) for an efficient WUC system. Cloud computing has been used for the management purposes in oil and gas industries, whereas fog computing has been used for reducing data traffic and analysis of data at edge [137]. In oil and gas industries, huge data generated by the upstream operations (e.g., drilling and seismic exploration) is a major challenge. Fog computing can be used for provision of localized data analytic being generated in real-time. It helps in minimizing communication delays and faster event response. Moreover, time-critical applications require efficient decision-making procedure because it is possible that decision-making opportunity is gone by the time data reaches the cloud. Hence, fog computing should be integrated with WUC systems [170].

Efficient Localization Methods
Localization can be done in many applications such as WUC monitoring, geo-tagged sensing, and optimized fracturing. There are limited studies which try to find location of buried nodes of MI-based WUC. In [114], authors developed a testbed for tracking objects in MI-based WUC. Abrudan et al. [2] study how mineral and rocks in underground environment affect the accuracy of localization. The accuracy of MI-based WUC is also investigated in [149, 163, 168]. It is important to note that localization work exists only for the MI-based and there is no such investigation done in EM-, acoustic-, and VLC-based WUC. Therefore, robust and accurate localization methods are required for these WUC systems.

References

1. Abrudan TE, Kypris O, Trigoni N, Markham A (2016) Impact of rocks and minerals on underground magneto-inductive communication and localization. IEEE Access 4:3999–4010
2. Abrudan TE, Xiao Z, Markham A, Trigoni N (2016) Underground incrementally deployed magneto-inductive 3-d positioning network. IEEE Trans Geosci Remote Sens 54(8):4376–4391
3. Adamo F, Andria G, Attivissimo F, Giaquinto N (2004) An acoustic method for soil moisture measurement. IEEE Trans Instrum Meas 53(4):891–898
4. Ahmad TJ, Noui-Mehidi M, Arsalan M (2014) Performance analysis of downhole acoustic communication in multiphase flow. In IECON 2014-40th annual conference of the IEEE industrial electronics society. IEEE, New York, pp. 3909–3913
5. Akkaş MA, Akyildiz IF, Sokullu R (2012) Terahertz channel modeling of underground sensor networks in oil reservoirs. In 2012 IEEE global communications conference (GLOBECOM). IEEE, New York, pp. 543–548
6. Akyildiz IF, Stuntebeck EP (2006) Wireless underground sensor networks: research challenges. Ad Hoc Netw J (Elsevier) 4:669–686
7. Akyildiz IF, Sun Z, Vuran MC (2009) Signal propagation techniques for wireless underground communication networks. Phys Commun J (Elsevier) 2(3):167–183
8. Akyildiz IF, Wang P, Lin S-C (2016) Softwater: software-defined networking for next-generation underwater communication systems. Ad Hoc Netw 46:1–11

9. Alenezi A, Abdi A (2017) A comparative study of multichannel and single channel accelerometer sensors for communication in oil wells. In 2017 international conference on communication and signal processing (ICCSP). IEEE, New York, pp. 0153–0153
10. Algeroy J, Lovell J, Tirado G, Meyyappan R, Brown G, Greenaway R, Carney M, Meyer JH, Davies JE, Pinzon ID (2010) Permanent monitoring: taking it to the reservoir. Oilfield Rev 22(1):34–41
11. Anand N, Lee S-J, Knightly EW (2012) Strobe: actively securing wireless communications using zero-forcing beamforming. In 2012 Proceedings of the IEEE INFOCOM, pp. 720–728
12. Arnautovski-Toseva V, Grcev L (2016) On the image model of a buried horizontal wire. IEEE Trans Electromagn Compat 58(1):278–286
13. Aryafar E et al. (2013) ADAM: an adaptive beamforming system for multicasting in wireless LANs. IEEE/ACM Trans Netw 21(5):1595–1608
14. Atzori L, Iera A, Morabito G (2010) The internet of things: a survey. Comput Netw 54(15):2787–2805
15. Baldwin C (2018) Fiber optic sensors in the oil and gas industry: current and future applications. In Opto-mechanical fiber optic sensors. Elsevier, Amsterdam, pp 211–236
16. Bandyopadhyay LK, Chaulya SK, Mishra PK (2010) Wireless communication in underground mines: RFID-based sensor networking. Springer, Berlin
17. Banos A (1966) Dipole radiation in the presence of a conducting halfspace. Pergamon Press, New York
18. Behari J (2005) Microwave dielectric behavior of wet soils. Springer, Berlin
19. Bicen AO, Sahin AB, Akan OB (2012) Spectrum-aware underwater networks: cognitive acoustic communications. IEEE Veh Technol Mag 7(2):34–40
20. Biggs AW (1968) Dipole antenna radiation fields in stratified Antarctic media. IEEE Trans Antennas Propag 16(4):445–448
21. Bobrov PP, Repin AV, Rodionova OV (2015) Wideband frequency domain method of soil dielectric property measurements. IEEE Trans Geosci Remote Sens 53(5):2366–2372
22. Bogena HR, Huismana JA, Meierb H, Rosenbauma U, Weuthena A (2009) Hybrid wireless underground sensor networks: quantification of signal attenuation in soil. Vadose Zone J 8(3):755–761
23. Bogena HR, Herbst M, Huisman JA, Rosenbaum U, Weuthen A, Vereecken H (2010) Potential of wireless sensor networks for measuring soil water content variability. Vadose Zone J 9(4):1002–1013
24. Boyle KR, Yuan Y, Ligthart LP (2007) Analysis of mobile phone antenna impedance variations with user proximity. IEEE Trans Antennas Propag 55(2):364–372
25. Brekhovskikh LM (1980) Waves in layered media, 2 edn. Academic Press, New York
26. Castorina G, Di Donato L, Morabito AF, Isernia T, Sorbello G (2016) Analysis and design of a concrete embedded antenna for wireless monitoring applications. IEEE Antennas Propag Mag 58(6):76–93
27. Chen L, Thombre S, Järvinen K, Lohan ES, Alén-Savikko A, Leppäkoski H, Zahidul M, Bhuiyan H, Bu-Pasha S, Ferrara GN, Honkala S et al. (2017) Robustness, security and privacy in location-based services for future IoT: a survey. IEEE Access 5:8956–8977
28. Cisco Visual Networking Index. https://www.cisco.com/c/en/us/solutions/service-provider/visual-networking-index-vni/index.html
29. Chukhlantsev A (2006) Microwave radiometry of vegetation canopies. Springer, Berlin
30. Crossbow Mica2, Micaz, and IRIS motes (2020). http://www.xbow.com
31. Curtis JO (2001) A durable laboratory apparatus for the measurement of soil dielectric properties. IEEE Trans Instrum Meas 50(5):1364–1369
32. Dissanayake T, Esselle KP, Yuce MR (2009) Dielectric loaded impedance matching for wideband implanted antennas. IEEE Trans Microwave Theory Tech 57(10):2480–2487
33. Dobson MC et al. (1985) Microwave dielectric behavior of wet soil—Part II: dielectric mixing models. IEEE Trans. Geoscience and Remote Sensing GE-23(1):35–46
34. Dong X, Vuran MC (2011) A channel model for wireless underground sensor networks using lateral waves. In: Proceedings of the IEEE Globecom '11, Houston, TX

35. Dong X, Vuran MC (2013) Impacts of soil moisture on cognitive radio underground networks. In: Proceedings of the IEEE BlackSeaCom, Georgia
36. Dong X, Vuran MC, Irmak S (2012) Autonomous precision agriculture through integration of wireless underground sensor networks with center pivot irrigation systems. Ad Hoc Netw (Elsevier) 11(7):1975–1987
37. Dong X, Vuran MC, Irmak S (2013) Autonomous precision agriculture through integration of wireless underground sensor networks with center pivot irrigation systems. Ad Hoc Netw 11(7):1975–1987
38. Dong S, Yao A, Meng F (2015) Analysis of an underground horizontal electrically small wire antenna. J Electr Comput Eng 2851:9
39. Du Y et al. (2014) iBeam: intelligent client-side multi-user beamforming in wireless networks. In: Proceedings of the IEEE INFOCOM 2014
40. Du D, Zhang H, Yang J, Yang P (2017) Propagation characteristics of the underground-to-aboveground communication link about 2.4 GHz and 433 MHz radio wave: an empirical study in the pine forest of Guizhou province. In 2017 3rd IEEE international conference on computer and communications (ICCC). IEEE, New York, pp. 1041–1045
41. Evans D, Eyers DM (2012) Efficient data tagging for managing privacy in the internet of things. In: 2012 IEEE international conference on green computing and communications. IEEE, New York, pp 244–248
42. FCC order no. da 16-307 dated: Mar 24, 2016. https://apps.fcc.gov/edocs_public/attachmatch/DA-16-307A1.pdf
43. Fitzgerrell RG, Haidle, LL (1972) Design and performance of four buried UHF antennas. IEEE Trans. Antennas Propagation 20(1):56–62
44. Foth HD (1990) Fundamentals of soil science, 8 edn. Wiley, New York
45. Franz TE, Wahbi A, Vreugdenhil M, Weltin G, Heng L, Oismueller M, Strauss P, Dercon G, Desilets D (2016) Using cosmic?ray neutron probes to monitor landscape scale soil water content in mixed land use agricultural systems, Appl Environ Soil Sci 2016 432–742. https://doi.org/10.1155/2016/4323742
46. Freire R, de Abreu MHM, Okada RY, Soares PF, GranhenTavares CR (2015) Sound absorption coefficient in situ: an alternative for estimating soil loss factors. Ultrason Sonochem 22:100–107
47. Galejs J (1969) Antennas in inhomogeneous media. Pergamon Press, Oxford
48. Garcia-Morchon O, Keoh SL, Kumar S, Moreno-Sanchez P, Vidal-Meca F, Ziegeldorf JH (2013) Securing the IP-based internet of things with hip and DTLS. In: Proceedings of the sixth ACM conference on Security and privacy in wireless and mobile networks, pp 119–124
49. Gardner WR, Hyden RE, Linyaev EJ, Gao L, Robbins C, Moore J, et al. (2006) Acoustic telemetry delivers more real-time downhole data in underbalanced drilling operations. In: IADC/SPE drilling conference. Society of Petroleum Engineers, London
50. Gharbieh M, ElSawy H, Bader A, Alouini M-S (2017) Spatiotemporal stochastic modeling of IoT enabled cellular networks: Scalability and stability analysis. IEEE Trans Commun 65(8):3585–3600
51. Ghazanfari E, Pamukcu S, Yoon S-U, Suleiman MT, Cheng L (2012) Geotechnical sensing using electromagnetic attenuation between radio transceivers. Smart Mater Struct 21(12):125017
52. Gosalia K, Humayun MS, Lazzi G (2005) Impedance matching and implementation of planar space-filling dipoles as intraocular implanted antennas in a retinal prosthesis. IEEE Trans Antennas Propag 53(8):2365–2373
53. Goyal R, Kennedy R, Kelsey B, Whelan M, Janoyan K (2014) Underground wireless sensor networks using 2nd generation RF transceivers. In: Geo-congress 2014: Geo-characterization and modeling for sustainability, pp 2619–2629
54. Guo H, Sun Z (2014) Channel and energy modeling for self-contained wireless sensor networks in oil reservoirs. IEEE Trans Wireless Commun 13(4):2258–2269.

55. Guo H, Sun Z (2018) Full-duplex metamaterial-enabled magnetic induction networks in extreme environments. In: Proceedings of the IEEE INFOCOM 2018-IEEE conference on computer communications. IEEE, New York, pp 558–566
56. Guo H, Sun Z, Zhou C (2017) Practical design and implementation of metamaterial-enhanced magnetic induction communication. IEEE Access 5:17213–17229
57. Gutiarrez J, Villa-Medina JF, Nieto-Garibay A, Porta-Gandara MA (2014) Automated irrigation system using a wireless sensor network and GPRS module. IEEE Trans Instrum Meas 63(1):166–176
58. Gutierrez-Estevez MA, Krüger U, Krueger KA, Manolakis K, Jungnickel V, Jaksch K, Krueger K, Mikulla S, Giese R, Sohmer M et al (2013) Acoustic broadband communications over deep drill strings using adaptive OFDM. In Proceedings of the 2013 IEEE wireless communications and networking conference (WCNC). IEEE, New York, pp 4089–4094
59. Hahn D, Peters V, Rouatbi C, Scholz E (2008) Reciprocating pulser for mud pulse telemetry, US Patent 7,417,920
60. Hajirahimova MS (2015) Opportunities and challenges big data in oil and gas industry. In: Proceedings of the national supercomputer forum (NSKF 2015), Russia, Pereslavl-Zalessky, pp. 24–27
61. Han W, Wang Y, Ooi SK, Guo C (2013) Signal transmission and node deployment of a 2.4 Ghz wireless sensor network: a case study in a persimmon orchard. In: 2013 Kansas City, Missouri, July 21-July 24, 2013. American Society of Agricultural and Biological Engineers, St. Joseph, p 1
62. Hansen R (1963) Radiation and reception with buried and submerged antennas. IEEE Trans Antennas Propag 11(3):207–216
63. Harrell J, Brooks AG, Morsy HS (2000) Method and apparatus for mud pulse telemetry in underbalanced drilling systems, US Patent 6,097,310
64. Harrison WH, Mazza RL, Rubin LA, Yost AB et al. (1990) Air-drilling, electromagnetic, MWD system development. In: SPE/IADC drilling conference. Society of Petroleum Engineers, London
65. Hernandez M, MacNeill DW, Reeves M, Kirkwood AD, Ruszka JP, Zaeper R, Lemke SR, et al. (2008) High-speed wired drillstring telemetry network delivers increased safety, efficiency, reliability, and productivity to the drilling industry. In: SPE Indian oil and gas technical conference and exhibition. Society of Petroleum Engineers, London
66. Hipp JE (1974) Soil electromagnetic parameters as functions of frequency, soil density, and soil moisture. Proc IEEE 62(1):98–103
67. Hislop G (2015) Permittivity estimation using coupling of commercial ground penetrating radars. IEEE Trans Geosci Remote Sens 53(8):4157–4164
68. Hopkins J (2016) USDA ERS - ARMS farm financial and crop production practices: tailored reports: Crop production practices. ers.usda.gov. http://www.ers.usda.gov/data-products/arms-farm-financial-and-crop-production-practices/tailored-reports-crop-production-practices.aspx
69. Horvat G, Vinko D, Vlaović J (2016) Impact of propagation medium on link quality for underwater and underground sensors. In: 2016 39th international convention on information and communication technology, electronics and microelectronics (MIPRO). IEEE, New York, pp 129–134
70. Huang S (1979) An antenna for underground radio communication. Master's thesis, University of Houston, Houston
71. Huang H, Zheng YR (2018) 3-D localization of wireless sensor nodes using near-field magnetic-induction communications. Phys Commun 30:97–106
72. Huang Q, Zhang X, Ma J (2015) Underground magnetic localization method and optimization based on simulated annealing algorithm. In: 2015 IEEE 12th international conference on ubiquitous intelligence and computing and 2015 IEEE 12th international conference on autonomic and trusted computing and 2015 IEEE 15th international conference on scalable computing and communications and its associated workshops (UIC-ATC-ScalCom). IEEE, New York, pp 168–173

73. Hutin R, Tennent RW, Kashikar SV et al. (2001) New mud pulse telemetry techniques for deepwater applications and improved real-time data capabilities. In: SPE/IADC drilling conference. Society of Petroleum Engineers, London
74. Hutin R (2016) Zero sum pressure drop mud telemetry modulator, US Patent 9,228,432
75. Iizuka K (1964) An experimental investigation on the behavior of the dipole antenna near the interface between the conducting medium and free space. IEEE Trans Antennas Propag 12(1):27–35
76. Jiang S, Georgakopoulos SV, Jonah O (2012) RF power harvesting for underground sensors. In: Proceedings of the 2012 IEEE international symposium on antennas and propagation. IEEE, New York, pp 1–2
77. Jianhui Z, Liyan W, Fan L, Yanlei L (2007) An effective approach for the noise removal of mud pulse telemetry system. In: Proceedings of the 2007 8th international conference on electronic measurement and instruments. IEEE, New York, pp 1–971
78. Jonard F, Weihermüller L, Schwank M, Jadoon KZ, Vereecken H, Lambot S (2015) Estimation of hydraulic properties of a sandy soil using ground-based active and passive microwave remote sensing. IEEE Trans Geosci Remote Sens 53(6):3095–3109
79. Kang Z, Yu Y, Hou C (2011) Study on stress and strain and characteristics of acoustic emission in the process of rock failure. In: 2011 second international conference on mechanic automation and control engineering. IEEE, New York, pp 7737–7740
80. Kennedy GA, Foster PJ (2006) High resilience networks and microwave propagation in underground mines. In:2006 European conference on wireless technology. IEEE, New York, pp 193–196
81. Kersey AD (2000) Optical fiber sensors for permanent downwell monitoring applications in the oil and gas industry. IEICE Trans Electron 83(3):400–404
82. Kesar AS, Weiss E (2013) Wave propagation between buried antennas. IEEE Trans Antennas Propag 61(12):6152–6156
83. Khan US, Al-Nuaimy W, Abd El-Samie FE (2010) Detection of landmines and underground utilities from acoustic and GPR images with a cepstral approach. J Vis Commun Image Represent 21(7):731–740
84. Kim Y, Evans RG, Iversen WM (2008) Remote sensing and control of an irrigation system using a distributed wireless sensor network. IEEE Trans Instrum Meas 57(7):1379–1387
85. Kim SB, Ouellette JD, van Zyl JJ, Johnson JT (2016) Detection of inland open water surfaces using dual polarization l-band radar for the soil moisture active passive mission. IEEE Trans Geosci Remote Sens 54(6):3388–3399
86. King RWP, Smith G (1981) Antennas in matter. MIT Press, Cambridge
87. King R, Smith GS, Owens M, Wu TT (1981) Antennas in matter - fundamentals, theory, and applications. MIT Press, Cambridge
88. King RWP, Owens M, Wu TT (1992) Lateral electromagnetic waves. Springer, Berlin
89. Kisseleff S, Gerstacker W, Sun Z, Akyildiz IF (2013) On the throughput of wireless underground sensor networks using magneto-inductive waveguides. In:2013 IEEE global communications conference (GLOBECOM). IEEE, New York, pp. 322–328
90. Kisseleff S, Akyildiz IF, Gerstacker W (2013) Interference polarization in magnetic induction based wireless underground sensor networks. In: 2013 IEEE 24th international symposium on personal, indoor and mobile radio communications (PIMRC Workshops). IEEE, New York, pp 71–75
91. Kisseleff S, Sackenreuter B, Akyildiz IF, Gerstacker W (2015) On capacity of active relaying in magnetic induction based wireless underground sensor networks. In: Proceedings of 2015 IEEE international conference on communications (ICC). IEEE, New York, pp 6541–6546
92. Kisseleff S, Akyildiz IF, Gerstacker W (2015) Beamforming for magnetic induction based wireless power transfer systems with multiple receivers. In 2015 IEEE GLOBECOM
93. Kisseleff S, Akyildiz IF, Gerstacker WH (2015) Digital signal transmission in magnetic induction based wireless underground sensor networks. IEEE Trans Commun 63(6):2300–2311

94. Kisseleff S, Chen X, Akyildiz IF, Gerstacker W (2017) Localization of a silent target node in magnetic induction based wireless underground sensor networks. In: 2017 IEEE international conference on communications (ICC). IEEE, New York, pp 1–7
95. Kisseleff S, Akyildiz IF, Gerstacker WH (2018) Survey on advances in magnetic induction-based wireless underground sensor networks. IEEE Internet Things J 5(6):4843–4856
96. Klotz C, Bond PR, Wassermann I, Priegnitz S et al. (2008) A new mud pulse telemetry system for enhanced MWD/LWD applications. In: IADC/SPE drilling conference. Society of Petroleum Engineers, London
97. Koike H, Kamiya Y (2016) A new approach for subsurface wireless sensor networks. In: Intelligent interactive multimedia systems and services 2016. Springer, Berlin, pp 201–211
98. Konda A, Rau A, Stoller MA, Taylor JM, Salam A, Pribil GA, Argyropoulos C, Morin SA (2018) Soft microreactors for the deposition of conductive metallic traces on planar, embossed, and curved surfaces. Adv Funct Mater 28(40):1803020
99. Kragas TK, Williams BA, Myers GA, et al. (2001) The optic oil field: deployment and application of permanent in-well fiber optic sensing systems for production and reservoir monitoring. In: SPE annual technical conference and exhibition. Society of Petroleum Engineers, London
100. Kulkarni A, Kumar V, Dhok SB (2018) Enabling technologies for range enhancement of mi based wireless non-conventional media communication. In: Proceedings of the 2018 9th international conference on computing, communication and networking technologies (ICCCNT). IEEE, New York, pp 1–7
101. Lakshmanan S, Sundaresan K, Kokku R, Khojestepour A, Rangarajan S (2009) Towards adaptive beamforming in indoor wireless networks: an experimental approach. In: Proceedings of the INFOCOM 2009. IEEE, New York
102. Li M, Liu Y (2007) Underground structure monitoring with wireless sensor networks. In: Proceedings of the 2007 6th international symposium on information processing in sensor networks. IEEE, New York, pp 69–78
103. Lin Y, Kong X, Qiu Y, Yuan Q (2013) Calculation analysis of pressure wave velocity in gas and drilling mud two-phase fluid in annulus during drilling operations. Math Prob Eng 2013 318–912, http://dx.doi.org/10.1155/2013/318912
104. Lin S-C, Akyildiz IF, Wang P, Sun Z (2014) Optimal energy-throughput efficiency for magneto-inductive underground sensor networks. In: Proceedings of the 2014 IEEE international black sea conference on communications and networking (BlackSeaCom). IEEE, New York, pp 22–27
105. Lin S, Akyildiz IF, Wang P, Sun Z (2015) Distributed cross-layer protocol design for magnetic induction communication in wireless underground sensor networks. IEEE Trans Wireless Commun 14(7):4006–4019
106. Lin S-C, Akyildiz IF, Wang P, Sun Z (2015) Distributed cross-layer protocol design for magnetic induction communication in wireless underground sensor networks. IEEE Trans Wireless Commun 14(7):4006–4019
107. Lin S-C, Alshehri AA, Wang P, Akyildiz IF (2017) Magnetic induction-based localization in randomly deployed wireless underground sensor networks. IEEE Int Things J 4(5):1454–1465
108. Luo D, Qiu T, Deonauth N, Zhao A (2015) A small world model for improving robustness of heterogeneous networks. In: 2015 IEEE global conference on signal and information processing (GlobalSIP). IEEE, New York, pp 849–852
109. Lytle RJ (1974) Measurement of earth medium electrical characteristics: techniques, results, and applications. IEEE Trans Geosci Electron 12(3):81–101
110. Lytle RJ, Lager DL (1976) The Yosemite experiments: HF propagation through rock. Radio Sci 11(4):245–252
111. Ma J, Zhang XT, Huang QW (2014) Near-field magnetic induction communication device for underground wireless communication networks. Sci China Inf Sci 57(12):1–11

112. Ma J, Zhang X, Huang Q, Cheng L, Lu M (2015) Experimental study on the impact of soil conductivity on underground magneto-inductive channel. IEEE Antennas and Wirel Propag Lett 14:1782–1785
113. Ma D, Shi Y, Zhang W, Liu G (2018) Design of acoustic transmission along drill strings for logging while drilling data based on adaptive NC-OFDM. AEU-Int J Electron Commun 83:329–338
114. Markham A, Trigoni N (2012) Magneto-inductive networked rescue system (miners): taking sensor networks underground. In: Proceedings of the 11th international conference on information processing in sensor networks, IPSN '12. ACM, New York, pp 317–328
115. Markham A, Trigoni N, Ellwood SA, Macdonald DW (2010) Revealing the hidden lives of underground animals using magneto-inductive tracking. In: Proceedings of the 8th ACM conference on embedded networked sensor systems, pp 281–294
116. Markham A, Trigoni N, Macdonald DW, Ellwood SA (2011) Underground localization in 3-d using magneto-inductive tracking. IEEE Sens J 12(6):1809–1816
117. McCulloch J, McCarthy P, Guru SM, Peng W, Hugo D, Terhorst A (2008) Wireless sensor network deployment for water use efficiency in irrigation. In: Proceedings of the workshop on Real-world wireless sensor networks, pp 46–50
118. Mijarez R, Pascacio D, Guevara R, Pacheco O, Tello C, Rodríguez J (2013) Communication system for down-hole measurement tools based on real-time SNR characterization in coaxial cable used as communication channel. Additional Papers Presentations 2013(HITEN):000174–000183
119. Mishra PK, Kumar M, Kumar S, Mandal PK, et al. (2018) Wireless real-time sensing platform using vibrating wire-based geotechnical sensor for underground coal mines. Sens Actuators A Phys 269:212–217
120. Moore RK, Blair WE (1961) Dipole radiation in conducting half space. J Res Nat Bur Stand 65(6):547–563
121. Muduli L, Mishra DP, Jana PK (2018) Application of wireless sensor network for environmental monitoring in underground coal mines: a systematic review. J Netw Comput Appl 106:48–67
122. Nassar EM, Lee R, Young JD (1999) A probe antenna for in situ measurement of the complex dielectric constant of materials. IEEE Trans Antennas Propag 47(6):1085–1093
123. Nawandar NK, Satpute VR (2019) IoT based low cost and intelligent module for smart irrigation system. Comput Electron Agric 162:979–990
124. Neff JM, Camwell PL (2007) Field test results of an acoustic telemetry MWD system. In Proceedings of the SPEIADC drilling conference. Society of Petroleum Engineers, London
125. Nicolson AM, Ross GF (1970) Measurement of the intrinsic properties of materials by time-domain techniques. IEEE Trans Instrum Meas 19(4):377–382
126. Nikolidakis SA, Kandris D, Vergados DD, Douligeris C (2015) Energy efficient automated control of irrigation in agriculture by using wireless sensor networks. Comput Electron Agric 113:154–163
127. Nitsche T et al. (2015) Steering with eyes closed: Mm-wave beam steering without in-band measurement. In: IEEE INFOCOM
128. Norton KA (1937) The physical reality of space and surface waves in the radiation field of radio antennas. In: Proceedings of the institute of radio engineers, vol 25(9), pp 1192–1202 .
129. Oelze ML, O'Brien WD, Darmody RG (2002) Measurement of attenuation and speed of sound in soils. Soil Sci Soc Am J 66(3):788–796
130. Parameswaran G, Sivaprasath K (2016) Arduino based smart drip irrigation system using internet of things. Int J Eng Sci 5518
131. Parameswaran V, Zhou H, Zhang Z (2012) Irrigation control using wireless underground sensor networks. In: Proceedings of the 2012 sixth international conference on sensing technology (ICST). IEEE, New York, pp 653–659

132. Parameswaran V, Zhou H, Zhang Z (2013) Wireless underground sensor network design for irrigation control: simulation of RFID deployment. In: Proceedings of the 2013 seventh international conference on sensing technology (ICST). IEEE, New York, pp 842–849
133. Pathak V, Kumar V, Barik RK (2018) Magnetic induction communication based transceiver coil and waveguide structure modeling for non-conventional WSNS. In: 2018 9th international conference on computing, communication and networking technologies (ICCCNT). IEEE, New York, pp 1–7
134. Pelekanakis K, Chitre M, Kumar LS, Guan YL (2014) Performance of channel coding and equalization for acoustic telemetry along drill strings. In: Proceeding of the 2014 IEEE international conference on communication systems. IEEE, New York, pp 610–614
135. Peplinski N, Ulaby F, Dobson M (1995) Dielectric properties of soil in the 0.3–1.3 GHz range. IEEE Trans Geosci Remote Sens 33(3):803–807
136. Perera C, Zaslavsky A, Christen P, Georgakopoulos D (2013) Context aware computing for the internet of things: a survey. IEEE Commun Surv Tutorials 16(1):414–454
137. Perrons RK, Hems A (2013) Cloud computing in the upstream oil and gas industry: a proposed way forward. Energy Policy 56:732–737
138. Pompili D, Akyildiz IF (2009) Overview of networking protocols for underwater wireless communications. IEEE Commun Mag 47(1):97–102
139. Powell J, Chandrakasan A (2004) Differential and single ended elliptical antennas for 3.1-10.6 Ghz ultra wideband communication. In: Antennas and propagation society international symposium, Vol 2. Sendai, Japan
140. Prevedel B, Bulut F, Bohnhoff M, Raub C, Kartal RF, Alver F, Malin PE (2015) Downhole geophysical observatories: best installation practices and a case history from turkey. Int J Earth Sci 104(6):1537–1547
141. Puente Fernández JA, García Villalba LJ, Kim T-H (2018) Software defined networks in wireless sensor architectures. Entropy 20(4):225
142. Qu F, Zhang Z, Hu J, Xu J, Wang S, Wu Y (2018) Adaptive dual-sensor noise cancellation method for continuous wave mud pulse telemetry. J Pet Sci Eng 162:386–393
143. Quitin F et al. (2013) A scalable architecture for distributed transmit beamforming with commodity radios: design and proof of concept. IEEE Trans Wirel Commun
144. Ranjan A, Sahu HB, Misra P (2020) Modeling and measurements for wireless communication networks in underground mine environments. Measurement 149:106980
145. Reckmann H (2014) Downhole noise cancellation in mud-pulse telemetry. US Patent 8,811,118
146. Ritsema CJ, Kuipers H, Kleiboer L, Elsen E, Oostindie K, Wesseling JG, Wolthuis J, Havinga P (2009) A new wireless underground network system for continuous monitoring of soil water contents. Water Resour Res J 45:1–9
147. Roman R, Najera P, Lopez J (2011) Securing the internet of things. Computer 44(9):51–58
148. Saeed N, Al-Naffouri TY, Alouini M-S (2019) Towards the internet of underground things: a systematic survey. arXiv preprint:1902.03844
149. Saeed N, Alouini MS, Al-Naffouri TY (2019) On achievable accuracy of localization in magnetic induction-based internet of underground things for oil and gas reservoirs. arXiv preprint:1901.09556
150. Saeed N, Alouini M-S, Al-Naffouri TY (2019) Toward the internet of underground things: a systematic survey. IEEE Commun Surv Tutorials 21(4):3443–3466
151. Salam A (2019) A comparison of path loss variations in soil using planar and dipole antennas. In: 2019 IEEE international symposium on antennas and propagation. IEEE, New York
152. Salam A (2019) Design of subsurface phased array antennas for digital agriculture applications. In: Proceedings of the 2019 IEEE international symposium on phased array systems and technology (IEEE Array 2019), Waltham, MA, USA
153. Salam A (2019) A path loss model for through the soil wireless communications in digital agriculture. In: Proceeding of the 2019 IEEE international symposium on antennas and propagation. IEEE, New York, pp 1–2

154. Salam A (2019) Sensor-free underground soil sensing. In: ASA, CSSA and SSSA international annual meetings (2019) (ASA-CSSA-SSSA)
155. Salam A (2019) Subsurface MIMO: a beamforming design in internet of underground things for digital agriculture applications. J Sensor Actuator Netw 8(3):41
156. Salam A (2019) Underground environment aware MIMO design using transmit and receive beamforming in internet of underground things. Springer, Cham, pp 1–15
157. Salam A (2019) An underground radio wave propagation prediction model for digital agriculture. Information 10(4):147
158. Salam A (2019) Underground soil sensing using subsurface radio wave propagation. In: Proceedings of the 5th global workshop on proximal soil sensing, Columbia
159. Salam A (2020) Internet of Things for environmental sustainability and climate change. Springer, Cham, pp 33–69
160. Salam A (2020) Internet of Things for sustainability: perspectives in privacy, cybersecurity, and future trends. Springer, Cham, pp 299–327
161. Salam A (2020) Internet of Things for sustainable community development, 1 edn. Springer Nature, Berlin
162. Salam A (2020) Internet of Things for sustainable community development: introduction and overview. Springer, Cham, pp 1–31
163. Salam A (2020) Internet of Things for sustainable forestry. Springer, Cham, pp 147–181
164. Salam A (2020) Internet of Things for sustainable human health. Springer, Cham, pp 217–242
165. Salam A (2020) Internet of Things for sustainable mining. Springer, Cham, pp 243–271
166. Salam A (2020) Internet of Things for water sustainability. Springer, Cham, pp 113–145
167. Salam A (2020) Internet of Things in agricultural innovation and security, pp 71–112. Springer, Cham
168. Salam A (2020) Internet of Things in sustainable energy systems. Springer, Cham, pp 183–216
169. Salam A (2020) Internet of Things in water management and treatment. Springer, Cham, pp 273–298
170. Salam A (2020) Wireless underground communications in sewer and stormwater overflow monitoring: radio waves through soil and asphalt medium. Information, 11(2), 98
171. Salam A, Karabiyik U (2019) A cooperative overlay approach at the physical layer of cognitive radio for digital agriculture. In: Third international Balkan conference on communications and networking 2019 (BalkanCom'19), Skopje, Macedonia, the former Yugoslav Republic of Macedonia
172. Salam A, Shah S (2019) Internet of things in smart agriculture: Enabling technologies. In: Proceedings of the 2019 IEEE 5th World Forum on Internet of Things (WF-IoT). IEEE, New York, pp 692–695
173. Salam A, Vuran MC (2016) Impacts of soil type and moisture on the capacity of multi-carrier modulation in internet of underground things. In: Proceeding of the 25th ICCCN 2016. Waikoloa, Hawaii
174. Salam A, Vuran MC (2016) Impacts of soil type and moisture on the capacity of multi-carrier modulation in internet of underground things. In: 2016 25th international conference on computer communication and networks (ICCCN). IEEE, New York, pp 1–9
175. Salam A, Vuran MC (2017) Smart underground antenna arrays: a soil moisture adaptive beamforming approach. In: IEEE INFOCOM 2017-IEEE conference on computer communications. IEEE, New York, pp 1–9
176. Salam A, Vuran MC (2017) Wireless underground channel diversity reception with multiple antennas for internet of underground things. In: Proceedings of the 2017 IEEE international conference on communications (ICC). IEEE, New York, pp 1–7
177. Salam A, Vuran MC (2017) Em-based wireless underground sensor networks. In: Pamukcu S, Cheng L (eds) Underground sensing: monitoring and hazard detection for environment and infrastructure, 1st edn, chapter 5. Elsevier, Amsterdam
178. Salam A, Vuran MC (2017) EM-based wireless underground sensor networks. Academic Press, New York, pp 247–285

179. Salam A, Vuran MC (2017) Wireless underground channel diversity reception with multiple antennas for internet of underground things. In: Proceedings of the IEEE ICC 2017, Paris, France
180. Salam A, Vuran MC (2017) Smart underground antenna arrays: A soil moisture adaptive beamforming approach. In: Proceedings of the 36th IEEE INFOCOM 2017, Atlanta, USA
181. Salam A, Vuran MC, Irmak S (2016) Pulses in the sand: impulse response analysis of wireless underground channel. In: IEEE INFOCOM 2016-The 35th annual IEEE international conference on computer communications. IEEE, New York, pp 1–9
182. Salam A, Vuran MC, Irmak S (2016) Pulses in the sand: impulse response analysis of wireless underground channel. In: The 35th annual IEEE international conference on computer communications (INFOCOM 2016), San Francisco, USA
183. Salam A, Vuran MC, Irmak S (2017) Towards internet of underground things in smart lighting: a statistical model of wireless underground channel. In: Proceedings of the 14th IEEE international conference on networking, sensing and control (IEEE ICNSC), Calabria, Italy
184. Salam A, Hoang AD, Meghna A, Martin DR, Guzman G, Yoon YH, Carlson J, Kramer J, Yansi K, Kelly M et al. (2019) The future of emerging IoT paradigms: architectures and technologies. https://doi.org/10.20944/preprints201912.0276.v1
185. Salam A, Vuran MC, Dong X, Argyropoulos C, Irmak S (2019) A theoretical model of underground dipole antennas for communications in internet of underground things. IEEE Trans Antennas Propag 67(6):3996–4009
186. Salam A, Vuran MC, Irmak S (2019) Di-sense: In situ real-time permittivity estimation and soil moisture sensing using wireless underground communications. Comput Netw 151:31–41
187. Schnitger J, Macpherson JD et al. (2009) Signal attenuation for electromagnetic telemetry systems. In: SPE/IADC drilling conference and exhibition. Society of Petroleum Engineers, London
188. Schroeder RJ (2002) The present and future of fiber optic sensors for the oilfield service industry: where is there a role? In: Proceedings of the 2002 15th optical fiber sensors conference technical digest. OFS 2002 (Cat. No. 02EX533). IEEE, New York, pp 39–42
189. Sharma RK, Gupta AK (2010) Continuous wave acoustic method for determination of moisture content in agricultural soil. Comput Electron Agric 73(2):105–111
190. Sicari S, Rizzardi A, Grieco LA, Coen-Porisini A (2015) Security, privacy and trust in internet of things: the road ahead. Comput Netw 76:146–164
191. Silva AR (2010) Channel characterization for wireless underground sensor networks. Master's thesis, University of Nebraska-Lincoln, Nebraska-Lincoln
192. Silva AR, Moghaddam M (2015) Design and implementation of low-power and mid-range magnetic-induction-based wireless underground sensor networks. *IEEE Trans Instrum Meas* 65(4):821–835
193. Silva AR, Vuran MC (2009) Empirical evaluation of wireless underground-to-underground communication in wireless underground sensor networks. In: Proceedings of the IEEE DCOSS '09, Marina del Rey, CA, pp 231–244
194. Silva AR, Vuran MC (2009) Empirical evaluation of wireless underground-to-underground communication in wireless underground sensor networks. In: International conference on distributed computing in sensor systems. Springer, Berlin, pp 231–244
195. Silva AR, Vuran MC (2009) Empirical evaluation of wireless underground-to-underground communication in wireless underground sensor networks. In: Proceedings of the IEEE DCOSS '09, Marina Del Rey, CA
196. Silva AR, Vuran MC (2010) Channel contention in wireless underground sensor networks. In: to appear in Proceedings of the III international conference on wireless communications in underground and confined areas (ICWCUCA' 10), Val-d, Canada
197. Silva AR Vuran MC (2010) Communication with aboveground devices in wireless underground sensor networks: an empirical study. In: Proceedings of the IEEE ICC'10, Cape Town, South Africa, pp 1–6

198. Silva AR, Vuran MC (2010) Development of a testbed for wireless underground sensor networks. EURASIP J Wirel Commun Netw, 2010(1):1–14
199. Silva AR, Vuran MC (2010) $(CPS)^2$: integration of center pivot systems with wireless underground sensor networks for autonomous precision agriculture. In: Proceedings of the ACM/IEEE international conference on cyber-physical systems. Stockholm, Sweden, pp 79–88
200. Silva AR, Vuran MC (2010) Development of a testbed for wireless underground sensor networks. EURASIP J Wirel Commun Netw 2010:1–14
201. Singer A, Yang S, Oelze M (2017) Acoustic communications: through soils, sands, water, and tissue. J Acoust Soc Am 141(5):3986–3987
202. Sivaprasad K, King RWP (1963) A study of arrays of dipoles in a semi-infinite dissipative medium. IEEE Trans Antennas Propag 11(3):240–256
203. Sivaprasad K, Stotz KC (1973) Reflection of electromagnetic pulses from a multilayered medium. IEEE Trans Geosci Electron 11(3):161–164
204. Small EE, Larson KM, Chew CC, Dong J, Ochsner TE (2016) Validation of GPS-IR soil moisture retrievals: Comparison of different algorithms to remove vegetation effects. IEEE J Sel Top Appl Earth Obs Remote Sens 9(10):4759–4770
205. Smith GS, King RWP (1974) The resonant linear antenna as a probe for measuring the in situ electrical properties of geological media. J Geophys Res 79(17):2623–2628
206. Smith G, Nordgard J (1985) Measurement of the electrical constitutive parameters of materials using antennas. IEEE Trans Antennas Propag 33(7):783–792
207. Solimene R, D'Alterio A, Gennarelli G, Soldovieri F (2014) Estimation of soil permittivity in presence of antenna-soil interactions. IEEE J Sel Top Appl Earth Obs Remote Sens 7(3):805–812
208. Sommerfeld A (1909) Über die Ausbreitung der Wellen in der drahtlosen Telegraphie. Ann Phys 28:665–737
209. Stuntebeck E, Pompili D, Melodia T (2006) Underground wireless sensor networks using commodity terrestrial motes. In: Poster presentation at IEEE SECON 2006, Reston, USA
210. Su D, Miro JV, Vidal-Calleja T (2015) Modelling in-pipe acoustic signal propagation for condition assessment of multi-layer water pipelines. In: Proceedings of the 2015 IEEE 10th conference on industrial electronics and applications (ICIEA). IEEE, New York, pp 545–550
211. Suherman S, Rambe A, Tanjung A (2018) Underground radio propagation on frequency band 97 MHz–130 MHz. Int J Eng Technol 7(3):722–726
212. Sun Z, Akyildiz IF (2009) Underground wireless communication using magnetic induction. In: Proceedings of the 2009 IEEE international conference on communications. IEEE, New York, pp 1–5
213. Sun Z, Akyildiz IF (2010) Magnetic induction communications for wireless underground sensor networks. IEEE Trans Antennas Propag 58(7):2426–2435
214. Sun Z, Akyildiz IF (2010) Channel modeling and analysis for wireless networks in underground mines and road tunnels. IEEE Trans Commun
215. Sun Z, Akyildiz IF (2013) Optimal deployment for magnetic induction-based wireless networks in challenged environments. IEEE Trans Wirel Commun 12(3):996–1005
216. Sun L, Li Y (2010) Acoustic emission sound source localization for crack in the pipeline. In: Proceedings of the 2010 Chinese control and decision conference. IEEE, New York, pp 4298–4301
217. Sun Z et al. (2011) MISE-PIPE: MI based wireless sensor networks for underground pipeline monitoring. Ad Hoc Netw 9(3):218–227
218. Sun Z, Wang P, Vuran MC, Al-Rodhaan MA, Al-Dhelaan AM, Akyildiz IF (2011) Border patrol through advanced wireless sensor networks. Ad Hoc Netw 9(3):468–477
219. Sun Z, Akyildiz IF, Kisseleff S, Gerstacker W (2013) Increasing the capacity of magnetic induction communications in RF-challenged environments. IEEE Trans Commun 61(9):3943–3952

220. Swathi S, Santhanam SM (2017) An efficient mi waveguide based underground wireless communication for smart irrigation. In: Proceedings of the 2017 14th IEEE India council international conference (INDICON). IEEE, New York, pp 1–6
221. Tai CT, Collin RE (2000) Radiation of a Hertzian dipole immersed in a dissipative medium. IEEE Trans Antennas Propag 48(10):1501–1506
222. Tan X, Sun Z, Akyildiz IF (2015) Wireless underground sensor networks: Mi-based communication systems for underground applications. IEEE Antennas Propag Mag 57(4):74–87
223. Tan X, Sun Z, Wang P, Sun Y (2020) Environment-aware localization for wireless sensor networks using magnetic induction. Ad Hoc Netw 98:102030
224. Temel S, Vuran MC, Lunar MMR, Zhao Z, Salam A, Faller RK, Stolle C (2018) Vehicle-to-barrier communication during real-world vehicle crash tests. Comput Commun 127:172–186
225. Tiusanen MJ (2005) Attenuation of a Soil Scout radio signal. Biosyst Eng 90(2):127–133
226. Tiusanen MJ (2006) Wideband antenna for underground Soil Scout transmission. IEEE Antennas Wirel Propag Lett 5(1):517–519
227. Tiusanen MJ (2008) Wireless Soil Scout prototype radio signal reception compared to the attenuation model. Precis Agric 10(5):372–381
228. Tiusanen MJ (2013) Soil scouts: description and performance of single hop wireless underground sensor nodes. Ad Hoc Netw 11(5):1610–1618
229. Toftgard J, Hornsleth SN, Andersen JB (1993) Effects on portable antennas of the presence of a person. IEEE Trans Antennas Propag 41(6):739–746
230. Tooker J, Vuran MC (2012) Mobile data harvesting in wireless underground sensor networks. In: Proceedings of the 2012 9th annual IEEE communications society conference on sensor, mesh and Ad Hoc communications and networks (SECON). IEEE, New York, pp 560–568
231. Tokan F et al. (2014) The lateral wave antenna. IEEE Trans Antennas Propag 62(6):2909–2916
232. Toro-Vazquez J, Rodriguez-Solis RA, Padilla I (2012) Estimation of electromagnetic properties in soil testbeds using frequency and time domain modeling. IEEE J Sel Topics Appl Earth Obs Remote Sens 5(3):984–989
233. Trang HTH, Hwang SO et al. (2018) Connectivity analysis of underground sensors in wireless underground sensor networks. Ad Hoc Netw 71:104–116
234. Ulaby FT (2007) Fundamentals of applied electromagnetics, 5 edn. Pearson Prentice Hall, Upper Saddle
235. Ulaby FT, Long DG (2014) Microwave radar and radiometric remote sensing. University of Michigan Press, Michigan
236. van der Velde R, Salama MS, Eweys OA, Wen J, Wang Q (2015) Soil moisture mapping using combined active/passive microwave observations over the east of the Netherlands. IEEE J Sel Topics Appl Earth Obs Remote Sens 8(9):4355–4372
237. Van Hieu B, Choi S, Kim YU, Park Y, Jeong T (2011) Wireless transmission of acoustic emission signals for real-time monitoring of leakage in underground pipes. KSCE J Civil Eng 15(5):805
238. Vaziri F, Huang SCF, Long SA, Shen LC (1980) Measurement of the radiated fields of a buried antenna at VHF. Radio Sci 15(4):743–747
239. Vereecken H, Schnepf A, Hopmans JW, Javaux M, Or D, Roose T, Vanderborght J, Young MH, Amelung W, Aitkenhead M et al. (2016) Modeling soil processes: review, key challenges, and new perspectives. Vadose Zone J 15(5):1–57
240. Vresk T, Čavrak I (2016) Architecture of an interoperable IoT platform based on microservices. In: Proceedings of the 2016 39th international convention on information and communication technology, electronics and microelectronics (MIPRO). IEEE, New York, pp 1196–1201
241. Vuran MC, Akyildiz IF (2010) Channel model and analysis for wireless underground sensor networks in soil medium. Phys Commun 3(4):245–254
242. Vuran MC, Silva AR (2009) Communication through soil in wireless underground sensor networks—theory and practice. Springer, Berlin, pp 309–347

243. Vuran MC, Salam A, Wong R, Irmak S (2018) Internet of underground things in precision agriculture: architecture and technology aspects. Ad Hoc Netw 81:160–173
244. Vuran MC, Salam A, Wong R, Irmak S (2018) Internet of underground things: Sensing and communications on the field for precision agriculture. In: Proceedings of the 2018 IEEE 4th world forum on Internet of Things (WF-IoT) (WF-IoT 2018), Singapore
245. Wait JR (1961) The electromagnetic fields of a horizontal dipole in the presence of a conducting half-space. Canadian J Phys 39(7):1017–1028
246. Wait J, Fuller J (1971) On radio propagation through earth. IEEE Trans Antennas Propag 19(6):796–798
247. Wang JR, Schmugge TJ (1980) An empirical model for the complex dielectric permittivity of soils as a function of water content. IEEE Trans Geosci Remote Sens GE-18(4):288–295
248. Wei Z, Yibing S, Yanjun L (2013) Design of acoustic wireless remote transmission system for logging-while-drilling data. In: Proceedings of the 2013 IEEE 11th international conference on electronic measurement and instruments, Vol 1. IEEE, New York, pp 53–57
249. Weir WB (1974) Automatic measurement of complex dielectric constant and permeability at microwave frequencies. Proceed IEEE 62(1):33–36
250. Wheeler HA (1961) Useful radiation from an underground antenna. J Res 65:89–91
251. Widrow B et al. (1967) Adaptive antenna systems. Proceed IEEE 55(12):2143–2159
252. Wong R (2017) Towards cloud-based center pivot irrigation automation based on in-situ soil information from wireless underground sensor networks. Master's thesis, University of Nebraska-Lincoln, Lincoln
253. Wu TT (1961) Theory of the dipole antenna and the two-wire transmission line. J Math Phys 2(4):550–574
254. Yan L, Wei D, Pan M, Chen J (2018) Downhole wireless communication using magnetic induction technique. In: Proceedings of the 2018 United States national committee of URSI national radio science meeting (USNC-URSI NRSM). IEEE, New York, pp 1–2
255. Yang S, Baltaji O, Hashash YM, Singer A (2018) SoilComm: a miniaturized through-soil wireless data transmission system. J Acoust Soc Am 144(3):1872–1872
256. Yoon S-U, Cheng L, Ghazanfari E, Wang Z, Zhang X, Pamukcu S, Suleiman MT (2012) Subsurface monitoring using low frequency wireless signal networks. In: Proceedings of the 2012 IEEE international conference on pervasive computing and communications workshops. IEEE, New York, pp 443–446
257. Yu X, Han W, Wu P, Zhang Z (2015) Experiment of propagation characteristics based on different frequency channels of wireless underground sensor network in soil. Trans Chin Soc Agric Mach 46(4):252–260
258. Zemmour H, Baudoin G, Hamouda C, Diet A, Biancheri-Astier M (2015) Impact of soil on UWB buried antenna and communication link in IR-UWB WUSN applications. In: Proceedings of the 2015 European Radar conference (EuRAD), pp 353–356
259. Zemmour H, Baudoin G, Diet A (2016) Soil effects on the underground-to-aboveground communication link in ultrawideband wireless underground sensor networks. IEEE Antennas Wirel Propag Lett 16:218–221
260. Zhang Y, Ning J, Yang S, Cui H-L (2007) Field test investigation of fiber optic seismic geophone in oilfield exploration. In: Fiber optic sensors and applications V, Vol 6770, pp 677005. International Society for Optics and Photonics, Bellingham
261. Zheng Z, Hu S (2008) Research challenges involving cross-layered communication protocol design for underground WSNS. In: Proceedings of the 2008 2nd international conference on anti-counterfeiting, security and identification. IEEE, New York, pp 120–123
262. Zhou X, Yu Q, Peng W (2012) Simultaneous measurement of down-hole pressure and distributed temperature with a single fiber. Meas Sci Technol 23(8):085102
263. Zia H, Harris NR, Merrett GV, Rivers M, Coles N (2013) The impact of agricultural activities on water quality: a case for collaborative catchment-scale management using integrated wireless sensor networks. Comput Electron Agric 96:126–138
264. Zrelli A, Ezzedine T (2018) Design of optical and wireless sensors for underground mining monitoring system. optik 170:376–383

Chapter 2
Electromagnetic Characteristics of the Soil

2.1 Introduction

Subsurface electromagnetic (EM) wave propagation is governed by the structure of earth crust and its electrical properties. Earth crust is a relatively cool area above the earth's core and mantle. Its thickness is up to 30 miles. The characteristic of this layer, as a transmission medium, determines the feasibility of propagation of EM waves. Particular characteristics of interest are thickness, different formations, type of the rock, and electrical conductivity.

An ideal earth crust is composed of 5% sedimentary rocks and 95% igneous rock. Sedimentary rocks are shale, sandstone, and limestone, whereas igneous rocks are granite and basalt. Most sedimentary rocks are composed of deposits from ocean bottom; therefore, pore spaces have high conductivity due to conducting seawater and it is not suitable for EM waves propagation. However, terrestrial and chemical deposited rocks (e.g., chemically precipitated dense limestones) are different and this rule does not apply to them.

Under the highly conductive sedimentary rock lies igneous rock. Igneous rock is formed by crystallization of molten rock material. The igneous rock has thickness ranging from 1000 ft to 30 miles below the surface. Owing to its origin, it can accumulate the ocean water and is very suitable for the communication experiments. This rock is homogeneous formation of granite and basalt between the mantle and sedimentary surface. However, sedimentary and upper granite are relatively less homogeneous.

EM propagation in earth depends upon the conductivity σ and dielectric constant ϵ of the earth medium (soil); however, upper portion (a few thousand feet above) is rather a complex medium for radio propagation with continuous and sudden changes in the condition. The propagating surface has varying conductivity and dielectric constants for up to a few miles. After the sedimentary section, the rock structure is more relatively uniform in the higher depths. There have been many measurements done for dielectric constants and conductivity of the rock.

A. Salam, U. Raza, *Signals in the Soil*,
https://doi.org/10.1007/978-3-030-50861-6_2

As per minéralogie character, conductivities show variations over large magnitude; however, dielectric constants do not change much. Conductivity of a rock is an important parameter as path loss is dependent upon it. The conductivity of a rock is majorly determined by the water content and minorly by mineral content. Conductivity varies greatly for the dry and wet rocks. Granite is the most suitable rock for communication due to its low conductivity. Water content in the rock has a great effect on the performance of the communication; therefore, antennas are buried deeply into the ground to reach low conductivity rock (granite). The conductivity value of granite in Appalachian Mountain chain from Maine to Pennsylvania was measured as 10 5 mhos/m by Dr. G. V. Keller of the United States Geological Survey. The conductivity decreases with an increase in depth. However, this trend is observed only for the depth of a few miles. The temperature induced conductivity comes in effect as the depth goes beyond several miles. The conductivity value increases significantly (same as conductivity values in the crust) with the increase in temperature. If the in-homogeneity of the rock is ignored, the ideal model of rock conductivity is obtained as follows: the upper layer of the earth is relatively thin and has high conductivity, whereas lower layer is thick (up to 20 miles) and has low conductivity values. This acts as the middle layer between the upper thin layer and very low high temperature and high conductivity layers (Fig. 2.1).

The knowledge of dielectric properties of rocks, both solid and in powdered form, comes very handy while working on geophysical and geological applications. Rock are the collection of mineral grains bounded together by molecular interaction forces. Dielectric properties can be measured in many ways. For example, in [47, 60, 68], the authors perform a broadband measurement for dielectric permittivity. The

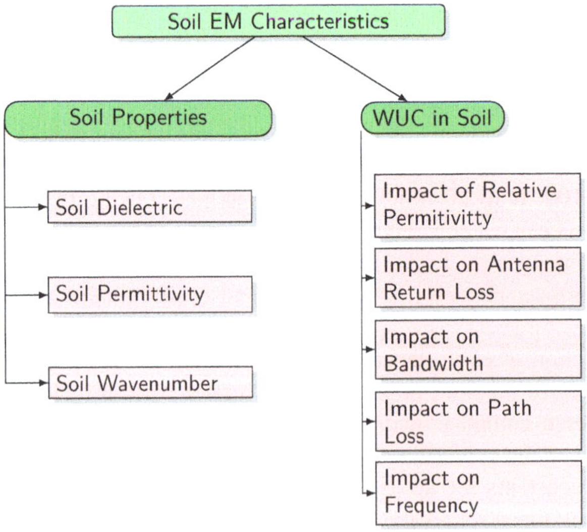

Fig. 2.1 Organization of the chapter

real part of the permittivity, between 450 MHz and 35 GHz, is around 5–6 for granite rocks. It does not depend on the frequency as the loss factor is very low, i.e., around 0.01–0.05. In another study [38, 48, 61], Parkhomenko lists very detailed dielectric property from various minerals in the frequency region of MHz and below.

2.1.1 Porous Bedrock

Bedrock is very heterogeneous ranging from microscopic hair cracks to very large scale of heterogeneity. They are porous allowing water to flow through them. The level of porosity varies from less than 1% (certain granite) to even 20–40% (shale and sandstone) [20, 32, 43]. The dielectric property is water dependent; therefore, it also depends upon the rock's porosity and micro-geometry.

At lower frequencies, the conductivity is found to be greater than the dielectric property of the brine-saturated water. Therefore, a relation between the porosity and conductivity of the medium can be found. Archie's law [3, 41, 42] gives one such approximate relation between water conductivity σ_w and mixture conductivity σ_{eff}:

$$\frac{\sigma_w}{\sigma_{eff}} = f^m, \tag{2.1}$$

where f is the porosity, m is the cementation exponent and predicts the conduction behavior based on various factors present in the rock. Its value for clay-free sedimentary rock is between 1.3 and 4. For decades, Archie's law is being used to estimate water saturation from electrical conductivity (EC) in borehole logging.

There are other different models for the percolation process, e.g., Bruggeman model, Maxwell Garnett model, or Coherent potential model. These models predict a non-zero porosity value f_c, also known as percolation threshold. At f_c, macroscopic conductivity changes suddenly, and above f_c, the conductivity is proportional to $(f - f_c)^m$. However, Archie's law gradually measures the increasing conductivity and starts from the percolation threshold value of zero. Therefore, Archie's law is not compatible with the mentioned models of percolation process [33, 46, 58].

There are some models which are compatible with the Archie's law. For example, the model presented in Sen et al. [60] measures effective permittivity as:

$$\frac{\epsilon_{eff} - \epsilon_e}{\epsilon_i - \epsilon_e} = f\left(\frac{\epsilon_{eff}}{\epsilon_i}\right)^{1/3}, \tag{2.2}$$

where host (ϵ_e) is the rock matrix, the inclusions (ϵ_i) represent water phase having volume fraction f same as porosity.

If we apply Eq. (2.2) to water-saturated rock matrix, then ϵ_e will be real because of insulating environment, and inclusion conductivity will become $\epsilon_i = \epsilon_i' - j\sigma_w/\omega$. Imaginary part will be dominant at lower frequencies, hence, $\epsilon_i \approx j\sigma_w/\omega$, and the effective permittivity will be

$$\left(\frac{\sigma_{eff}}{\sigma_w}\right)^{2/3} = f, \tag{2.3}$$

where the real part is negligible and imaginary part will be: $\epsilon_{eff} = \epsilon_{eff}^{'} - j\sigma_{eff}/\omega \approx -j\sigma_{eff}/\omega$ which leads to Archie's law: $\sigma_{eff}/\sigma_w = f^{2/3}$ having value of exponent $m = 1.5$. The exponent value of 1/3 in Sen et al. model is because of considering different sized spheres; for ellipsoids with depolarization factor N, they will become as:

$$\frac{\epsilon_{eff} - \epsilon_e}{\epsilon_i - \epsilon_e} = f\left(\frac{\epsilon_{eff}}{\epsilon_i}\right)^N \tag{2.4}$$

which predicts conductivity at low-frequency limit as per: $\epsilon_{eff}/\epsilon_w = f^{1/(1-N)}$. It gives the freedom to choose the cementation exponent value as depolarization factor, in case of ellipsoid, is between 0 and 1.

In the light of current research, Archie's law seems to be an oversimplification [28, 51, 61] because it fails in carbonate rocks which have very inhomogeneous structure and complex pore structure. It has economical significance when it is applied in important application of estimating hydrocarbon reservoirs of the world.

Anisotropy plays an important role in rock material. Figure 2.2 shows that special directions are visible in sedimented rocks and stones. Anisotropy is highly dependent upon the scale, e.g., the dielectric anisotropy for Gypsum can be around 1.5–2 at lower frequencies.

2.1.2 Soil

Soil is a complex substance made up of different components. It contains different solid materials, air, free and bound water. There are different types of soil covering the surface of the earth. It is useful to separate the dry and wet components of soil while estimating soil's dielectric properties. The permittivity of dry soil is better understood than the wet soil because dielectric losses can lead to water contribution [38, 45, 52].

Dry soil is a two-component mixture. The inclusion of rock material takes a certain volume in air.[1] Soil can be classified into three types: clay, silt, and sand. This classification is on the basis of the size of the soil's inclusions and increasing order particles coarseness. However, the dielectric properties of the soil are independent of the particle. size. Soil permittivity ϵ_s can be approximately related to soil density ρ_s.

[1]This oversimplified assumption for modeling the dielectric. Soil is a complex fragmented rock consisting of plants microscopic organisms, fungi, plant, and other inorganic matter.

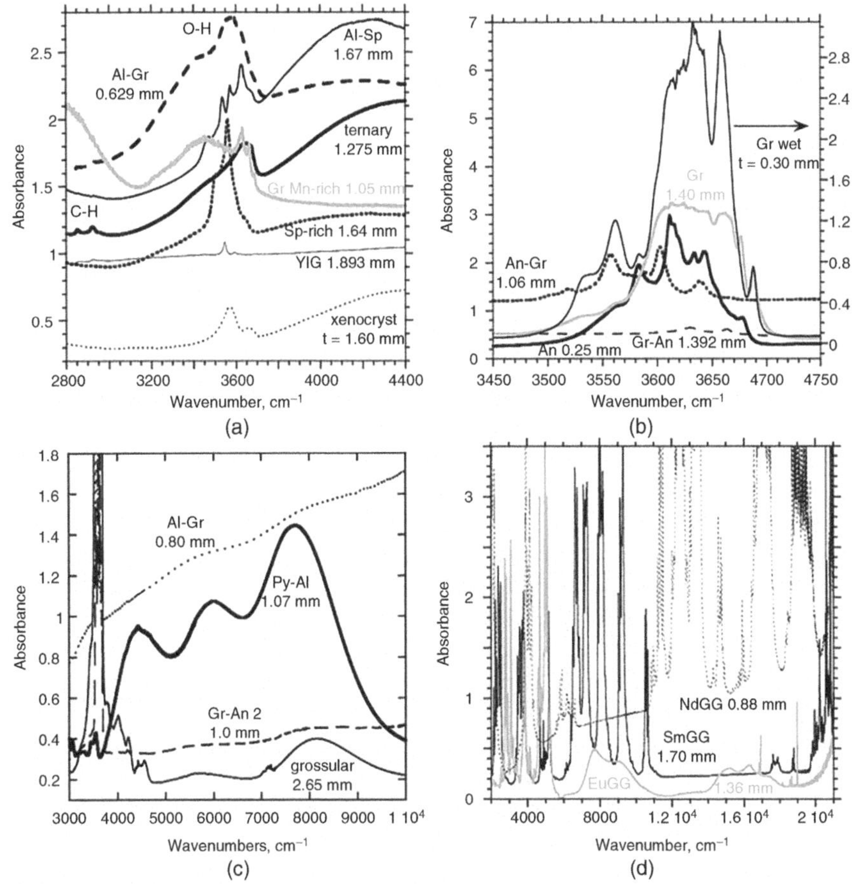

Fig. 2.2 The near-IR spectra of representative samples [16, 48, 56]

In literature, the relation between the soil density and soil permittivity is given as a polynomial, e.g., in [62] it is given as follows:

$$\epsilon_s/\epsilon_0 = (1 + 0.5\rho_s)^2, \tag{2.5}$$

where the soil density is measured as g/cm^3. Dobsen et al. [9] perform the modeling on the basis of measurements taken in the range of 1.4 and 18 GHz. It gives slightly lower prediction as:

$$\epsilon_s/\epsilon_0 = (1.01 + 0.44\rho_s)^2 - 0.062. \tag{2.6}$$

The above models are the empirical models. Birchak refractive index model averages out the square roots of permittivity of all components and can be connected to the above-mentioned models. Of course, the soil density is directly proportional to the amount of solid matter present in the soil and it is raised to second power while calculating permittivity [34, 50, 59].

Campbell and Ulrichs [7, 37, 39] studied different types of terrestrial rocks and includes extensive measurements. They also proposed a classical mixing model for predicting dielectric properties of powdered rocks. The result was that Maxwell Garnett mixing model, having spherical inclusions, matches the measurements from rocks like olivine peridotite powder and olivine basalt. However, it does not work well with some other rock types such as granite and aplite. Bruggeman model worked better for these rocks. All measurements were done at 450 MHz.

The permittivity was measured for dry Sahara Desert [23, 49, 55] between the frequencies of 245 MHz and 6 GHz. The real component (around $\epsilon_s/\epsilon_0 \approx 2.53$) of the permittivity remained constant above 1 GHz. Along with the sand density of 1.45 g/cm^3, these measurements follow the model presented in Eqs. (2.5) and (2.6). In microwave region, the imaginary part was shown to be decreasing. The Debye-type spectrum suggests relaxing the frequency around 270 MHz.

The permittivity of the dry soil is not affected much by the dry soil; however, it is not true in case of wet or moist soil. A clay surface area is much greater than the sand because clay has much larger, in many order of magnitude, number of particles as compared to sand. Therefore, the liquid water distribution is completely different for both of the cases. Moreover, the surface absorbs more liquid water than the bound water; hence, more dependency of soil texture on permittivity can be expected [29, 30, 66].

In Hallikainen et al. [15, 31, 50, 53], the effect of temperature, soil type, frequency, and wetness, on the wet soil permittivity are investigated. It concludes that the permittivity is highly dominated by the water phase. For example, values for real and imaginary part of relative permittivity are 25 and 10 for 1 GHz frequency and water fraction of 0.4. The imaginary part is over two orders of magnitude as compared to what it is for dry sand. The results also show a clear soil type dependency [35, 67]. This dependency is more evident for lower frequencies. Even with the same moisture, coarse-structured soil (sand and loam) has higher permittivity as compared to fine-grained soil (silt and clay). This may be because of greater specific surface area, fine-grained soil attaches more water to the surface, and sand has more free water. The permittivity of the free water is greater than the bound water. The reason for this is because water molecules are bounded in bound water and do not respond freely to the electric field.

2.1.3 Rain Attenuation

Fog is very transparent at microwave frequencies; however, rain is not. For a rainy troposphere, the attenuation in radio signals can be calculated by measuring the loss of air–precipitation mixture effective permittivity; however, water fraction in air must be known for that.

The size of a raindrop, within the rain, depends on the intensity of the rainfall. Mostly Marshall–Palmer drop size distribution is used in the literature which calculates number of drops having diameter d per unit volume from the following distribution:

$$p(d) = N_0 e^{-bd}, \tag{2.7}$$

where $N_0 = 8.0 \times 10^6\,\mathrm{m}^{-4}$ and exponent b is calculated as follows:

$$b = 4100 R_r^{-0.21} [1/\mathrm{m}], \tag{2.8}$$

where rain rate is measured in $\mathrm{mm\,h}^{-1}$. The relation between rate and fraction volume f of water phase in air is given by integration as:

$$f = 8.894 \times 10^{-8} R_r^{0.84}. \tag{2.9}$$

The attenuation A is calculated using the imaginary part of effective permittivity of dilute water–air mixture:

$$A = \frac{8686\pi}{\lambda} \epsilon''_{eff,r} [\,\mathrm{dB\,km}^{-1}], \tag{2.10}$$

where wavelength λ is in meters, and imaginary part of relative effective permittivity is given as:

$$\epsilon''_{eff,r} = -3 f Im \left\{ \frac{\epsilon_w - \epsilon_0}{\epsilon_w - 2\epsilon_0} \right\} \tag{2.11}$$

and raindrops are assumed to be spherical. Small raindrops are spherical for sure; however, as the size increases they start to resemble oblate ellipsoid [24, 57]. Figure 2.3 plots attenuation with rain rate at 1 GHz of frequency. It shows that attenuation is directly proportional to the rain rate. Only the absorption was considered for the calculation. However, this assumption was undermined with increasing frequency mainly because of increase in scattering and, secondly, it is not possible to calculate absorption while assuming that internal field of raindrop remains the same throughout the volume (a quasi-static assumption). However, even in the cases where rain rate is higher, the dilute mixture assumption remains valid because inclusion fraction is below the order of 10^{-6}.

An experiment was performed to estimate the safety of neglecting contribution due to high frequency in Fig. 2.3. The experiment compared absorption due to quasi-static assumption with extinction efficiency Q_{ext} of a spherical raindrop. Q_{ext} is estimated by full Mie theory.

$$Q_{abs,qs} = 4x Im \left\{ \frac{\epsilon_w - \epsilon_0}{\epsilon_w + 2\epsilon_0} \right\}, \tag{2.12}$$

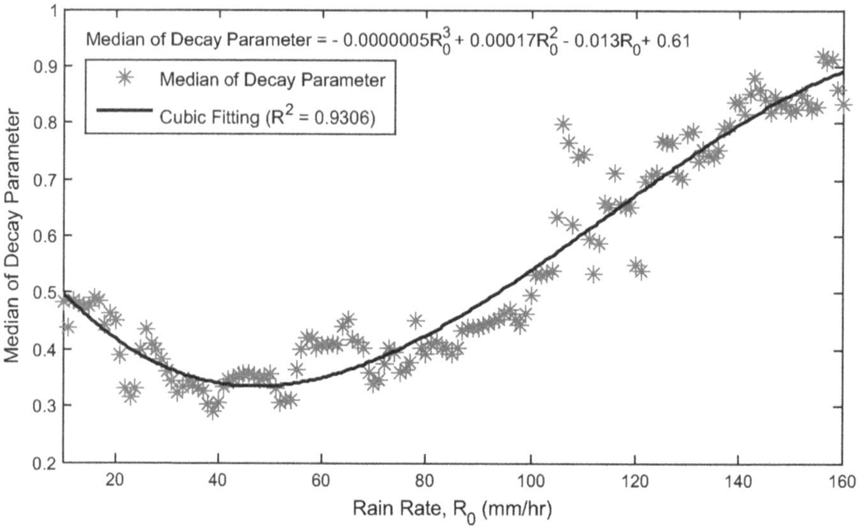

Fig. 2.3 Variation of the median values of decay parameter with rain rate [27]

where $x = k_0 a$ is the size of sphere.

Figure 2.4 plots the comparison where the extinction is plotted with the frequency. Here, the extinction is calculated with the raindrop having a diameter of 1 mm. After 5 GHz, the quasi-static assumption about absorption fails. There are different drop sizes in the rain and the average drop size is directly proportional to the rain rate as shown by Marshall–Palmer distribution function.

Oguchi [25] presents a survey on scattering from hydrometers. The review of mixing approach and connection of backscattering and mixing approach can be seen in [5, 6].

2.2 Dielectric Soil Properties Model

Soil is considered as a dielectric material which is composed of bound water, air, free water, and bulk soil. It is mainly described by its relative permittivity or dielectric constant. EM waves are affected by the permittivity of the medium through which they pass. Relative permittivity has a direct relation with EM wave propagation, i.e., smaller the permittivity, better will be EM propagation. Soil with smaller bulk density will be more porous having more air in the medium, therefore, causing lower attenuation in EM waves. The presence of water in the soil has a direct relation with the attenuation in the soil and soil parameters like the volumetric water content (VWC) are the main causes of attenuation in EM wave propagation [1, 2, 47, 63–65].

VWC is made up of the bound water and the free water. The bound water is a thin layer of water which strongly bounded to the surface of the soil. The free water is

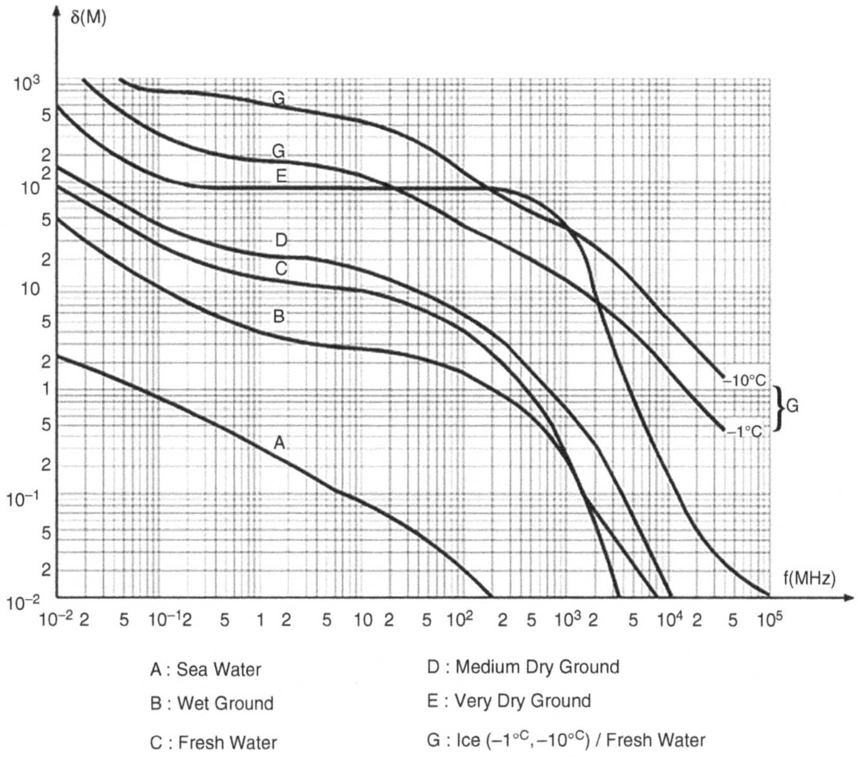

Fig. 2.4 Variations of penetration depth vs. frequency [14]

loosely or freely held water molecule in the soil [4, 48]. Therefore, even if the VWC value of two soil samples is the same, one sample may contain more free water as compared to others because of different sample texture. Bound water is determined by the texture and clay quantity in the soil [4, 13, 40]. This will further be explained in Sect. 2.4, where power is varied more than 20 dB for samples having the same VWC but different textures. This is because of the difference in dielectric properties of bound and free water. Therefore, the dielectric model weighs the bound and free water differently.

A soil with predominantly clay particles is worse for underground communication as compared to sandy soil. A well-known fact is that one gram of clay retains more water as compared to similar amount of sand or silt [13, 36]. A surface area of soil particle is directly proportional to the water absorption in soil and inversely proportional to the size of soil particles. Clay particle has a high surface area and small particle size. Therefore, they absorb more water as compared to sand and silt.

Soil permittivity also depends on the frequency of the signal; hence, in addition to VWC, soil permittivity is also affected by the frequency of the signal [4, 22, 45]. Therefore, it can be said that frequency indirectly affects the attenuation of EM waves. Soil conductivity, along with certain frequency ranges, causes the signal attenuation in soil [4, 21]. This negates the general belief of lower signal attenuation under low frequencies [4, 26]. It is true for most of the cases; however, there are some exceptional frequency ranges which do not support this argument. Even if values for all soil parameters are available, there is no direct method or formula for calculating soil permittivity using these parameters and, without soil permittivity, it is not possible to estimate signal attenuation in soil.

There has been many frequency specific dielectric model proposed to lessen the severity of the above problem [4, 8, 44]. As these models were for some specific applications with restricted frequency ranges, many assumptions can be made to simplify the model. Thus, the model is simplified without losing its accuracy. Consequently, it is important to determine what frequencies must be used in WUSNs while developing dielectric model of Zenneck waves (ZW).

Frequencies around 1 GHz are used for practical wireless communication and microwave remote sensing and give smaller values of soil permittivity [2, 17, 26, 54]. Signal can have smaller attenuation under the frequencies lower than 300 MHz; however, smaller frequencies cause an increase in wavelength and antenna size. Therefore, it is not recommended to use frequencies below 300 MHz in WUSNs. Considering these facts about frequency, ZW model employs a balanced approach of achieving practicality with accuracy. Therefore, semi-empirical soil dielectric model presented in [26] is selected for ZW model. This model limits the frequency range to 300–1300 MHz.

2.3 Permittivity and Wavenumber in Soil

This section discusses the analysis of the constitutive parameters of the soil at different soil moisture levels. It also studies the dielectric constant in silt loam and sandy soil types at different frequency and water content levels. A detailed analysis of the impact of soil moisture variations on permittivity and wavenumber in soil has also been carried out. This section also presents the effect of change in soil moisture on wavenumber and soil permittivity. Phase constant in the soil is given by β_s and that of air is represented by β_0. Similarly, wavelength in the soil is represented by λ_s, and λ_0 is the wavelength in air. Complex wavenumber $k_s = \beta_s + i\alpha_s$, where α_s represents the attenuation constant in soil, is used to calculate β_s/β_0 and λ_s/λ_0. Figure 2.5 plots the change in these parameters with the change in soil moisture for sandy, silt loam, and silty clay loam soil types. It can be observed in Fig. 2.5a increasing the soil moisture results in increase of β_s/β_0. For example, when soil moisture is 40%, the phase shift increases 5% more than what it is in free space β_0. Figure 2.5b plots the change in wavelength λ_s/λ_0 with OTA wavelength. It is observed that frequency shift is comparatively less than that of OTA and is in direct relation with soil moisture levels, i.e., decreases with decreasing soil moisture level and vice versa. Moreover, the difference in frequency shift for different soil types shrinks as the soil moisture levels increase. This behavior is because of high permittivity of the soil (Fig. 2.6).

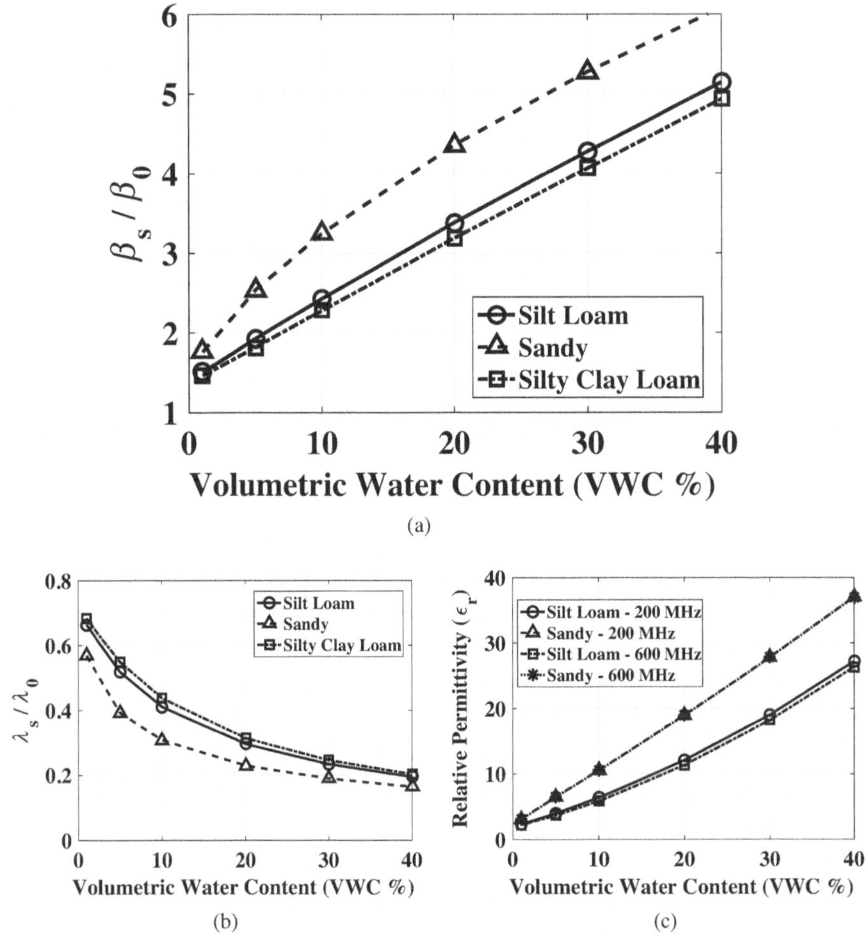

Fig. 2.5 (**a**) Change in β_s/β_0 [39], (**b**) Effect of soil moisture on the parameter λ_s/λ_0 in three different soil types [39], (**c**) Effect of soil moisture on the relative permittivity in three different soil types at different frequencies [39]

Figure 2.5c plots the relative permittivity with soil moisture for silt loam and sandy soil. The experiment uses frequency of 200 and 600 MHz. It can be observed that: (a) relative permittivity of sandy soil is affected more than that of silt loam soil and (b) frequency does not have any effect on soil permittivity. This is due to dielectric and conduction losses. Dielectric losses cause soil moisture to change because of water held by the soil is released [9, 31].

Underground wireless signal can propagate through three different paths, namely reflected, directed, and lateral as shown in Fig. 2.7. Directed and reflected are the possible signal paths in the soil and they remain completely in the soil. However,

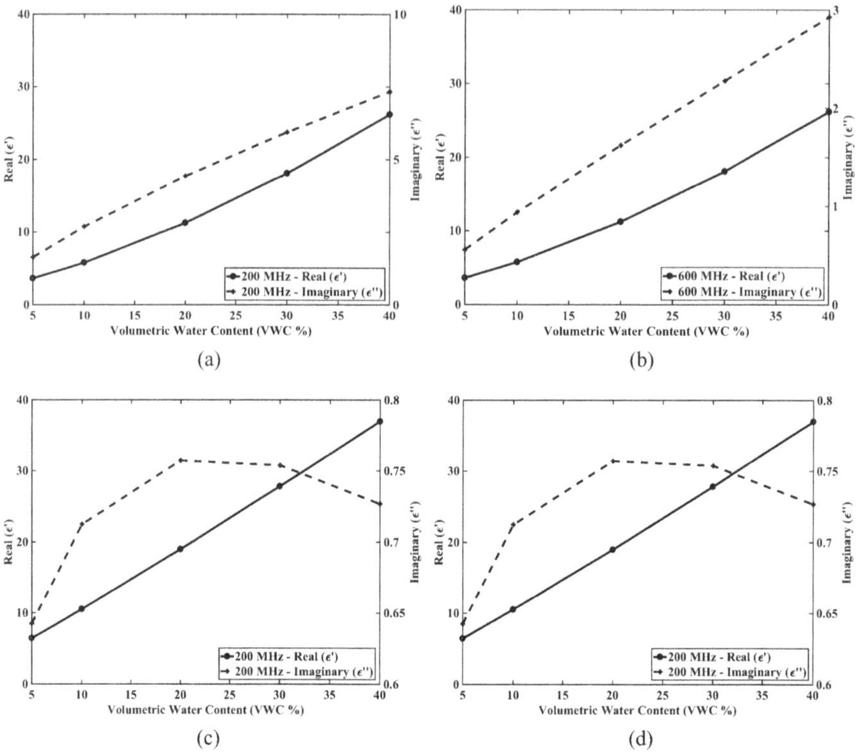

Fig. 2.6 Soil moisture effect on real and imaginary part of dielectric constant different frequencies. (**a**) Silt loam—200 MHz. (**b**) Silt loam—600 MHz. (**c**) Sand—200 MHz. (**d**) Sand—600 MHz

the lateral component can pass along soil–air interface right above the soil surface. All three UG channel components are studied with great detail in [10, 45, 46, 55].

Figure 2.6 plots the dielectric constant with soil moisture levels. The experiments are performed for sandy and silt loam soil at frequency of 200 and 600 MHz. It can be seen that ϵ_s' is directly proportional to VWC, i.e., increasing linearly with increase in VWC.

2.4 Impact of Soil on Wireless Underground Communication

This section discusses how wireless communication is affected by the soil. There are mainly three ways in which soil can impact the communication: return loss of the antenna, the path loss of the propagation, and bandwidth of the antenna. Soil permittivity is affected greatly by change in soil moisture and frequency. In the upcoming sections, all three impacts are studied separately [50, 58].

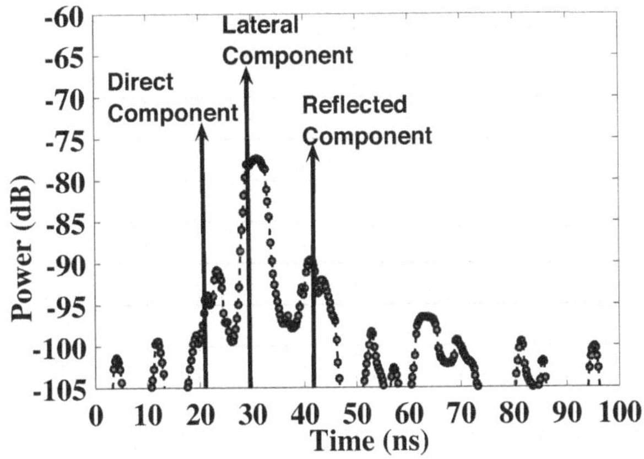

Fig. 2.7 The three components of the wireless subsurface propagation channel in soil [55]

2.4.1 *Relative Permittivity of Soil*

The permittivity of soil is higher than the air due to which it changes the wavelength of EM waves incident upon it. Soil permittivity relies on many soil parameters such as: soil texture, bulk density, salinity, soil moisture (volumetric water content), and temperature. There has been intensive investigation for modeling the characteristics of relative permittivity [9, 26, 66]. These models explain the relative permittivity of various soil–water mixture components: free water, soil, air, and bounded water [9]. The effective permittivity of soil–water mixture is the overall permittivity of air, soil, and water and is a complex number (Fig. 2.8).

2.4.2 *The Impact of Soil on the Return Loss of an Antenna*

The return loss of buried antenna changes because of permittivity of the soil. Moreover, change in soil moisture causes a change in soil permittivity and antenna return loss. Figure 2.9a shows the result of the experiments performed with a 70 mm monopole antenna with different soil moisture levels which confirms that increasing soil moisture causes resonant frequency, corresponding to return loss, to shift in lower spectrum.

The return loss occurs due to impedance mismatch; therefore, it is important to calculate the impedance first. However, there is no closed form representation of an arbitrary antenna but a good impedance approximation for dipole antenna is provided in [19, 21, 47, 69]. The analysis of impedance model in [19, 37] is analyzed below. As per [19], the input impedance of a dipole can be approximated by using

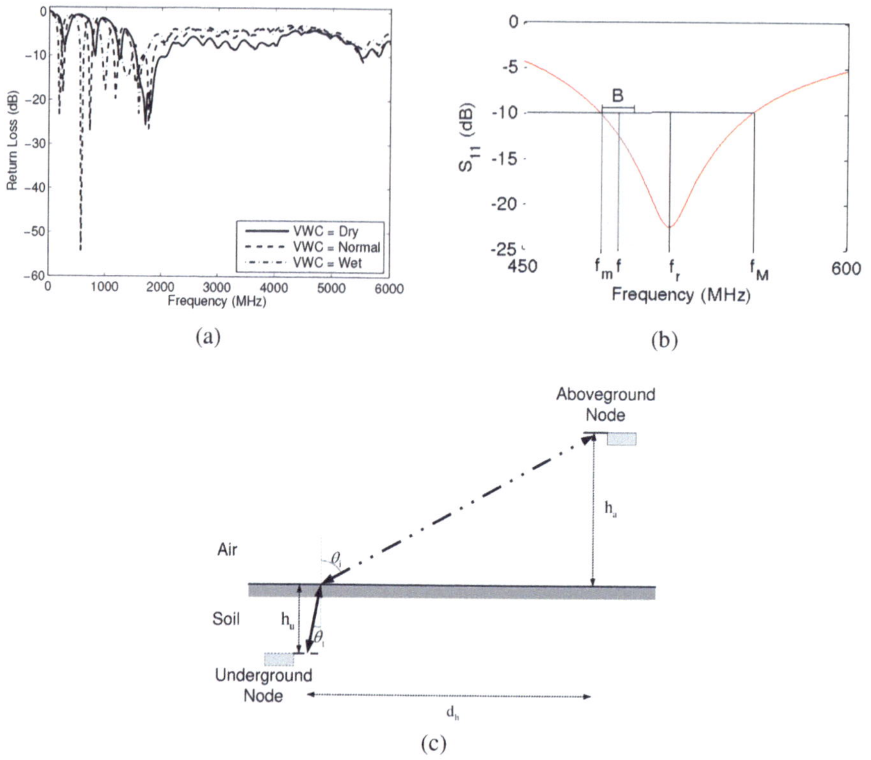

Fig. 2.8 Models: (**a**) Return loss of an antenna estimated at different soil moisture values, (**b**) Bandwidth at non-resonant frequency, (**c**) Channel model [11]

induced-EMF method as follows:

$$Z_a \approx f_1(\beta l) - i\left(120\left(\ln\frac{2l}{d} - 1\right)\cot(\beta l) - f_2(\beta l)\right), \tag{2.13}$$

where

$$f_1(\beta l) = -0.4787 + 7.3246\beta l + 0.3963(\beta l)^2 + 145.6131(\beta l)^3,$$

$$f_2(\beta l) = -0.4456 + 17.0082\beta l - 8.6793(\beta l)^2 + 9.6031(\beta l)^3$$

β is a real part of wavenumber, diameter of d is given as d, and half of the length of the dipole is given as l. βl is expressed as :

$$\beta l = \frac{2\pi l}{\lambda_0} Re\sqrt{\epsilon_s}, \tag{2.14}$$

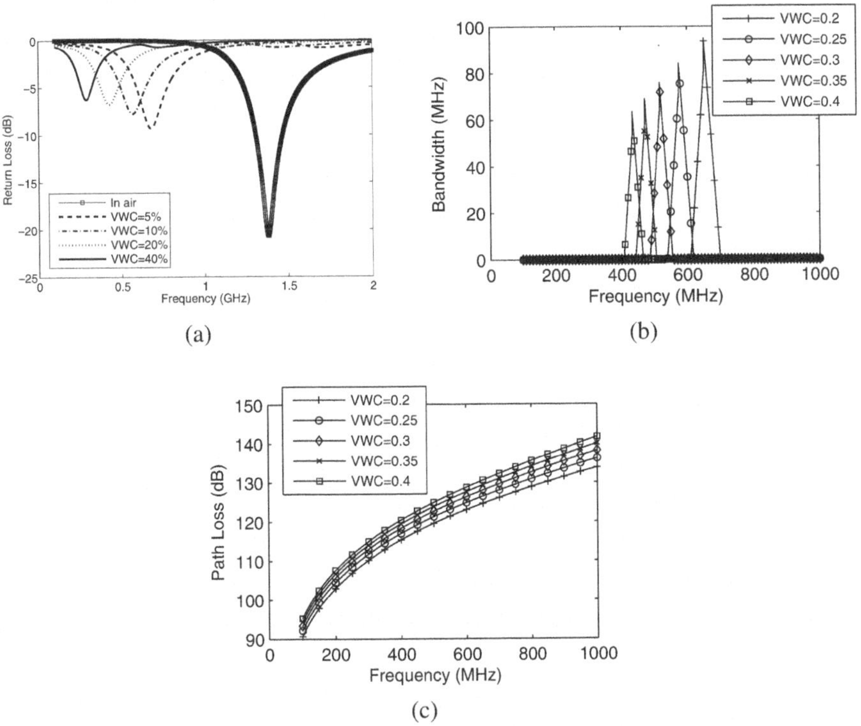

Fig. 2.9 (**a**) Return loss, (**b**) Bandwidth, (**c**) Path loss in UG2AG channel [11]

where λ_0 is the wavelength in the air, and ϵ_s is the soil relative permittivity [26]. As the ϵ_s relies on frequency so βl is not linear function of l/λ_0. Therefore, the resonant frequency changes which in turn changes the impedance at those frequency values when the antenna is moved from air to soil [30, 39].

The return loss is because of mismatch in impedance. The return loss of antenna is express in dB and is calculated as:

$$\mathrm{RL_{dB}} = 20\log_{10}\left|\frac{\mathrm{Z_s} + \mathrm{Z_a}}{\mathrm{Z_s} - \mathrm{Z_a}}\right|, \tag{2.15}$$

which approximately matches the experiment results.

2.4.3 *The Impact of Soil on Bandwidth*

Channel capacity increases with the increase in bandwidth; hence, bandwidth also affects the performance of underground communication. In wireless communication, bandwidth is affected by antenna and return loss of antenna is calculated from bandwidth. Return loss RL (see Fig. 2.9) is the function of frequency ($RL = R(f)$). When antenna is excited at resonant frequency, the bandwidth refers to the spectrum where negative of the return loss $< \delta$. Bandwidth always remains smaller than the resonant frequency when no resonant frequency is being used. Bandwidth is calculated as:

$$B = \begin{cases} 0 & if - R(f) > \delta, \\ 2(f - f_m) & if - R(f) \leq \delta \; and \; f < f_r, \\ 2(f_M - f) & if - R(f) \leq \delta \; and \; f \geq f_R, \end{cases} \tag{2.16}$$

where f_r is the resonant frequency, f_m and f_M represent the lowest and highest frequency at which $R(F) \leq \delta$.

Figure 2.9b plots negative of RL, i.e., S_{11} with frequency f. Operational frequency is higher and there is a difference of 24 MHz between operational frequency and resonant frequency, and $\delta = -10$ dB. It can be seen that $S_1 1 < \delta$ for whole spectrum and bandwidth is 14 MHz.

2.4.4 The Impact of Soil on Path Loss

In [10, 12, 41, 46], aboveground-to-underground (AG2UG) and underground-to-aboveground (AG2UG) channel in WUSN are studied in detail. Figure 2.9c depicts the communication path where attenuation in EM waves passing through the soil is the function of distance, soil moisture, and soil type. Irrespective of the direction, path loss is given as:

$$L = (L_{ug}(d_{ug}) + L_{ag}(d_{ag}) + L_{(R,\rightarrow)}), \tag{2.17}$$

where $L_{ag}(d_{ag})$ and $L_{ug}(d_{ug})$ are the loss at aboveground and underground portions, respectively. $L_{(R,\rightarrow)}$ is the refraction loss on the basis of direction of the propagation, $\rightarrow$,, i.e., $ug2ag$ or $ag2ug$.

The UG and AG losses from Eq. (2.17) can be calculated as:

$$L_{ug}(d_{ug}) = 6.4 + 20\log d_{ug} + 20\log\beta + 8.69\alpha d_{ug}, \tag{2.18}$$

$$L_{ag}(d_{ag}) = -147.6 + 10\eta\log d_{ag} + 20\log f, \tag{2.19}$$

respectively, η represents the attenuation coefficient in air, β is the phase shifting constant, f is the operational frequency, and α denotes the attenuation constant. $\eta > 2$ because of ground reflection. The η range is 2.8–3.3 and is estimated by empirical experiments [7]. Last two terms in Eq. (2.18) give the effect of soil properties on attenuation, where α and β are calculated as:

$$k_s = \alpha + i\beta = i\omega\sqrt{\mu_0 \epsilon_s}, \tag{2.20}$$

where k_s denotes the propagation constant in soil, ϵ denotes the effective soil permittivity, and μ_0 denotes permeability in free space. EM waves can only enter the soil–air interface if the incident angle is small; otherwise, it is refracted and reflected at the interface. It is mainly because of high permittivity value of the soil [32, 42]. Only waves having small incident angle θ_t, as shown in Fig. 2.9c, will transmit to air in UG2AG propagation. In AG2UG propagation, the propagation in soil is in the vertical direction where refracted angle is approximately close to zero. Therefore, the underground distance in both links can be approximated as $d_{ug} \simeq h_u$, where h_{ug} denotes the burial depth and aboveground distance in both links can be approximated as $d_{ug} = \sqrt{h_a^2 + d_a^2}$, where h_a denotes the height of the AG node and d_g horizontal distance between the node.

Maximum power path is considered for the AG2UG link where incident angle $\theta_i \to 0$. Therefore, refraction loss $L_{(R,\to)}$ from Eq. (2.17) can be approximated as [18]:

$$L_{(R,ag2ug)} \simeq 20\log\frac{n+1}{4}, \tag{2.21}$$

where refractive index of soil is represented by n and calculated by Dong et al. [12] as:

$$n = \sqrt{\frac{\sqrt{\epsilon'^2 + \epsilon''^2} + \epsilon'}{2}}. \tag{2.22}$$

The signal in UG2AG link propagates from higher density medium to lower density medium in a perpendicular direction; hence, all energy is refracted, i.e., $L_{(R,ug2ag)=0}$.

2.4.5 *The Impact of Soil on Frequency*

In soil, higher frequencies suffer more attenuation because when EM waves propagate their wavelength shortens due to higher permittivity of soil than the air. Hence, due to less effects of permittivity of soil on lower frequency spectrum, it is more suitable for UG2UG communication as larger communication distances can be achieved. In order to have minimum attenuation, an operation frequency should be selected, for each distance and depth, such that attenuation is minimized. This is important from IOUT topology design perspective because deployment needs to be customized to the soil type and frequency range of sensors being used for deployment.

In Fig. 2.10, attenuation with frequency at different distances of up to 12 m is presented. Transmitter and receiver depths are set to 20 cm. At 2 m distance,

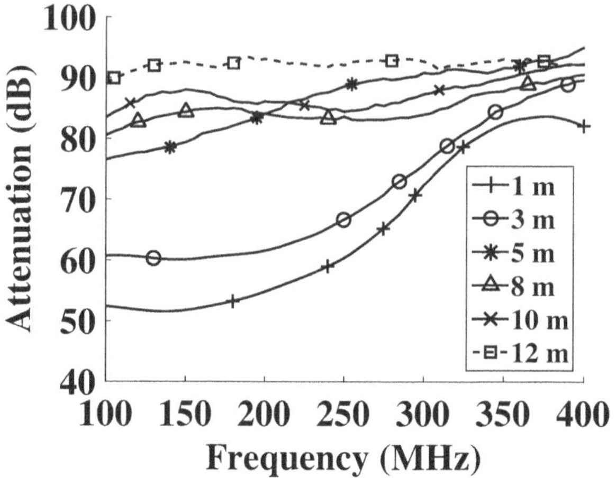

Fig. 2.10 Attenuation with frequency [55]

attenuation increases by 24 dB when frequency increases from 200 to 400 MHz. Similarly, for 200 MHz, attenuation is increased from 51 to 92 dB (80%) when distance increases from 50 cm to 12 m.

References

1. Akyildiz IF, Stuntebeck EP (2006) Wireless underground sensor networks: research challenges. Ad Hoc Netw J 4:669–686
2. Akyildiz IF, Sun Z, Vuran MC (2009) Signal propagation techniques for wireless underground communication networks. Phys Commun J 2(3):167–183
3. Archie GE et al. (1942) The electrical resistivity log as an aid in determining some reservoir characteristics. Trans AIME 146(1):54–62
4. Behari J (2005) Microwave dielectric behavior of wet soils. Springer, Berlin
5. Bohren CF, Battan LJ (1980) Radar backscattering by inhomogeneous precipitation particles. J. Atmos Sci 37(8):1821–1827
6. Bohren CF, Battan LJ (1982) Radar backscattering of microwaves by spongy ice spheres. J Atmos Sci 39(11):2623–2628
7. Campbell MJ, Ulrichs J (1969) Electrical properties of rocks and their significance for lunar radar observations. J Geophys Res 74(25):5867–5881
8. Chukhlantsev A (2006) Microwave radiometry of vegetation canopies. Springer, Berlin
9. Dobson M et al. (1985) Microwave dielectric behavior of wet soil—part II: dielectric mixing models. IEEE Trans Geosci Remote Sens GE-23(1):35–46. https://doi.org/10.1109/TGRS.1985.289498
10. Dong X, Vuran MC (2011) A channel model for wireless underground sensor networks using lateral waves. In: Proceedings of IEEE GLOBECOM '11, Houston
11. Dong X, Vuran MC (2013) Impacts of soil moisture on cognitive radio underground networks. In: 2013 first international black sea conference on communications and networking (Black-SeaCom). IEEE, Piscataway, pp 222–227

12. Dong X, Vuran MC, Irmak S (2013) Autonomous precision agriculture through integration of wireless underground sensor networks with center pivot irrigation systems. Ad Hoc Netw 11(7):1975–1987
13. Foth HD (1990) Fundamentals of soil science, 8th edn. Wiley, London
14. Ghasemi A, Abedi A, Ghasemi F (2016) Radiowave propagation in troposphere. Springer, Cham, pp 57–126. https://doi.org/10.1007/978-3-319-32783-9_3
15. Hallikainen MT, Ulaby FT, Dobson MC, El-Rayes MA, Wu LK (1985) Microwave dielectric behavior of wet soil—part 1: empirical models and experimental observations. IEEE Trans Geosci Remote Sens GE-23(1):25–34
16. Hofmeister AM (2006) Thermal diffusivity of garnets at high temperature. Phys Chem Miner 33(1):45–62
17. Holmes TRH (2003) Measuring surface soil parameters using passive microwave remote sensing. the ELBARA field campaign 2003. Master's thesis, Vrije Universiteit Amsterdam
18. Johnk CT (1988) Engineering electromagnetic fields and waves, 2nd edn. Wiley, London
19. Johnson RC (ed) (1993) Antenna engineering handbook, 3rd edn. McGraw-Hill, New York
20. Keller GV (2017) Electrical properties of rocks and minerals. In: Handbook of physical properties of rocks (1982). CRC Press, Boca Raton, pp 217–294
21. King R, Smith GS, Owens M, Wu TT (1981) Antennas in matter—fundamentals, theory, and applications. MIT Press, Cambridge
22. Konda A, Rau A, Stoller MA, Taylor JM, Salam A, Pribil GA, Argyropoulos C, Morin SA (2018) Soft microreactors for the deposition of conductive metallic traces on planar, embossed, and curved surfaces. Adv Funct Mater 28(40):1803020. https://doi.org/10.1002/adfm.201803020
23. Matzler C (1998) Microwave permittivity of dry sand. IEEE Trans Geosci Remote Sens 36(1):317–319
24. Matzler C, Strozzi T, Weise T, Floricioiu DM, Rott H (1997) Microwave snowpack studies made in the Austrian Alps during the SIR-C/X-SAR experiment. Int J Remote Sens 18(12):2505–2530
25. Oguchi T (1981) Scattering from hydrometeors: a survey. Radio Sci 16(5):691–730
26. Peplinski N, Ulaby F, Dobson M (1995) Dielectric properties of soil in the 0.3–1.3 GHz range. IEEE Trans Geosci Remote Sens 33(3):803–807
27. Rakshit G, Adhikari A, Maitra A (2017) Modelling of rain decay parameter for attenuation estimation at a tropical location. Ad Space Res 59(12):2901–2908. https://doi.org/10.1016/j.asr.2017.03.028
28. Salam A (2018) Pulses in the sand: Long range and high data rate communication techniques for next generation wireless underground networks. ETD collection for University of Nebraska–Lincoln (AAI10826112). http://digitalcommons.unl.edu/dissertations/AAI10826112
29. Salam A (2019) A comparison of path loss variations in soil using planar and dipole antennas. In: 2019 IEEE international symposium on antennas and propagation. IEEE, Piscataway
30. Salam A (2019) Design of subsurface phased array antennas for digital agriculture applications. In: Proceedings of the 2019 IEEE international symposium on phased array systems and technology (IEEE Array 2019), Waltham
31. Salam A (2019) A path loss model for through the soil wireless communications in digital agriculture. In: 2019 IEEE international symposium on antennas and propagation. IEEE, Piscataway, pp 1–2
32. Salam A (2019) Sensor-free underground soil sensing. In: ASA, CSSA and SSSA international annual meetings (2019), ASA-CSSA-SSSA
33. Salam A (2019) Subsurface MIMO: a beamforming design in internet of underground things for digital agriculture applications. J Sensor Actuator Netw 8(3). https://doi.org/10.3390/jsan8030041
34. Salam A (2019) Underground environment aware MIMO design using transmit and receive beamforming in internet of underground things. Springer, Cham, pp 1–15

35. Salam A (2019) An underground radio wave propagation prediction model for digital agriculture. Information 10(4). https://doi.org/10.3390/info10040147
36. Salam A (2019) Underground soil sensing using subsurface radio wave propagation. In: Fifth global workshop on proximal soil sensing, Columbia
37. Salam A (2020) Internet of Things for environmental sustainability and climate change. Springer, Cham, pp 33–69. https://doi.org/10.1007/978-3-030-35291-2_2
38. Salam A (2020) Internet of Things for sustainability: perspectives in privacy, cybersecurity, and future trends. Springer, Cham, pp 299–327. https://doi.org/10.1007/978-3-030-35291-2_10
39. Salam A (2020) Internet of Things for sustainable community development, 1st edn. Springer, Berlin. https://doi.org/10.1007/978-3-030-35291-2
40. Salam A (2020) Internet of Things for sustainable community development: introduction and overview. Springer, Cham, pp 1–31. https://doi.org/10.1007/978-3-030-35291-2_1
41. Salam A (2020) Internet of Things for sustainable forestry. Springer, Cham, pp 147–181. https://doi.org/10.1007/978-3-030-35291-2_5
42. Salam A (2020) Internet of Things for sustainable human health. Springer, Cham, pp 217–242. https://doi.org/10.1007/978-3-030-35291-2_7
43. Salam A (2020) Internet of Things for sustainable mining. Springer, Cham, pp 243–271. https://doi.org/10.1007/978-3-030-35291-2_8
44. Salam A (2020) Internet of Things for water sustainability. Springer, Cham, pp 113–145. https://doi.org/10.1007/978-3-030-35291-2_4
45. Salam A (2020) Internet of Things in agricultural innovation and security. Springer, Cham, pp 71–112. https://doi.org/10.1007/978-3-030-35291-2_3
46. Salam A (2020) Internet of Things in sustainable energy systems. Springer, Cham, pp 183–216. https://doi.org/10.1007/978-3-030-35291-2_6
47. Salam A (2020) Internet of Things in water management and treatment. Springer, Cham, pp 273–298. https://doi.org/10.1007/978-3-030-35291-2_9
48. Salam A (2020) Wireless underground communications in sewer and stormwater overflow monitoring: radio waves through soil and asphalt medium. Information 11(2):98
49. Salam A, Karabiyik U (2019) A cooperative overlay approach at the physical layer of cognitive radio for digital agriculture. In: Third international Balkan conference on communications and networking 2019 (BalkanCom'19), Skopje
50. Salam A, Shah S (2019) Internet of things in smart agriculture: enabling technologies. In: 2019 IEEE 5th world forum on Internet of Things (WF-IoT) (WF-IoT 2019), Limerick
51. Salam A, Vuran MC (2016) Impacts of soil type and moisture on the capacity of multi-carrier modulation in internet of underground things. In: Proceedings of the 25th ICCCN 2016, Waikoloa, Hawaii
52. Salam A, Vuran MC (2017) EM-based wireless underground sensor networks. In: Underground sensing: monitoring and hazard detection for environment and infrastructure, pp 247–285. https://doi.org/10.1016/B978-0-12-803139-1.00005-9
53. Salam A, Vuran MC (2017) Smart underground antenna arrays: a soil moisture adaptive beamforming approach. In: Proceedings of the IEEE INFOCOM 2017, Atlanta
54. Salam A, Vuran MC (2017) Wireless underground channel diversity reception with multiple antennas for internet of underground things. In: Proceedings of the IEEE ICC 2017, Paris
55. Salam A, Vuran MC, Irmak S (2016) Pulses in the sand: impulse response analysis of wireless underground channel. In: The 35th annual IEEE international conference on computer communications (INFOCOM 2016), San Francisco
56. Salam A, Vuran MC, Irmak S (2017) Towards internet of underground things in smart lighting: a statistical model of wireless underground channel. In: Proceedings of the 14th IEEE international conference on networking, sensing and control (IEEE ICNSC), Calabria
57. Salam A, Hoang AD, Meghna A, Martin DR, Guzman G, Yoon YH, Carlson J, Kramer J, Yansi K, Kelly M, et al. (2019) The future of emerging IoT paradigms: architectures and technologies. https://doi.org/10.20944/preprints201912.0276.v1

58. Salam A, Vuran MC, Dong X, Argyropoulos C, Irmak S (2019) A theoretical model of underground dipole antennas for communications in internet of underground things. IEEE Trans Antennas Propag 67(6):3996–4009
59. Salam A, Vuran MC, Irmak S (2019) Di-sense: in situ real-time permittivity estimation and soil moisture sensing using wireless underground communications. Comput Netw 151:31–41. https://doi.org/10.1016/j.comnet.2019.01.001
60. Sen P, Scala C, Cohen M (1981) A self-similar model for sedimentary rocks with application to the dielectric constant of fused glass beads. Geophysics 46(5):781–795
61. Sen PN (1997) Resistivity of partially saturated carbonate rocks with microporosity. Geophysics 62(2):415–425
62. Shutko AM (1982) Microwave radiometry of lands under natural and artificial moistening. IEEE Trans Geosci Remote Sens GE-20(1):18–26
63. Silva AR, Vuran MC (2009) Empirical evaluation of wireless underground-to-underground communication in wireless underground sensor networks. In: Proceedings of the IEEE DCOSS '09, Marina Del Rey
64. Silva AR, Vuran MC (2010) Communication with aboveground devices in wireless underground sensor networks: an empirical study. In: Proceedings of the IEEE ICC '10, Cape Town
65. Silva AR, Vuran MC (2010) $(CPS)^2$: integration of center pivot systems with wireless underground sensor networks for autonomous precision agriculture. In: Proceedings of the ACM/IEEE international conference on cyber-physical systems, Stockholm, pp 79–88. https://doi.org/10.1145/1795194.1795206
66. Temel S, Vuran MC, Lunar MM, Zhao Z, Salam A, Faller RK, Stolle C (2018) Vehicle-to-barrier communication during real-world vehicle crash tests. Comput Commun 127:172–186
67. Vuran MC, Salam A, Wong R, Irmak S (2018) Internet of underground things in precision agriculture: architecture and technology aspects. Ad Hoc Netw 81:160–173 https://doi.org/10.1016/j.adhoc.2018.07.017
68. Vuran MC, Salam A, Wong R, Irmak S (2018) Internet of underground things: sensing and communications on the field for precision agriculture. In: 2018 IEEE 4th world forum on Internet of Things (WF-IoT) (WF-IoT 2018), Singapore
69. Wu TT (1961) Theory of the dipole antenna and the two-wire transmission line. J Math Phys 2(4):550–574

Chapter 3
Wireless Underground Channel Modeling

3.1 Introduction

Wireless underground sensor networks (WUSNs) can be considered as one of the emerging technologies having broad set of applications. The potential applications of WUSNs include environment and infrastructure monitoring [32, 53, 69], precision agriculture [14, 61, 62], and border patrol [1]. An efficient and a reliable UG channel plays a very important role in implementing these applications. There is a need to theoretical explain and empirically validate the models of underground channel for further advancement of WUSNs. This chapter focuses on characterization of these channels and measures the effect of different parameters on UG communication. The discussion in this chapter is divided into two major parts: (1) The first part presents the effect of soil on communication components, i.e., UG channel, UG signal attenuation, return loss in soil and channel capacity, etc., and (2) The second part discusses different channel models under various soil factors and their effect on UG communication. Figure 3.1 shows the general structure of the chapter.

3.2 Effect of Soil Properties on UG Communication

Wireless communication is highly medium-dependent and wireless UG communication is accomplished using soil as a medium. Therefore, soil components play an important role in determining the efficiency of UG communication. These components are described below:

(i) ***Soil Effective Permittivity*****:** It is a complex number that adds extra attenuation to electromagnetic (EM) waves in addition to diffusion attenuation. The extra

A. Salam, U. Raza, *Signals in the Soil*,
https://doi.org/10.1007/978-3-030-50861-6_3

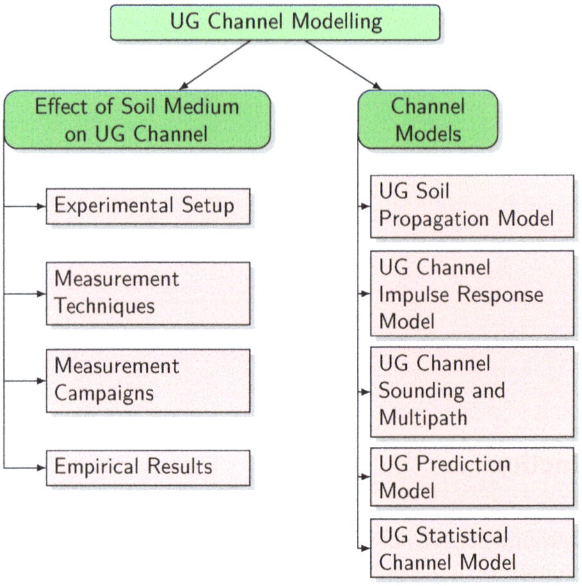

Fig. 3.1 Chapter organization

attenuation is due to absorption of EM waves by soil and is also frequency-dependent [57].

(ii) ***Permittivity of Soil*:** Permittivity of soil has much greater value than the permittivity of air which reduces the wavelength of EM waves propagating through soil. Therefore, aboveground antennas do not perform well in underground environment [56].

(iii) ***Temporal Variation of Soil Moisture*:** Permittivity of soil keeps on changing with the change in soil moisture which in turn changes the wavelength. Soil moisture mostly changes due to irrigation and precipitation, therefore, for a given frequency, wavelength is always varying with the time. This soil behavior has great effect on the return loss and bandwidth of antenna [49].

From the above discussion, it can be concluded that soil properties significantly impact the performance of the underground wireless communication. Therefore, it is necessary to measure the effect of soil on wireless communication. To that end, extensive experimentation is done over the period of 2–3 years to study UG communication and its response to different soil-related deployment (e.g., burial depth) and operational parameters (e.g., antenna, soil moisture). In the following sections, the methodology and results from these experiments are discussed in detail.

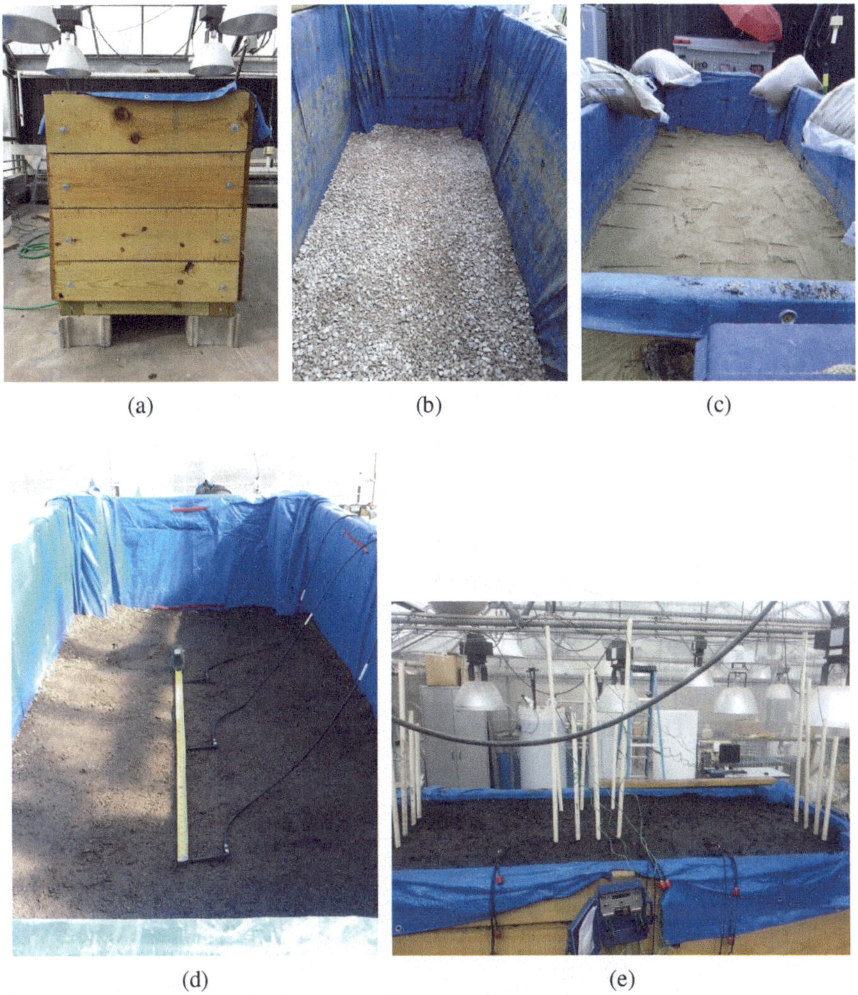

Fig. 3.2 Phases in development of an indoor testbed: (**a**) wooden box of the testbed, (**b**) gravel in the testbed (placed at the bottom), (**c**) soil in the testbed, (**d**) placement of antenna in testbed, (**e**) final testbed look [61]

3.2.1 Experimental Setup

3.2.1.1 The Indoor Testbed

Outdoor experimentation in WUSN is not an easy task. Outdoor settings face challenges of extreme weather and temperature conditions. It is very important to get timely results of experiments and getting the different soil moisture level in a short span of time is difficult in outdoor settings. Furthermore, it lacks dynamic soil moisture control, difficult to get changing soil types and deployment of equipment is also cumbersome. An indoor testbed can overcome these challenges. Figure 3.2

shows an indoor testbed. This indoor testbed is developed in a greenhouse setting. It is a wooden box with a dimension of 100 in×36 in×48 in (see Fig. 3.3b) having 90 ft^3 of soil in it (see Fig. 3.2a). It is equipped with a drainage system at the bottom and has waterproof sides. The sides are waterproofed using waterproof tarp to prevent leaks from the side. The box has a 3 in layer of gravel underneath to allow free water drainage (see Fig. 3.2b). Figure 3.2c shows the box with soil in it.

It uses Watermark sensors to monitor the soil moisture level. A total of 8 sensors are installed on the sides of the box. These sensors are placed at the depth of 10cm, 20cm, 30cm, and 40cm. Two Watermark dataloggers are connected to these sensors. For each installed antenna, a tamper tool is used to pack soil every 30cm. It is done to

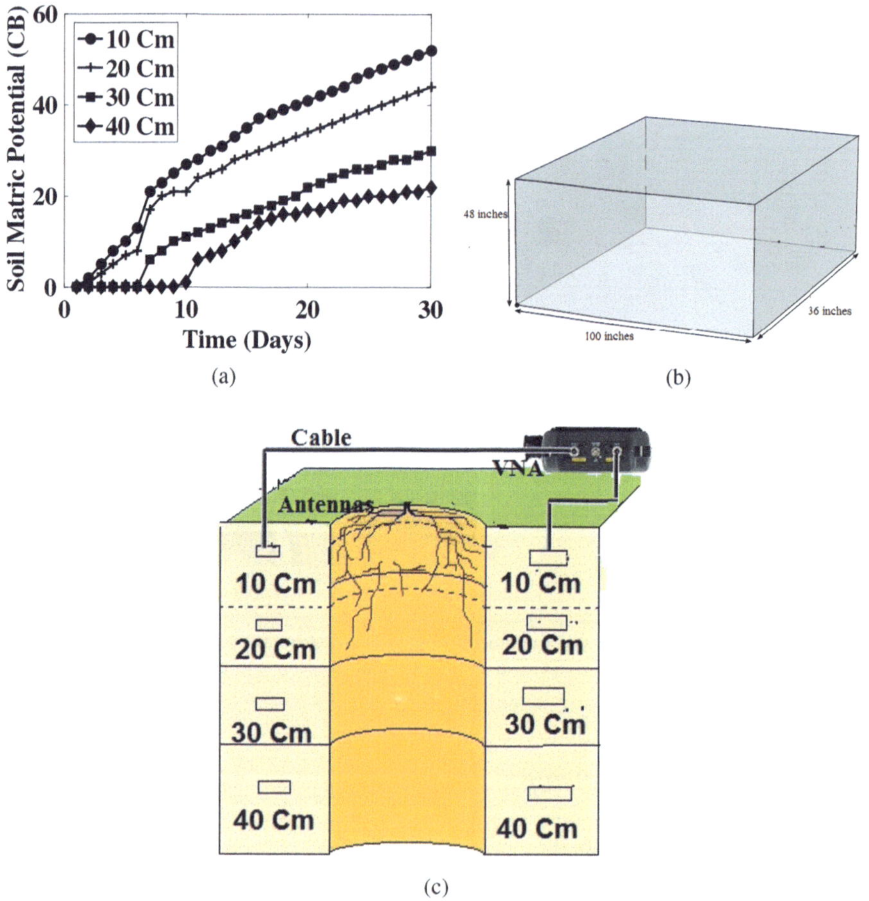

Fig. 3.3 (**a**) Change in soil moisture vs. time, (**b**) box dimensions, (**c**) layout of antennas in the experiment [61]

Table 3.1 Particle size distribution and classification of testbed soils [61]

Textural class	% Sand	% Silt	% Clay
Sandy soil	86	11	3
Silt loam	33	51	16
Silty clay loam	13	55	32

achieve the bulk density[1] to simulate the real-world scenario. A total of 12 antennas, divided into sets of three, are installed (see Fig. 3.2d) at increasing depths of 10cm, 20cm, 30cm, and 40cm (Fig. 3.2d). These sets are separated at 50cm from each other. Figure 3.2e shows the final testbed.

Two different soil types were used for the experimentation: silt loam and sandy soil. Table 3.1 lists the soil type used and distribution of the particle size within the soil type. Experiments were performed to investigate the effect of soil texture on underground communications. To that end, soil having sand content from 13% to 86% and clay content from 3% to 32% was chosen. For experimentation, a nearly saturated soil is used as an input to achieve maximum volumetric water content (VWC) level. Afterwards, results are gathered as the soil water potential decreases from saturated state to field capacity[2] and then finally to wilting point.[3] Figure 3.3a shows the moisture level change in silt loam soil.

3.2.1.2 The Field Testbed

A field testbed consists of dipole antennas, buried at 20cm depth, with silty clay loam soil (see Fig. 3.5a). Figure 3.4 shows the different stages in which field testbed is developed. The purpose of this testbed is to perform underground-to-aboveground (UG2AG) experiments. Field testbed is also used to compare the results from the indoor testbed. For UG2AG experiments, an adjustable height pole is used. The experiments are conducted with the radii of 2m, 4m, 5.5m, and 7m. The distance of max 73 is used due to cable limitations of antenna used for vector network analyzer (VNA). Receiver angles of 0°, 30°, 45°, 60° and 90° are used. The testbed is shown in Fig. 3.5.

3.2.1.3 UG Software-Defined Radio (SDR)

Another testbed is used for conducting experiments with underground software-defined radios. This testbed consists of a total of 16 dipole antennas, divided into 4 sets, in silt loam soil. For each set, the antenna's burial distance is 50cm, 2m, and 4m from the first antenna. In each set, antennas are buried at the depth 10cm,

[1]Ratio of dry soil weight to the volume of soil (also including volume of pores in particles).

[2]Water content in the soil after the drainage of excess water.

[3]Minimum level of water in the soil.

Fig. 3.4 Field testbed development in the silty loam soil: (**a**) testbed layout, (**b**) antenna placement, (**c**) outlook after antenna installation, (**d**) antenna cables out of soil at different depths, (**e**) USRPs and datalogger for soil moisture measurements [33]

(a) (b)

Fig. 3.5 The field testbed [61]

20cm, 30cm, and 40cm. This dipole antenna uses the resonant frequency of 433MHz for over-the-air communication. Figure 3.4 shows step by step development of the testbed (starting from Fig. 3.4a till Fig. 3.4e). Figure 3.4a shows the layout,

Fig. 3.4b,c shows the testbed after placement and installation of the antenna, Fig. 3.4d shows the cable coming out of the antennas buried at different depths. Finally, Fig. 3.4e shows the testbed with USRPSs and datalogger used for measuring soil moisture.

3.2.1.4 Soil Moisture Logging

As discussed previously, soil moisture has an impact on the communication. Therefore, it is important to log the soil moisture data after each experiment for correct characterization of channel. Oven drying method can be used for determining soil moisture, however, it requires removal of soil from the testbed. Watermark sensors are used because of their ability of logging soil moisture data with timestamp. Watermark sensors are fast, efficient, and less error prone. It also overcomes the challenge of oven drying method. If there is metallic object in the soil within the vicinity of buried antennas, it can cause interference during the communication. Therefore, to avoid interference, sensors are deployed at the edges of the testbed.

3.2.2 Measurement Techniques and Experiments Description

This testbed uses Keysight Technologies N9923A, FeildFox, and VNA for taking measurements. The layout of the measurements is shown in Fig. 3.3c. In the following section, measurements taken for the indoor testbed are explained in detail.

3.2.2.1 Path Loss Measurements

VNA uses a UG transmitter and receiver (T-R) pair for measuring the signal loss. It transmits a known signal and comparison of received and incident signal is performed to calculate the loss. Path loss is defined as the ratio of power of a signal at sender end P_t to power of the signal reached at receiver end P_r. Path loss is calculated by the following equation:

Table 3.2 Underground channel measurement parameters [61]

Parameter	Value
Start frequency	10 MHz
Stop frequency	4 GHz
Number of frequency points	401
Transmit power	5 dBm
Vector network analyzer	Agilent FieldFox

$$PL = P_t - P_r = 10. \log 10(P_t/P_r) \tag{3.1}$$

where PL denotes the path loss of the system. It also includes the effect of transmitting antenna gain G_t and receiving antenna gain G_r. The frequency range used for the measurement of path loss is 10 Mhz to 4 GHz. A total of 401 discrete frequency points were taken at varying distance and depths. These measurements were used to study how physical properties of the soil impact the attenuation of UG channel.

3.2.2.2 Power Delay Profile (PDP) Measurements

Power delay profile (PDP) is a way to measure signal intensity through a multipath channel. The intensity is measured as a time delay. Time delay is the difference of time taken by signal to travel through multiple paths. The multipath characteristic of a wireless underground channel is studied by channel sounding experiment. VNA accurately characterizes the UG channel by transmitting multiple sine waves at transmitter and measuring them at receiver end. These sine waves are from low to high frequency. Impulse response of UG channel is measured by VNA using one frequency at a time. It measures the impulse response in frequency domain instead of time-domain.

3.2.3 Measurement Campaigns

This section presents the measurements parameters for experiments on each type of the soil.

3.2.3.1 Sandy Soil Experiments

For a distance of 50cm between transmitter and receiver, antenna was buried at the depth of 10cm, 20cm, 30cm, and 40cm in a sandy soil. For a soil moisture range of 0–250 CB, they are buried at the depth of 1m.

3.2.3.2 Silty Clay Loam Experiments

For a distance of 50cm between transmitter and receiver, antenna was buried at the depth of 20cm in a silty clay loam soil. In field testbed, it is buried 1m for a soil moisture range of 0–50 CB.

3.2.3.3 Silt Loam Experiments

For a distance of 50cm between transmitter and receiver, antenna was buried at the depth of 10cm, 20cm, 30cm, and 40cm in a silt loam soil. In indoor testbed, for a soil moisture range of 0–50 CB, they are buried at the depth of 1m.

3.2.3.4 Planar Antenna Experiments

Indoor testbed and two planar antennas are used for the planar antenna experiments. These experiments are conducted in silty clay loam and sandy soils. The return loss and path loss are measured without obstructions and with obstruction between these two antennas. For former experiment, the antenna is kept 1m apart and buried at the depth of 20cm. For latter, another antenna is placed in the middle (50cm) of the two antennas acting as an obstruction and depth for all the antennas is kept same (20cm). These experiments were also conducted with silty clay loam soil using the same empirical parameters and return loss and path loss are compared for both types of the soil.

3.2.4 Empirical Results

This section explores the channel transfer function (S_{12}) for different type of soil using varying values for distances, soil moisture, and depth.

3.2.4.1 Channel Transfer Function Measurement

3.2.4.2 Impact of Burial Depth and Antennas Distance on Attenuation

Figure 3.6 plots the attenuation at different depth and distance in silt loam soil. Figure 3.6a and b shows the attenuation, when the antennas are 50cm apart with varying frequency and depths (10cm, 20cm, 30cm, and 40cm). It can be observed in these figures that at 200MHz path loss differs by 5dB at 10cm (where path loss = 40.37dB) from 40cm (where path loss = 45.26dB). The difference further increases to 8dB at 250 MHz. Figure 3.6c and d shows the attenuation, when the antennas are 1m apart with frequency and depth (10cm, 20cm, 30cm, and 40cm).

At 200 MHz, path loss for the depth of 10cm is 44.37 dB and that for 40cm is 60.12 dB. The difference in path loss is 16dB which is higher as compared to difference at 50cm. The difference increases to 20 dB at the frequency of 250 MHz.

Results for similar experiments with same depths and frequencies, but with different soil type (silt loam soil), are shown in Figs. 3.7 and 3.8. Figure 3.7 shows the results with transmitter and receiver 50cm apart and Fig. 3.8 shows for the

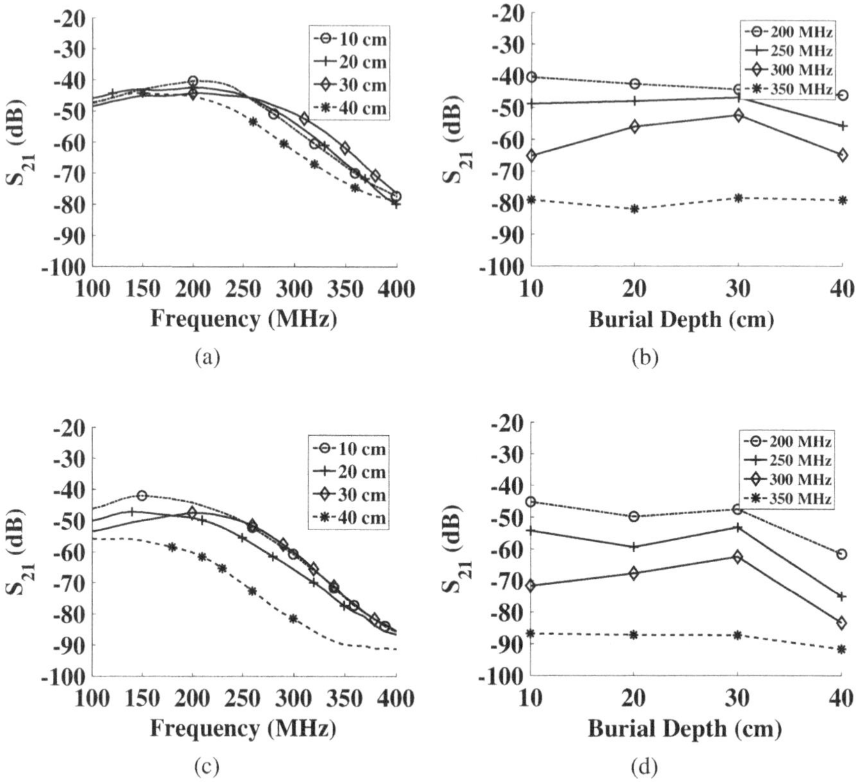

Fig. 3.6 Attenuation in silt loam soil at 50 cm distance: (**a**) with frequency, (**b**) at different depths. Attenuation in silt loam soil at 1 m distance: (**c**) with frequency, (**d**) at different depths [33]

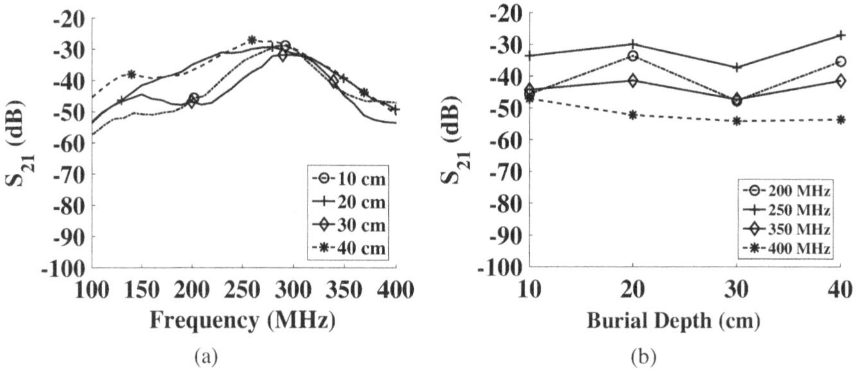

Fig. 3.7 Attenuation in sandy soil at 50 cm distance: (**a**) with frequency, (**b**) at different depths

distance of 1m. For both distances attenuation is highest at the frequency of 400 MHz for all depths.

Another important impact to consider is that of distance between the antennas on attenuation. From Figs. 3.6a (50cm) and 3.6c (1m), it can be observed that path loss increases as the distance between antennas is increased from 50cm to 1m. For example, when antennas distance increases from 50cm to 1m at 200 MHz, path loss is increased by 5 dB, 7 dB, 3 dB, and 15 dB at depths of 10cm, 20cm, 30cm, and 40cm, respectively. Similar trend can be observed for the increased distance in sandy soil (see Figs. 3.7 and 3.8), however, the difference is very low as compared to silt loam soil.

Hence, it can be concluded that path loss is affected by frequency, depth, and distance, i.e., increases with increasing frequency, distance, and depth of antennas. This variation because of frequency is due to soil permittivity. There are three major paths in underground–underground communication: (1) direct, (2) reflected, and (3) lateral. Direct wave is the line of sight path; reflected path is the one taken by wave when it is reflected from soil–air interface; and lateral path is taken because of the propagating along the soil–air interface. The reason for the path loss variation due to depth is because of multipath effect. The direct path has no effect due to the depth increase, however, it affects the lateral and reflected path. So when the depth is increased, path loss due to lateral and reflected wave increases which in turn increases the overall path loss.

3.2.4.3 Impact of Soil Type on Attenuation

Figure 3.9 shows the comparison of attenuation in different soil type for varying depths and frequency. Figure 3.9a shows the result at transmitter–receiver distance of 50cm and Fig. 3.9b shows for the distance of 1m. For 50cm, sandy soil has 10–30 dB, 13 dB, 18 dB, and 20 dB lower path loss value at the depth of 10cm, 20cm,

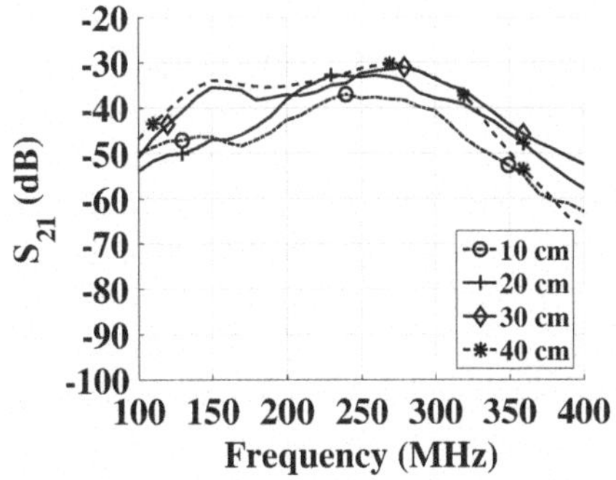

(a)

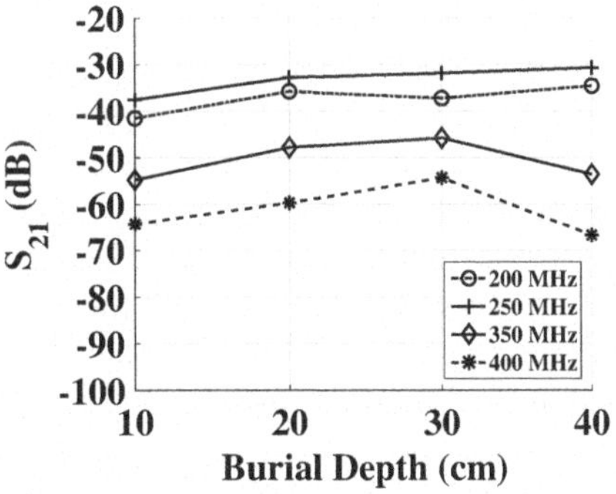

(b)

Fig. 3.8 Attenuation in sandy soil at 1 m distance: (**a**) with frequency, (**b**) at different depths [33]

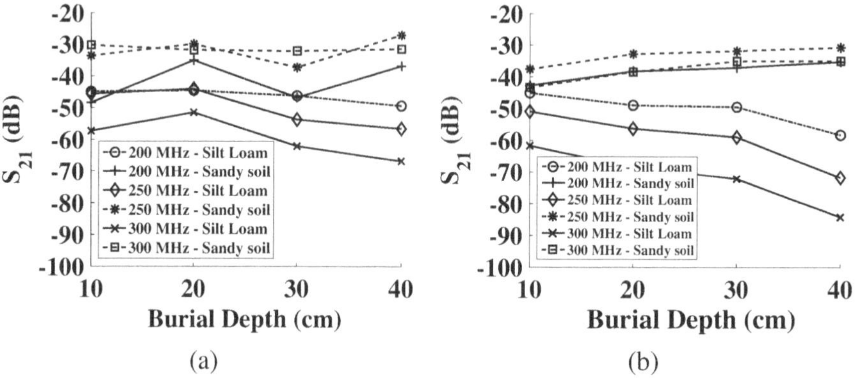

Fig. 3.9 (**a**) Comparison of attenuation in silt loam and sandy soils at 50 cm distance at different frequencies, (**b**) comparison of attenuation in silt loam and sandy soils at 1 m distance at different frequencies [33]

30cm, 40cm, respectively. Similarly, for 1m, sandy soil has 12 dB, 20 dB, 18 dB, and 38 dB lower path loss value at the depth of 10cm, 20cm, 30cm, 40cm, respectively.

It can be observed that, for both distances, path loss in sandy soil is less as compared to silt loam soil for all depths. The reason for this is that electromagnetic waves are absorbed by the bound water present in the soil and sandy soil absorbs less bound water as compared to silt loam soil. The water holding power of medium textured soil, e.g., silt loam and silty clay loam, is much higher than that of coarse soils such as sand and sandy loam. This is because coarse soils have lower size of pores hence low resistance against gravity and no aggregation.

3.2.4.4 Impact of Soil Moisture on Attenuation

This section evaluates the experiments done to study the effect of soil moisture on the signal attenuation. Different soil moisture levels, ranging from 0 to 50 CB, are used for these experiments. Figure 3.10 shows the experiment results at transmitter–receiver distance of 50cm in silt loam soil. It can be seen that in Fig. 3.10a there is a decrease of 5 dB as moisture level goes from 0 to 50 and frequency increases. Figure 3.10b uses the frequency of 200 MHz to show the soil moisture effect at varying depth levels. There is a consistent path loss decrease at each depth value as the soil moisture level increases from 0 to 50 CB. For example, path loss decreases 5 dB, 3 dB, 5dB, and 3dB at depths of 10cm, 20cm, 30cm, and 40 cm, respectively, with increasing soil moisture. Figure 3.10c plots the path loss by keeping the fixed depth value, i.e., 10cm, and varying frequency. For fixed depth of 10cm, path loss decreases by 2dB (at 250 MHz) and 3dB (at 250 MHz).

Figure 3.11 shows the experiment results at transmitter–receiver distance of 1m in silt loam soil. In Fig. 3.11a, a 3 dB decrease in path loss can be observed at 250 MHz as soil moisture goes from 0 to 50 CB. In Fig. 3.11c, experiments with varying

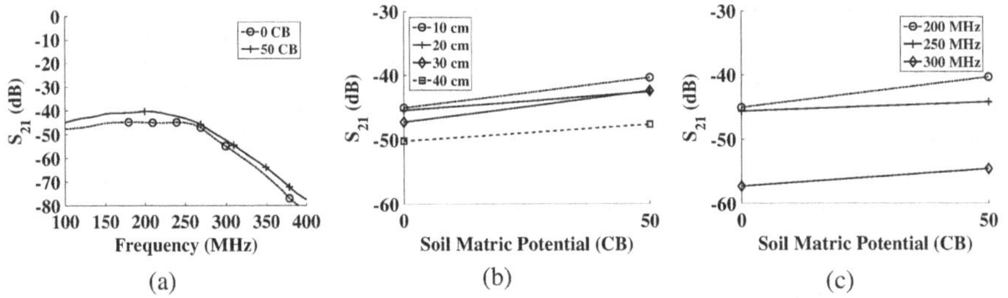

Fig. 3.10 Attenuation with soil moisture in silt loam soil at 50 cm distance: (**a**) S_{21}, (**b**) effect of change in soil moisture at different depths at 200 MHz frequency, (**c**) effect of change in soil moisture at different frequencies at 10 cm depth [33]

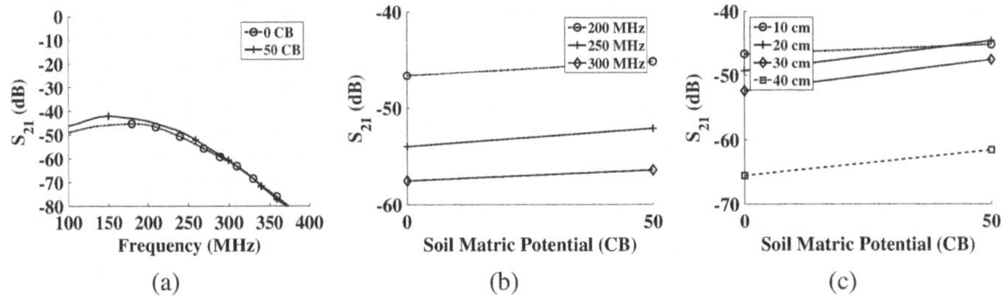

Fig. 3.11 Attenuation with soil moisture in silt loam soil at 1 m distance: (**a**) S_{21}, (**b**) effect of change in soil moisture at different frequencies at 10 cm depth, (**c**) effect of change in soil moisture at different depths at 200 MHz frequency

frequency and fixed depth of 10cm show that path loss decreases by 2–3 dB as soil moisture goes from 0 to 50 CB. This difference increases further as frequency increases. Figure 3.11b uses the frequency of 200 MHz to show the soil moisture effect at varying depth levels. The trend is similar to the one observed in Fig. 3.10b, i.e., path loss decreases with increasing depth. For example, path loss decreases 1 dB, 5 dB, 5dB, and 4dB at depths of 10cm, 20cm, 30cm, and 40 cm, respectively, with increasing soil moisture.

Figures 3.12 and 3.13 show the effect of soil moisture on path loss for sandy soil at transmitter–receiver distance of 50cm and 1m, respectively. Values of soil moisture considered for this experiment are in the range of 0 to 255 CB. In Fig. 3.12a, it can be seen that the path loss is minimum at 300 MHz and after 300 MHz it starts increasing. Figure 3.12b shows that, for a fixed depth of 10cm, path loss is showing negative trend (increasing) at 200 MHz and positive trend (decreasing) at 250 MHz and 300 MHz. Figure 3.12c plots the effect of soil moisture with fixed frequency but varying depths. It can be observed that for all depth levels path loss is increasing with decreasing soil moisture. The reason for this is the change of resonant frequency to high spectrum with decreasing soil moisture.

Figure 3.13a shows similar trends as of Fig. 3.12a, however, for 1m distance path is minimum at 250 MHz at 0 CB soil moisture level and it is minimum 300

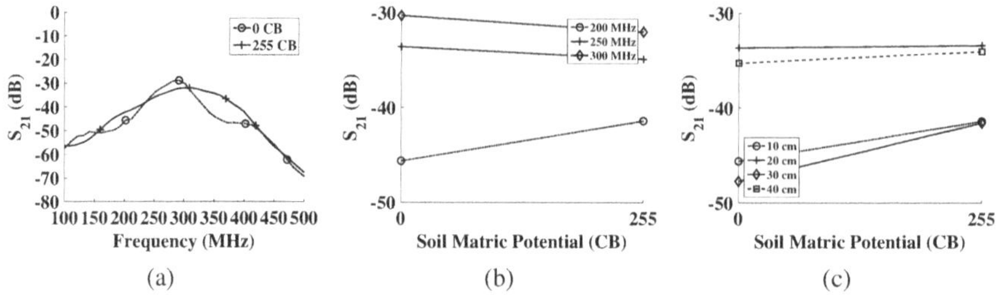

Fig. 3.12 Attenuation with soil moisture in sandy soil at 50 cm distance: (**a**) S_{21}, (**b**) effect of change in soil moisture at different frequencies at 10 cm depth, (**c**) effect of change in soil moisture at different depths at 200 MHz frequency [33]

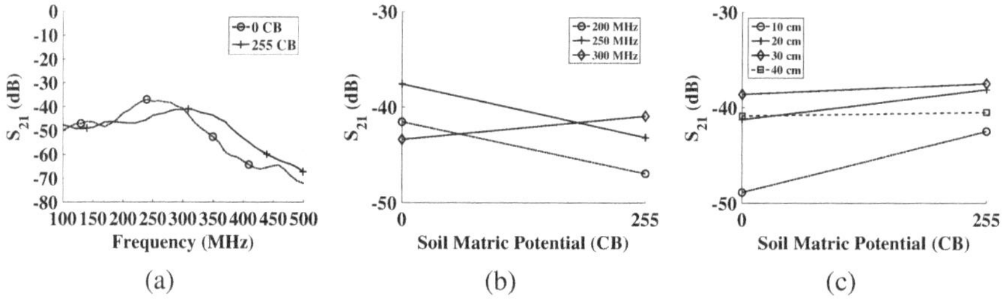

Fig. 3.13 Attenuation with soil moisture in sandy soil at 1 m distance: (**a**) S_{21}, (**b**) effect of change in soil moisture at different frequencies at 10 cm depth (**c**) effect of change in soil moisture at different depths at 330 MHz frequency [33]

MHz at 255 CB soil moisture level. In Fig. 3.13b, path loss is decreasing at 200 and 250 MHz and decreasing at 300 MHz as soil moisture goes from 0 to 250 CB. Figure 3.13c shows decrease in path loss for all depths at 200 MHz.

These experiments conclude that soil moisture is inversely proportional to path loss, i.e., path loss increases with decrease in soil moisture for both soil types. This is because of high permittivity at high soil moisture.

3.2.4.5 Power Delay Profile Measurements

Figure 3.14 shows the result of power delay profiling (PDP) for the experiments at 50cm and 1m.

Speed of the wave in soil is given as: $S = c/n$, where n is the refractive index and c is the speed of light 3×10^8 m/s. Since the permittivity of soil is a complex number, the refractive index of soil is calculated as:

$$n = \sqrt{\frac{\sqrt{\epsilon'^2 + \epsilon''^2} + \epsilon'}{2}}, \tag{3.2}$$

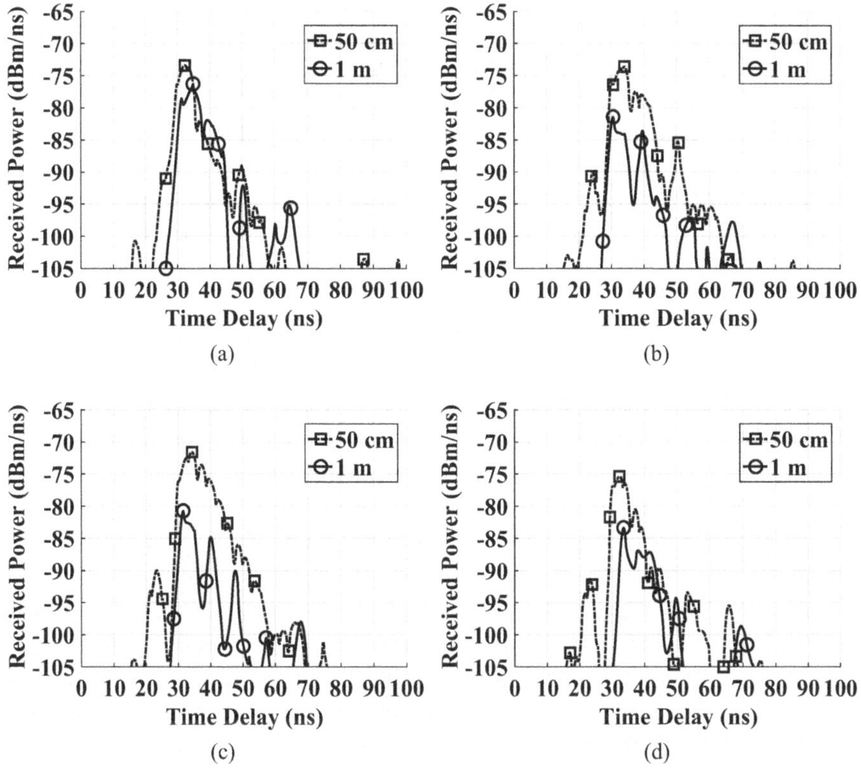

Fig. 3.14 Power delay profiles (PDP) measured at 50 cm and 1 m distance, at different depths in silt loam soil at near-saturation: (**a**) 10 cm, (**b**) 20 cm, (**c**) 30 cm, (**d**) 40 cm [33]

where ϵ' and ϵ'' are the real and imaginary parts of the relative permittivity of the soil.

The wave speed is calculated in silt loam soil using refractive index. Refractive index is calculated as $5.6 \times 10_7$ on the basis of properties given in Table 3.2. It is 5 times slower and 19% of speed of light.

Figure 3.14 shows the results from the experiments done at the distance of 50cm and 1m for all depths. First multipath component is the direct wave. It is present at delay of 18–28 ns at 50cm and is not appearing for 1m distance. The reason for this is that direct wave is more attenuated at 1m than 50cm. Lateral waves can be seen at time delay of 30–40 ns as the strongest component among all PDPs. The delay for lateral wave is similar for both, 1m and 50cm, because it propagates faster in air. In general, power of lateral wave is 10–15 dB faster than direct wave.

Figure 3.15 compares PDP for all four depths at the distances of 50cm (see Fig. 3.15a) and 1m (see Fig. 3.15b) between transmitter and receiver. It can be observed from the results that if distance is kept same, increase in depth causes decrease in power and time delay of received signal. For example, at 1m, peak power

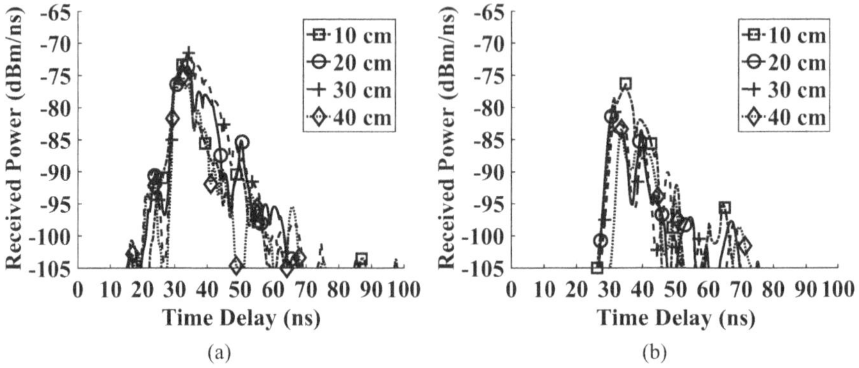

Fig. 3.15 Power delay profile in silt loam soil at different depths at: (**a**) 50 cm T-R distance, (**b**) 1 m T-R distance [33]

of lateral wave is −75 dB and −83 dB at the depth of 10cm and 40cm, respectively. Similarly, same trend is seen for the time delay, i.e., in Fig. 3.15a lateral wave arrives at 29 ns and 32 ns at the depth of 10cm and 40cm, respectively.

Figure 3.16 shows the PDP measurement results at distance of 50cm and 1m and depth of 20cm. The effect of soil moisture is measured by varying the soil moisture values. For both distances, the received signal strength is increasing with decreasing soil moisture. An important observation to note is that direct component vanishes with increasing distance due to higher soil attenuation.

Figure 3.17 shows the measurement of PDP with different type of soil. It is observed that, due to its low water holding capacity, the power of received signal is highest in sandy soil as compared to silt loam and clay loam soil.

3.2.4.6 UG Antenna Return Loss Measurements

The return loss of the antenna (in dB) is calculated as:

$$RL_{dB} = 20\log_{10}\left|\frac{Z_a - Z_0}{Z_a + Z_0}\right|, \tag{3.3}$$

where Z_a = the antenna impedance and Z_0 = the characteristics impedance of the transmission line.

The reflection coefficient Γ is calculated using return loss as: $|\Gamma| = 10^{\frac{RL}{20}}$. Reflection coefficient is converted to impedance using: $Z_a = Z_0\frac{1+\Gamma}{1-\Gamma}$. Standing wave ratio (SWR) is calculated as: $SWR = \frac{1+|\Gamma|}{1-|\Gamma|}$

The frequency where the antenna's input impedance is pure resistance is known as resonant frequency (f_r). At resonant frequency f_r:

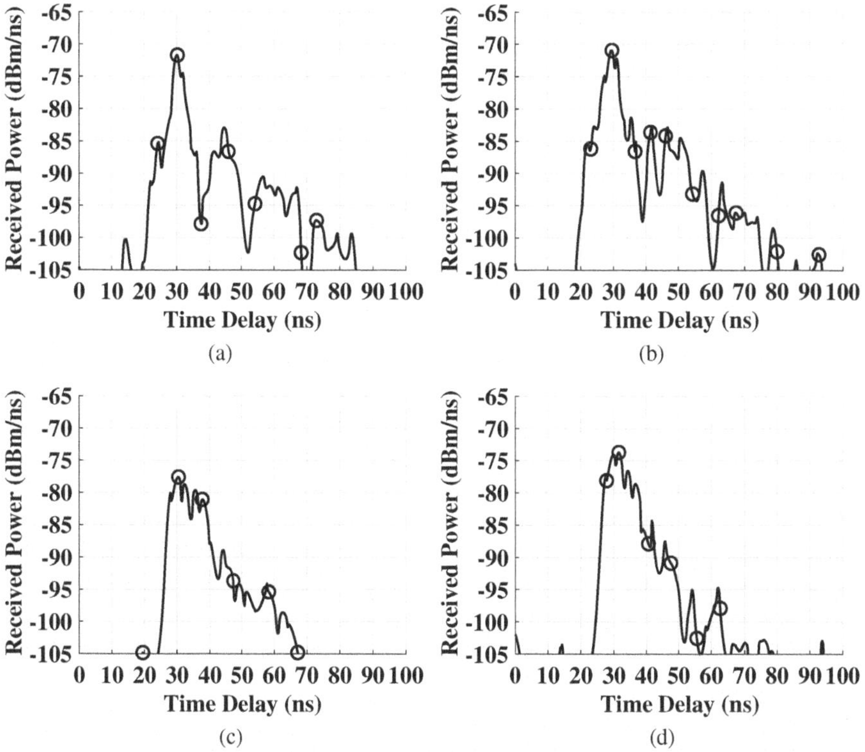

Fig. 3.16 Power delay profiles (PDP) measured at 50 cm and 1 m distance, at 20 cm depths for different soil moisture levels: (**a**) 0 CB–50cm, (**b**) 50 CB–50cm, (**c**) 0 CB–1m, (**d**) 50 CB–1m [33]

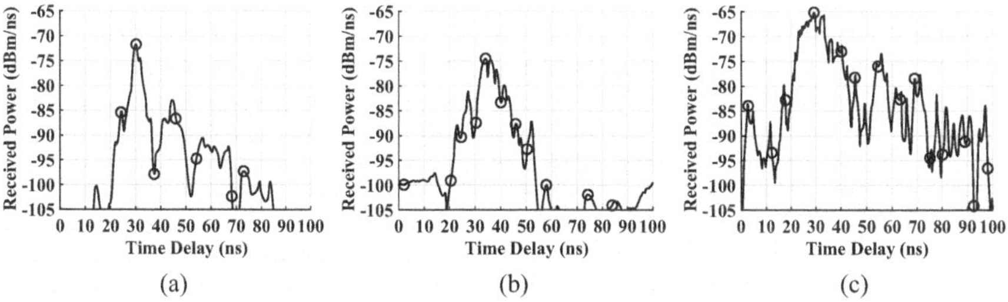

Fig. 3.17 Power delay profiles (PDP) measured in different soils: (**a**) silt loam, (**b**) silty clay loam, (**c**) sandy soil

$$Z_a|_{f=f_r} = Z_r = R_a. \tag{3.4}$$

It is the frequency where return loss is maximum such that:

$$f_r = \max(RL_{dB}). \tag{3.5}$$

The return loss experiments are done to measure the effect of soil type, soil moisture, and burial depth on the resonant frequency of the UG antenna.

3.2.4.7 Effects of Soil Type

Figures 3.18 shows the effect of soil type on antenna return loss. There is a difference of 63 MHz, 59 MHz, 55 MHz, and 43 MHz at the depths of 10cm, 20cm, 30cm, and 40cm, respectively.

3.2.4.8 Impact of Change in Soil Moisture

Figure 3.19 shows different parameters effecting the return loss of antenna. It uses the soil matric potential values from 0 to 50 CB for all experiments. The soil used for this experiment is silt loam soil. Figure 3.19a shows the return loss for 10cm depth at varying frequencies. It can be seen in Fig. 3.19b, for a soil moisture values of 0 CB and 50 CB, resonant frequency changes from: 211 MHz to 219 MHz at the depth of 10cm, 221 MHz to 227 MHz at the depth of 20cm, 221 MHz to 231 MHz at the depth of 30cm, and 201 MHz to 231 MHz at the depth of 40cm.

In Fig. 3.19c, at shifted frequency, reflection coefficient changes from: −15 dB to −17 dB at the depth of 10cm, −18 dB to −15 dB at the depth of 20cm, −20 dB to −17 dB at the depth of 30cm, and −20 dB to −18 dB at the depth of 40cm. Figure 3.19c plots change in antenna bandwidth at soil moisture values of 0 CB and 255 CB. It can be observed that with decreasing soil moisture bandwidth changes from: 40 MHz to 29 MHz at the depth of 10cm, 30 MHz to 34 MHz at the depth of 20cm, 39 MHz to 34 MHz at the depth of 30cm, and 37 MHz to 30 MHz at the depth of 40cm.

Figure 3.20 plots the return loss for sandy soil with change in soil moisture. It uses the soil matric potential values from 0 to 255 CB for all experiments. Figure 3.20a shows the antenna return loss for 10cm depth. The resonant frequency

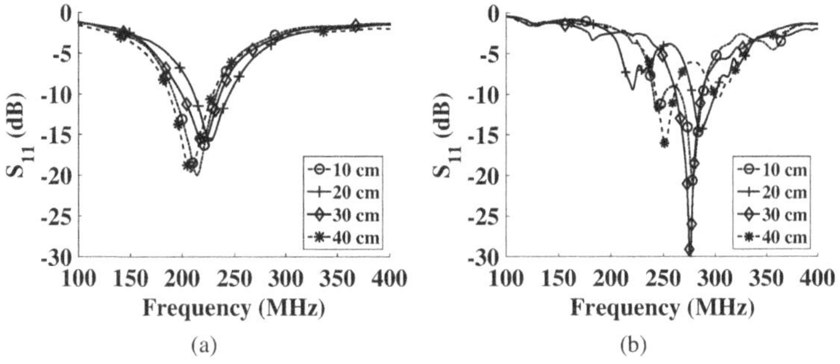

Fig. 3.18 Change in return loss in different soil: (**a**) silt loam, (**b**) sandy soil [33]

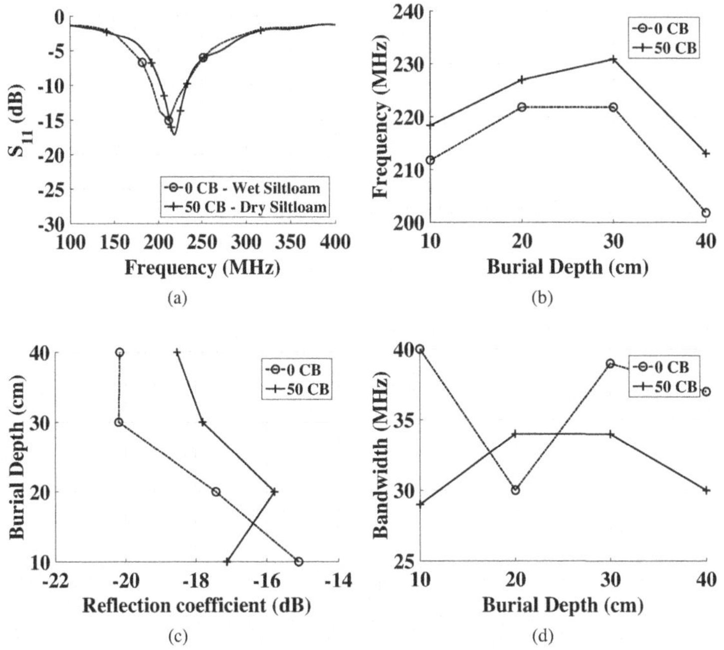

Fig. 3.19 Return loss in silt loam soil: (**a**) S_{11} at different frequencies, (**b**) change in resonant frequency with burial depth, (**c**) reflection coefficient (dB) at different burial depths, (**d**) antenna bandwidth at different burial depths [33]

changes from 278 MHz to 305 MHz with decrease in soil moisture. It can be seen in Fig. 3.20b, for soil moisture values of 0 CB and 255 CB, resonant frequency changes from: 276 MHz to 301 MHz at the depth of 20cm and 30cm, and 251 MHz to 279 MHz at the depth of 40cm.

In Fig. 3.20c, reflection coefficient changes from: −20 dB to −16 dB at the depth of 10cm, −14 dB to −12 dB at the depth of 20cm, −31 dB to −15 dB at the depth of 30cm, and −16 dB to −15 dB at the depth of 40cm. Figure 3.20d plots change in antenna bandwidth at soil moisture values of 0 CB and 255 CB. It can be observed that as the soil moisture decreases bandwidth changes from: 22 MHz to 8 MHz at the depth of 10cm, 23 MHz to 15 MHz at the depth of 20cm, 25 MHz to 16 MHz at the depth of 30cm, and 18 MHz to 16 MHz at the depth of 40cm.

The analysis has shown that antenna return loss is affected by soil moisture for sandy and silt loam soils. As the soil moisture increases, the resonant frequency goes to lower range. Contrary to over-the-air communications, resonant frequency and optimal frequency (where maximum capacity is achieved) of antenna are different. Finding an optimal frequency for UG antenna in soil is a potential area to work on.

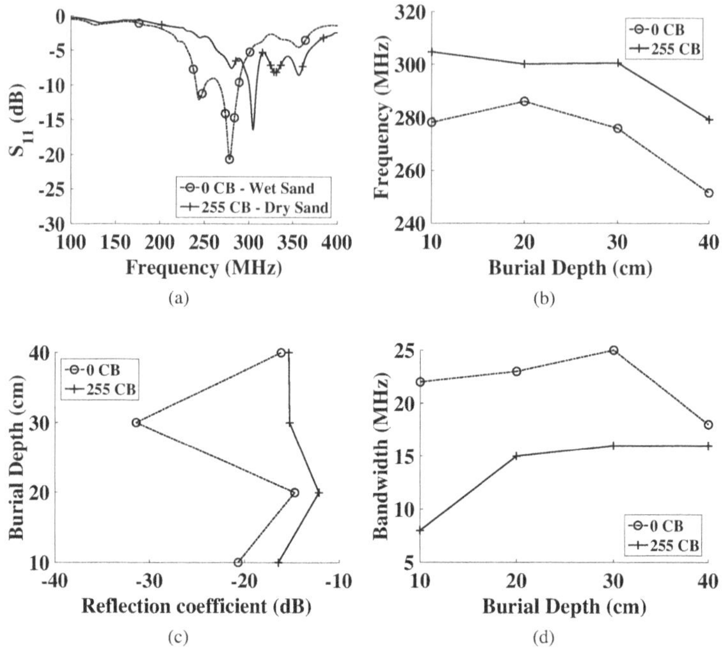

Fig. 3.20 Return loss in sandy soil: (**a**) S_{11} at different frequencies, (**b**) change in resonant frequency with burial depth, (**c**) reflection coefficient (dB) at different burial depths, (**d**) antenna bandwidth at different burial depths [33]

3.2.4.9 Impact of Burial Depth

Figure 3.18 shows the effect of burial depth on return loss in sandy soil (Fig. 3.18b) and silt loam (Fig. 3.18a). The chosen depths are: 10cm, 20cm, 30cm, and 40cm. Resonant frequency increases from 215 MHz (at 10cm depth) to 227 MHz (at 20cm depth) and decreases to 220 MHz (at 30cm) and 208 MHz (at 40cm). Reflection coefficient is -19.92 dB, -15.76 dB, -16.04 dB, and -19.57 dB at the depths of 10cm, 20cm, 30cm, and 40cm, respectively. To analyze the bandwidth of antenna, -10 dB is chosen as threshold power value. Bandwidth is measured as 40MHz, 32 MHz, 37MHz, and 42 MHz at the depths of 10cm, 20cm, 30cm, and 40cm, respectively.

Figure 3.18b shows the effect of sandy soil on return loss. As in the case of silt loam soil, resonant frequency increases from 278 MHz (at 10cm depth) to 286 MHz (at 20cm depth) and then decreases to 275 MHz (at 30cm) and 251 MHz (at 40cm). Reflection coefficient is -19.92 dB, -15.76 dB, -16.04 dB, and -19.57 dB at the depths of 10cm, 20cm, 30cm, and 40cm, respectively. Reflection coefficient is -20.66 dB (for 10cm depth), -14.57 dB (for 20cm depth), -32 dB (for 30cm depth), and -16.07 dB (for 40cm depth). Bandwidth was measured as 22MHz,

21 MHz, 26 MHz, and 17 MHz at the depths of 10cm, 20cm, 30cm, and 40cm, respectively.

It is very important to characterize underground channel models to design IoUT applications [27, 33, 41, 55–59, 61, 62, 64, 65, 71, 72] In this chapter, design of different types of testbeds for underground channel modeling is presented. It also presented a detailed measurement campaign which was completed over several years. Empirical results for the underground antenna performance and underground channel transfer functions were presented using different types of soil (sandy and silt loam). Experiments show that indoor testbed gives speedy, efficient, and improved characterization of underground communications. From experiments, it is shown how soil moisture, soil texture, and burial depth of the antenna effect the performance of underground channel and communication. These empirical measurements can act as the preliminary step in designing next generation IoUT communication protocols for the development of improved wireless underground communications [34–40, 42–54, 63, 68].

3.3 UG Channel Models

Ideally, a generalized UG channel model is needed for an optimized and integrated UG applications. However, it is a very challenging task because of many factors involved in an underground ecosystem. These factors include: unpredictable environment, continuously changing topology, capacity of the channel, operational frequency, and antenna parameters. The capacity of a channel is an important metric for analysis of the performance of communication channel. Optimal operational frequency can be defined as the frequency at which system achieves maximum bandwidth. For a given antenna design, change in bandwidth, return loss, path loss because of soil moisture and frequency also result as a change in optimal operational frequency. Underground devices must adapt, in real-time, to changes that may occur because of soil moisture.

In [25, 59], the electromagnetic fields of antennas in half space and infinite dissipative medium have been theoretically derived. However, it assumes that the dipole antennas are matched and do not consider the return loss. [22, 71, 72] measures the impedance of dipole antenna in a solution. It discusses the effect of antenna depth on the length of the dipole, solution surface, and the complex permittivity of the solution. However, that does not apply to WUSN because the soil permittivity is not same as of solution and it does not consider the effect on permittivity due to change in soil moisture.

The permittivity model in [11, 30, 39] provides the basis for developing path loss model in WUSNs channel. References [28, 70] were first to model the underground-to-underground (UG2UG) communication channel. In [13, 65], the authors develop a three-path channel model for the explanation of electromagnetic waves propagation in soil. They use electromagnetic analysis and testbed verification for the model. References [6, 14, 67] develops underground-to-aboveground communication models and [14] verifies them through testbed experiments.

Cognitive radio networks are made up of software-defined radios and have recently gained popularity [3, 27]. Most of the literature uses them to deal with scarcity of the spectrum. While there are studies on using cognitive radio to develop underwater solutions [4, 13, 76], there are no solutions, to the best of our knowledge, that consider cognitive radio for wireless underground communication. Moreover, impact of soil moisture on the WUSN communication in terms of path loss, antenna return, loss and bandwidth, has not been yet analyzed.

In the following sections, different UG channel models are discussed with experimental setup and results.

3.4 Underground Soil Propagation Models

There are three major components of a wireless signal: direct wave (DW), reflected wave (RW), and lateral wave (LW) (Fig. 3.23). A signal, propagating through the soil, can exhibit any of the three paths involved in an underground channel propagation. Among these three paths, DW and RW are through soil only. However, LW propagates through, both, soil and air [34, 58]. This section presents the discussion on the characterization of the propagation of these three waves through various channel models.

3.4.1 Path Loss Model

In this section, a path loss model is developed to predict the impact of soil type, soil moisture, operation frequency, distance, and burial depth of sensors for through-the-soil wireless communications channel. The soil specific model is developed based on empirical measurements [61] in a testbed and field settings. The model can be used in different soils for a frequency range of 100 MHz to 1 GHz. The standard deviation between measured and predicted path loss is from 4 dB to 6 dB in the silt loam, sandy, and silty clay loam soil types. The model leads to development of sensor-guided irrigation system in the field of digital agriculture [71].

Digital agriculture [48, 64, 65, 71] is an emerging field in which technology is used to effectively manage agriculture by understanding the temporal and spatial changes in soil, crop, production, and management through innovative techniques. In literature, the Hata-Okumura models are widely used for prediction and simulation of signal strength in cellular environments. Such empirical models do not exist in digital agriculture to predict path loss in through-the-soil wireless communications channel. The wireless communications in wireless underground channel is effected by different factors [40, 61] (e.g., soil type, soil moisture, operation frequency, T-R distances, and burial depths). The prediction of path loss in through-the-soil communications is vital for digital agriculture sensing and communication system design. In this section, a path loss model is developed to predict the impact of soil type, soil moisture, operation frequency, distance, and burial depth of sensors.

3.4.2 Model Parameters

The standard formula for the model is given as [61]:

$$P_r(f, \delta, \phi, \rho, \nu) = -58.8 - 20 \times log10(r1) - 20 \times log10(r2) - 20 \times log10(\delta) + ((\nu \times -10) + \gamma) \times 3 - \xi \times r1 - \xi \times r2 - \xi \times (\phi + \rho) + F + K \quad (3.6)$$

where f is the operational frequency, δ is the distance between transmitter and receiver, ϕ and ρ are the transmitter and receiver depths, respectively, ν is the volumetric water content in percentage unit, $r1 = \sqrt{(\phi - \rho)^2 + \delta^2)}$, $r2 = \sqrt{(\phi + \rho)^2 + \delta^2}$, K is the soil dependent constant, and F is the frequency dependent constant. The γ and ξ, soil moisture and soil attenuation factor, respectively, are given in (3.7) and (3.8):

$$\begin{aligned} \gamma &= p1 * f^3 + p2 * f^2 + p3 * f + p4 \\ \text{where } p1 &= -1.6748e^{-26}, \\ p2 &= 3.8512e^{-17}, \\ p3 &= -3.6971e^{-08}, \\ p4 &= -4.9007, \end{aligned} \quad (3.7)$$

$$\begin{aligned} \xi &= ((\nu * 10) * (p1 * f) + p2 + \nu) * 8.7 + (\nu * 20) \\ \text{where } p1 &= 4.1355e^{-10} \\ p2 &= 2.1161 \end{aligned} \quad (3.8)$$

and the path loss PL is given as:

$$PL(f, \delta, \phi, \rho, \nu) = P_t + G_t + G_r - P_r(f, \delta, \phi, \rho, \nu) \quad (3.9)$$

where P_t is the transmitted power, G_t is the transmitter antenna gain, and G_r is the receiver antenna gain. The values of soil dependent constants are given in Table 3.3.

The frequency dependent constants for silty clay loam, sandy, and silt loam soil types are given in Tables 3.4, 3.5, and 3.6, respectively.

Table 3.3 Soil dependent constant

Param	Silty clay loam	Sandy	Silt loam
K	1	21	6

Table 3.4 Numerical values of frequency dependent constant for silty clay loam soil

Frequency	$\delta < 1$ m	$\delta \geq 1$ m
$f < 300$ MHz	5	15
300 MHz $< f <$ 600 MHz	−10	5
$f > 600$ MHz	−25	1

Table 3.5 Numerical values of frequency dependent constant for sandy soil

Frequency	δ<1 m	$\delta \geq 1$ m
$f < 400$ MHz	1	15
400 MHz $< f <$ 600 MHz	−15	15
$f > 600$ MHz	−15	1

Table 3.6 Numerical values of frequency dependent constant for silt loam soil

Frequency	$\delta < 1$ m	$\delta \geq 1$ m
$f < 400$ MHz	1	15
400 MHz $< f <$ 600 MHz	−15	1
$f > 600$ MHz	−25	1

3.4.3 Model Validation with Empirical Data

The model is validated through data collected during an empirical campaign in an indoor testbed and outdoor field settings [48, 61]. The measured and model results are compared in Figs. 3.21 and 3.22. The experiments are performed in silt loam, sandy, and silty clay loam soil at 50 cm and 1 m distances. It can be observed that the path loss changes with change in the soil type. The sandy soil has 19 dB less path loss as compared to the silt loam soil. The soil types rich in clay content suffers from higher propagation path loss because of the higher water holding capacity [16, 43]. Overall, the developed model matches with an empirical measured data with maximum prediction difference of 5 dB.

It can also be observed that the path loss in through-the-soil wireless communications is high. A 50 cm increase in the communication distance, i.e., T-R separation, leads to 21 dB increase in the propagation path loss due to complex permittivity of the soil [30, 36]. The short communication range is a major challenge in development of in-soil communications system for digital agriculture field operation. Therefore, an advanced signal processing technique (e.g., moisture adaptive beamforming, multi-carrier) is needed for long-range communications in soil for sensor-guided variable-rate irrigation applications [34, 58, 71].

1. Path differences because of soil type.
2. Soil moisture impact on Path Loss.
3. Effect of operation frequency.
4. Distance and depth

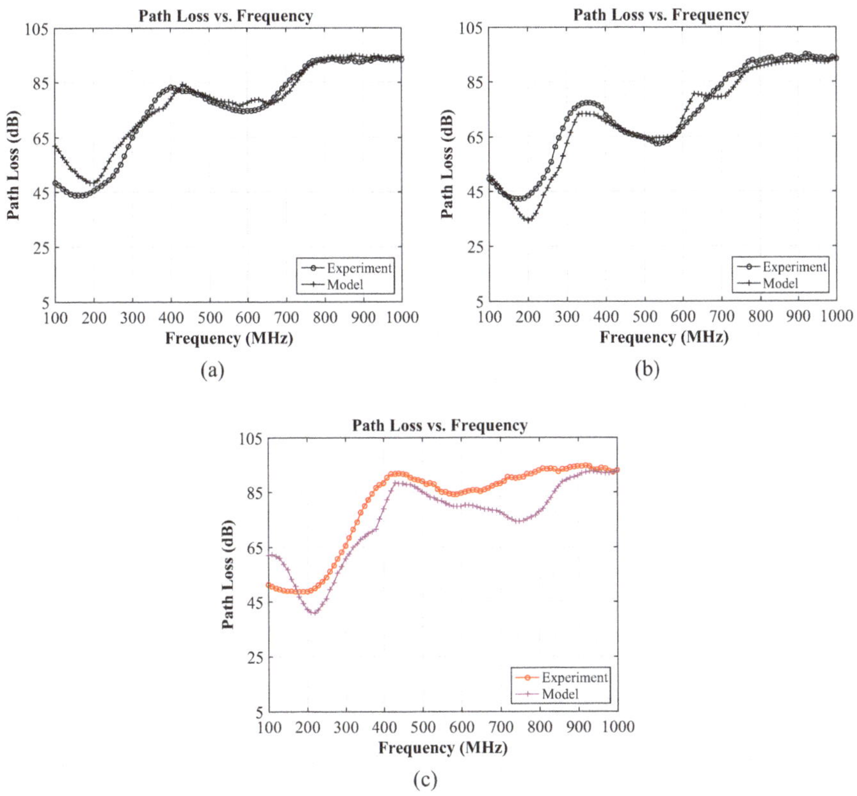

Fig. 3.21 The model comparison with measurement data: (**a**) in silty clay loam at 1 m, (**b**) in silty clay loam at 50 cm (**c**) in silt loam at 1 m

Based on an extensive set of measurements, a model has been developed to investigate the impact of different parameters on communications in silt loam, sandy, and silty clay loam soils. The model is useful to predict the propagation path loss in digital agriculture through-the-soil wireless communications channel. For in-soil system design, the model allows path loss prediction in various soil types under different soil moisture levels without conducting extensive measurements.

3.4.4 Two-Ray Underground Reflection Model

The two-ray UG reflection model considers two signal paths for channel modeling, i.e., DW and RW (Fig. 3.23) [36, 39, 70]. DW is direct between sender and receiver whereas RW is the path of the wave reflected from the ground. Reflected wave is completely negligible at high depths because of longer propagation path. Therefore,

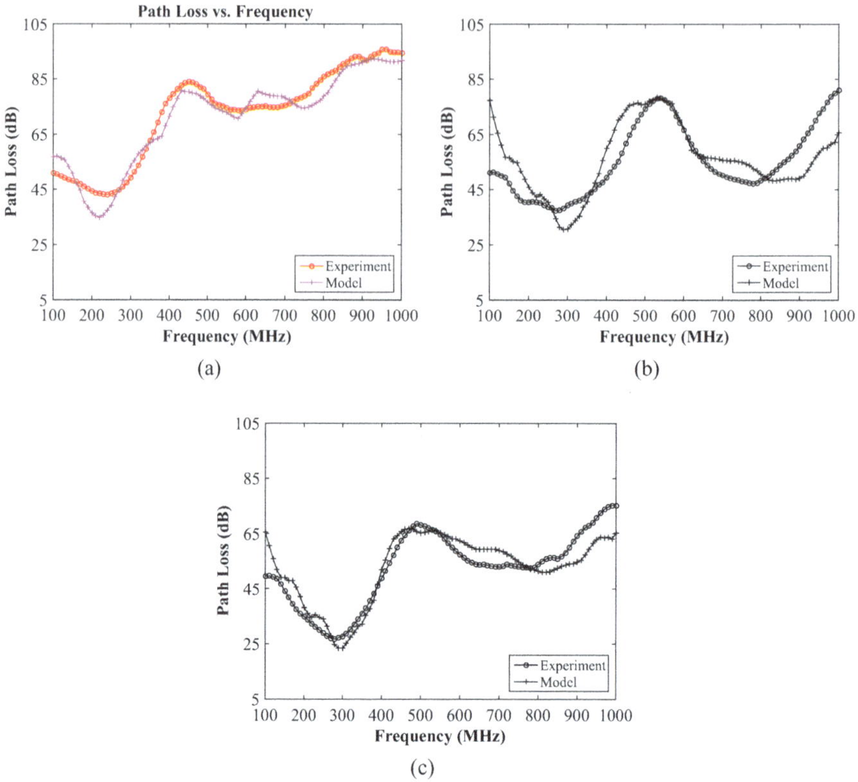

Fig. 3.22 The model comparison with measurement data: (**a**) in silt loam at 50 cm, (**b**) in sandy soil at 1 m (**c**) in sandy soil at 50 cm

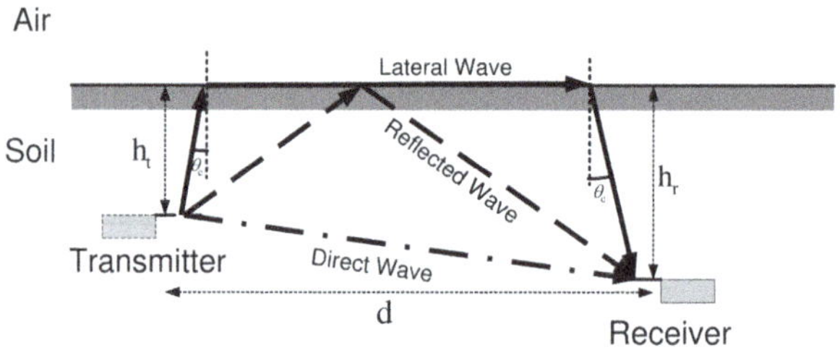

Fig. 3.23 EM wave propagation in soil [61]

only direct path is dominant in the communication. Hence, an adjusted Friis free space path loss equation for the soil is given as:

$$P_r = G_t + G_r + P_t - L_p, \tag{3.10}$$

where P_r is the received power, G_t is the transmitter gain, G_r is the receiver antenna gain, $L_p = L_0 + L_s$ where L_0 is the free space path loss, and L_s is the path loss exhibited in soil. $L_s = L_\beta + L_\alpha$, where L_α is the transmission loss and L_β is the attenuation loss due to difference of wavelength in soil and free space. L_p is given as (in dB):

$$L_p = 20\log_{10}(d) + 20\log_{10}(B) + 8.69 \cdot A \cdot d \ + 6.4, \tag{3.11}$$

where d is the distance in meters, A is the attenuation constant, B is the phase shift constant. Path loss results are shown in Fig. 3.24.

Figure 3.24b compares the path loss in soil with the path loss in the free space for soil moisture. It is observed that path loss increases with the increase in soil water content. For example, at 5 % volumetric water content (VWC), path loss in soil exceeds free space path loss by 64 dB. It becomes worse as the VWC reaches to 25 %.

3.4.5 Three-Ray Underground Reflection Model

Three-wave channel model also considers lateral waves [12, 27]. Lateral waves propagate through soil and air medium (Fig. 3.23). These waves are modeled as follows:

Direct Wave The direct wave has a line-of-sight path between the sender and receiver, i.e., no obstruction is present between both ends and wave completely travels through the soil. It is expressed as:

$$\mathcal{P}^d_{av} = \left(\frac{D_d}{4\pi r_1^2}\right) e^{i2k_s r_1} G_t \ , \tag{3.12}$$

where $\mathcal{P}^d_{av}$ is the averaged Poynting vector, distance between the transmitter and receiver is given by r_1; D_d is a constant factor and its value relies on the permittivity of the soil, wave number in soil is $k_s = \beta_s + i\alpha_s = \omega\sqrt{\mu_0 \hat{\epsilon}_s}$, and β_s is the phase shifting constant, α_s attenuation constant; $\omega = 2\pi f$, where f is the wave frequency; μ_0 is the permeability and $\hat{\epsilon}_s$ is the soil permittivity.

Reflected Wave Soil and air have different permittivity values. This difference in permittivity causes a wave to reflect and is known as reflected wave. Similar to direct waves, complete travel path of reflected waves is also through soil. The reflected wave is expressed as:

$$\mathcal{P}^r_{av} = D_r \left(\frac{1}{4\pi r_2^2}\right) e^{i2k_s r_2} G_t \Gamma^2 \ , \tag{3.13}$$

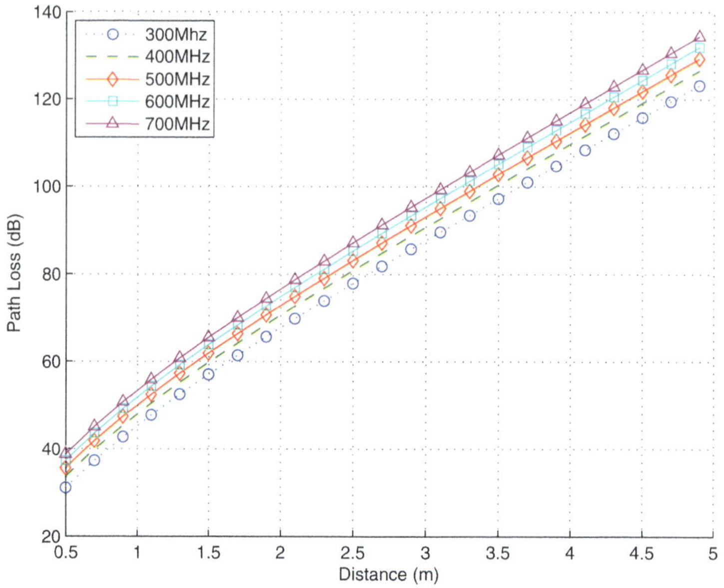

(a)

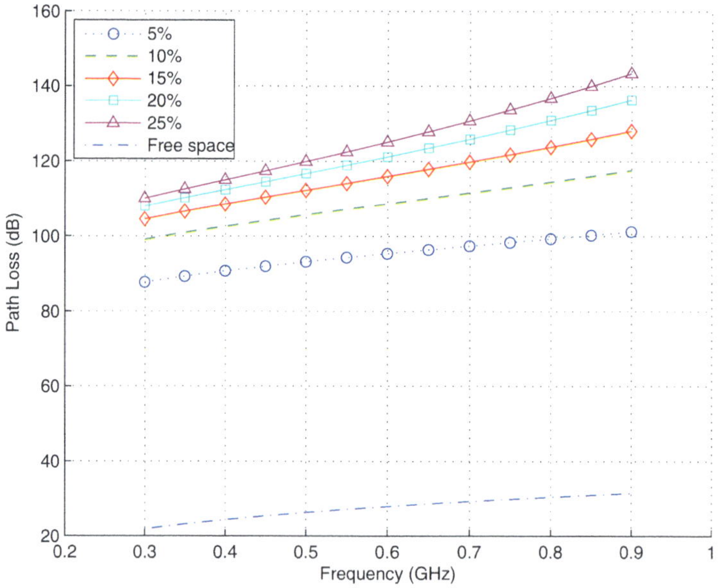

(b)

Fig. 3.24 Two-wave path loss model [70]: (**a**) path loss with variation in distance at different frequencies, (**b**) path loss with variation in frequencies at different VWC where D_r represents the permittivity dependent constant. The length of reflection path r_2 is given as:

$$r_2 = \sqrt{(h_t + h_r)^2 + d^2} \tag{3.14}$$

and Γ is the reflection coefficient given by [24]:

$$\Gamma = \frac{\frac{1}{n}\cos\theta_{ri} - \cos\theta_{rt}}{\frac{1}{n}\cos\theta_{ri} + \cos\theta_{rt}}, \tag{3.15}$$

where θ_{ri} and θ_{rt} are the incident and refracted angle, respectively. n represents the refractive index of soil. As per Snell's law,

$$\sin\theta_{ri} = \frac{d}{r_2}, \quad \cos\theta_{ri} = \frac{h_t + h_r}{r_2}, \sin\theta_{rt} = n\sin\theta_{ri}, \quad \cos\theta_{rt} = \sqrt{1 - \sin^2\theta_{rt}}.$$

To determine the refractive index of soil, real imaginary parts of soil relative permittivity are used as follows:

$$n = \sqrt{\frac{\sqrt{\epsilon'^2 + \epsilon''^2} + \epsilon'}{2}}, \tag{3.16}$$

Lateral Wave Unlike DW and RW, travel path of LW is through soil and air. It propagates horizontally, along soil–air interface and reaches the receiver through soil. Therefore, propagation model for the lateral wave is the mix of both soil and air. The lateral wave is given as:

$$\mathcal{P}^L_{av} = \left(\frac{D_l}{4\pi d^4}\right) e^{i2k_s(h_t+h_r)} e^{i2k_0 d} T^2 G_t, \tag{3.17}$$

where D_l is based on permittivity of the soil and is a constant value, T is the refraction coefficient calculated as :

$$T = \frac{2\cos\theta_{li}}{n\cos\theta_{li} + \cos\theta_{lt}}, \tag{3.18}$$

The power received by DW component is given as $P_r^D = P_t + 20\log_{10}\lambda_s - 20\log_{10} r_1 - 8.69\alpha_s r_1 - 45$. The power received by reflected component is given by: $P_r^R = P_t + 20\log_{10}\lambda_s - 20\log_{10} r_2 - 8.69\alpha_s r_2 + 20\log_{10}\Gamma - 45$. Similarly, for lateral wave it is: $P_r^L = P_t + 20\log_{10}\lambda_s - 40\log_{10} d - 8.69\alpha_s(h_t + h_r) + 20\log_{10} T - 30$,

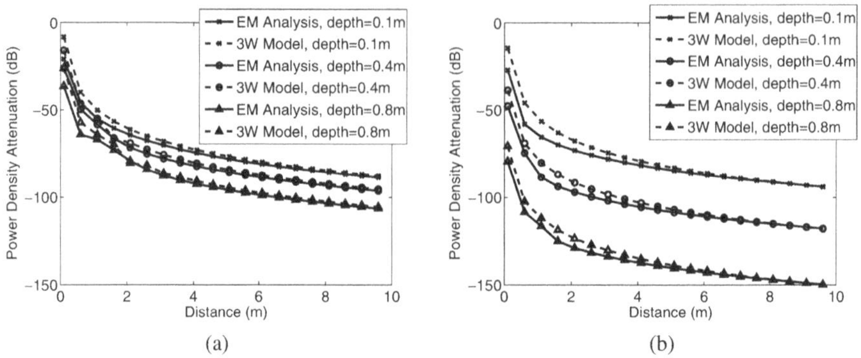

Fig. 3.25 Comparative results for the models at: (**a**) volumetric water content of 10 %; (**b**) volumetric water content of 35 % [58]

$$
\begin{aligned}
P_r^D &= P_t + 20\log_{10}\lambda_s - 20\log_{10} r_1 - 8.69\alpha_s r_1 - 45, \\
P_r^R &= P_t + 20\log_{10}\lambda_s - 20\log_{10} r_2 - 8.69\alpha_s r_2 + 20\log_{10}\Gamma - 45, \\
P_r^L &= P_t + 20\log_{10}\lambda_s - 40\log_{10} d - 8.69\alpha_s (h_t + h_r) + 20\log_{10} T - 30,
\end{aligned}
\tag{3.19}
$$

and the total received power is expressed as:

$$
P_r = 10\log_{10}(10^{\frac{P_r^D}{10}} + 10^{\frac{P_r^R}{10}} + 10^{\frac{P_r^L}{10}}) . \tag{3.20}
$$

Figure 3.25 gives a comparison graph between attenuation of the model and EM wave. The graph is for multiple distances and two different soil moisture levels of 10 % and 30 %. The parameters include depths of 10 cm, 40 cm, and 80 cm, and distance of $D_l = 0.15$, and $D_d = D_r = 0.005$. Maxwell equation is used for EM analysis. EM analysis also uses electromagnetic field analysis presented in [26, 33, 65]. With increase in depth of 0.1 m to 0.81 m, up to 21 dB of attenuation is observed at 10 % VWC. Similarly, for same values of distance and depth, the attenuation increases by 130 dB, at 35 % VWC.

The 3-wave model shows better match with empirical measurements than two-wave model (Fig. 3.26). Empirical measurements are obtained using Mica2 motes with 10 dBm transmit power and burial depth of 40 cm. The minor difference between empirical measurements and 3-wave model is because of inaccuracies in received signal strength of Mica2 motes.

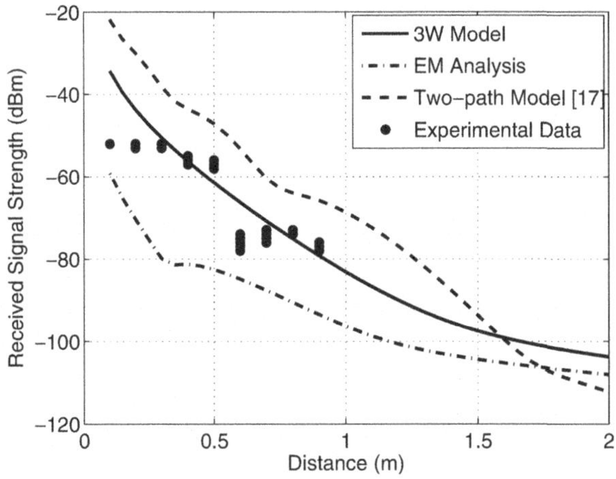

Fig. 3.26 Comparing with measured data at burial depth=0.4 m

3.5 UG Channel Impulse Response Model

An efficient communication scheme requires an in-depth investigation of an underground channel model. To that end, a UG channel impulse response model characterizes the underground channel, helps in designing the physical and link layer protocols, and ensures that the required bandwidth is provided for the effective communication. This section presents the UG channel impulse response model [35, 44, 61]. Moreover, outdoor experiments for validation of UG channel models is a challenging task due to changing weather conditions and highly dynamic properties of the soil medium. To that end, an indoor testbed can be developed to simulate the outdoor environment for underground channel characterization.

The results show that indoor testbed is an efficient way to validate UG channel models. It also gives the flexibility of using different types of soil which is very difficult to do in outdoor setup. The effect of soil properties and implementation parameters, e.g., soil moisture, texture, and burial depth, has been analyzed on UG communication and performance of UG antenna. Moreover, power delay profile of UG channel was also analyzed. Results from the experiments facilitate the development of a mature WUSN platform for the practical implementation.

UG channel impulse response of a wireless underground channel is given as [61]:

$$h(t) = \sum_{l=0}^{N-1} \alpha_n \delta(t - \tau_l) \; , \tag{3.21}$$

where the total number of multipaths is given as N, gain and arrival delay for the n-th multipath are represented by α_n and τ_n, respectively.

Impulse response of the UG model with all three wave components, i.e. RW, DW, and LW, is given as:

$$h_{ug}(t) = \sum_{l=0}^{L-1} \alpha_l \delta(t - \tau_l) + \sum_{d=0}^{D-1} \alpha_d \delta(t - \tau_d) + \sum_{r=0}^{R-1} \alpha_r \delta(t - \tau_r) , \quad (3.22)$$

where L, D, and R represent the lateral, direct, and refracted wave multipaths, respectively.

The magnitude square method gives the total received power of these components [37, 63]. This method sums up the power of all three components over the total delays of the components.

$$P_r = \sum_{l=0}^{L-1} |\alpha_l|^2 + \sum_{d=0}^{D-1} |\alpha_d|^2 + \sum_{r=0}^{R-1} |\alpha_r|^2. \quad (3.23)$$

After calculating the total received power, path loss is given as:

$$PL(\text{dBm}) = G_r(\text{dBi}) + G_t(\text{dBi}) + P_t(\text{dBm}) - P_r(\text{dBm}), \quad (3.24)$$

G_r is the receiver gain, G_t is the transmitter gain, P_t is the transmitted power, and P_r is the received power.

Metrics for Impulse Response Characterization

The metrics for the analysis of the impulse response model are: mean excess delay and RMS delay spread [55, 64]. Time between the first and last component arriving at receiver is known as excess delay. Mean excess delay (τ) is defined as the first power delay profile (PDP) definition [31]:

$$\tau = \frac{\sum_k P_k \tau_k}{\sum_k P_k}, \quad (3.25)$$

where τ_k is the k-th bin delay, and absolute instantaneous power, of k-th bin, is given by P_k. It is important to mention that multipath component is only considered when it is within 30 dB of the strongest component.

In wireless channel, the inter-symbol interference (ISI) is given by Root Mean Square (RMS) delay spread as follows [31]:

$$\tau_{rms} = \sqrt{(\tau^2) - (\tau)^2} , \quad (3.26)$$

where $(\tau^2) = \sum_k P_k \tau_k^2 / \sum_k P_k$,

$$\langle \tau^2 \rangle = \sum_k P_k \tau_k^2 \Big/ \sum_k P_k v \,, \tag{3.27}$$

where τ_k is the delay of the k-th bin, and P_k is the absolute instantaneous power at k-th bin.

3.5.1 Testbed Setup

Testbed shown in Fig. 3.2 is used to conduct experiments for wireless channel modeling. It is an indoor facility housed in a greenhouse setting [46, 61]. UG communication devices have been used with burial depth of 50 cm which is close to the earth surface [5, 43]. Four dipole antennas are buried at the four different depths of: 10 cm, 20 cm, 30 cm and 40 cm, at horizontal distances of 50 cm and 1 m [49, 61]. The maximum distance is kept at 1 m due to the fact that signal attenuates more in the soil. It is helpful in analyzing the channel symmetry by reversing node roles, i.e., switching transmitter and receiver. Furthermore, it also helps to verify accuracy and repeatability of the experiments. Using the multiple burial depth will allow to measure the effect of the burial depth on channel performance. Water is drained through the pipes at the bottom of the testbed. Soil is packed to simulate the same bulk density as of the field.

Soil moisture has a large effect on the performance of UG communication channel. Hence, the effect of this important property of soil should also be measured with channel characterization. To that end, Watermark sensors are used to measure the time-varying soil moisture values as it is less prone to error. An outdoor testbed has also been prepared. It consists of dipole antennas buried at 20 cm depth at distances of up to 12 m. Table 3.1 shows properties of the soils used in the testbed.

3.5.2 Time-Domain Parameters

The impulse response metrics used for the characterization of UG channel are: Excess delay, mean access delay given by Eq. (3.25), RMS delay spread [7, 31, 50, 51, 66] given by Eq. (3.26), and coherence bandwidth [21, 53].

The difference between flat fading and frequency selective channels lies in the extent of frequency distortion of different components. In frequency selective channel, each component has different distortions whereas distortion is same for all components in flat fading channel. In flat fading channel, delay spread is always less than the symbol period [40, 54]. In over-the-air channel (OTA), channels become frequency selective due to mobility whereas in UG channel soil properties (texture, moisture, soil–air interface, etc.) and other organic matter present in the soil cause major disturbance in the channel.

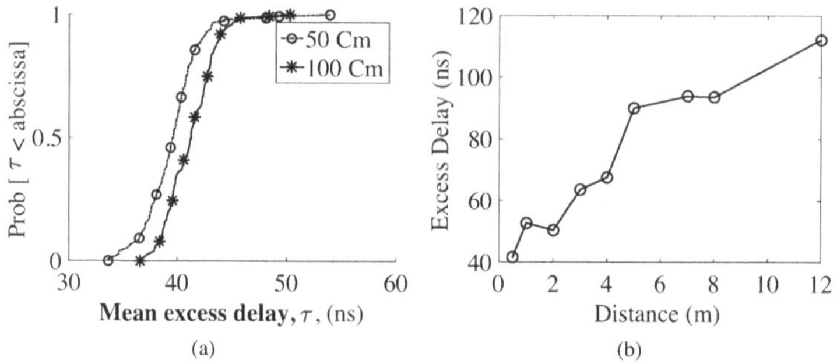

Fig. 3.27 (**a**) Mean excess delay in silt loam soil for the indoor experiments, (**b**) excess delay in silty clay loam soil for the field experiments at the depth of 20 cm [61]

Table 3.7 Silt loam experiment for indoor experiment: standard deviation (σ) and mean (μ) (ns) [61]

	Mean excess delay τ				RMS delay spread τ_{rms}			
	50 cm		1 m		50 cm		1 m	
Depth	μ	σ	μ	σ	μ	σ	μ	σ
10 cm	33.53	1.24	36.09	0.80	20.05	2.24	21.94	2.32
20 cm	34.66	1.07	37.12	1.00	24.93	1.64	25.10	1.77
30 cm	35.87	0.72	37.55	0.65	24.84	2.17	25.34	3.41
40 cm	36.43	0.74	40.18	0.94	23.91	2.84	25.62	1.87

Excess delay of the UG channel is related to the transmitter–receiver separation (distance). It is also shown in Fig. 3.27a, where mean excess delay is plotted at the distance of 50 cm and 1 m at varying depths. Mean excess delay increases by 2 ns to 3 ns for increase of 50 cm in a distance, i.e., 50 cm to 1 m. Table 3.7 shows the values for mean μ and standard deviation, σ, at varying depths. It is observed that mean excess delay increases not only with increase in distance but also with increase in depth. Figure 3.27b plots excess delays along with the distance in silty clay loam soil at the depth of 20 cm. The graph shows that the excess delay is 116 ns for distance of up to 12 m.

Figure 3.28b plots RMS delay spread at transmitter–receiver (T-R) distance of 50 cm and 1 m. Similar to mean excess delay, RMS delay is also dependent upon the distance and burial depths. The increased delay spread is due to reflected and lateral waves. Table 3.7 presents the detailed RMS delay spread statistics for varying burial depths and distances.

Another important metric for analyzing the impulse response of the channel is the coherence bandwidth [41, 68]. It is defined as the inverse of RMS delay spread of UG channel. Figure 3.28d plots coherence bandwidth with T-R distances of 50 cm and 1 m. It is shown that coherence bandwidth decreases from 1 MHz to 600 kHz and then 400 kHz. As the RMS delay value increases, coherence bandwidth of the UG channel decreases. The coherence bandwidth is an important factor to consider

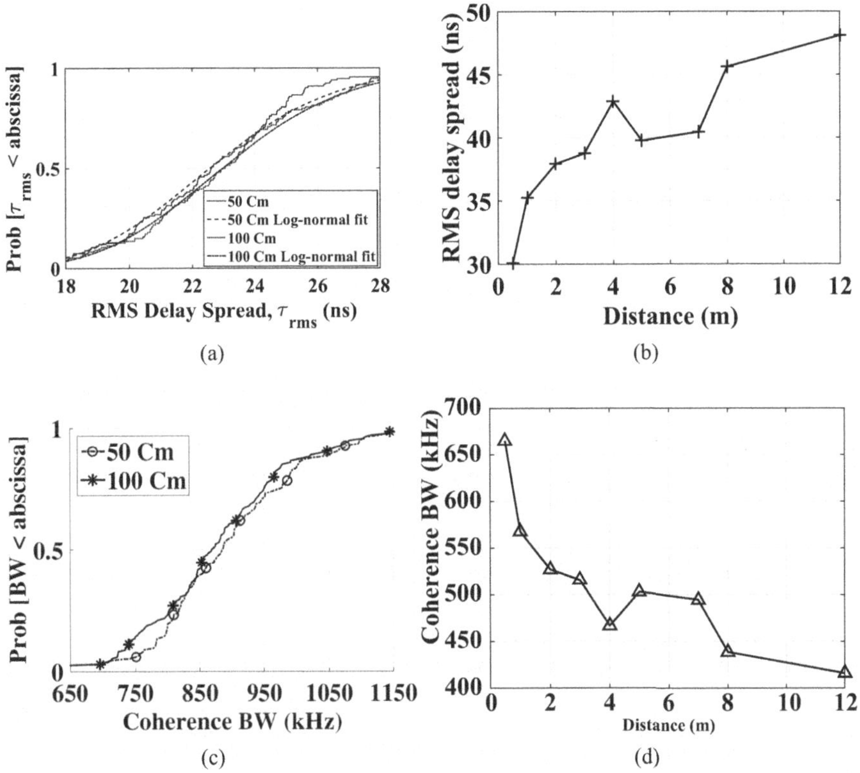

Fig. 3.28 (**a**) RMS delay spread distribution, τ_{rms} in silt loam soil (indoor testbed), (**b**) RMS delay spread, τ_{rms} silty clay loam (field experiment), (**c**) coherence bandwidth distribution in silt loam soil (indoor testbed), (**d**) coherence bandwidth in silty clay loam (field experiment) [61]

while designing practical UG communication systems. Section 3.5.2.1 discusses the effect of soil moisture, and Sect. 3.5.2.3 discusses the effect of change in operation frequency on underground communication system.

3.5.2.1 Capturing the Water Content Changes

Soil water can be classified as bound and free water. Bound water is present in the initial soil layer and is strongly held by soil particles, thus the named *bound* [16, 48]. The strong bonding is because of osmotic and matric forces in the initial layers of soil and forces start decreasing after these layers. EM waves scatter because of the bound water in the soil [42, 45, 52]. Soil dielectric constant is dependent upon the soil moisture, therefore, it changes with the time as the soil moisture changes. The change in dielectric constant causes the wavelength of the signal to change which effects EM attenuation in the soil. Figure 3.29a shows the soil moisture effect on

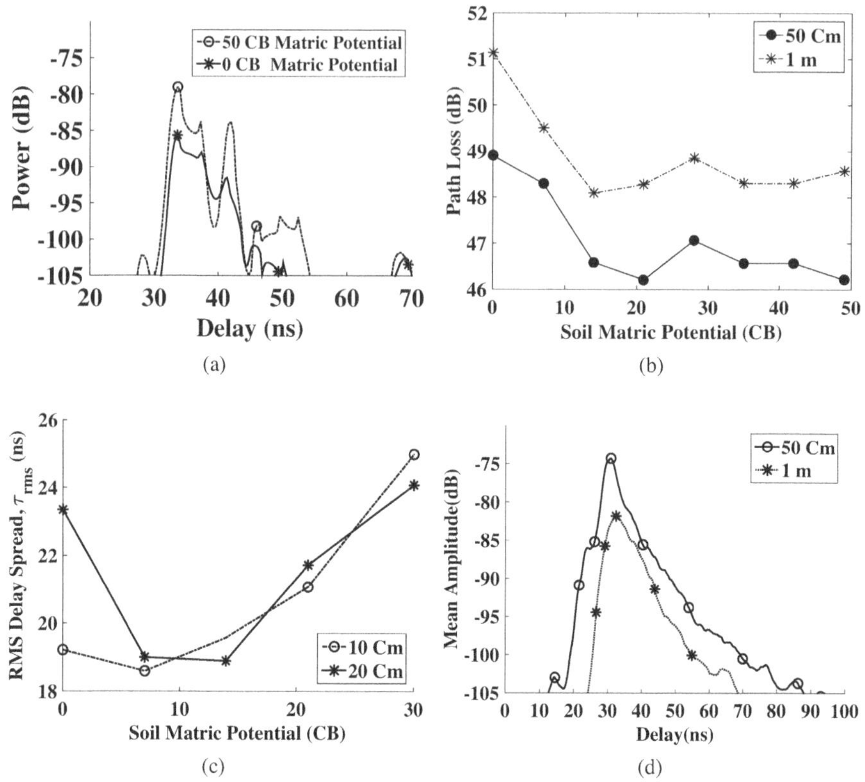

Fig. 3.29 Experiment from indoor testbed using silt loam soil: (**a**) variation in power with delay (Power Delay Profile (PDP)) [61], (**b**) variation in Path loss at different soil moisture values, (**c**) variation in RMS delay spread at different soil moisture values at distance of 50 cm, (**d**) mean amplitudes at all depths at distances of 50 cm and 1 m [61, 61]

delay profiles. The results are for silt loam soil at the distance of 50 cm. It can be observed that gain consistently decreases for all delay range from 5 dB to 8 dB for complete power delay profile.

Figure 3.29b plots the path loss with changing soil moisture. The experiments were performed at the distances of 50 cm and 1 m and burial depth of 10 cm in silt loam soil. It can be seen that path loss decreases by 7 % as the soil moisture changes from 0 cbar to 50 cbar (Centibar).

The variation in amplitude and RMS delay spread are key factors in designing wireless communication system. It is due to the fact that parameters of the system which are suitable for low soil moisture value might not be effective for high soil moisture values. Therefore, a wireless communication system should have the ability to use the optimal values of the system parameters [3, 38].

3.5.2.2 Soil Type Considerations

Texture of a soil is determined by their particle size [20, 57]. Soil texture also plays an important role in the performance of communication system. Therefore, it is important to understand the impact of soil texture on communication systems. To that end, this section studies the impact of soil types on the metrics, i.e., RMS delay spread, mean excess delay, and path loss. The three different soils used for this experiment are: silt loam, silty clay loam, and sandy (see Table 3.8). The change in soil texture leads to significant changes in pore size, and space between the particles. Due to these differences, water retaining capacity of a soil also changes from one type to another [16, 47]. A soil with high water retention capabilities will suffer from high attenuation and vice versa. Furthermore, there is a trade-off between RMS delay and attenuation. Hence, soil retaining high water will have high attenuation but lower RMS delay and vice versa.

3.5.2.3 Radio Frequency Selection

Figure 3.30 studies the effect of frequency on attenuation of the signal. The experiment is done at the T-R separation distances of up to 12m and burial depth of 20 cm is used for both transmitter and receiver. For a distance of 2 m, a 24 dB of attenuation is experienced for a frequency increase of 200 MHz, i.e., from

Table 3.8 Standard deviation (σ) and mean (μ) for all three soil types (ns)

	Mean excess delay				RMS delay spread				Path loss	
	Distance				Distance				Distance	
	50 cm		1 m		50 cm		1 m		50 cm	1 m
Soil type	μ	σ	μ	σ	μ	σ	μ	σ		
Silty clay loam	34.77	2.44	38.05	0.74	25.67	3.49	26.89	2.98	49 dB	52 dB
Silt loam	34.66	1.07	37.12	1.00	24.93	1.64	25.10	1.77	48 dB	51 dB
Sandy soil	34.13	1.90	37.87	0.80	27.89	2.76	29.54	1.66	40 dB	44 dB

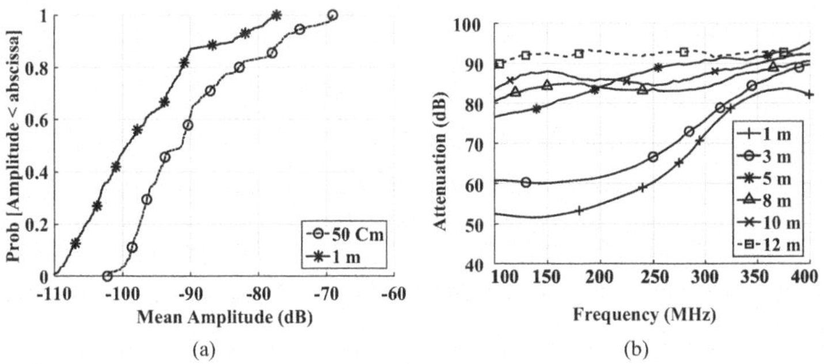

Fig. 3.30 Attenuation with frequency [61]

200 MHz to 400 MHz. Similarly, an 80 % increase in attenuation is observed when distance is increased from 1 m to 12 m at a fixed frequency of 200 MHz.

EM waves with higher frequency attenuate quickly in the soil because of its shorten wavelength due to high soil permittivity. Hence, to get large communication range, lower frequency is more desirable for communication in underground channel [47]. An optimal operation frequency is required for the optimal operation of a particular distance and depth combination. This shows that deployment of such system needs to be customized for a number of factors such as distances, depths, frequency range, and soil texture.

3.5.3 Empirical Validation

This section calculates the wave speed in soil to compare the results of models. The following equation calculates the speed of EM waves in the soil [10]:

$$S = c/\eta, \tag{3.28}$$

where c is the speed of light measured in $\mathrm{m\,s^{-1}}$. The value of $c = 3 \times 10^8\ \mathrm{m\,s^{-1}}$. η is the refractive index of the soil. It is calculated as: $\eta = \sqrt{\sqrt{\epsilon''^2 + \epsilon'^2} + \frac{\epsilon'}{2}}. \epsilon'$ and ϵ'' are the real and imaginary parts of soil's permittivity, respectively.

The three components of EM wave (DW τ_d, LW τ_l, and RW τ_r) may arrive at different time. Therefore, time of arrival for all three components is given as:

$$\tau_d = (\delta_s/S) + 2 \times (L/S_c)\,, \tag{3.29}$$

$$\tau_r = 2 \times (\delta_s/S) + 2 \times (L/S_c)\,, \tag{3.30}$$

$$\tau_l = 2 \times (\delta_s/S) + (\delta_a/c) + 2 \times (L/S_c)\,. \tag{3.31}$$

Some of the components may travel through the soil, therefore, δ_s and S in Eqs. (3.29), (3.30), and (3.31) represent the distance and speed of wave in the soil, respectively. S_c is the speed of the EM waves, and L is the length of EM waves in coaxial cable.

For testbeds (Table 3.1), soil's refractive indices (Table 3.9) are calculated using Eqs. (3.29), (3.30), and (3.31). Moreover, exact arrival time of all components is also measured for comparison with experimental results of these components.

Figure 3.31 compares the results for sandy, silt loam, and silty clay loam soils. The results from the three-wave model for UG channel are shown in dotted lines. The burial depth used for these results is 40 cm. These results are similar to the measurement results from the model. Owing to the high speed of lateral waves, it always arrives before all other components except at depth of 10 cm and T-R separation of 50 cm.

Table 3.9 Wave speed using refractive index (n) of soil [61]

Soil type	Speed in soil m/s	% of c	Refractive index n
Silt loam	5.66×10^7	18.89	5.28
Sandy soil	5.01×10^7	16.71	5.98
Silty clay loam	5.67×10^7	18.91	5.29

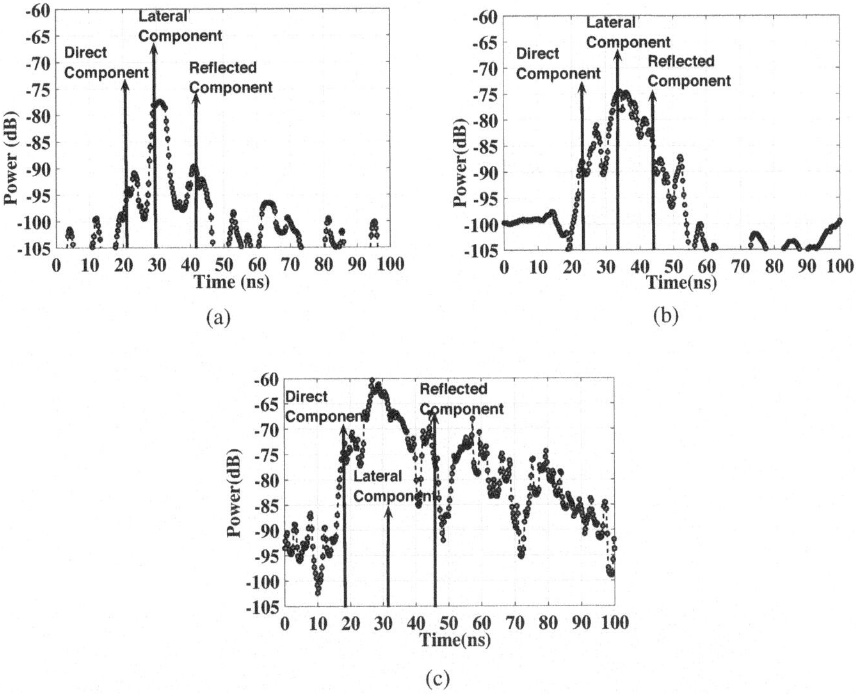

Fig. 3.31 Comparison of impulse response (represented by dotted lines) and impulse response model (represented by solid lines) for soil types of: (**a**) silt loam, (**b**) silty clay loam soil, and (**c**) sandy soil [61]

It can be observed from Fig. 3.31 that the lateral wave is the dominant component of EM wave and suffers from least attenuation. The reason for this is that lateral waves travel from air and soil whereas direct and reflected components completely travel from the soil.

3.6 Underground Channel Sounding and Multipath Characteristics

This section discusses the characterization of a wireless underground channel in terms of path loss, channel sounding, multipaths, and BER analysis. The experiments were conducted in underground testbed in SCAL (South Central Agricultural Laboratory). Theoretical foundation of these experiments along with results is explained in the following sections.

3.6.1 Path Loss Experiments

The testbed consists of 4 sets of buried dipole antennas. Each set contains four antennas buried at a distances of 50 cm, 200 cm, 400 cm, respectively, from each other. Burial depths are 10 cm, 20 cm, 30 cm and 40 cm, respectively. Figure 3.32 shows the testbed.

A wideband Gaussian signal (2 MHz) RF waveform is transmitted from underground dipole antenna buried to 40 cm depth by using one USRP. Signal is received on other buried dipole antenna by using USRPs connected to antennas buried at four different depths (10 cm, 20 cm, 30 cm and 40 cm) with a fixed T-R distance of 50 cm. Experiment is repeated for all these depths by varying the distance to 200 cm and 400 cm. For each frequency, sender transmits data for one second. Receivers collect

Fig. 3.32 The cable layout in the testbed [58]

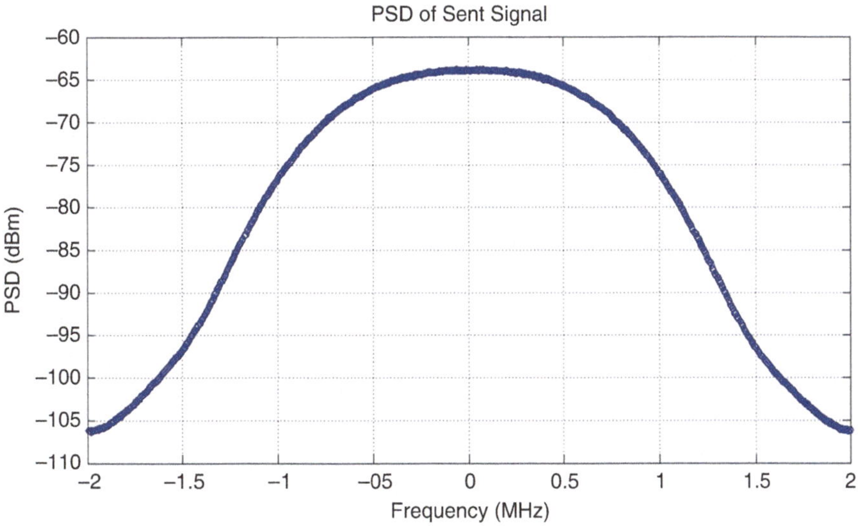

Fig. 3.33 The power spectral density of the transmitted signal

IQ data of 4 Mega Samples. Receivers acknowledge sender after finishing and then transmitter moves to the next frequency. Three measurements are taken for each depth and distance. Analysis is done in Matlab [29]. GNU Radio [17] and USRPs [15] are used to conduct experiments. Figure 3.33 shows PSD of a sent signal.

For spectral estimation and path loss analysis Welch's method [73] has been used. This method is an enhanced form of periodogram analysis. By using a computationally efficient discrete Fourier transforms, data is divided into fixed blocks to calculate periodograms and modified periodograms. These modified periodograms are averaged to calculate the power spectrum. The analysis follows details from [73].

Collected data X[n], for n = 0, ..., $N-1$, is divided into K blocks. Length of these overlapping block is L with a difference of D units from each other. Then

$$\mathbf{X_1}(j) = X(j), \qquad j = 0 \ldots L-1 \tag{3.32}$$

Similarly,

$$\mathbf{X_2}(j) = X(j+D), \qquad j = 0 \ldots L-1 \tag{3.33}$$

and

$$\mathbf{X_K}(j) = X(j+(K-1)D), \qquad j = 0 \ldots L-1. \tag{3.34}$$

For each block of length L, modified periodogram is calculated by selecting a data window $W(j)$ for $j = 0 \ldots L-1$, in order to form the sequence $X_i(j)W(j), \ldots, X_K W(j)$. Fourier transforms of these sequences $A_i(n), \ldots, A_k(n)$ are taken as follows:

$$\mathbf{A_k}(n) = \frac{1}{L}\sum_{j=0}^{L-1} X_k(j)W(j)e^{-2kijn/L} \tag{3.35}$$

where $i = (-1)^{\frac{1}{2}}$. K modified periodograms are obtained as follows:

$$\mathbf{I_k}(f_n) = \frac{L}{U}|A_k(n)|^2 \tag{3.36}$$

where

$$f_n = \frac{n}{L} n = 0, \ldots, L/2 \tag{3.37}$$

and

$$U = \frac{1}{L}\sum_{j=0}^{L-1} W^2(j) \tag{3.38}$$

Power spectral density (PSD) is the average of these periodograms and is given by:

$$\mathbf{P}(f_n) = \frac{1}{K}\sum_{k=0}^{K} I_k(f_n) \tag{3.39}$$

Figure 3.34 shows PSD of a received signal at 50 cm distance and depths of 10 cm, 20 cm, 30 cm, and 40 cm. Sender's burial depth is 40 cm. It is clearly evident that with an increase in burial depth PSD decreases significantly.

Figure 3.35 shows received signal strength for depths of 10 cm, 20 cm, 30 cm, and 40 cm. Received signal strength decreases significantly as burial depth is increased. Figure 3.36 shows path loss with varying frequencies at depths of 10 cm, 20 cm, 30 cm, and 40 cm. Our results show that path loss increases significantly with frequency. This shows that lower frequencies (less than 500 MHz) are more suitable for underground communications. Results of path loss model are compared with theoretical model [12]. Received power for isotropic antennas with $G_t = G_r = 1$ and for all three components is given by Eq. (3.19) and total power is given by Eq. (3.20) [12].

Figure 3.37 compares experimental results with theoretical model of Eq. (3.20). Experimental results for 30 cm and 40 cm depth are in complete match with the theoretical model. In order to use wireless underground channel for effective and efficient communication, a characterization of its impulse response and multipath is needed. These characteristics like delay spread, power delay profile, and rms delay spread are crucial to characterize a wireless channel [33, 41, 58]. We

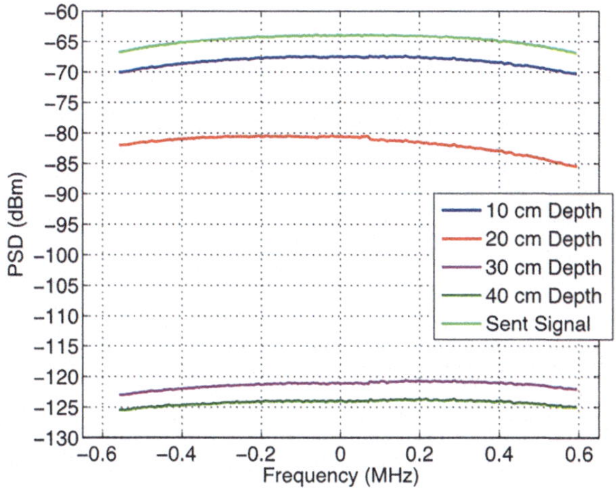

Fig. 3.34 PSD of received signal.$d = 50\,\text{cm}\,f = 225\,\text{MHz}$ [58]

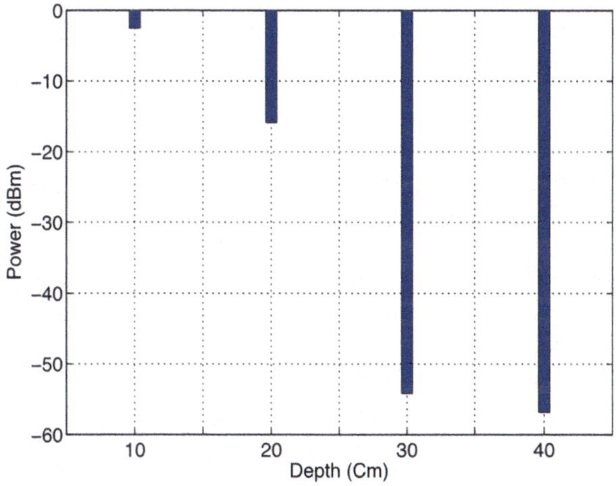

Fig. 3.35 Received signal strength $d = 50\,\text{cm}\,f = 225\,\text{MHz}$

have done underground channel measurement experiments to estimate multipath characteristics of underground channel. Experiment methodology is explained first followed by results of UG channel impulse responses as function of transmitter–receiver distance, burial depth, frequency, and soil moisture.

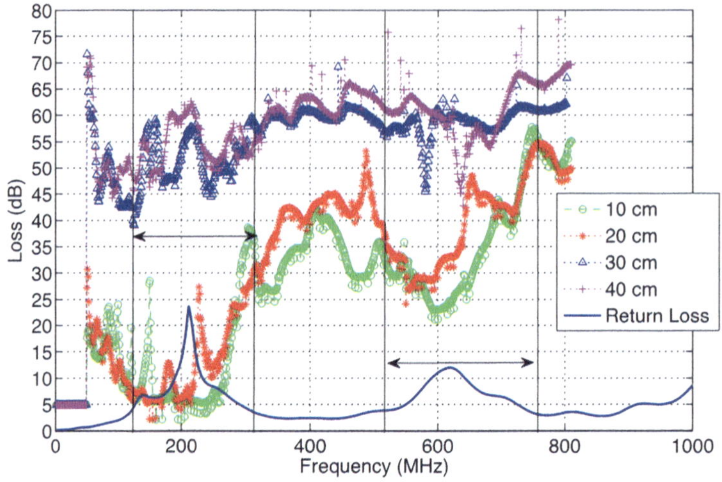

Fig. 3.36 The path loss with frequency

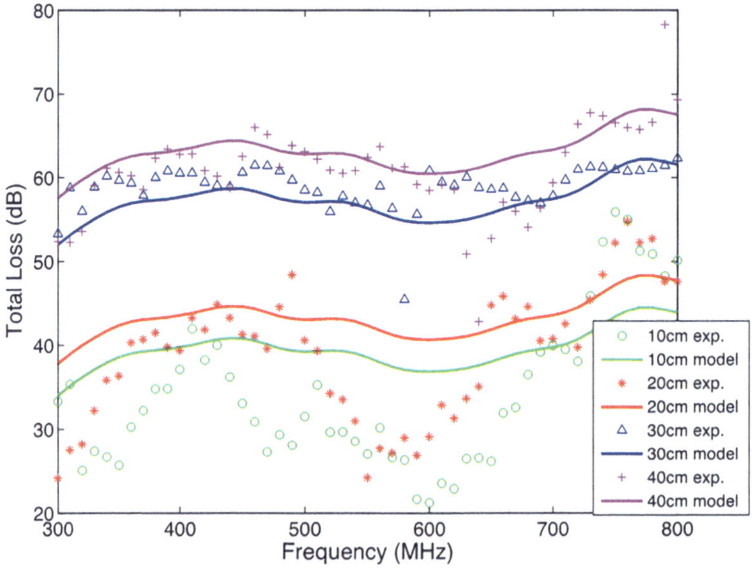

Fig. 3.37 Comparative analysis between experiment and theoretical results [58]

3.6.2 *Experiment Methodology*

A channel sounding system, designed in [23, 38], is used in experiments which follows channel measurement techniques of [9]. In this design, pseudo-random sequences are used for measurement of channel impulse response. These sequences consist of series of 1 and -1 and are generated by shift register. Auto-correlation function of these sequences $s(t)$ is given by [9, 54]:

$$\mathbf{R_s}(\tau) = \frac{1}{T}\int_0^{T} s(t)s(t-\tau) \tag{3.40}$$

Correlation output of the N-length x and y sequences is given as follows[18]:

$$\mathbf{R_{xy}[n]} = \begin{cases} 1. & \text{if } n = 0, N, -N, 2N, -2N, \ldots \\ \frac{-1}{N}, & \text{otherwise} \end{cases} \tag{3.41}$$

Let $s(t)$ and $r(t)$ be the sent and received signals, respectively. $\mathrm{h}(\tau)$ is an impulse response of the channel where τ represents a multipath delay. $r(t)$ can be written as a convolution of the $s(t)$ with $\mathrm{h}(\tau)$.

$$\mathbf{r(t)} = s(t) \otimes h(\tau) = \int_{\inf}^{-\inf} s(t)h(\tau)d\tau \tag{3.42}$$

Important system parameters are summarized below [8]:

1. Power delay profile (PDP): is described as power associated with each component delayed in time. It is given as [8]:

$$\mathbf{p}(\tau) = E\left\{|c(\tau)^2|\right\} \tag{3.43}$$

2. *Multipath Resolution*: It is the capability to estimate the maximum detectable component. It is smaller than rate at which sequence starts repeating itself.
3. *Minimum Multipath Resolution*: Difference between two contagious multipath components is called minimum multipath resolution and it depends on chip rate.
4. *Dynamic Range*: Dynamic range is the ratio of the magnitude of the correlation peak and the magnitude of the maximum correlation value [8, 63]:

$$\mathbf{DR} = 10\log_{10}(L^2) \tag{3.44}$$

where L is the m-sequence length. It represents the ability to detect weak signals.

3.6.2.1 Transmitter

Our experiment uses Ettus Research's Universal Software Radio Peripheral (USRP) [15] along with a GNURadio platform [17]. GNURadio and USRPs provide a very flexible, configurable functionality for implementation of this algorithm. The transmitter and receiver are synchronized by using MIMO cable of [15]. A shift register is used with angle of 10°, and 17 MHz bandwidth to give a resolution of 60 ns. The sequence repeats after every 61 μs. This sequence is then amplitude modulated using root raised cosine filter (RRC) and forwarded to USRP block for transmission. Sent signal is given by [23, 45]

$$\mathbf{s(t)} = \sum_{n} x[n]p(t - nT_s) \tag{3.45}$$

The effects channel dispersion can be modeled as follows [23]:

$$\mathbf{h(t)} = \sum_{l=0}^{L-1} \alpha_l \delta(t - \tau_l) \tag{3.46}$$

Here L, α_l, τ_l are the number of multipaths, complex gain of multipaths, and delays associated with multipaths, respectively.

3.6.2.2 Receiver

Received signal in time-domain is given by [23]:

$$\mathbf{r(t)} = (x * h)(t) \sum_{l=0}^{L-1} \alpha_l x(t - \tau_l) \tag{3.47}$$

Properly synchronized discrete symbol is given by:

$$\mathbf{r[j]}) = \sum_{l=0}^{L-1} \alpha_l \sum_{n} x[n]p(j - nT_s) - \tau_l \tag{3.48}$$

With the assumption that τ_l is the integer multiple of T_s, then $\tau_l = c_l T_s$ where cl is a nonnegative integer. Hence,

$$r[j] = \sum_{l=0}^{L-1} \alpha_l \sum_{n} x[n]p[(j - n - c_1)T_s] \tag{3.49}$$

$$= \sum_{l=0}^{L-1} \alpha_l x[j - c_1]$$

Correlator's output is given as:

$$R_{cy}[n] = corr(c, y) = \sum_{l=0}^{L-1} \alpha_l R_{cx}[n - c_l] \tag{3.50}$$

in this equation R_{cx} is $corr(c, y)$. Complex multipath gains α_l are found using Eqs. (3.41) and (3.50). Taking magnitude squared of these complex multipath gains α_l gives us multipath delay profile. Path loss is obtained by subtracting combined power of all multipath components from the transmitted power.

The system uses sliding correlator based channel sounding technique in literature and has been widely used by researchers for doing channel sounding in different environments. Equipment consists of USRPs connected with MIMO synchronization cables. The transmitter rate is 25 MHz, which gives 40 ns time resolution. With 40 nanoseconds chip rate all the components arriving 40 ns apart at receiver can be easily identified by the measurement system. Symbol level synchronization is achieved by use of 10 training sequences. Each sequence consists of 1000 bits. Gold Sequences of 10° are used for correlation.

Experiments are conducted in 800 MHz and Power Delay Profiles (PDP) are taken. Each PDP shows normalized correlation value over time. The LOS case is simple, however, OLOS scenarios are complicated where the line-of-sight has been blocked by inserting two metal plates of 8 mm thickness and then further blocking by two 15 mm thick cabinet covers.

3.6.3 Experimental Results and Analysis

A measurement system has been developed for wireless underground channel modeling and to characterize underground wireless channel. This system expands path loss modeling and bit error rate analysis experiments to measure underground channel multipath delays and channel attenuation. This section presents results of channel sounding experiments conducted in Cyber Physical Networking (CPN) laboratory for both Line-of-Sight (LOS) and Non-Line-of-Sight scenarios (OLOS). These experiments are conducted to verify the functionality of sliding correlator based channel sounding system developed in lab and to validate system's ability to detect multipath in different environments. These experiments were done in underground testbed in SCAL (South Central Agricultural Laboratory). Theoretical foundation of these experiments along with results is explained in the following sections.

All the power delay profile figures depict the normalized correlation power of first arriving signal at the receiver relative to the peak power of the first arriving signal. X-axis shows propagation delay. Delay components within 15 dB are categorized as multipaths and over 15 dB can be categorized as noise.

Figures 3.38, 3.39, 3.40 show the power delay profile for LOS scenario. There is a one peak that is the strongest one and indicates the line of sight. Its propagation delay is approximately 300 ns. Low peaks, followed by the strongest one, are the weak multiple components and its difference from the strongest one is more than 15 dB.

These figures also show the one component at 300 ns and then multiple components within 15 dB range and last till 500 ns with delay spread of total 200 ns.

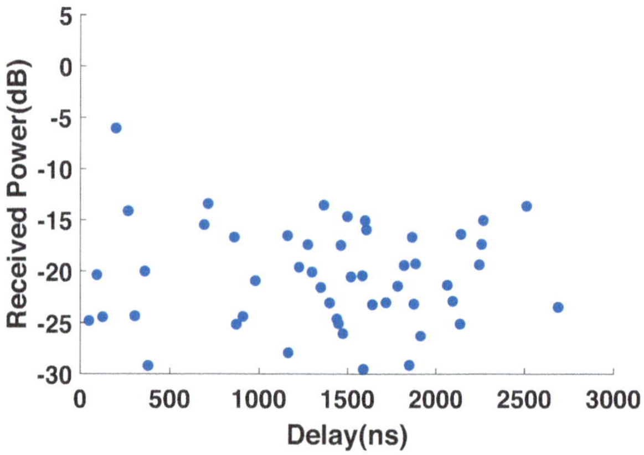

Fig. 3.38 Power delay profile—line of sight

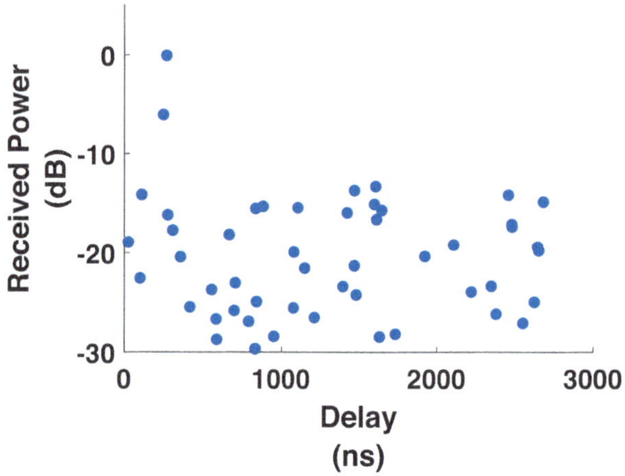

Fig. 3.39 Power delay profile—obstructed LOS—Case 1

Since line-of-sight is obstructed, delayed components experience delay due to more distance covered. These delayed multipaths can contribute to intersymbol interference.

The system's ability to detect multipath with delay resolution of 40 ns has been validated. For indoor LOS scenario delay spread is less than 80 ns and for OLOS scenario delay spread range is 100 ns to 150 ns. This system can be used for taking underground channel measurements and these results along with path loss experiments and BER results are presented in subsequent chapters.

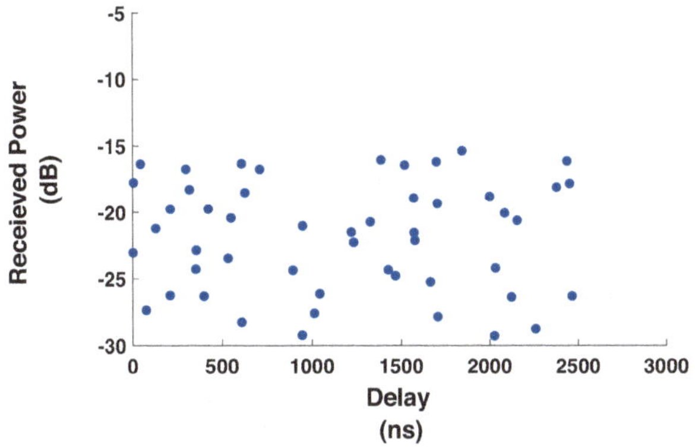

Fig. 3.40 Power delay profile—obstructed LOS—Case 5

3.7 UG Prediction Model for Subsurface Attenuation

Electromagnetic signals suffer attenuation with increase in distance [2] because of which the wireless communication has a limited range. It is very rare to achieve long ranges in WUC and if it is achieved in some cases, it is due to randomness of wireless channel. Wireless signals change their direction because of reflection and scattering phenomenon. The arrival time of these electromagnetic signals is different because of multipath effect. Therefore, a channel model considering the multipath effect in UG communication is required. To that end, Eq. (3.51) takes multipath effect and attenuation into consideration in wireless communication. In this section, the effect of distance and angle on path loss and multipath effect of the channel is investigated using empirical evaluations [59, 71, 72].

$$P_{rx}\left(d_{x,\theta}\right) = P_{tx}\left(d_{x,\theta}\right) - PL\left(d_{0,\theta}\right) - 10n\log_{10}\left(\frac{d}{d_0}\right) + X_\sigma \tag{3.51}$$

where

$d_{x,\theta}$: is the T-R separation with angle θ
$d_{0,\theta}$: is the reference distance with angle θ
P_{rx}: is the receiving power
P_{tx}: is the transmission power
$PL(d_{0,\theta})$: is the path loss at the reference distance
n: is the path loss exponent
X_σ: is a normal random variable with zero mean and standard deviation σ

For a normal random variable, the path loss exponent and standard deviation are measured using the empirical data by the following method:

1. Packets are classified into multiple bins on the basis of communication links. For example, there can be a separate bin for East UG to counterclockwise AG node and similarly separate bins for other links.
2. All items within the bins are classified on the basis of communication angles. Due to symmetric radiation pattern of aboveground (AG) antenna, packets with same angle but opposite direction, e.g., 30° and −30°, can be considered same and placed together.
3. A reference distance with at least hundred entries is selected.
4. Next step is to calculate the mean of the receiving power. This mean is calculated at reference distance $d_{0,\theta}$. It is because averaging over hundred readings will cause the random variable to approach zero. Equation (3.51) can be re-written as Eq. (3.53).

$$\bar{P}_{rx}\left(d_{0,\theta}\right) = P_{tx}\left(d_{x,\theta}\right) - PL\left(d_{0,\theta}\right) - 10n\log_{10}\left(\frac{d_{0,\theta}}{d_{0,\theta}}\right) + X_{\sigma} \tag{3.52}$$

$$P_{rx}\left(d_{x,\theta}\right) = \bar{P}_{rx}\left(d_{0,\theta}\right) + 10n\log\left(\frac{d_{x,\theta}}{d_{0,\theta}}\right) + X_{\sigma} \tag{3.53}$$

5. Only those entries are selected from the leftovers where distance > $d_{0,\theta}$.
6. Calculate $P_{rx}(d_{x,\theta}) - \bar{P}_{d_{0,\theta}}$ and $\frac{dx,\theta}{d_{0,\theta}}$.
7. After plotting the values, path loss coefficient n is given by the slope of best fit line. Standard deviation is the root mean squared error of the normal random variable. The communication process between UG node at east and the UG node at opposite direction is shown using AG2UG communication link in Fig. 3.41.

The path loss coefficients and the corresponding log normal variable standard deviation values are used to perform this procedure.

3.8 UG Statistical Channel Model

A standard OTA modeling approach given in [19, 31, 66, 74] is adopted, with a slight modification, to statistically model the UG wireless channel. Some assumptions for the analysis are given below:

- multipath components (LW, RW, and DW) are not correlated practically, however, individual component is affected by the strongest one. Therefore, within each component, a uniformly spaced tap-delay-lines are assumed.
- The random phases of the components are assumed at the receiver. The phases follow the uniform distribution over $[0, 2\pi]$.

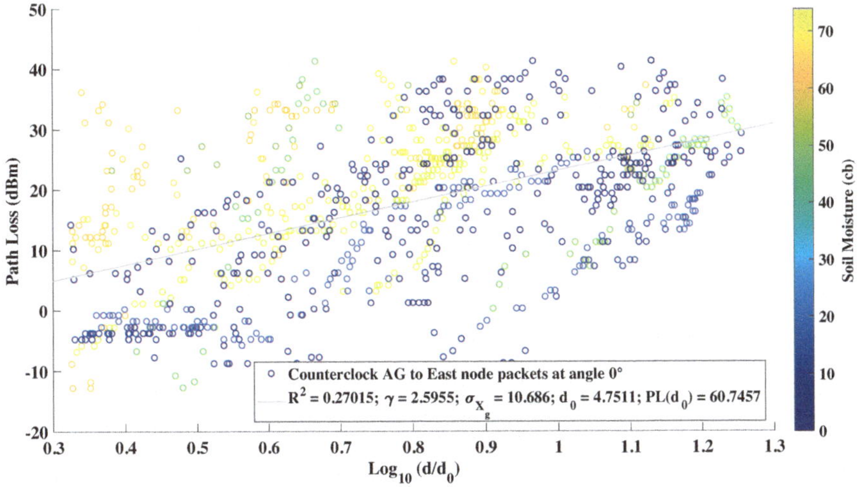

Fig. 3.41 Empirical evaluation of channel path loss [75]

To reduce the model complexity, each component has constant arrival rate of delay and statistically independent amplitude. It is helpful in analyzing the UG channel along with the delay statistics. Different factors are involved in deciding which component will arrive first. These factors include T-R separation (distance) and burial depth. It is due to the fact that propagation speed is different in soil and air. Hence, order of arrival will vary based on the characteristics of these mediums. Direct component arrives earlier for T-R distances of up to 50 cm. For all other distances, lateral component is dominant because of its high speed in the air. Component overlapping does not occur because of the big difference between component's speed, and gain of each components decays before the other one arrives. In the given measurements, no components were detected after the delay of 100 ns.

Next, the amplitudes of individual component are derived. The delays of each component's amplitude, i.e., Lateral (α_{li}), Direct (α_{dj}), and Reflected (α_{rk}) are τ_{li}, τ_{di}, τ_{ri}, respectively. Figure 3.42 plots the mean amplitude of a profile with time decay at a distance of 50 cm. It is observed that the gain of each component exhibits an exponential decay, hence, amplitude of a component is modeled as exponentially decaying. For lateral waves, arrival time τ_L, decay rate γ_L, and amplitude α_L are used for the calculation of amplitude and it is given as [66]:

$$\alpha_{li} = \alpha_{l0} e^{-(i-\tau_L)/\gamma_L} \; \forall \, i > \tau_l \; and \; i < \tau_l + L. \tag{3.54}$$

Similarly, same parameters, i.e., decay rate γ_D, arrival time τ_D, and amplitude α_D, are used for calculating the amplitude of direct component α_{dj}. It is expressed as:

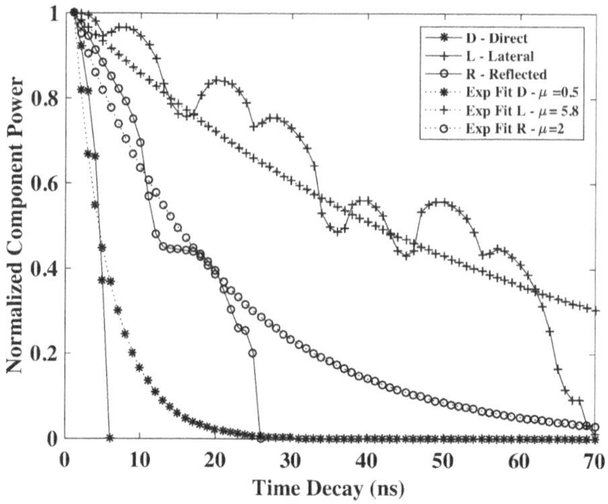

Fig. 3.42 Decay of L,D, and R wave-components [62]

$$\alpha_{dj} = \alpha_{d0} e^{-(j-\tau_D)/\gamma_D} \ \forall \ j > \tau_d \ and \ j < \tau_d + D. \tag{3.55}$$

Following the same method for the reflected component, α_{rk} is calculated as:

$$\alpha_{rk} = \alpha_{r0} e^{-(k-\tau_R)/\gamma_R} \ \forall \ k > \tau_r \ and \ k < \tau_d + R. \tag{3.56}$$

To calculate the gain of first multipaths (α_{d0}, α_{l0}, and α_{r0}) both model and measured values can be used. First multipath gain is given as: α_{d0}, α_{l0}, and α_{r0}. Each multipath is calculated as follows [12]:

$$\begin{aligned}
\alpha_{d0} =& P_t + 20\log_{10}\lambda_s - 20\log_{10} r_1 - 8.69\alpha_s r_1 - 22 + 10\log_{10} D_{rl}, \\
\alpha_{r0} =& P_t + 20\log_{10}\lambda_s - 20\log_{10} r_2 - 8.69\alpha_s r_2 + 20\log_{10}\Gamma - 22 + \\
& 10\log_{10} D_{rl}, \\
\alpha_{l0} =& P_t + 20\log_{10}\lambda_s - 40\log_{10} d - 8.69\alpha_s (h_t + h_r) + 20\log_{10} T - 22 + \\
& 10\log_{10} D_{rl},
\end{aligned} \tag{3.57}$$

where P_t is the transmitted power, Γ is the reflection coefficient, T is the transmission coefficients [12], r_2 is the length of the reflection path, $r_1 = \sqrt{(h_t - h_r)^2 + d^2}$, $r_2 = \sqrt{(h_t + h_r)^2 + d^2}$, where h_t is the transmitter depth and h_r is the receiver burial depth, and λ_s represents the wavelength in soil [60].

The exponential decay seems reasonable because of the different time delays due to different travel paths. Moreover, path gains are affected by the soil, hence, multipath gains are also dependent upon the delays. Other important factors effecting the multipath gains (α_{li}, α_{dj}, α_{rk}) are soil moisture and soil type, e.g.,

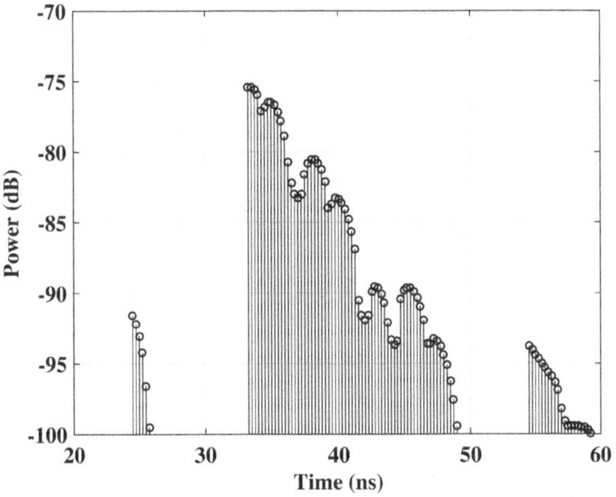

Fig. 3.43 Impulse response of wireless underground channel [62]

sandy soil has higher path gain as compared to the silty clay loam and silt loam soils. This is because of sand holds less water as compared to later soil types. However, this effect of soil type on multipath gains does not need to be remodeled in Eqs. (3.54)–(3.56) and is captured in main components α_{l0}, α_{d0}, α_{r0}. As multipath gains are dependent upon these components, the effect will be reflected in gains (Eqs. 3.54–3.56) as well.

A threshold value for the gain is selected for estimating the number of multipaths L, D, and R in each component and after that, multipath generation is stopped for that particular component. The threshold value is usually within 30 dB from peak. As discussed earlier, the effect of soil texture will result in relatively large number of components for sandy soil than the silt loam and silty clay loam soils. This shows a very good matching with empirical results. Furthermore, the number of components also represent the channel spread which I also inversely proportional to the soil moisture values, i.e., low soil moisture will result in large spread and vice versa. It is because of the decreased attenuation for low soil moisture values. Figure 3.43 shows the underground impulse response model and model parameters are given in Table 3.10.

To summarize, delays within multipath components are used to calculate α_l, α_d, α_r. These delays rely on the exponential decay of multipaths. However, for normalized path gains, the amplitudes become independent of their corresponding delays [66]. The normalized delays are calculated using average such that $\alpha_{li}/\bar{\alpha_{li}}$, $\alpha_{dj}/\bar{\alpha_{dj}}$, and $\alpha_{rk}/\bar{\alpha_{rk}}$. Figure 3.44 shows normalized path gains, calculated from similar process, following a Weibull probability distribution.

3.8.1 Model Parameters and Evaluation

Table 3.10 lists all parameters required to evaluate the statistical model. First, the component's gains (α_{li}, α_{dj}, α_{rk}) and their corresponding delays (τ_{li}, τ_{di}, τ_{ri})

Table 3.10 The impulse response model parameters [62]

Parameter	Description	Model	Values
S	Speed of wave in soil [61]	C/η	$C = 3 \times 10^8$
η	Refraction index [61]	$\eta = \sqrt{\sqrt{\epsilon'^2 + \epsilon''^2} + \epsilon'/2}$	ϵ', ϵ''
ϵ'	Real part of relative permittivity of the soil [30]	$\epsilon'_s = \begin{cases} 1.15\left[1 + \rho_b/\rho_s\,(\epsilon_s^{\delta} - 1) + (m_v)^{v'}(\epsilon'_{fw})^{\delta} - m_v\right]^{1/\delta} - 0.68 & 0.3\,\text{GHz} \le f \le 1.4\,\text{GHz}\,, \\ \left[1 + \rho_b/\rho_s\,(\epsilon_s^{\delta} - 1) + (m_v)^{v'}(\epsilon'_{fw})^{\delta} - m_v\right]^{1/\delta} & 1.4\,\text{GHz} \le f \le 18\,\text{GHz}\,, \end{cases}$	S = Sand in %, C= Clay in %, $\delta = 0.65$, $v' = 1.2748 - 0.519S - 0.152C$, $v'' = 1.33797 - 0.603S - 0.166C$ $\epsilon'_{fw}, \epsilon''_{fw}$
ϵ''	Imaginary part of relative permittivity of the soil [30]	$\epsilon''_s = \left[(m_v)^{v''}(\epsilon''_{fw})^{\delta},\right]^{1/\delta}$	
ϵ'_{fw}	Real part of relative permittivity of the free water [30]	$\epsilon'_{fw} = e_{w\infty} + \frac{\epsilon_{w0} - \epsilon_{w\infty}}{1 + (2\pi f \tau_w)^2}$	$\epsilon_{w\infty} = 4.9$ is the limit of ϵ'_{fw} when $f \to \infty$, ϵ_{w0} is the static dielectric constant for water, τ_w is the relaxation time for water, and ϵ_0 is the permittivity of free space. At room temperature, $2\pi\tau_w = 0.58 \times 10^{-10} s$ and $\epsilon_{w0} = 80.1$, effective conductivity, δ_{off}
ϵ''_{fw}	Imaginary part of relative permittivity of the free water [30]	$\epsilon''_{fw} = \frac{2\pi.f\tau_w(\epsilon_{w0} - \epsilon_{w\infty})}{1 + (2\pi.f\tau_w)^2} + \frac{\delta_{eff}}{,2\pi\epsilon_0 f}\frac{(\rho_s - \rho_b)}{\rho_s m_v}$	
δ_{eff}	Effective conductivity of soil [30]	$\delta_{eff} = \begin{cases} 0.0467 + 0.2204\rho_b - 0.4111S + 0.6614C & 0.3\,\text{GHz} \le f \le 1.4\,\text{GHz}\,. \\ -1.645 + 1.939\rho_b - 2.25622S + 1.594C & 1.4\,\text{GHz} \le f \le 18\,\text{GHz} \end{cases}$	ρ_b is the bulk density

τ_d	Arrival time of direct component	$\tau_d = (\delta_s / S)$	S is the speed of wave in soil
τ_r	Arrival time of reflected component	$\tau_r = 2 \times (\delta_s / S)$	S is the speed of wave in soil
τ_l	Arrival time of reflected component	$\tau_l = 2 \times (\delta_s / S) + (\delta_a / c)$	S is the speed of wave in soil C is the speed of wave in air
$\alpha_{d0}, \alpha_{r0}, \alpha_{l0},$	Gains of the three main components	$\alpha_{d0} = P_t + 20\log_{10}\lambda_s - 20\log_{10}r_1 - 8.69\alpha_s r_1 - 22 + 10\log_{10}D_{rl}$ $\alpha_{r0} = P_t + 20\log_{10}\lambda_s - 20\log_{10}r_2 - 8.69\alpha_s r_2 + 20\log_{10}\Gamma - 22 + 10\log_{10}D_{rl}$ $\alpha_{l0} = P_t + 20\log_{10}\lambda_s - 40\log_{10}d - 8.69\alpha_s(h_t + h_r) + 20\log_{10}T - 22 + 10\log_{10}D_{rl}$, See also analysis from Table VI.	μ and σ
$\alpha_{di}, \alpha_{rj}, \alpha_{lk}$	Path amplitudes of the three components	$\alpha_{li} = \alpha_{l0}e^{-(i-\tau_L)/\gamma_L}\ \forall,\ i > \tau_l\ and\ i < \tau_l + L$ $\alpha_{dj} = \alpha_{d0}e^{-(j-\tau_D)/\gamma_D}\ \forall,\ j > \tau_d\ and\ j < \tau_d + D$ $\alpha_{rk} = \alpha_{r0}e^{-(k-\tau_R)/\gamma_R}\ \forall,\ k > \tau_r\ and\ k < \tau_d + R$	

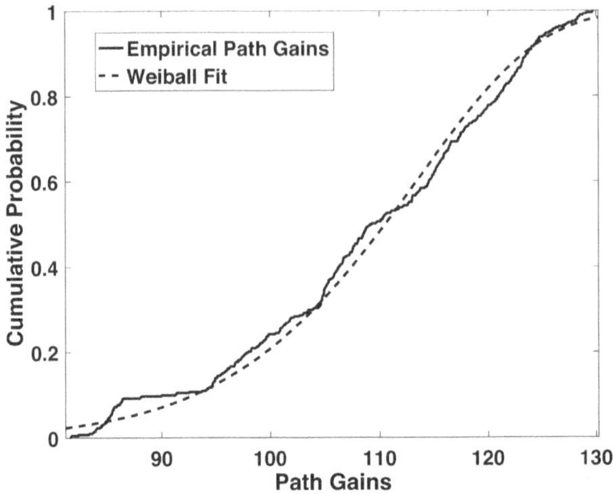

Fig. 3.44 Amplitude gains resulted from the Weibull distribution [62]

Algorithm 1 UG channel impulse response simulation

1: *Initialization* :
2: Input soil parameters
3: Obtain the soil moisture level
4: **BEGIN**
5: Generate decay exponents for the lateral, direct, and reflected components
6: Determine the arrival time
7: Calculate the first multipath gain of each of the three components
8: Generate multipaths and impulse response
9: **END**

are calculated numerically. After that, measured data is compared with the other impulse response parameters. Algorithm 1 is used to generate the UG channel impulse response.

The different soil parameters, e.g., soil texture and moisture, act as an input to the algorithm. Equations (3.29) to (3.31) are used to calculate the arrival time of direct component (τ_d), lateral component (τ_l), and reflected component (τ_r). Peak power gains for the components τ_{d0}, τ_{r0}, τ_{l0} are calculated considering soil type. These gains are estimated from [61]

Table 3.9 shows the comparison between statistical parameters from measured and numerically computed channel model. The error rate of RMS delay spread is 14.67 % and for coherence bandwidth is 14.08 %. This difference in theoretical and experimental values is because of erroneous observation or model uncertainty, nonetheless, the overall error rate stays below 15 %. Therefore, it can be said that statistical model results agree with the empirical results.

Table 3.9 Comparison of parameters for impulse response model [62]

Impulse response parameter	Measured	Modeled
RMS delay spread (τ_{rms})	45.52 ns	38.84 ns
Coherence bandwidth	439 kHz	514 kHz

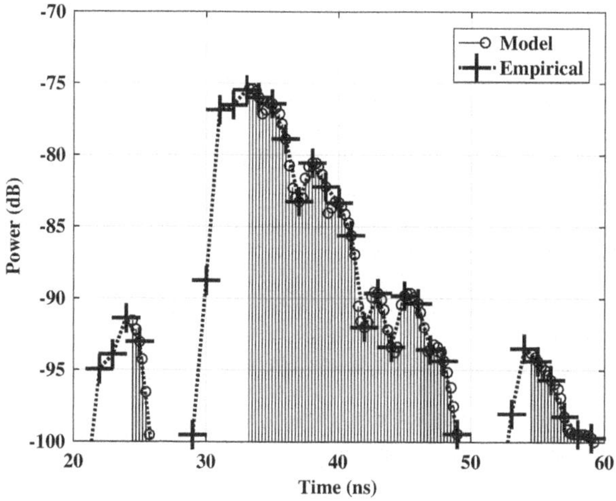

Fig. 3.45 Comparative analysis of results for impulse response from model and empirical evaluation experiments in silt loam [62]

3.8.2 *Empirical Validation*

A desirable property of a good statistical model is that it should simulate the empirical measurements with high accuracy, and that too with a very little difference between the characteristics of the measured results. This section validates the multipath component arrival. To that end, results from the indoor testbed experiments are used for validation [61].

Equations (3.29), (3.30), and (3.31) are used to estimate the refractive indices of each component. The refractive index differs for each soil type as given in [61] along with the results. Reference [61] gives the value of 5.6×10^7 m s^{-1} which is approximately one-fifth of speed of light.

The comparative graph measures empirical PDP values in silt loam soil as shown in Fig. 3.45. The depth is 40 cm and T-R separation is kept at 50 cm. It was observed that, for a fixed T-R separation of 50 cm and varying depths, lateral component is arriving (at receiver) before direct component at 10 cm depth only. For all other depths, measured and model values are similar.

It is also observed from Fig. 3.45 that lateral component is the strongest of all other components. The reason is that the direct and reflected components propagate radially outward because of their spherical nature. On the contrary, lateral waves exhibit two different paths: vertical path in upward direction (source to boundary) and horizontal path as cylindrical wave. Smart lighting is the feasible application for the proposed model. However, it is important to improve this model for customized smart lighting in IOUT environments with specific sensing and communication systems deployed in the IOUT.

References

1. Akyildiz IF, Stuntebeck EP (2006) Wireless underground sensor networks: Research challenges. Ad Hoc Netw 4(6):669–686
2. Akyildiz IF, Vuran MC (2011) Wireless sensor networks (advanced texts in communications and networking). Wiley, Chichester
3. Akyildiz IF, and et al (2006) Next generation/dynamic spectrum access/cognitive radio wireless networks: A survey. Comput Netw J (Elsevier) 50:2127–2159
4. Bicen AO, Sahin AB, Akan OB (2012) Spectrum-aware underwater networks: Cognitive acoustic communications. Veh Technol Mag IEEE 7(2):34–40
5. Bogena HR, Herbst M, Huisman JA, Rosenbaum U, Weuthen A, Vereecken H (2010) Potential of wireless sensor networks for measuring soil water content variability. Vadose Zone J 9(4):1002–1013
6. Bogena HR, Huismana JA, Meierb H, Rosenbauma U, Weuthena A (2009) Hybrid wireless underground sensor networks: Quantification of signal attenuation in soil. Vadose Zone J 8(3):755–761
7. Cassioli D, Win MZ, Molisch AF (2002) The ultra-wide bandwidth indoor channel: from statistical model to simulations. IEEE JSAC 20(6):1247–1257
8. Chen Q (2013) Wideband channel sounding techniques for dynamic spectrum access networks. Ph.D. thesis, University of Kansas
9. Cox D (1972) Delay Doppler characteristics of multipath propagation at 910 Mhz in a suburban mobile radio environment. IEEE Trans Antennas Propag 20(5):625–635
10. CRC Handbook (2014) CRC handbook of chemistry and physics, 95th edn. CRC Press, Boca Raton
11. Dobson MC, Ulaby FT, Hallikainen MT, El-Rayes MA (1985) Microwave dielectric behavior of wet soil-part ii: Dielectric mixing models. IEEE Trans Geosci Remote Sens (1):35–46
12. Dong X, Vuran MC (2011) A channel model for wireless underground sensor networks using lateral waves. In: Proc. of IEEE Globecom '11, Houston, TX, December 2011
13. Dong X, Vuran MC (2011) A channel model for wireless underground sensor networks using lateral waves. In: 2011 IEEE global telecommunications conference-GLOBECOM 2011, pp. 1–6. IEEE, Houston, TX, USA
14. Dong X, Vuran MC, Irmak S (2013) Autonomous precision agriculture through integration of wireless underground sensor networks with center pivot irrigation systems. Ad Hoc Netw 11(7):1975–1987
15. Ettus research website (2016)
16. Foth HD (1990) Fundamentals of soil science, 8th edn. Wiley, Chichester
17. GNU radio website (2020)
18. Goldsmith A (2005) Wireless communications. Cambridge University Press, New York, NY, USA
19. Hashemi H (1993) Impulse response modeling of indoor radio propagation channels. IEEE JSAC 11(7):967–978
20. Hillel D (2003) Introduction to environmental soil physics. Elsevier Science, Amsterdam
21. Howard SJ, Pahlavan K (1990) Measurement and analysis of the indoor radio channel in the frequency domain. IEEE Trans Instrum Meas 39(5):751–755
22. Iizuka K (1964) An experimental investigation on the behavior of the dipole antenna near the interface between the conducting medium and free space. IEEE Trans Antennas Propag 12(1):27–35

23. Islam MN, Kim BJ, Henry P, Rozner E (2013) A wireless channel sounding system for rapid propagation measurements. In: Proc. IEEE ICC' 2013, pp. 5720–5725
24. Johnk CTA (1988) Engineering electromagnetic fields and waves, 2nd edn. Wiley, Chichester
25. King RWP, Smith G (1981) Antennas in matter. MIT Press, Cambridge, MA
26. King RWP, Owens M, Wu TT (1992) Lateral electromagnetic waves. Springer, New York
27. Konda A, Rau A, Stoller MA, Taylor JM, Salam A, Pribil GA, Argyropoulos C, Morin SA (2018) Soft microreactors for the deposition of conductive metallic traces on planar, embossed, and curved surfaces. Adv Funct Mater 28(40):1803020
28. Li L, Vuran MC, Akyildiz IF (2007) Characteristics of underground channel for wireless underground sensor networks. In: Proc. Med-Hoc-Net 07, Corfu, Greece, June 2007
29. Matlab (2020). www.matlab.com. Accessed on 27 July 2020
30. Peplinski N, Ulaby F, Dobson M (1995) Dielectric properties of soil in the 0.3–1.3 Ghz range. IEEE Trans Geosci Remote Sens 33(3):803–807
31. Rappaport TS, Seidel SY, Takamizawa K (1991) Statistical channel impulse response models for factory and open plan building radio communicate system design. IEEE Trans Commun 39(5):794–807
32. Ritsema CJ, Kuipers H, Kleiboer L, Elsen E, Oostindie K, Wesseling JG, Wolthuis J, Havinga P (2009) A new wireless underground network system for continuous monitoring of soil water contents. Water Resour Res J 45:1–9
33. Salam A (2018) Pulses in the sand: Long range and high data rate communication techniques for next generation wireless underground networks. ETD collection for University of Nebraska - Lincoln, (AAI10826112)
34. Salam A (2019) A comparison of path loss variations in soil using planar and dipole antennas. In: 2019 IEEE international symposium on antennas and propagation. IEEE, Houston, TX, USA
35. Salam A (2019) Design of subsurface phased array antennas for digital agriculture applications. In: Proc. 2019 IEEE international symposium on phased array systems and technology (IEEE Array 2019), Waltham, MA, USA
36. Salam A (2019) A path loss model for through the soil wireless communications in digital agriculture. In: 2019 IEEE international symposium on antennas and propagation, pp 1–2. IEEE, Houston, TX, USA
37. Salam A (2019) Sensor-free underground soil sensing. In: ASA, CSSA and SSSA international annual meetings (2019). ASA-CSSA-SSSA, 2019
38. Salam A (2019) Subsurface MIMO: A beamforming design in internet of underground things for digital agriculture applications. J Sens Actuator Netw 8(3):41
39. Salam A (2019) Underground environment aware MIMO design using transmit and receive beamforming in internet of underground things, pp 1–15. Springer International Publishing, Cham
40. Salam A (2019) An underground radio wave propagation prediction model for digital agriculture. Information 10(4):147
41. Salam A (2019) Underground soil sensing using subsurface radio wave propagation. In: 5th global workshop on proximal soil sensing, Columbia, MO, May 2019
42. Salam A (2020) Internet of things for environmental sustainability and climate change, pp 33–69. Springer International Publishing, Cham
43. Salam A (2020) Internet of things for sustainability: perspectives in privacy, cybersecurity, and future trends, pp 299–327. Springer International Publishing, Cham
44. Salam A (2020) Internet of things for sustainable community development, 1st edn. Springer Nature, Basel
45. Salam A (2020) Internet of things for sustainable community development: introduction and overview, pp 1–31. Springer International Publishing, Cham
46. Salam A (2020) Internet of things for sustainable forestry, pp 147–181. Springer International Publishing, Cham
47. Salam A (2020) Internet of things for sustainable human health, pp 217–242. Springer International Publishing, Cham

48. Salam A (2020) Internet of things for sustainable mining, pp 243–271. Springer International Publishing, Cham
49. Salam A (2020) Internet of things for water sustainability, pp 113–145. Springer International Publishing, Cham
50. Salam A (2020) Internet of things in agricultural innovation and security, pp 71–112. Springer International Publishing, Cham
51. Salam A (2020) Internet of things in sustainable energy systems, pp 183–216. Springer International Publishing, Cham
52. Salam A (2020) Internet of things in water management and treatment, pp 273–298. Springer International Publishing, Cham
53. Salam A (2020) Wireless underground communications in sewer and stormwater overflow monitoring: Radio waves through soil and asphalt medium. Information 11(2):98
54. Salam A, Karabiyik U (2019) A cooperative overlay approach at the physical layer of cognitive radio for digital agriculture. In: Third international Balkan conference on communications and networking 2019 (BalkanCom'19), Skopje, Macedonia, the former Yugoslav Republic of, June 2019
55. Salam A, Shah S (2019) Internet of things in smart agriculture: Enabling technologies. In 2019 IEEE 5th world forum on internet of things (WF-IoT), pp 692–695. IEEE
56. Salam A, Vuran MC (2017) Smart underground antenna arrays: A soil moisture adaptive beamforming approach. In: Proc. IEEE INFOCOM 2017, Atlanta, USA, May 2017
57. Salam A, Vuran MC (2016) Impacts of soil type and moisture on the capacity of multi-carrier modulation in internet of underground things. In: Proc. of the 25th ICCCN 2016, Waikoloa, Hawaii, USA, Aug 2016
58. Salam A, Vuran MC (2017) EM-based wireless underground sensor networks, pp 247–285 Academic Press, Cambridge
59. Salam A, Vuran MC (2017) Wireless underground channel diversity reception with multiple antennas for internet of underground things. In: Proc. IEEE ICC 2017, Paris, France, May 2017
60. Salam A, Vuran MC (2017) Smart underground antenna arrays: A soil moisture adaptive beamforming approach. Technical report
61. Salam A, Vuran MC, Irmak S (2016) Pulses in the sand: Impulse response analysis of wireless underground channel. In: The 35th annual IEEE international conference on computer communications (INFOCOM 2016), San Francisco, USA, April 2016
62. Salam A, Vuran MC, Irmak S (2017) Towards internet of underground things in smart lighting: A statistical model of wireless underground channel. In: Proc. 14th IEEE international conference on networking, sensing and control (IEEE ICNSC), Calabria, Italy, May 2017
63. Salam A, Hoang AD, Meghna A, Martin DR, Guzman G, Yoon YH, Carlson J, Kramer J, Yansi K, Kelly M, et al (2019) The future of emerging IoT paradigms: Architectures and technologies. https://doi.org/10.20944/preprints201912.0276.v1
64. Salam A, Vuran MC, Dong X, Argyropoulos C, Irmak S (2019) A theoretical model of underground dipole antennas for communications in internet of underground things. IEEE Trans Antennas Propag 67(6):3996–4009
65. Salam A, Vuran MC, Irmak S (2019) Di-sense: In situ real-time permittivity estimation and soil moisture sensing using wireless underground communications. Comput Netw 151:31–41
66. Saleh AAM, Valenzuela RA (1987) A statistical model for indoor multipath propagation. IEEE J Sel Areas Commun 5(2):128–137
67. Sun Z, Akyildiz IF (2010) Connectivity in wireless underground sensor networks. In: Proc. of IEEE communications society conference on sensor mesh and Ad Hoc communications and networks (SECON '10), Boston, MA, 2010
68. Temel S, Vuran MC, Lunar MMR, Zhao Z, Salam A, Faller RK, Stolle C (2018) Vehicle-to-barrier communication during real-world vehicle crash tests. Comput Commun 127:172–186
69. Tiusanen MJ (2006) Wideband antenna for underground soil scout transmission. IEEE Antennas Wirel Propag Lett 5:517–519
70. Vuran MC, Akyildiz IF (2010) Channel model and analysis for wireless underground sensor networks in soil medium. Phys Commun 3(4):245–254

71. Vuran MC, Salam A, Wong R, Irmak S (2018) Internet of underground things in precision agriculture: Architecture and technology aspects. Ad Hoc Netw 81:160–173
72. Vuran MC, Salam A, Wong R, Irmak S (2018) Internet of underground things: Sensing and communications on the field for precision agriculture. In: 2018 IEEE 4th world forum on internet of things (WF-IoT) (WF-IoT 2018), Singapore, February 2018
73. Welch PD (1967) The use of fast Fourier transform for the estimation of power spectra: A method based on time averaging over short, modified periodograms. IEEE Trans Audio Electroacoust 15(2):70–73
74. Win MZ, Scholtz RA (2002) Characterization of ultra-wide bandwidth wireless indoor channels: a communication-theoretic view. IEEE J Sel Areas Commun 20(9):1613–1627
75. Wong R (2017) Towards cloud-based center pivot irrigation automation based on in-situ soil information from wireless underground sensor networks. Master's thesis, University of Nebraska-Lincoln
76. Yonggang W, Jiansheng T, Yue P, Li H (2008) Underwater communication goes cognitive. In: MTS/IEEE Kobe techno-ocean OCEANS 2008, pp 1–4, 2008

Part II
Underground Antenna and Radio Interface Technologies

Chapter 4
Modulation Schemes and Connectivity in Wireless Underground Channel

4.1 Introduction

Wireless underground communications are implemented with heterogeneous set of communication devices [74], i.e., various UG sensor nodes for sensing and different AG nodes for supporting and maintaining infrastructure [41, 49]. In WUC communication range depends on the locations of nodes with respect to the ground surface. Hence, deployment depth should also be considered for deployment analysis in WUSNs [46, 47]. In the following section, we present an in-depth analysis of the underground modulation schemes, deployment, and connectivity.

4.2 Multi-Carrier Modulation: Sub-carriers in UG Channel

High data rate is one of the important requirement of IOUT. Data rate is highly dependent on IOUT communication channel; however, IOUT channel characteristics are not modeled. Therefore, extensive experimentation is required to characterize IOUT wireless channel. Soil properties (moisture and type) and antenna properties (burial depth and distance) impact capacity of communication channel. Therefore, there is a need to develop a theoretical model to estimate the channel capacity of IOUT channel considering all these factors [56, 64] (Fig. 4.1).

4.2.1 Capacity Model

The channel capacity depends upon the bandwidth and change in transfer function with the change in bandwidth [62]. Therefore, two properties of UG channel are considered for channel capacity model: (1) bandwidth of the sender and receiver

A. Salam, U. Raza, *Signals in the Soil*,
https://doi.org/10.1007/978-3-030-50861-6_4

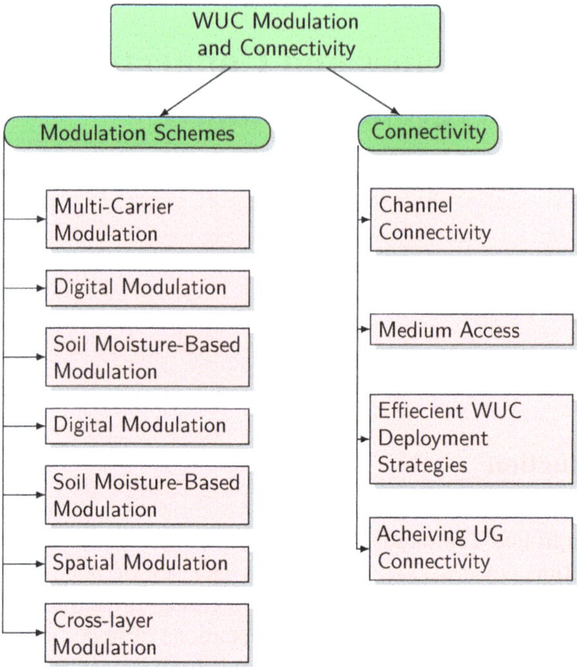

Fig. 4.1 Organization of the chapter

antenna, and (2) channel transfer function. Bandwidth is determined from return loss of an antenna. Return loss occurs due to impedance mismatch which can be calculated as:

$$RL_{\text{dB}} = 20\log_{10}\left|\frac{Z_s + Z_a}{Z_s - Z_a}\right|, \tag{4.1}$$

where Z_s represents the transmission line and Z_a represents impedance of antenna.

Figure 4.2a plots return loss of antenna at zero (0) soil matric potential. Bandwidth is calculated using a threshold value δ for antenna return loss. Generally, a value of -10 dB is used for δ in existing literature [4, 35]. It is assumed that the receiver and sender will have same return loss. The reason for this assumption is that short distance and spatial homogeneity of soil reduce the probability of difference in resonant frequency and return loss of antennas at sender and receiver. Bandwidth of an antenna is maximum at its resonant frequency, therefore, second assumption is that the system is operating at resonant frequency to maximize the bandwidth [14, 38]. Bandwidth of an underground system with the antenna operating at resonant frequency is given as:

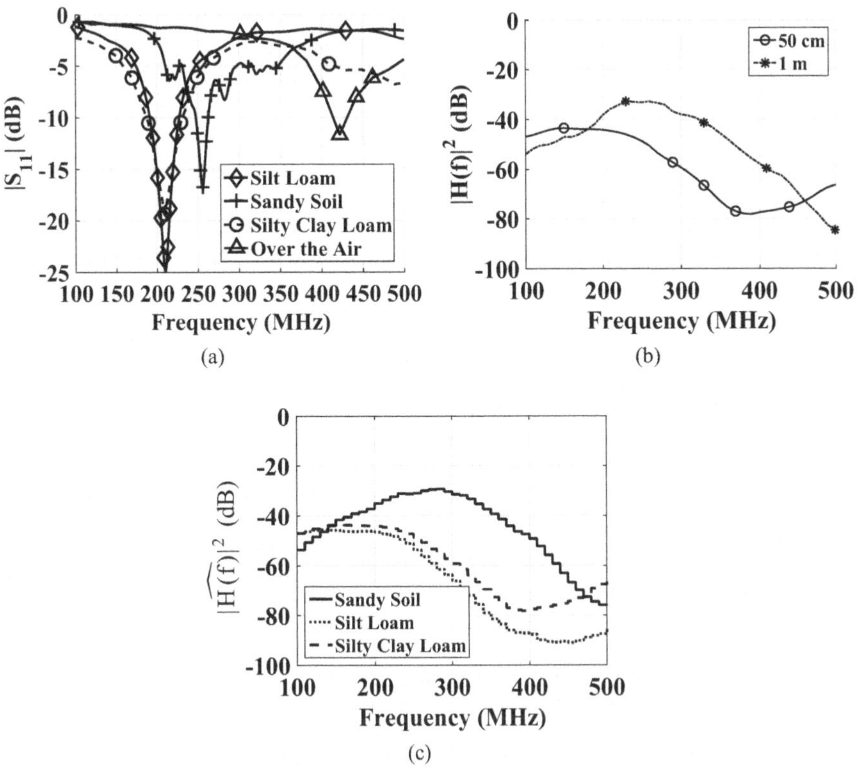

Fig. 4.2 (**a**) Antenna return loss at 0 CB [56], (**b**) channel transfer function [56], (**c**) approximating channel transfer function at the distance and depth of 50 and 20 cm, respectively [56]

$$B_s = \begin{cases} 0 & \text{if } -R(f) > \delta, \\ 2(f - f_m) & \text{if } -R(f) \leq \delta \text{ and } f < f_r, \\ 2(f_M - f) & \text{if } -R(f) \leq \delta \text{ and } f \geq f_r, \end{cases} \tag{4.2}$$

where f_r is the resonant frequency, f_m is lowest, and f_M is the highest frequency at which $R(f) \leq \delta$.

The total number of sub-carriers are calculated in a multi-carrier modulation. Here, a total number of sub-carriers are defined as the least number required to prevent inter-symbol interference (ISI) due to on B_s (system bandwidth). The total number of sub-carriers are given as:

$$N_c = \left\lceil \frac{B_s}{B_{cb}} \right\rceil, \tag{4.3}$$

where B_{cb} is the channel coherence bandwidth.

The spectral efficiency of m-ary quadrature amplitude modulation (MQAM) is high. Therefore, MAQM is considered as a modulation method for each sub-carrier [27, 47]. Total bit rate of the UG channel is calculated as:

$$R_{ug} = \sum_{i=1}^{N_c} r_i B_{cb}, \tag{4.4}$$

where N_c is the total number of sub-carriers in a multi-carrier transmission system, B_{cb} is the bandwidth of an individual sub-carrier, and r_i is the bits per symbol for each carrier.

At high signal-to-noise ratio (SNR), the symbol-error probability, P_{sc}, for each of the i-th carrier is calculated as:

$$P_{sc_i} = K_{ri} Q\left(\sqrt{\frac{3En}{(M_i - 1)N_0}}\right). \tag{4.5}$$

The value of a constant K_{ri} is dependent upon the total bits per symbol. Its value lies in the range of $2 \leq K_{ri} < 4$.

$$\sum_{i=1}^{N_c} \gamma_i \cdot P = P, \qquad \gamma_i > 0. \tag{4.6}$$

Power allocation among sub-carriers can be optimized based on target probability of symbol error for each sub-carrier, $P_{sc}^{*} > P_{sc_i}, \forall i$, and a constant power constraint P. It also maximizes the bit rater of UG channel, R_{ug}.

Figure 4.2b plots the empirical channel transfer function for sandy soil at a distance of 50 cm and 1 m. For the individual lower bandwidths, B_{cb}, staircase function gives close approximation of channel transfer function $\mid H(f) \mid^2$. Therefore, the transfer function is approximated using a staircase function, $\mid \hat{H(f)} \mid^2$. Figure 4.2c shows the staircase approximation of channel transfer function in sandy soil at 20 cm depth and 50 cm T-R distance. The total bit rate of UG channel is given as summation of all sub-carriers [22]:

$$R_{ug} = \sum_{i=1}^{N_c} B_{cb} \log_2 \left\{ 1 + \frac{\frac{3\gamma_i P}{(N_0 B_{cb}) \mid H_i(f) \mid^2}}{\left[Q^{-1}\left\{\frac{P_{sc}^{*}}{K_{ri}}\right\}\right]^2} \right\}, \tag{4.7}$$

where N_c is from Eq. 4.3, P is transmit power constraint, and γ's are given such that:

R_{ug} is maximized by optimized power distribution among sub-carriers. This optimization problem [31, 39, 58, 82] is solved similar to as water filling problem of [8, 10, 55] was solved. Lagrangian multiplier is used to obtain the optimized power allocation, γ_i^*, leading to water filling allocation as [10, 41]:

$$\gamma_i^* = \begin{cases} \lambda - \frac{1}{K_0 |H(f)_i|^2)} & \text{if } \gamma_i > 0, \\ 0 & \text{otherwise,} \end{cases} \tag{4.8}$$

where $\sum_{i=1}^{N_c} \gamma_i^* = 1$, and $K_0 = 3P/(N_0/[Q^{-1}[.]]^2)$. An updated maximum bit rate, R_{ug}^{max} is calculated by jointly solving equations 4.6 and 4.8 [22, 36]. R_{ug}^{max} is given as:

$$R_{ug}^{max} = \sum_{i=1}^{N_c} B_{cb} \cdot \log_2 \left\{ \lambda K_0 \mid H_i(f) \mid^2 \right\}. \tag{4.9}$$

This rate is a high-SNR optimal because it is for the cases having high-SNR. Another sub-optimal solutions for an equal power allocation to sub-carriers has capacity close to that of an optimal solution [9, 31, 34, 43]. R_{ug} for equal power allocation is calculated as:

$$R_{ug}^{eq} = \sum_{i=1}^{N_c} B_{cb} \log_2 \left\{ 1 + \frac{\frac{3P/N_c}{(N_0 B_{cb}) \mid H_i(f) \mid^2}}{\left[Q^{-1} \left\{ \frac{P_{sc}^*}{K_{ri}} \right\} \right]^2} \right\}. \tag{4.10}$$

4.2.2 Results and Discussion

4.2.2.1 Effect of Soil Texture

Figure 4.3 plots the multi-carrier capacity of UG channel. Three different types of soil (silty loam, silty clay loam, and sandy) are used for the experiments at the distances of 50 cm (Fig. 4.3a) and 1 m (Fig. 4.3b). The effect of soil type was measured on capacity and system bandwidth. The system bandwidth was calculated as 20 MHz. Sandy soil had the highest capacity at both distances of 50 cm (30% higher) and 1 cm (39% higher). This is because attenuation of EM waves in soil is dependent upon the water holding capacity of a soil and sandy soil has lowest water holding capacity as compared to silty loam and silty clay loam. Therefore, soil with lower clay content (sandy soil) experiences minimum attenuation. Channel capacity decreases for each type with increase in distance. For sandy soil it went from 233 to 180 Mbps, 195 to 137 Mbps for silty clay loam soil, and 178 to 129 Mbps for silty loam soil. This is because EM signals attenuates with the increase in distance [37].

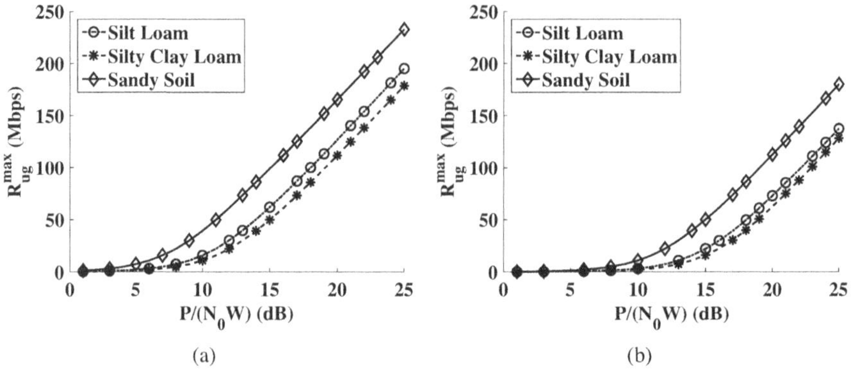

Fig. 4.3 Channel capacity for 200 MHz bandwidth and 20 cm depth, at distance of [56]: (**a**) 50 cm, total number of sub-carriers are 25 (sandy), 25 (silty clay loam), and 24 (silt loam), (**b**) 1 cm, total number of sub-carriers are 27 (sandy), 21 (silty clay loam), and 25 (silt loam)

4.2.2.2 Effect of Soil Moisture

Figure 4.4 plots the effect of soil moisture on various parameters of multi-carrier capacity of UG channel. Figure 4.4a shows antenna bandwidth against soil moisture in a silt loam soil. Antenna bandwidth increases up to 80% (i.e., from 20 to 36 MHz) with the decrease in soil moisture. Figure 4.4b shows the coherence bandwidth and number of sub-carriers plotted against soil moisture in a silt loam soil. The coherence bandwidth decreases up to 69% (i.e., from 55 to 17 kHz) and number of sub-carriers increased by 175% (i.e., from 20 to 55) as the soil moisture decreases. The impact on sub-carrier can be minimized by adjusting sub-carrier bandwidth [32].

Figure 4.4c and d shows the maximum bit rate from Eq. 4.9, with soil moisture in silt loam and sandy soil, respectively. For both soils, with increasing value of (P/N_0W), the bit rate increases with decrease in soil moisture, e.g., for (P/N_0W) = 18 dB, rate increases from 39 to 194 Mbps, and for (P/N_0W) = 25 dB it goes from 127 to 362 Mbps. Similarly for sandy soil, data rate is increased from 126 to 213 Mbps. EM waves are absorbed by soil water content causing more attenuation. Change in soil moisture changes soil permittivity resulting in fluctuating wavelength which further attenuates the signal. Therefore, decrease in soil moisture lowers signal attenuation, increasing data rate [40].

4.2.2.3 Effect of Distance

Experiments were conducted in an outdoor testbed to study the effect of distance on the channel capacity. Silty clay loam soil and burial depth of 20 cm are used for experiment. Figure 4.5b shows the number of sub-carriers against distance. It shows that for 20 MHz antenna bandwidth, the coherence bandwidth decreases from 678

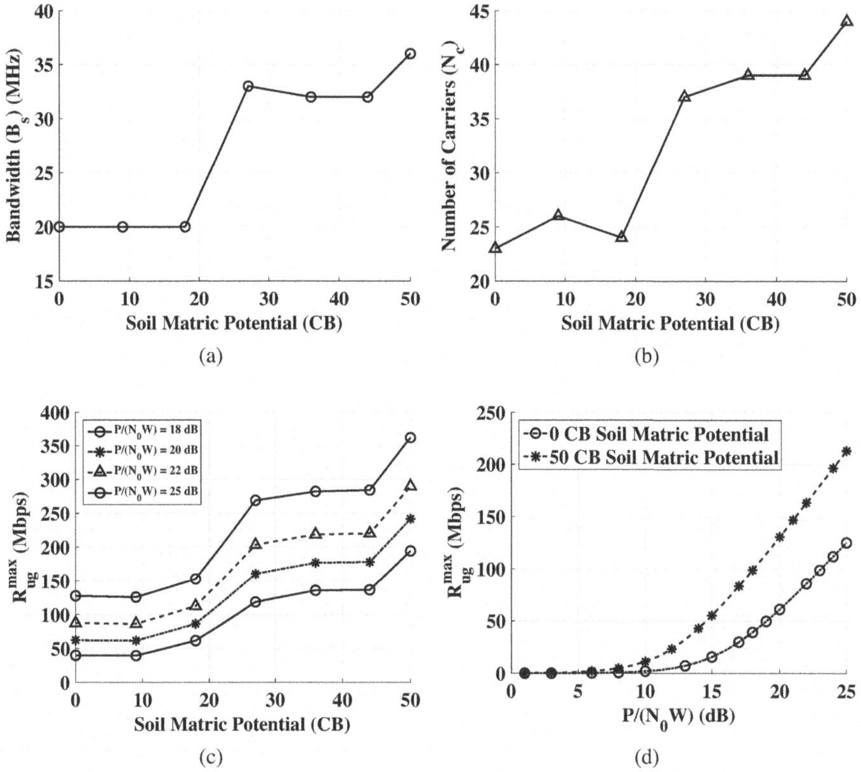

Fig. 4.4 Soil moisture impact in silt loam soil [56]: (**a**) system bandwidth, (**b**) number of sub-carriers, (**c**) data rate, (**d**) channel capacity depth and distance of 20 cm and 50 cm, respectively, in sandy soil

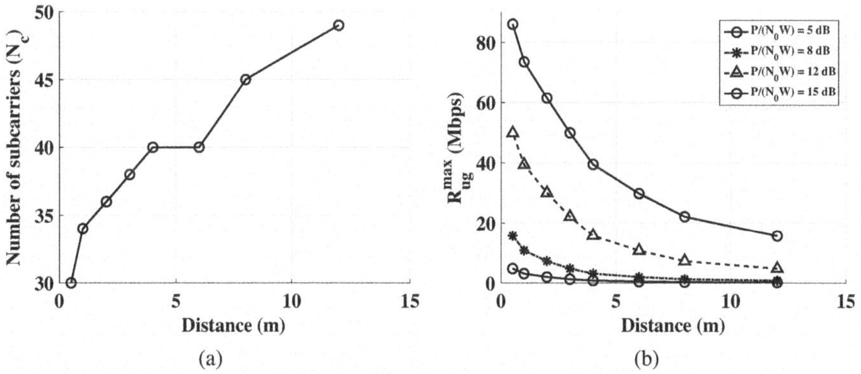

Fig. 4.5 Effect of distance on [56]: (**a**) underground channel capacity, (**b**) number of sub-carriers (N_c)

to 411 kHz with decrease in distance. This is because RMS varies with change in distance. Figure 4.5b plots T-R distance with channel capacity [49, 52]. It can be seen that a bit rate of 80 Mbps can be achieved at distance up to 12 m. Underground communication is carried out by three component waves: direct, reflected, and lateral. As the distance is increased, direct and reflected component fades and the only significant component of the received signal is lateral wave. As the distance is increased further, lateral waves also attenuate, hence, decreasing the data rate.

4.3 Digital Modulation: Wireless UG Receiver Design Based on Diversity

The delay spread of underground channel has adverse effect on the performance of the channel. It results in frequency selective fading [36, 63]. Frequency selective fading limits the data rate of the channel and induces an irreducible bit error rates (BER). Therefore, impact of delay spread and resulting frequency selective fading on UG channel is an important issue to investigate [34, 57, 63, 66]. An ideal UG channel must dynamically adapt to soil parameters while achieving high data rate and low BER. To that end channel capacity has already been investigate in [14, 41, 43]; however, impact of digital modulation techniques on UG channels is still not investigated. Therefore, this section explains these effects using the impulse response of UG channel.

4.3.1 Diversity Model

This section develops a model which uses channel impulse response to analyze the impact of RMS delay on conventional modulation techniques, i.e., pulse-amplitude modulation (PAM), differential phase shift keying (DPSK), quadrature phase shift keying (QPSK), Gaussian minimum-shift keying (GMSK), and m-ary quadrature amplitude modulation (MQAM). The aim is to optimize the IOUT design parameters, i.e., modulation techniques and BER.

The EM-based communication in IOUT consists of three components (see Fig. 4.6: direct, reflected, and lateral, out of which lateral wave is the strongest.

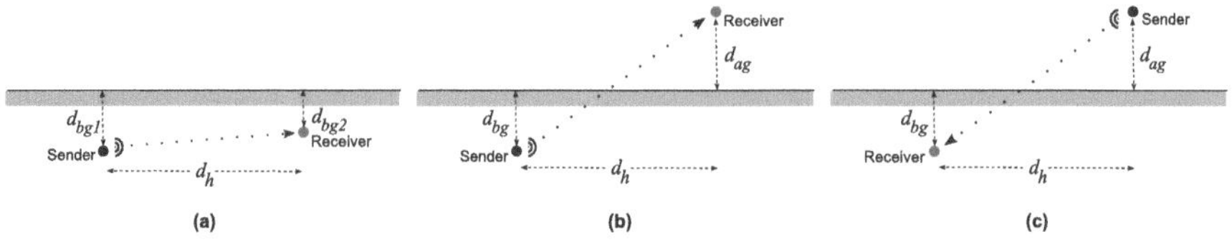

Fig. 4.6 Different communication links of WUSNs [72]

The impulse response of UG channel is expressed as sum of all these components [49, 63]:

$$h_{ug}(t) = \sum_{l=0}^{L-1} \alpha_l \delta(t - \tau_l) + \sum_{d=0}^{D-1} \alpha_d \delta(t - \tau_d) + \sum_{r=0}^{R-1} \alpha_r \delta(t - \tau_r), \tag{4.11}$$

where L, D, and R are number of multiple paths; α_l, α_d, and α_r are complex gains; and τ_l, τ_d, τ_r are delays for lateral, direct, and reflected wave component. These measurements were taken in indoor and outdoor testbeds. The experiment setup details are given in [45, 48, 63].

The output waveform of the received signal is given by the convolution of the baseband input to UG channel, $u(t)$ and channel impulse response, h_{ug} as follows:

$$z(t) = u(t) * h_{ug}, \tag{4.12}$$

Using Eq. 4.11, the waveform can expressed as follows:

$$z(t) = \sum_{l=0}^{L-1} \alpha_l u(t - \tau_l) + \sum_{d=0}^{D-1} \alpha_d u(t - \tau_d) + \sum_{r=0}^{R-1} \alpha_r u(t - \tau_r). \tag{4.13}$$

The delay spread of the UG channel, τ_d is normalized using sample period T and RMS delay spread τ_{rms} as follows:

$$\tau_d = \frac{\tau_{rms}}{T}. \tag{4.14}$$

Bandwidth is calculated as $B = 1/T$. The input signal, $u(t)$ is convolved with signaling waveform $u(t)$ for all modulation schemes. For signaling, raised cosine pulses and rectangular waveform were used. The advantage of raised cosine waveform is that raised cosine filter reduces the ISI and can be realized through raised cosine spectrum with β as a roll-off factor. BER performance is calculated at receiver [46, 50].

The impulse response h_{ug} has been calculated from PDPs measured using different soil parameters (moisture level, depths, and distances).

UG 3W-Rake Receiver The purpose of designing UG 3W-Rake Receiver is to minimize the multipath fading in the system; however, it is done without considering spatial diversity. The approach is based on RAKE [30, 51] and resolves the fading of three components of waves, namely, direct, reflected, and lateral. It exploits the high diversity of all three components. UG 3W-RAKE has a branch for each component which correlates the corresponding component of the received signal, thus, separating all components.

Signal-to-Noise ratio of an underground received signal is random. This is because of multipath fading happening in underground. Therefore, error probability of Additive White Gaussian Noise (AWGN) is averaged over the probability density function (pdf) of the SNR$_{\gamma b}$. Average BER probability, $P_b(\bar{\gamma} b)$

$$P_b(\bar{\gamma}) = \int_0^{\infty} P_{e|\gamma_b} p(\gamma_b) d_{\gamma_b}, \tag{4.15}$$

where $\bar{\gamma}_b$ is average SNR/bit, $p(\gamma_b)$ is pdf of SNR, and $P_{e|\gamma_b}$ is conditional AWGN error probability. There is no close form solution available for the pdf of γ_b, therefore, $p(\gamma_b)$ is measured from UG channel response experiments [63]. It is done, for each measured response, by averaging $P_{e|\gamma_b}$ over instantaneous SNR. As UG 3W-Rake is for the processing of multipaths in all three components of wave, the SNR per received bit γ_b is calculated as:

$$\gamma_b = \sum_{l=0}^{L-1} \gamma_l + \sum_{d=0}^{D-1} \gamma_d + \sum_{r=0}^{R-1} \gamma_r, \tag{4.16}$$

Equation 4.16 can be rewritten as:

$$\gamma_b = \frac{E_b}{N_0} \left[\sum_{l=0}^{L-1} \mid \gamma_l \mid^2 + \sum_{d=0}^{D-1} \mid \gamma_d \mid^2 + \sum_{r=0}^{R-1} \mid \gamma_r \mid^2 \right], \tag{4.17}$$

where $\frac{E_b}{N_0}$ is the ratio of energy per bit to noise power spectral density (PSD) and calculated as:

$$\frac{E_b}{N_0} = \frac{P_t T}{N_0 PL}, \tag{4.18}$$

where P_t is the power transmitted, N_0 is the noise density, T is the sample period, and PL is path loss.

This process is used to approximate a discrete $p(\gamma)$. After the calculation of $p(\gamma)$, average bit rate probability, $P_b(\bar{\gamma})$, is calculated using Eq. 4.15.

LDR Receiver Design This section discusses the development procedure of a novel LDR reception technique considering different parameters of UG communication such as delay spreads, angular angle, and travel path. It shows significant increase in performance as compared to existing techniques. The key characteristics of the technique are elimination of multipath fading and elimination of inter-symbol-interference between three components (lateral, direct, and reflected) of waves.

LDR Antenna Orientation Both, transmitter and receiver, are buried in considered IOUT system. Transmitter is equipped with single antenna and receiver is equipped with three antennas, one for each wave component [42]. For diversity reception of the UG channel, LDR antenna orientation is shown in Fig. 4.7. To receive direct wave (D-wave), corresponding receiving antenna is aligned at 90° from x-axis. To receive reflected wave (R-wave), corresponding receiving antenna

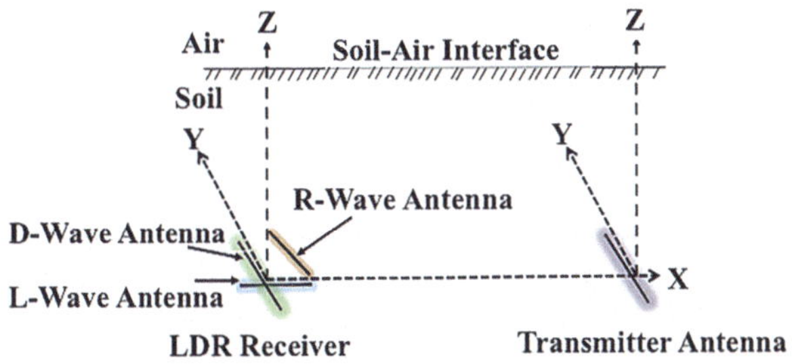

Fig. 4.7 The underground antenna positions [60]

is aligned at the connecting line of x-axis and z-axis with center at 45° from x-axis. To receive Lateral wave (L-wave), corresponding receiving antenna is aligned at 0° from x-axis. Same depth is assumed for transmitter and receiver antenna along the x-axis to avoid variations in axis of the receiver [44].

LDR System Model For the antenna orientation in Fig. 4.7, received signal is calculated as:

$$z = h_{ug}u + n, \tag{4.19}$$

where z is a 3×1 vector for received output, u is the transmitter's data symbol, h_{ug} is the vector for channel impulse response for wave components (L-, D-, R-wave channel response), and n is the 3×1 vector for noise. Channel response of each wave component can be separated as $h_d$for direct, $h_r$for reflected, and $h_l$for lateral. Instantaneous SNR for each receiving antenna is defined as:

$$\gamma_i = \frac{E_b \mid h_i \mid^2}{N_0}, \tag{4.20}$$

where $i = L, D, \text{and} R$ components.

Optimum Maximum Ratio Combining (MRC-LDR) SNR can be determined by using maximum ratio combining (MRC) [30, 47, 52]. This SNR is increased three times from the SNR of a single antenna matched filter UG receiver:

$$\gamma = \sum_{i=1}^{3} w_i \frac{E_b \mid h_i \mid^2}{N_0}, \tag{4.21}$$

where w_i is the combining weight. Although, MRC-LDR achieves the maximum gain, however, interference due to reflected component is still present. The suppression of this interference is discussed in the next section.

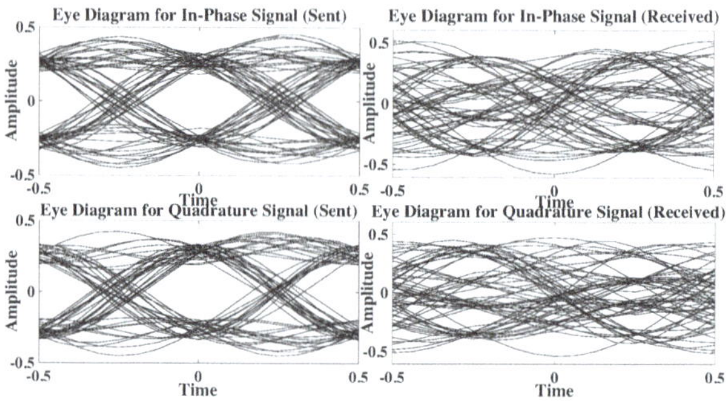

(a)

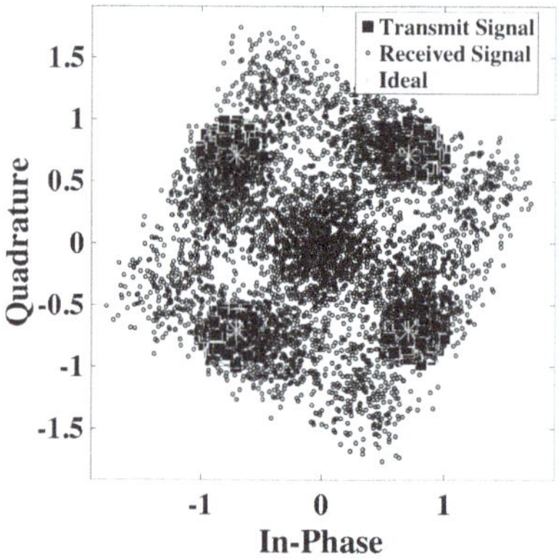

(b)

(c)

Fig. 4.8 (**a**) QPSK eye patterns, (**b**) QPSK constellation diagram, and (**c**) experiment setup for estimating BER [60]

Adaptive Combining (AC-LDR) Among the three components of the signal wave, one of either D-wave or L-wave is dominant at receiver end [64]. This dominance rely on the proximity of LDR receiver. AC-LDR uses this fact to, dynamically, switch and adapt strongest of the two components (L and D). R-component is not considered because of being the weakest among all and may cause the performance to degrade as:

$$\gamma = \begin{cases} \frac{E_b |h_L|^2}{N_0}, & \text{if } \mid h_L \mid^2 > \mid h_D \mid^2, \\ \frac{E_b |h_D|^2}{N_0}, & \text{otherwise.} \end{cases} \tag{4.22}$$

AC-LDR differs from MRC-LDR in that it removes interference while achieving channel gain. For both LDR approaches, the average BER, $P_b(\bar{\gamma_b})$, is calculated as [30, 64]:

$$P_b(\bar{\gamma}) = \int_0^{\infty} P_{e|\gamma_b} p(\gamma_b)\, d_{\gamma_b}. \tag{4.23}$$

4.3.2 *Performance Analysis*

Power delay profile of multipath UG channel is dependent upon the depth and moisture of the soil, and depths and distance of the UG antennas (transmitting and receiving). The UG channel is simulated by with the range of 0.4–0.002 for τ_d.

1. **Coherent Detection**: This section analyzes the performance of modulation techniques, namely, QAM, PSK, PAM, and MSK. The parameters used for the experiments are soil type is silty clay loam, soil moisture is 0 CB and τ_{rms} is 25 ns, distance between transmitter and receiver antenna is 50 cm, and burial depth of antenna is 20 cm. Very high error rates ($<10^{-1}$) were observed for all modulation techniques. This is because of the effect of multipath fading in UG channel. The coherent modulation technique requires the knowledge of exact state of UG channel; however, higher delay spread makes it difficult to track reference symbol for the channel. Moreover, it was observed that error rate is independent of the sample time because it does not change for the τ_d range of 0.002–0.4. Figure 4.8a and b shows the plot of constellation and eye diagram, respectively, for QPSK modulation. It can be seen from Fig. 4.8a, that inter-symbol interference and large delay spread between wave components cause severe detonation and complete eye closure which consequently result in high error rate. Hence, digital modulation in UG channel is highly affected by multipath fading and does not improve by increasing the power rate. The empirical validation of results is shown in coming sections [32, 55].
2. **Experimental Evaluation:** Experimental setup for BER performance analysis under UG channel is shown in Fig. 4.8c. It consist of a: GNU Radio [38], Ettus N210 USRPs [61], burial depth is 20 cm, distance is 50 cm and soil is silty clay

loam soil, moisture level is 50 Cb, $\tau_{rms} = 25.67$ ns [49, 63], transmit power is 10 dBm, range for operational frequency is 100–300 MHz and normalized delay spread, τ_d ranges from 0.005–0.43. The Amplitude shift keying (ASK) modulation technique is used for the experiment.

Experiments were conducted on software-defined radio (SDR). The results show very high bit rate further proving the vulnerability of UG channel to multipath fading and delay spread. In over-the-air (OTA), this can be minimized by using adaptive equalization.

3. **Differential Detection:** Differential quadrature phase shift keying (DQPSK) and differential binary phase shift keying (DBPSK) were used with $\tau_{rms} = 25$ ns to analyze the performance of UG channel. Figure 4.9a compares the BER performance of DQPSK and DBPSK. It can be seen that with τ_d values less than 0.1, error rate reduces by 10^{-3} which is significant improvement over the error rate of 10^{-1} for coherent modulation schemes. The error rate increases with increase in value of τ_d; however, differential schemes are relatively performing better [37, 65].
4. **3W-Rake Performance in UG Channel:** This section analyzes the performance of 3W-Rake receiver. In order to achieve the target BER thresholds, SNR values are analyzed under different modulation schemes. Different combination of soil types, soil moisture levels, burial depth, and distances between transmitter and receiver were considered for the experiments. Figure 4.9b plots the BER along with changing depth in silt loam soil at distance of 50 cm and 1 m. It can be seen that BER increases with the increase in burial depth; however, the increase is relatively greater when the distance between the antennas is greater. Figure 4.9c plots the BER along with distances less than 1 m in silty clay loam soil. It can be observed that BER increases with the increase in distances because propagation loss of all three components increases with the distance which causes attenuation leading to increase in BER. Similarly, Fig. 4.9d plots the BER along with soil moisture range from 0–50 CB at the distances of 50 cm and 1m in silt loam soil. The BER trend is similar for both distances values. It first decreases for initial values of soil moisture and start increasing afterwards. This is because of water repellency due to soil texture slowing down the water infiltration at high soil moisture levels [33, 62, 65].
5. **LDR Performance Analysis:** Fig. 4.9e compares the performance of 3W-Rake with MRC-LDR and AC-LDR. It can be observed that both LDRs are performing better than 3W-Rake. There is significant improvement for SNR > 13 dB. It is because LDR uses separate antenna for each wave component whereas 3W-Rake degrades due to bottleneck while correlating the components.
6. **LDR Implementation:** AC-LDR uses zero-forcing (ZF) precoding for it implementation. ZF precoding performs inversion of channel matrix for the removal of unwanted components. Optimum MRC combining can be implemented using extra digital signal processing (DSP) hardware and it can used as benchmark to analyze the theoretical performance of UG channel. However, LDR is considered more suitable to fulfill the high data rate and long-range of next generation IOUT systems [81].

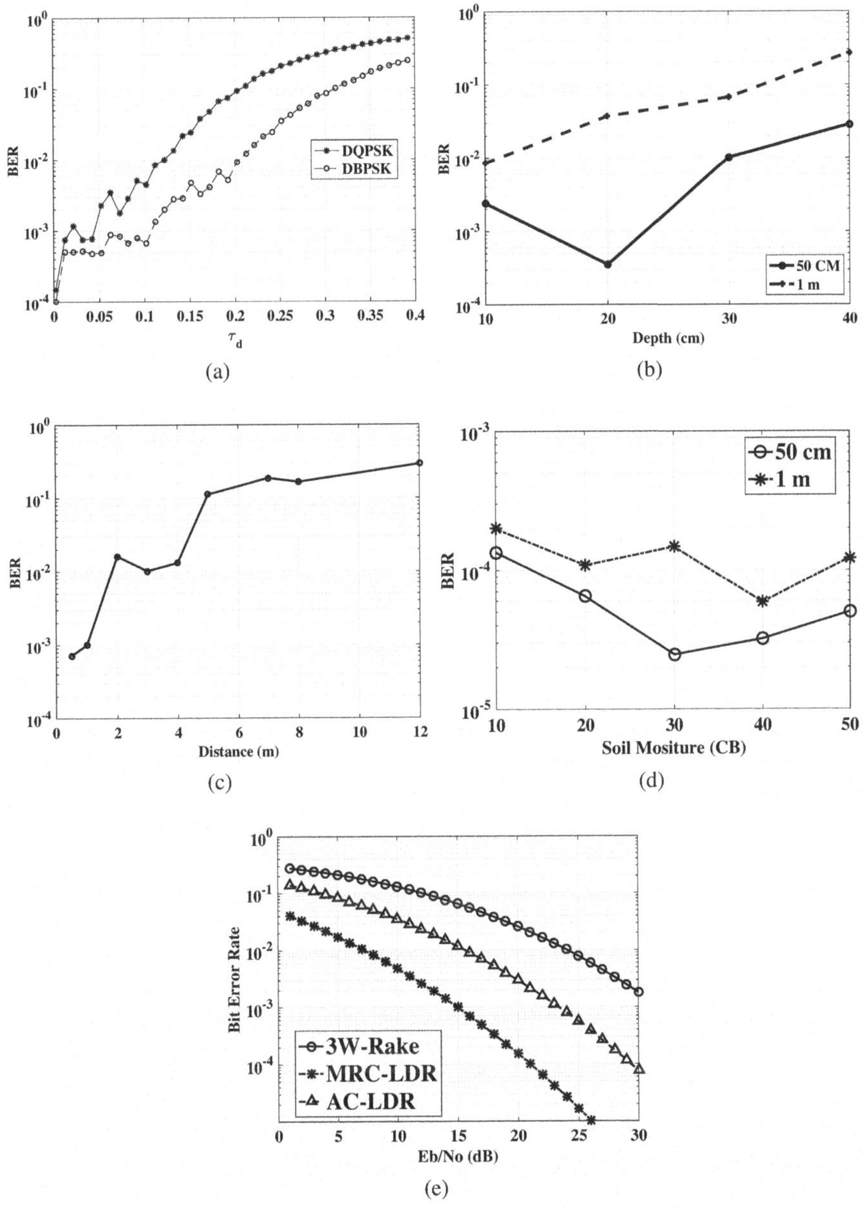

Fig. 4.9 Bit error rate [60]: (**a**) using different modulation techniques, (**b**) with varying depths in silt loam soil, (**c**) with varying distance in silty clay loam soil, (**d**) with varying soil moisture at two distance in silt loam soil, (**e**) comparison BER for AC-LDR, 3W-Rake, and MRC-LDR

4.4 Soil Moisture-Based Modulation

In UG communications soil moisture information can be used to design a unique soil moisture-based modulation scheme. The core idea is based on selecting a channel state using the propagation environment in the vicinity of transmit antenna(s), which in turn will change with soil moisture changes. This can be viewed as creating a multitude of channel states, each corresponding to a different transmission path, where the transmitter can select any of the channel states in each transmission.

In UG communication, these states are inherently available due to soil moisture changes. This approach enables transmitter to use the information to be transmitted as an index to select a particular channel state. Soil moisture modulation uses channel state variations to embed information in transmission media. It is different from conventional wireless communications where data is embedded into a radio frequency prior to transmission.

Media based modulation (MBM) is a similar technique used in conventional wireless systems. It randomizes a wireless channel by perturbing the propagation environment in the vicinity of transmit antennas resulting in change of overall transmission path [23, 70]. In contrast to MBM, different channel states are inherently available in UG communication due to the dependence of channel on soil moisture.

Soil moisture-based modulation will have M channel states created due to soil moisture variations. A total of $\log_2(M)$ bits can be transmitted by selecting any one of these states for transmission. In other words, the UG channel state depends on soil moisture. By selecting the most suitable channel state based on soil moisture, the transmitter not only maximizes the probability of communication success but also transmits the soil moisture information without incurring any additional cost. Consequently, the soil moisture-based modulation scheme allows *resource-free* transmission of soil moisture data. Each channel state and gain will be different based on the soil moisture variations, therefore, each state will have its own channel gain which is likely to follow a complex Gaussian distribution.

The strategy investigates practical methods to independently identify the transmission path for each soil moisture state. To convey a particular selection to the receiver, transmitter will then send a signal through the selected channel state.

4.5 Spatial Modulation: Subsurface MIMO

Underground communication is challenging because of impact of soil texture and water content. Range and energy of UG communication can be improved by using phased-array antenna in UG radio wave propagation [33, 67]. The antenna array is used to direct the wave power via Zenneck waves. This section explains the design of UG phased-array of antennas for digital agriculture and develops

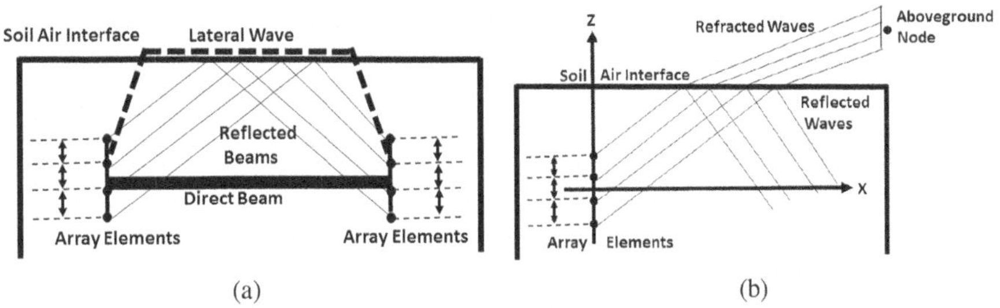

Fig. 4.10 UG MIMO communication [37]: (**a**) receive beamforming, (**b**) transmit beamforming

the underground (UG) multiple-input multiple-output(MIMO) using transmit and receive beamforming in IOUT communication systems.

Underground nodes communicate with other UG nodes via UG2UG link and with aboveground (AG) nodes via UG2AG link. Figure 4.10 shows the overview of UG MIMO communication system. An aboveground node can be a fixed or mobile sink. A mobile AG sink is attached to some moving infrastructure, e.g., tractors. Soil–air interface has an impact on wave propagation to receiver nodes. For example, it is refracted in AG communication and for UG communication lateral waves are used. Figure 4.10 shows the beam patterns for UG and AG communication. In Fig. 4.10b, the UG propagation is impacted by soil-air interface i.e., reflected and refracted EM waves. These EM waves effect the beam pattern which is being propagated towards AG node.

The main purpose of transmit beamforming [59, 61] is to focus energy towards desired direction. The wave can reach receiver from any of three available paths [45, 63] in UG soil medium. When UG receiver is getting the data only from the desired path, UG MIMO channel acts as a three-path interference channel, i.e., direct, reflected, and lateral. Therefore, capacity of MIMO channel and freedom degree is required to be carefully modeled.

There are three components of EM wave propagating in the soil, i.e., direct, lateral, and reflected. UG receiver is required to cancel interference from all these components. The UG MIMO focuses on arrival of the minimum interference signal at the receiver end. It does so by removing the undesired interference that may arise due to undesired components via receive beamforming. An underground MIMO technique is developed using the transmit and receive beamforming which is aware of the UG environment. Accordingly, experiments are performed to test UG MIMO techniques [62, 80].

4.5.1 UG MIMO System Models

Some assumptions made for designing MIMO system are (1) transmitter and receiver have multiple transmit and receiving antennas, respectively, along with the beamforming capabilities and, (2) transmitter antenna has ability to steer beam and receiving antenna can receive all three components of wave from soil. Receiver has ability to select and switch path using a selection method depending upon the strength of received path at the receiver. It is also assumed that power is equally allocated to all UG MIMO transmitters. Total power constraint is assumed to analyze the capacity of the system. Next, zero-forcing (ZF) UG MIMO transceiver technique is explained as below [44, 68, 81].

Contrary to over-the-air (OTA) MIMO methods, information about receiver channel state is not necessary for ZF UG MIMO transceiver technique. It improves the received signal strength by removing the interference between the components. **TR** is the channel between the transmitter and receiver represented as a complex number. The size of the channel is given as $N_t \times N_r$, where N_t is the number of transmitter antenna and N_r is the number of receiving antenna. A total of k spatial underground components are differentiated using $w_1, \ldots, w_k$. Each w is associated with a component. Interference between the component is denoted by $N_t \times N_r$ matrix I_k. With equal power constraint, signal at the receiver is given as [7, 24]:

$$y_k = w_k^* T R f_k x_k + w_k^* I_k f_i x_i + w_k^* n_k, \tag{4.24}$$

where x_k is the transmitted signal of the UG component k, n_k is additive white Gaussian noise (AWGN) vector, and w_k and f_k represent the transmit and receive beamforming vectors.

Next, capacity maximization expression is evaluated for cases having low SNR values. From Eq. 4.24, SINR at receiver for k_{th} component is given as:

$$SINR_k = \frac{w_k f_k f_k * TRTR^* w_k^*}{w_k^* (I_k I_k * f_i f_i^*) w_k^*}. \tag{4.25}$$

The capacity of three UG components of EM waves is given as:

$$C = \sum_{k=1}^{3} \log_2(1 + SINR_k). \tag{4.26}$$

Main objective of the technique is to improve gain and reduce inter-component interference, therefore, only those beamforming vectors are considered which comes under the lower bound capacity. Next, the method to completely remove the inter-component interference will be presented. For every sensed wave component, instantaneous SNR can be calculated as:

$$\gamma_i = \frac{E_b \mid h_i \mid^2}{N_0}, \tag{4.27}$$

where i = Lateral direct and reflected components; $\mid h_i \mid$ is impulse response, and E_b is energy per bit. In contrast to single antenna match filter-based, SNR can be increased three times via maximum ratio combining (MRC) approach:

$$\gamma = \sum_{i=1}^{3} w_i \frac{E_b \mid h_i \mid^2}{N_0}, \tag{4.28}$$

where w_i is the combining weight. Although SISO achieves the maximum gain, however, interference due to reflected component is still present. Therefore, transmit beamforming vectors are used to suppress unwanted interference. Hence, received signal is represented as [7, 53]:

$$y_k = w_k^* T R f_k x_k + w_k^* I_k f_i x_i + w_k^* n_k \tag{4.29}$$

$$y_k = \frac{w_k^* T R f_k x_k}{||T R f_i||} + \frac{w_k^* I_k f_i x_i}{||T R f_i||} + \frac{w_k^* n_k}{||T R f_i||}. \tag{4.30}$$

It is important for MRC approach to satisfy the following equation via transmit beamforming vector for complete removal of interference from Eq. 4.30:

$$w_1^* I_1 f_i = 0 \tag{4.31}$$

UG MIMO adjusts the weights according to soil moisture; however, in addition to the environment-aware weight adjustment, it also uses the feedback mechanism. Feedback signals adjust the MIMO weights via array gain feedback loops. The array gain is maximized using the pilot signals. UG MIMO array at transmitter adjusts its parameter by receiving a pilot signal in receive mode. The transmitter channel state is determined by varying the scan angles while transmitter is operating in receive mode. Using the best statistics of SNR, parameters are adjusted with change in soil moisture.

Far-field power density for the identical element is calculated as follows [19, 51]:

$$P_{den} = \frac{\mid E(\theta, \phi) \mid^2}{120\pi}, \tag{4.32}$$

where $E(\theta, \phi)$ denotes the intensity of electric field for each individual array and is calculated as:

$$\mid E(\theta, \phi) \mid = \sqrt{P_{et}} \sqrt{G_{et}} \frac{\sqrt{30}}{d}, \tag{4.33}$$

where P_{et} is the transmit power, G_{et} is the transmit gain, and d is the distance. Array gain G_a is calculated by getting sum of E-field contributions (E_a) from all elements [19, 46]. Therefore,

$$G_a(\theta, \phi) = \frac{d^2}{30} \frac{\mid E_a \zeta(\theta, \phi) \mid^2}{P_t}, \tag{4.34}$$

where ζ represents the element phase factor and

$$E_a = \frac{\sqrt{30}}{d} \sum_n \sqrt{P_{et}} \sqrt{G_{et}}. \tag{4.35}$$

Effective isotropic radiated power (EIRP) can be calculated as:

$$P_{rad} = G_t P_t, \tag{4.36}$$

where P_t is the power transmitted and G_t is the array gain. The far-field power density P_{av} is calculated as [12, 42]:

$$P_{av} = P_{av}^D + P_{av}^R + P_{av}^L, \tag{4.37}$$

where D is the power density of direct component, R is the power density of reflected component, and L is the power density of lateral component [63].

The received power is calculated by multiplying far-field power density P_{av} and antenna aperture $(\lambda_s^2/4\pi)$. The received power is expressed as [12, 50, 66]:

$$\begin{aligned} P_r^d &= P_t + 20\log_{10}\lambda_s - 20\log_{10}r_1 - 8.69\alpha_s r_1 - 22 + 10\log_{10}D_{rl}, \\ P_r^r &= P_t + 20\log_{10}\lambda_s - 20\log_{10}r_2 - 8.69\alpha_s r_2 - 20\log_{10}\Gamma - 22 + 10\log_{10}D_{rl}, \\ P_r^L &= P_t + 20\log_{10}\lambda_s - 4\log_{10}d - 8.69\alpha_s(h_t + h_r) + 20\log_{10}T - 22 \\ &\quad + 10\log_{10}D_{rl}, \end{aligned} \tag{4.38}$$

where Γ is the reflection coefficient, T is the transmission coefficient [12], and λ_s denotes wavelength in the soil. The received power of an isotropic antenna is calculated as [12]:

$$P_r = 10\log_1 0 \left(10^{\frac{P_r^d}{10}} + 10^{\frac{P_r^r}{10}} + 10^{\frac{P_r^L}{10}} \right). \tag{4.39}$$

4.5.2 Performance Analysis

This section analyzes and evaluates the performance of UG MIMO model.

Transmit Beamforming: Transmit MMSE, MRT, and ZFBF beamforming technique in [5] are used for the evaluation. Heuristic beamforming techniques, such as ZFBF, MRT, transmit MMSE/regularized ZFBF/SLNR-MAX beamforming, and the corresponding power allocation, is also based on the [5, 76]. UG impulse response has been investigated with details in [63, 65]. Therefore, instead of generating a random OTA channel, UG impulse response presented in [63] is used for UG MIMO application. Figure 4.11b shows a three-component model of the UG channel. SNR values are considered in range of −10 to 30 dB. For

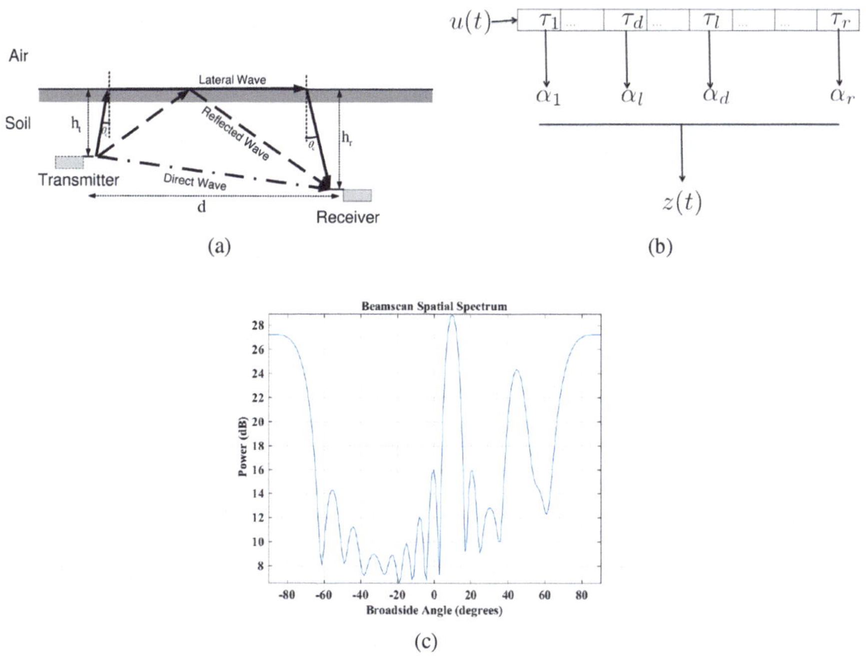

Fig. 4.11 (**a**) Three-wave components in underground channel [63], (**b**) UG channel model realization with direct, lateral, and reflected components, (**c**) spatial spectrum of the direct, lateral, and reflected components [37]

each approach, channel matrices are generated as a first step of simulation. From these channel matrices, normalized beamforming matrices are calculated for each technique. UG MIMO is evaluated for all three components of UG channel and the sum rate is computed for all beamforming approaches.

Figure 4.12 plots average sum rate with change in average SNR. Figure 4.12a, b and c shows average sum rate for one (direct), two (direct and reflected), and three (direct, lateral, and reflected) wave components, respectively. For single, there is no effect of beamforming and all three approaches have same average sum rate of 1.5 to 1.7. In case of two components, the sum rate increased from 1.6 to 3.1 at average SNR of 10 dB. All three approaches have minor difference of 0.1 in average sum rate at SNR of 0 dB. The difference between ZFBF and MMSE increases with the increase in SNR, thus, MIMO UG shows better performance than ZFBF and have improved power gains which is further shown in Fig. 4.12c. In case of two components, the sum rate increased from 3.1 to 6.6 at average SNR of 10 dB which increased to 8.4 at average SNR of 30 dB. This high performance gain shows that UG MIMO performs to its maximum when all three components are used for transmit beamforming in underground environment.

Receive Beamforming: To evaluate receive beamforming, a 16-element uniform linear array with an inter-element distance of half wavelength and 300 MHz of

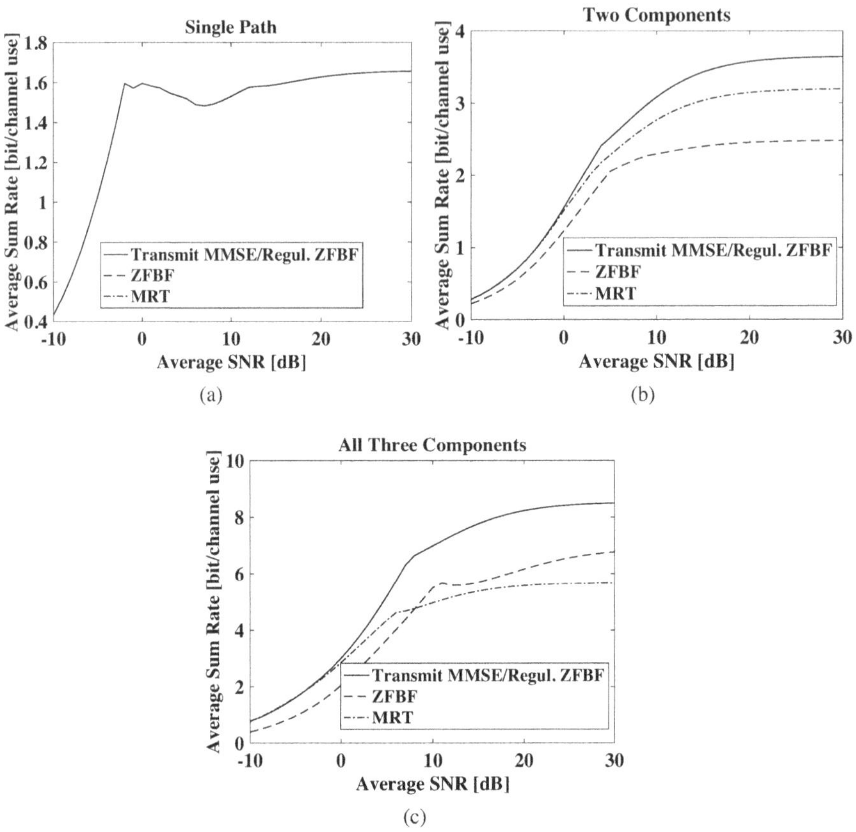

Fig. 4.12 Average sum rate v/s average SNR in UG MIMO [37]: (**a**) single-wave component path, (**b**) two-wave component paths, (**c**) three-wave component paths

operational frequency is used. Path loss of UG communication increases with the frequency, therefore, lower range of frequency spectrum is used to achieve long-range communication. The three components of UG communication are shown in Fig. 4.11a. The signal received at 10°–15° azimuth is the most powerful signal. The direct wave is received from 90° azimuth, and reflected signal reaches the receiver at 45° azimuth. Figure 4.11c shows a spatial spectrum of receive beamforming for all components. The highest power gain is shown at 10° for lateral wave, second highest is at 90° for direct wave, and third highest is at 45° for reflected waves [64].

Impact of soil–air interface: The separation between soil and air medium is called soil–air interface. Both mediums have different properties which cause waves refraction. This phenomenon is also termed as beam squint [21, 24]. Beam squint can cause an error of 5°–15°, depending on soil moisture and incidence angle at interface. The velocity of wave is also affected by refraction which is corrected

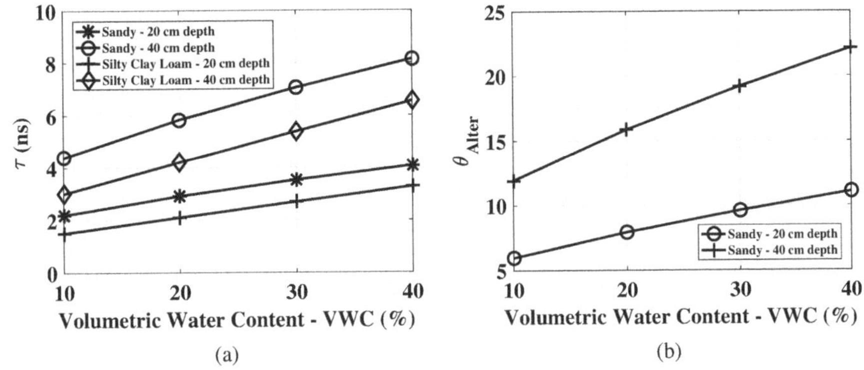

Fig. 4.13 (**a**) Time delay τ with change in soil moisture [37], (**b**) Phase shift adjustment with change in soil moisture [37]

by angle adjustments and time delay (τ). Figure 4.13a plots τ with change in soil moisture. It can be seen that τ increases with the increase in soil moisture and increasing depth further makes it worse. Figure 4.13b shows the same plot but with adjustments made in original phase. It can be observed that phase has to be adjusted as per increase in soil moisture level and depth [32]. For each approach, channel matrices are generated as a first step of simulation. From these channel matrices, normalized beamforming matrices are calculated for each technique. UG MIMO is evaluated for all three components of UG channel and the sum rate is computed for all beamforming approaches [56].

4.6 Cross-Layer Modulation: On the Use of Soil Properties

UG channel model shows that communication in WUSNs is effected by various environmental parameters such as seasonal change, soil type, tunnel size, soil moisture variations, and antenna position/polarization. Therefore, it is important to design environment-adaptive protocols for WUSNs. Moreover, physical layer is quite unpredictable and directly effect the quality of communication, therefore, for better performance, environment-adaptive cross-layer protocols are required. To that end, a packet size optimization framework is presented for WUSNs [78]. The model will use the empirical results towards improving underground channel model by complete characterization of the UG channel. It will also identify the networking related challenges in the underground environment, e.g., effects of soil on routing and medium access control. A simulation module for an underground communication will be given. This simulator will consist of rich and accurate models for underground communication. The purpose is to provide motivation for developing more accurate models for the UG communication.

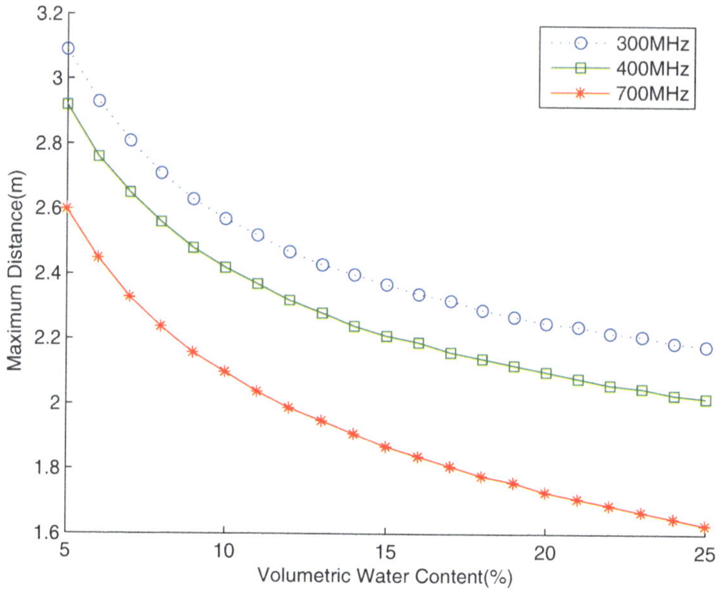

Fig. 4.14 Maximum inter-node distance with variation in volumetric water content (VWC)

4.6.1 *Cross-Layer Protocol Design*

As discussed earlier, WUSN communication is effected by dynamic environment parameters. Therefore, a cross-layer design should be able to adapt to those dynamically changing operating parameters to achieve efficient system performance.

Figure 4.14 shows that WUSN communication range is inversely proportional to the soil moisture. The achievable distance decreases with the increase in soil moisture content. Although, the model gives a thorough idea of UG communication, however, its effects on high-layer networking protocol are yet to be investigated.

There is a need of detailed empirical validation for the underground wireless channel. To that end, field experiments and software simulations are performed in a testbed setting. The testbed has 50 tested wireless sensor nodes. This testbed is used to develop a generic framework by evaluating following research issues:

4.6.1.1 Empirical Evaluations

In [35], authors have proposed an EM-based propagation model considering soil dielectric properties [29]. The model is empirically tested for signal strength of transmitting and receiving antenna, and packet loss rate in different testbeds (experiment sites). Results from these experiments are used to propose a more general channel model.

4.6.1.2 Medium Access Performance

Wireless communication medium is shared by multiple nodes. In shared medium, a node can experience interference from neighboring nodes sharing the medium. Therefore, it is important to understand the difference between the communication and interference by a node. For example, even if a node x is not able to communicate with some other node y, its communication with other nodes may still effect the y communication as an interference. To that end, [39, 76] investigates the performance of multiple communicating sensors in presence of a controlled background traffic.

4.6.1.3 Effects of Seasonal Changes

Weather changes directly effect the soil moisture and temperature, hence, causing the communication performance to change. To that end, seasonal effects are evaluated for a period of 12-months to study the effect of humidity, temperature, and precipitation on communication performance [35].

4.6.1.4 Development of a Cross-Layer Communication Software Module

Empirical results will be used to develop, modify, and improve the communication model to use the sensor information efficiently. The communication parameters such as signal strength, modulation scheme, route information, and packet size are determined using the real-time information from the sensor [38, 53].

4.6.2 Future Research Directions

WUSN paradigm can also be used for communication between underground and aboveground (AG) nodes, e.g., sinks, relay nodes, or control units. The underground sensor node can send their data to some central location using AG nodes. Hence, WUSN provides the combination of UG and AG communication. To that end, communication characteristics of UG to AG communication can be further explored for research purposes, e.g., difference in attenuation level of a signal in soil (UG) and air (AG). Investigation of this area can result in better understanding and further improvement of UG channel communication.

4.7 Medium Access in Wireless Underground Communications

Medium Access Control (MAC) is one of the important feature of wireless networks. Wireless Underground Sensor Networks (WUSN) architecture is distributed where each sensor attempts to communicate in a shared wireless channel [13]. Therefore, to implement an optimized and efficient WUSN, it is very important to improve the sharing mechanism by improving MAC protocols for WUSNs [11, 62]. These MAC protocols are analyzed by performing the underground-to-underground (UG2UG) communication experiments and simulations. The analysis shows that the behavior of communication parameters, e.g., interference and carrier ranges, is completely different for underground then what it is in over-the-air (OTA). This significant difference results in increased contentions leading to large amount of packet collisions. A model is developed to capture the effect of soil on the performance of medium access in an attempt to propose a novel MAC solution for underground sensor networks and eliminate possible shortcomings. The MAC protocols for (WUSNs) are described [13, 61, 65]. This section discusses the performance of different traditional MAC protocols for underground environment, and studies the feasibility of these MAC protocols for underground space and identifies shortcomings due to complex communication medium [13].

4.7.1 Traditional Modulation Scheme

Modulation schemes such as Amplitude Shift Keying (ASK), Frequency Shift Keying (FSK), and Phase Shift Keying (PSK) are studied in [79] for their effect on Bit Error Rate (BER) [2]. Figure 4.15a shows the plot between VWC and maximum inter-node distance (with BER target of 10^{-3}) under various modulation schemes. It can be seen that PSK achieves the largest communication range, therefore, PSK is selected for the analysis [2].

4.7.2 Operation Frequency and Deployment Depth

In Fig. 4.15b, path loss is plotted against the burial depth H at different operating frequencies [2]. Path loss can be reduced by finding an optimal frequency at optimum depth. This is an important fact in WUSN as the sensor deployment in WUSNs is highly dependent upon the operating frequency of sensors [2]. Moreover, Fig. 4.15b shows that path loss tends to become constant as the depth increases. Hence, underground channel tends to become single-path channel with the increase in depth and negligible reflection. Therefore, two-path channel should be considered for lower depths [2].

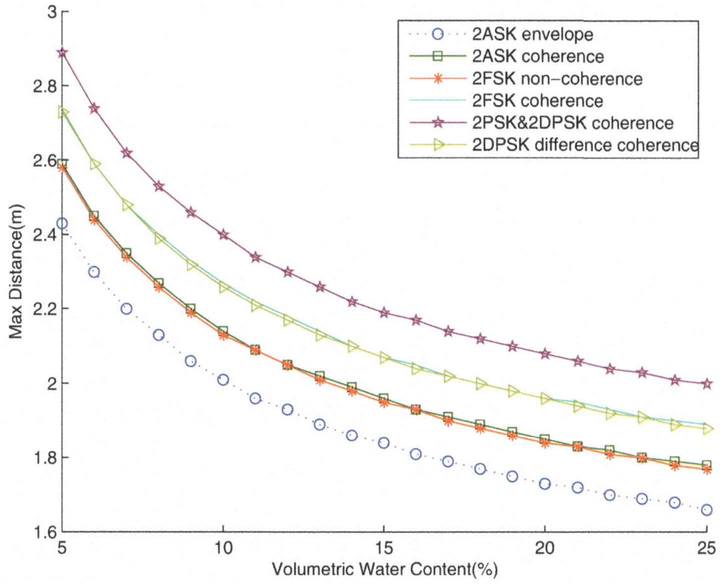

(a)

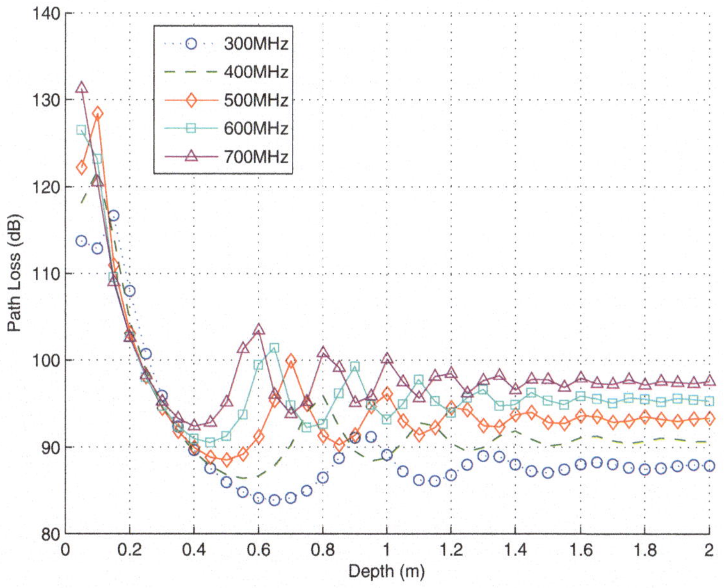

(b)

Fig. 4.15 (**a**) Maximum inter-node communication distance with variation in soil moisture [2], (**b**) Path loss with depth variations in two-path channel model [2]

4.8 Cost-Efficient Underground Deployment

A cost-efficient deployment of sensor nodes is a challenge in WUSN paradigm and needs to be addressed. While deploying sensors in three-dimensional environments, it is imperative to consider range and connectivity constraints. Communication range also depends upon the depth of sensor nodes, hence, deployment depth should also be considered while analyzing the deployments strategies in WUSNs. Therefore, UG deployment [2, 25, 71, 78] becomes an optimization problem using minimum number of sensor nodes within mentioned constraints. To achieve optimal performance, it needs novel optimization programming methods from networking paradigm. It is also important to consider the randomness of soil and weather conditions while deploying network of UG nodes. Deploying deterministic optimization technique to such random behaving network is a challenging task. Moreover, performance of base stations also suffers from the effect of soil–air interface, hence, a separate optimization model for base stations and an optimal routing protocol is also required for longer lifetime of the network. As sensor nodes are buried in underground at different depths, changing the battery of such sensors is also a challenging issue and needs to be addressed efficiently for longer life of network so that a certain level of performance can be guaranteed. Following research issues are discussed under UG deployments:

- *Deployment of sensors*—The network connectivity depends upon the connectivity of UG nodes between UG nodes and the sink nodes. The deployment of over-the-air wireless nodes has extensively been studied; however, UG development is yet to be explored. UG deployments is a NP-hard problem, therefore, heuristics are used to determine the optimal deployment strategies. Soil properties and effect of environmental changes are additional factors to consider while deploying UG nodes. This is analyzed through sensitivity analysis for development of more robust deployment strategies [42, 51].
- *Deployment of the base stations*— In UG networks, base stations are located aboveground, therefore, attenuation of signal attenuation between two AG nodes is different from that of between AG and a UG node due to different communication medium. To that end, an optimization model is given for determining the optimal number of base stations to deploy along with their optimal location. For multiple base station, deployment is modeled considering that they will communicate with each other [36, 44].
- *Routing policy*—Finally, after network deployment, an optimal routing protocol is determined. Randomness and uncertainty associated with the sensor lifetimes require a simulation modeling to determine how long a network can last using a certain routing protocol. As all these parameters will be evaluated using a simulation, a sampling error may also occur in measurement. To that end, simulation optimization technique in [43, 66] will be used.

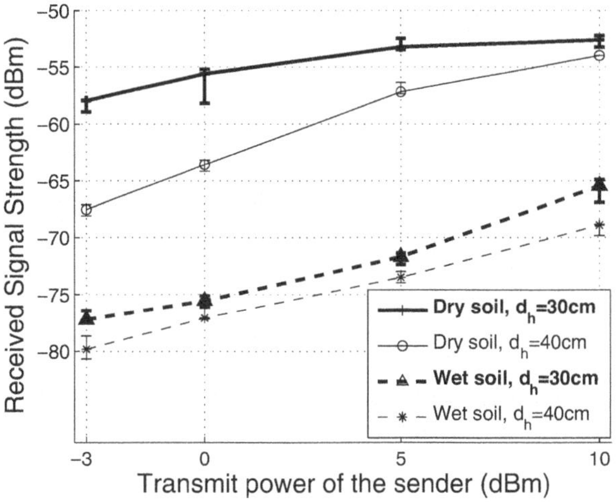

Fig. 4.16 RSS v/s transmitted power under various soil moisture levels [71]

4.8.1 Simulation Results

To understand the effect of soil properties on wireless communication and underground communication experiment is performed [71]. For this experiment, two sensors are used which are buried at 40 cm. Experiment is performed using the separation distance, between both sensors, of 30 and 40 cm. Furthermore, to incorporate the effect of environment, two types of soils are used for the experiment: dry soil and wet soil. The soils are taken from the location where 2.5% of precipitation was registered [28] giving the soil moisture of 11% for dry soil and 18% for wet soil. Figure 4.16 plots the received power with the transmitted power for the experiment. It can be seen the received signal strength is 20 dB lower for the wet soil as compared to dry soil. It results in twice the communication error rate significantly effecting the higher layer communication protocol.

Figure 4.17 shows the communication scenarios for underground transmitter and an aboveground transmitter. The connectivity region of the UG transmitter consists of both UG and AG components; however, communication range is very short due to high attenuation because of underground environment. On contrary, communication range of AG transmitter is relatively large as compared to that of UG node. AG range also consists of some *nulls* at the angles where UG signals get reflected. AG communication range in WUSN is similar to that of traditional over-the-air (OTA) wireless nodes. Short range of UG transmitter is mainly due to the fact that signals are highly reflected when enters from a medium of a lower density, i.e., air, to a medium of a higher density, i.e., soil [45, 50].

Most of the studies about network connectivity employ percolation theory, which was started by the pioneering work of Gilbert. The basic idea about percolation is that there is a critical density, λ_c, such that when the node density is greater than λ_c,

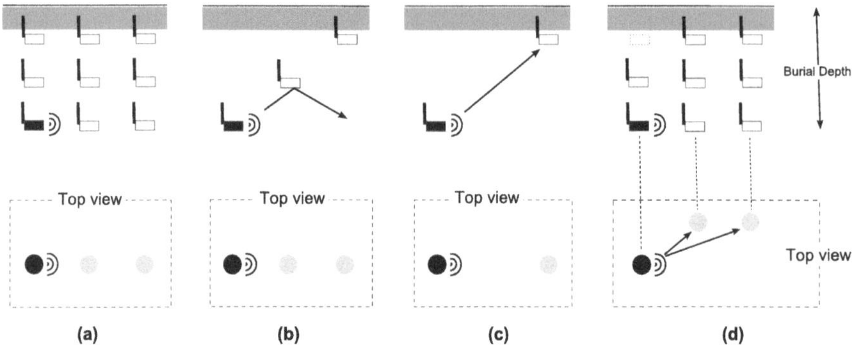

Fig. 4.17 The communication scenarios for underground transmitter and an aboveground transmitter [72]

there is a positive percolation probability that an arbitrary node belongs to a cluster of infinite size (supercritical phase); on the other hand, when the node density is smaller than λ_c, the percolation probability is zero (sub-critical phase) [56, 64].

Most of the literature, dealing with network connectivity utilize percolation theory, which will be the foundation of my research too. In [15, 24], the function of base station is similar to the aboveground nodes in my WUSN consideration. According to the author, for the infinite 2-D plane, base station does not improve connectivity. However, in the WUSN field, it is finite, which I think will be more like the 2-D strip case, thus the deployment of the above ground nodes may help improve network connectivity.

The interference model in [17, 33, 35] shows if the connectivity of the network needs to be kept with the impact of the interference, the orthogonality factor γ needs to be small. But, the conclusion is drawn under the assumption that the transmit power of each node is the same. According to [38, 54], designed transmit power assignment can be used to maintain network connectivity. The interference model in [39] is simplified to consider interfering nodes individually, which is not true in real situation.

The unreliable link issue considered in [37, 76] is quite similar in WUSNs. However, the modeling of the dynamic as a Markov on-off process is not suitable for WUSNs, where the links are quite stable during short period but dramatically change caused by soil moisture. This process may be modeled as several different networks with different path loss function and transmit power. Thus, the connectivity of the network may be modeled as a set of several different networks.

4.8.2 *Efficient Deployment for Full Connectivity of Ad-Hoc Networks*

In this section, the possibility of deploying base stations to ensure connectivity when the node density, λ, is small, is considered. The results show that because of percolation there is no benefit in terms of connectivity in the supercritical phase and the benefit remains marginal in the sub-critical phase, unless the node spatial distribution is close to one-dimensional [15, 62].

The transmission range of the nodes is modeled as a ball of radius $r/2$ and two nodes are neighbors if their representing balls intersect. Thus, the communication range is fixed at r. This model can be easily transformed to a graph [61].

The authors first consider the situation of a pure ad-hoc network for three difference cases: one-dimensional, two-dimensional plane, and two-dimensional strip. For the one-dimensional case, they calculate the connectivity probability, $P_c(x)$, of two nodes at distance x and find the bounds of the probability when $x \geq 2r$, which is

$$\begin{aligned} &(1 - e^{\lambda r})e^{\lambda(x-2r)e^{-\lambda r}} - \lambda e^{-\lambda r} \\ &\leq P_c(x) \leq (1 - e^{-\lambda r})e^{-\lambda(x-r)e^{-\lambda r}} . \end{aligned} \tag{4.40}$$

This leads to the conclusion that in one dimension, the network is almost surely (a.s.) divided into an infinite number of bounded clusters, between which no communication is possible [65, 81].

For the two-dimensional case, the percolation theory holds such that there is a critical density λ_c for the supercritical phase and a unique unbound cluster a.s. The probability that an arbitrary node belongs to this unbound cluster is called percolation probability θ. We have $P_c(x) \geq \theta^2$ and $\inf_{x>0} P_c(x) = \theta^2$.

For the two-dimensional trip with width d, percolation never occurs.

4.8.3 *Topology Control Network Connectivity*

The issue of topology control under the physical SINR model with the objective of maximizing network capacity is studied in this section. The intention of adopting the SINR model is that first the interference in the resulting topology is likely to be high because the node degree does not capture interference adequately and second a wireless link that exists in the communication graph may in practice not exist under the physical model because of high interference [58].

In this section, a centralized approach (MaxSR) is proposed for topology control. This approach consists two parts: T2P and P2T. T2P optimizes the assignment of transmission power given a fixed topology to minimize the average interference degree, and P2T constructs, based on the power assignment, a new topology by deriving a spanning tree that gives the minimal interference degree [68, 80].

In their model, the communication graph is defined based on the received power at the receiver without consideration of interference. Thus, a link (v_i, v_j) is said to exist if and only if:

$$p_t(i) \geq \frac{d_{i,j}^{\alpha} RX_{\min}}{g_{i,j}}, \tag{4.41}$$

where $RX_{\min}$ is the minimum received power requirement and $g_{i,j}$ is the path gain. Moreover, the interference is considered individually in the works [32, 67]. Interference node is defined such that the transmission of an interference node blocks the transmission on link (v_i, v_j). In other words, node k is an interference node for link (v_i, v_j) if

$$\frac{p_t(i) d_{i,j}^{-\alpha}}{N + p_t(k) d_{k,j}^{-\alpha}} < \beta\,. \tag{4.42}$$

The interference degree of a link (v_i, v_j) is defined as the number of interfering nodes for (v_i, v_j). The authors argue that interference degree is a better index than node degree in quantifying interference.

For *Topology to Power assignment*, and indicator function is employed to indicate if a link can transmit and the function is smoothened by the sigmoid function. The problem is then modeled as a linear programming problem. For *Power assignment to Topology*, interference degree is first calculated for each edge and exploited as the weight of the edge. Then, an algorithm similar to the minimum spanning treealgorithm is adopted to reconstruct the topology [40, 55]. Repeating those two algorithms alternately, the power assignment and the topology will converge to the optimal results.

4.9 Modeling the Effects of Interference

4.9.1 The Protocol Interference Model

Consider a pair of transmitter and receiver $\{X_i, X_{R(i)}\}$, where X_i and $X_{R(i)}$ denote a transmitting terminal and its corresponding receiving terminal, respectively. Communication over link $(X_i, X_{R(i)})$ is successful if the distance between the receiver $X_{R(i)}$ and any other terminal X_k transmitting on the same channel is larger than the distance between X_i and $X_{R(i)}$ by a factor $(1 + \Delta)$, that is

$$|X_k - X_{R(i)}| > (1 + \Delta)|X_i - X_{R(i)}|\,. \tag{4.43}$$

4.9.2 The Physical Interference Model

Consider a pair of transmitter and receiver $\{X_i, X_{R(i)}\}$ with a transmit power assignment P_i, the transmission is successful if the signal to interference plus noise ratio (SINR) at $X_{R(i)}$ is equal to or larger than a given threshold β, that is

$$\frac{\frac{P_i}{|X_i - X_{R(i)}|^{\eta}}}{N + \sum_{k \in V, k \neq i} \frac{P_k}{|X_k - X_{R(i)}|^{\eta}}} \geq \beta \,, \tag{4.44}$$

where N is the additive noise power.

4.10 Connectivity in Underground Environment

The fundamental property of a wireless network is to maintain connectivity between all network components (sensors and sinks) for proper functionality. However, connectivity definition for terrestrial wireless channel and underground wireless network is different. A terrestrial wireless network is considered *connected*, if a sensor node is connected to atleast one sink using multi-hop connection [16]. Connectivity of a UG node can be defined in terms of either mobile or fixed AG sink nodes. A UG wireless sensor network is considered connected if a UG sensor is connected to atleast one *fixed* AG sink node, or atleast one *mobile* AG sink node once every maximum tolerable latency time t [75].

Connectivity analysis of a WUSN is much more complicated as compared to that of terrestrial wireless network. WUSN framework is complicated as it consist of two types of sensor nodes (UG and AG nodes), and three types of transmission links, i.e., underground-to-underground (UG2UG), underground-to-aboveground (UG2AG), and aboveground-to-underground (AG2UG) links [1, 6, 77]. Among three links, the transmission range of UG2UG is lowest due to soil [26, 73]; transmission range of UG2AG is highest as most of the signal travels through air. AG2UG is lower than UG2AG because of reflection from soil–air interface [73]. Moreover, in WUSNs, there is a trade-off between connectivity and latency.

A WUSN system consists of multiple sensor nodes for sensing data, and sink nodes for collecting data from the sensor nodes. A sing sink node results in using large numbers of UG nodes, therefore, the high density is required to achieve full connectivity. However, high density of UG nodes increases the maintenance and deployment cost. Therefore, high density of UG nodes is not recommended, instead, number of AG nodes are increased [1, 6]. It is not recommended to deploy the AG nodes in field as it can effect the everyday operation of application where they are deployed, e.g., big machinery in agriculture cannot be used with high number of AG nodes. This constraint may cause the high density of sensor nodes on the border far from the field. To that end, AG nodes can be fixed on a mobile machinery and

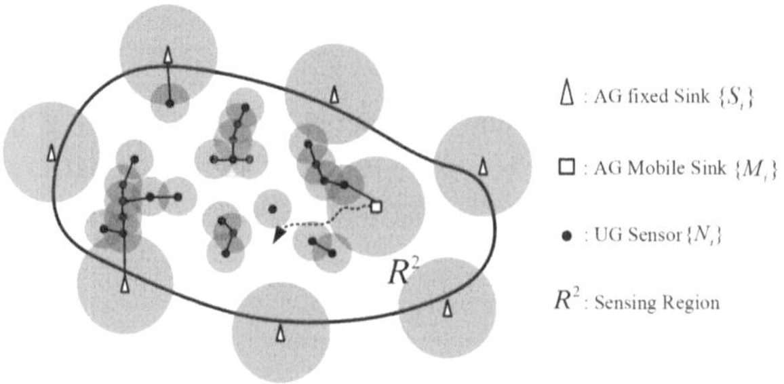

Fig. 4.18 WUSNs network topology [75]: gray shaded region represents the range of each node for connecting with the other node

sensors far from field can send their data to sinks when machinery comes to their transmission range.

There has been already work done in analyzing the connectivity of ad-hoc networks. For example, authors in [20], transmission ranges of nodes are scaled to get the connectivity, and in [3] maximum connectivity probability is computed using node density. Authors in [69] give the simulation results to study the connectivity issues in mobile ad-hoc network. In [16], authors study the connectivity in wireless network using single sink and [18] does it using multiple sinks. This section discusses the probability of connectivity in multi-hop wireless network using multiple sinks, environmental effect on connectivity, and latency-connectivity tradeoff in wireless channel.

4.10.1 Channel Models

In [26], authors developed a WUSN channel model which is validated in [6, 73]. As discussed in previous section, WUSN channel model consists of three channels. The transmission of three channels are given below:

- *UG–UG Channel*—If power transmitted by the UG sensors is given by P_t^u, receiver and transmitter antenna gains are g_r and g_t, respectively. The received power is calculated as $P_r^{U2U} = P_t^u + g_r + g_t - L_{UG-UG}$, where L_{UG-UG} is the path loss of this channel. Similarly, the transmission range of UG–UG channel is given as [75]:

$$R_{UG-UG} = \max\{d : P_r^{U2U}/P_n > SNR_{th}\}, \tag{4.45}$$

  where d is the distance between sending and transmitting sensor, noise is given as P_n, SNR_{th} represents minimum signal-to-noise ratio for receiver.
- *UG–AG Channel*—For a total received power of $P_r^{U-A} = P_t^u + g_r + g_t - L_{UG-AG}$, the transmission range for UG-AG channel is given as [75]:

$$R_{UG-AG} \simeq \max\{d_{AG} : P_r^{U-A}/P_n > SNR_{th}\}. \quad (4.46)$$

- *AG–UG Channel*—Similarly, for a total received power of $P_r^{A-U} = P_t^a + g_r + g_t - L_{AG-UG}$, the transmission range for AG–UG channel is given as [75]:

$$R_{AG-UG} \simeq \max\{d : P_r^{A-U}/P_n > SNR_{th}\}. \quad (4.47)$$

4.10.1.1 Network Model

Figure 4.18 shows the network topology used for the analysis. The network consists of multiple UG sensor nodes and n fixed AG nodes and m mobile AG nodes, carried by some machinery, distributed in a WUSN region $\mathbb{R}^2$ where transmission range of nodes is much less than the region $\mathbb{R}^2$. There are two phases of communication: data from UG sensors is collected in sensing phase (using UG2UG channel) and data from sensing phase is reported to AG nodes in control phase (using UG2AG & AG2UG channel).

4.10.1.2 Mobility Model

Mobile AG nodes are fixed on equipment and field machinery. The mobility of machinery is modeled as random walk [13] and its probability distribution function is given as follows:

$$P(t, (x_t, y_t)) = \frac{1}{2\pi t\sigma^2/\tau} \exp\left(- \frac{(x_t - x_0)^2 + (y_t - y_0)^2}{2t\sigma^2/\tau} \right), \quad (4.48)$$

where σ^2 is variance of the flight duration and τ is mean step time. A mobile AG sink will be more active when the value for σ^2 is larger and τ is smaller.

4.10.2 Lower and Upper Bound of the Connectivity Probability

Connectivity probability of a WUSN is dependent upon various dynamically changing environmental and system factors. Therefore, connectivity in WUSN can be given as a probability function. The lower bound on the probability function can be calculated from two quantities: (1) lower bound on a probability sensor deployed at location (x_i, y_i) connects with the sink at (x_j, y_i), $P(N_i^{(x_i,y_i)} \sim \text{Sink}_j^{(x_j,y_j)})$, in multi-hop manner and first hit time of the random walk given in mobility model. The lower bound of connectivity probability is given as follows [75]:

$$P_c \geq \exp\left\{\lambda S_{fix} - \lambda S_{\mathbb{R}^2} + \lambda \int_{S_{\mathbb{R}^2} - S_{fix}} [(1-F)\cdot G^m + 1 - G^m] dx_i dy_i\right\}, \tag{4.49}$$

where λ is the UG sensor node density, σ^2/τ is the mobility of mobile sinks, $S_{\mathbb{R}^2}$ is the area of region $\mathbb{R}^2$, S_{fix} is the area covered by fixed sink node in $\mathbb{R}^2$,

Similarly, upper bound on probability of not having an isolated UG node in WUSN is given as [75]:

$$P_c \leq \exp\{-\lambda \cdot e^{-\lambda\pi R^2_{UG2UG}} \cdot (S_{\mathbb{R}^2} - S_{fix}) \cdot H(m, R_{UG2AG}, \sigma^{-2}/\tau, t_s)\}, \tag{4.50}$$

where R_{UG2AG}, R_{UG2UG} are the transmission ranges, and t_s is the latency that can be tolerated during sensing phase.

4.10.3 Performance Evaluation

From Eqs. 4.49 and 4.50, it can be observed the lower and upper connectivity probability bound of a WUSN is dependent upon various parameters. These bounds are validated by performing a simulation experiment in [75]. The simulation parameters are as follows: 12 fixed and 10 mobile AG nodes are deployed in 100 m × 100 square grid. Fixed AG nodes follow uniform distribution and deployed at the border (3 at each corner), whereas mobile move as per random walk mode with $\sigma^2/\tau = 1\,\text{m}^2/\text{s}$. Tolerable latency in sensing and control phase are $t_s = 30\,\text{s}$ and $t_c = 20\,\text{min}$. Transmitting power is 10 mW, receiving and transmitting antenna gains are 5 dB, burial depth is kept 0.5 m, and height of antenna at sink is 1 m. Soil moisture (VWC) is kept at 10%, sand particle 50%, and clay is 15%. Each simulation is run for 500 times and results are averaged. The bounds are calculated using Eqs. 4.49 and 4.50. Figures 4.19 and 4.20 show the simulation results to study the effect of different parameters on connectivity.

1. *Soil Moisture*—Fig. 4.19a shows the connectivity probability with UG sensor density in wet (25% VWC) and dry soil (10% VWC). As the transmission range decreases in wet soil due to high water content, the UG nodes in wet soil scenario are doubled to get the equal connectivity probability. It can be seen that soil moisture significantly effects the connectivity of WUSN and due to continuous dynamic change in soil moisture, connectivity status of WUSN also keeps changing.

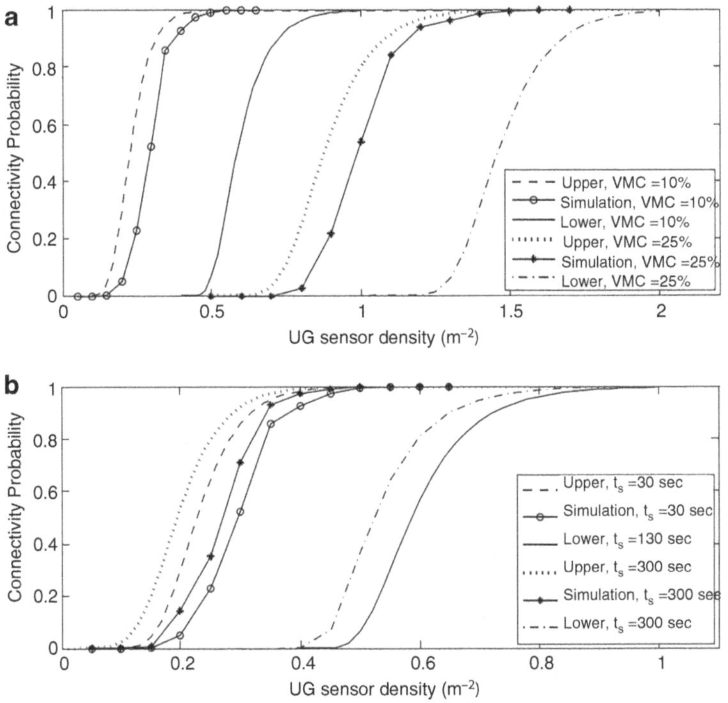

Fig. 4.19 Simulation results [75]: (**a**) effect of soil moisture, (**b**) latency effect

2. *Tolerable Latency*—Fig. 4.19b shows the connectivity probability with latency of the WUSN. With higher latency the connectivity of the WUSN increases. For this simulation, the latency was increased from 30 to 300. Therefore, there is a trade-off between latency and WUSN connectivity.
3. *Depth and Height*— Fig. 4.20 shows the effect of UG node depth and AG antenna height on connectivity probability of the WUSN. The depth of deployment significantly effects the UG2AG and AG2UG communication whereas height effects the AG2UG communication only. It can be seen in Fig. 4.20a, during sensing phase, that connectivity is significantly lower for depth over 2.5m. Conversely, connectivity probability increases with doubling the height in Fig. 4.20b.

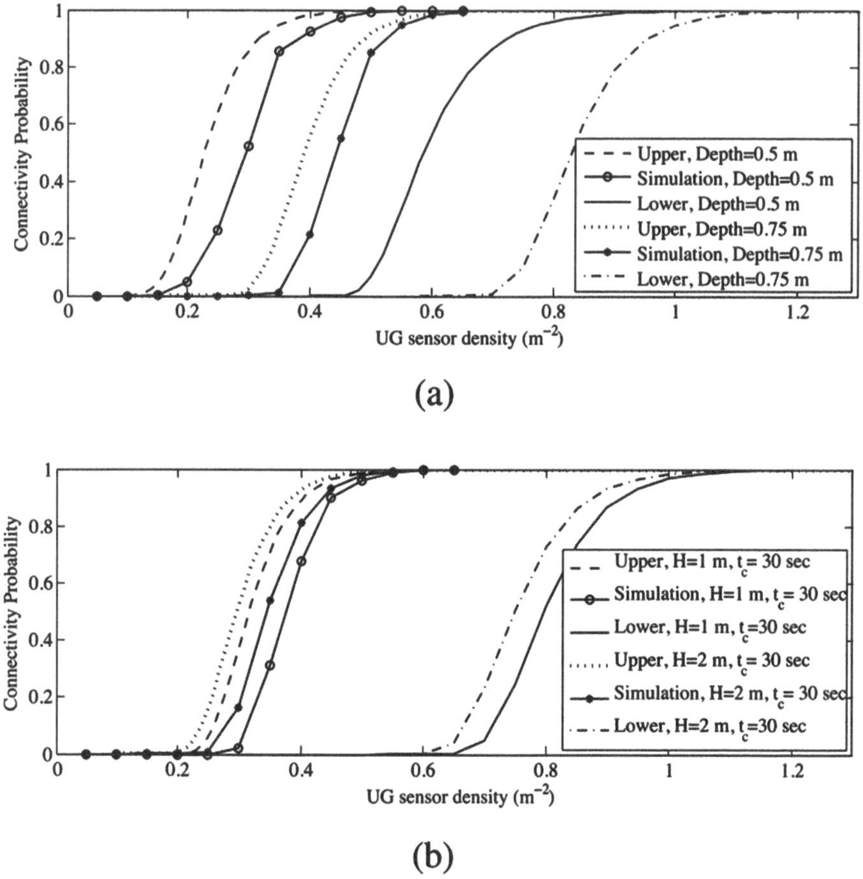

Fig. 4.20 Connectivity probabilities with (**a**) different UG sensor burial depth in sensing phase and (**b**) different AG sink antenna height in control phase [77]

References

1. Akyildiz IF, Stuntebeck EP (2006) Wireless underground sensor networks: research challenges. Ad Hoc Netw 4(6):669–686
2. Akyildiz IF, Sun Z, Vuran MC (2009) Signal propagation techniques for wireless underground communication networks. Phys Commun J 2(3):167–183
3. Bettstetter C (2002) On the minimum node degree and connectivity of a wireless multihop network. In: Proceedings of the 3rd ACM international symposium on mobile ad hoc networking & computing, pp 80–91
4. Bird TS (2009) Definition and misuse of return loss [report of the transactions editor-in-chief]. IEEE Antennas Propag Mag 51(2):166–167
5. Björnson E, Bengtsson M, Ottersten B (2014) Optimal multiuser transmit beamforming: a difficult problem with a simple solution structure [lecture notes]. IEEE Signal Process Mag 31(4):142–148

6. Bogena HR, Huisman JA, Meier H, Rosenbaum U, Weuthen A (2009) Hybrid wireless underground sensor networks: quantification of signal attenuation in soil. Vadose Zone J 8(3):755–761
7. Chae C-B, Hwang I, Heath RW, Tarokh V (2012) Interference aware-coordinated beamforming in a multi-cell system. IEEE Trans Wireless Commun 11(10):3692–3703
8. Cheng RS, Verdú S (1993) Gaussian multiaccess channels with ISI: capacity region and multiuser water-filling. IEEE Trans Inf Theory 39(3):773–785
9. Chow PS, Cioffi JM (1992) Bandwidth optimization for high speed data transmission over channels with severe intersymbol interference. In: [Conference record] GLOBECOM'92-communications for global users: IEEE. IEEE, Piscataway, pp 59–63
10. Cover TM, Thomas JA (2012) Elements of information theory. John Wiley & Sons, Hoboken
11. Demirkol I, Ersoy C, Alagöz F (2006) MAC protocols for wireless sensor networks: a survey. IEEE Commun Mag 44(4):115–121
12. Dong X, Vuran MC (2011) A channel model for wireless underground sensor networks using lateral waves. In: 2011 IEEE global telecommunications conference-GLOBECOM 2011. IEEE, Piscataway, pp 1–6
13. Dong X, Vuran MC (2012) Empirical analysis of the hidden terminal problem in wireless underground sensor networks. In: International conference on wireless communications in unusual and confined areas (ICWCUCA)
14. Dong X, Vuran MC (2013) Impacts of soil moisture on cognitive radio underground networks. In: 2013 first international black sea conference on communications and networking (BlackSeaCom). IEEE, Piscataway, pp 222–227
15. Dousse O, Thiran P, Hasler M (2002) Connectivity in ad-hoc and hybrid networks. In: Proceedings of INFOCOM 2002, New York, vol. 2, pp 1079–1088
16. Dousse O, Mannersalo P, Thiran P (2004) Latency of wireless sensor networks with uncoordinated power saving mechanisms. In: Proceedings of the 5th ACM international symposium on mobile ad hoc networking and computing, pp 109–120
17. Dousse O, Baccelli F, Thiran P (2005) Impact of interferences on connectivity in ad hoc networks. IEEE/ACM Trans. Networking 13(2):425–436
18. Fabbri F, Verdone R (2008) A statistical model for the connectivity of nodes in a multi-sink wireless sensor network over a bounded region. In: 2008 14th European wireless conference. IEEE, Piscataway, pp 1–6
19. Fenn AJ, Hurst PT (2015) Ultrawideband phased array antenna technology for sensing and communications systems. MIT Press, Cambridge
20. Gupta P, Kumar PR (1999) Critical power for asymptotic connectivity in wireless networks. In: Stochastic analysis, control, optimization and applications. Springer, Berlin, pp 547–566
21. Haupt RL (2015) Timed arrays. Wiley, Hoboken
22. Kalet I (1989) The multitone channel. IEEE Trans Commun 37(2):119–124
23. Khandani AK (2014) Media-based modulation: Converting static Rayleigh fading to AWGN. In: In 2014 IEEE international symposium on information theory (ISIT), pp 1549–1553
24. Konda A, Rau A, Stoller MA, Taylor JM, Salam A, Pribil GA, Argyropoulos C, Morin SA (2018) Soft microreactors for the deposition of conductive metallic traces on planar, embossed, and curved surfaces. Adv. Funct. Mater. 28(40):1803020
25. Li L, Vuran MC, Akyildiz IF (2007) Characteristics of underground channel for wireless underground sensor networks. In: Proceedings of Med-Hoc-Net 07, Corfu
26. Li L, Vuran MC, Akyildiz IF (2007) Characteristics of underground channel for wireless underground sensor networks. In: Proceedings of Med-Hoc-Net, vol 7, pp 13–15
27. Ma R, Zhang W (2015) Adaptive MQAM for energy harvesting wireless communications with 1-bit channel feedback. IEEE Trans Wirel Commun 14(11):6459–6470
28. NOAA's national weather service weather forecast office. http://www.crh.noaa.gov/gid/?n=neb_rainfall_observed#archive
29. Peplinski N, Ulaby F, Dobson M (1995) Dielectric properties of soils in the 0.3–1.3-ghz range. IEEE Trans. Geosci. Remote Sens. 33(3):803–807
30. Proakis JG, Salehi M (2001) Digital communications, vol 4. McGraw-Hill, New York

31. Rhee W, Cioffi JM (2000) Increase in capacity of multiuser OFDM system using dynamic subchannel allocation. In: VTC2000-spring. 2000 IEEE 51st vehicular technology conference proceedings (Cat. No. 00CH37026), vol 2. IEEE, Piscataway, pp 1085–1089
32. Salam A (2018) Pulses in the sand: Long range and high data rate communication techniques for next generation wireless underground networks. ETD collection for University of Nebraska – Lincoln (AAI10826112)
33. Salam A (2019) A comparison of path loss variations in soil using planar and dipole antennas. In: 2019 IEEE international symposium on antennas and propagation. IEEE, Piscataway
34. Salam A (2019) Design of subsurface phased array antennas for digital agriculture applications. In: Proceedings of 2019 IEEE international symposium on phased array systems and technology (IEEE Array 2019), Waltham
35. Salam A (2019) A path loss model for through the soil wireless communications in digital agriculture. In: 2019 IEEE international symposium on antennas and propagation. IEEE, Piscataway, pp 1–2
36. Salam A (2019) Sensor-free underground soil sensing. In: ASA, CSSA and SSSA international annual meetings (2019). ASA-CSSA-SSSA
37. Salam A (2019) Subsurface MIMO: a beamforming design in internet of underground things for digital agriculture applications. J Sens Actuator Netw 8(3):41
38. Salam A (2019) Underground environment aware MIMO design using transmit and receive beamforming in internet of underground things. Springer International Publishing, Cham, pp 1–15
39. Salam A (2019) An underground radio wave propagation prediction model for digital agriculture. Information 10(4), 147
40. Salam A (2019) Underground soil sensing using subsurface radio wave propagation. In: 5th global workshop on proximal soil sensing, Columbia
41. Salam A (2020) Internet of things for environmental sustainability and climate change. Springer International Publishing, Cham, pp 33–69
42. Salam A (2020) Internet of things for sustainability: perspectives in privacy, cybersecurity, and future trends. Springer International Publishing, Cham, pp 299–327
43. Salam A (2020) Internet of things for sustainable community development, 1 edn. Springer Nature, Heidelberg
44. Salam A (2020) Internet of things for sustainable community development: introduction and overview. Springer International Publishing, Cham, pp 1–31
45. Salam A (2020) Internet of things for sustainable forestry. Springer International Publishing, Cham, pp 147–181
46. Salam A (2020) Internet of things for sustainable human health. Springer International Publishing, Cham, pp 217–242
47. Salam A (2020) Internet of things for sustainable mining. Springer International Publishing, Cham, pp 243–271
48. Salam A (2020) Internet of things for water sustainability. Springer International Publishing, Cham, pp 113–145
49. Salam A (2020) Internet of things in agricultural innovation and security. Springer International Publishing, Cham, pp 71–112
50. Salam A (2020) Internet of things in sustainable energy systems. Springer International Publishing, Cham, pp 183–216
51. Salam A (2020) Internet of things in water management and treatment. Springer International Publishing, Cham, pp 273–298
52. Salam A (2020) Wireless underground communications in sewer and stormwater overflow monitoring: radio waves through soil and asphalt medium. Information 11(2):98
53. Salam A, Karabiyik U (2019) A cooperative overlay approach at the physical layer of cognitive radio for digital agriculture. In: Third international Balkan conference on communications and networking 2019 (BalkanCom'19), Skopje
54. Salam A, Karabiyik U (2019) A cooperative overlay approach at the physical layer of cognitive radio for digital agriculture. Faculty Publications, West Lafayette

55. Salam A, Shah S (2019) Internet of things in smart agriculture: enabling technologies. In: 2019 IEEE 5th world forum on internet of things (WF-IoT). IEEE, Piscataway, pp 692–695
56. Salam A, Vuran MC (2016) Impacts of soil type and moisture on the capacity of multi-carrier modulation in internet of underground things. In: Proceedings of the 25th ICCCN 2016, Waikoloa
57. Salam A, Vuran MC (2016) Impacts of soil type and moisture on the capacity of multi-carrier modulation in internet of underground things. In: 2016 25th international conference on computer communication and networks (ICCCN). IEEE, Piscataway, pp 1–9
58. Salam A, Vuran MC (2017) EM-based wireless underground sensor networks. Academic Press, Cambridge, pp 247–285
59. Salam A, Vuran MC (2017) Smart underground antenna arrays: a soil moisture adaptive beamforming approach. In: IEEE INFOCOM 2017-IEEE conference on computer communications. IEEE, Piscataway, pp 1–9
60. Salam A, Vuran MC (2017) Wireless underground channel diversity reception with multiple antennas for internet of underground things. In: 2017 IEEE international conference on communications (ICC). IEEE, Piscataway, pp 1–7
61. Salam A, Vuran MC (2017) Smart underground antenna arrays: a soil moisture adaptive beamforming approach. In: Proceeding of IEEE INFOCOM 2017, Atlanta
62. Salam A, Vuran MC (2017) Wireless underground channel diversity reception with multiple antennas for internet of underground things. In: Proceedings of IEEE ICC 2017, Paris
63. Salam A, Vuran MC, Irmak S (2016) Pulses in the sand: Impulse response analysis of wireless underground channel. In: IEEE INFOCOM 2016-The 35th annual IEEE international conference on computer communications. IEEE, Piscataway, pp 1–9
64. Salam A, Vuran MC, Irmak S (2016) Pulses in the sand: Impulse response analysis of wireless underground channel. In: The 35th annual IEEE international conference on computer communications (INFOCOM 2016), San Francisco
65. Salam A, Vuran MC, Irmak S (2017) Towards internet of underground things in smart lighting: a statistical model of wireless underground channel. In: Proceedings of 14th IEEE international conference on networking, sensing and control (IEEE ICNSC), Calabria
66. Salam A, Hoang AD, Meghna A, Martin DR, Guzman G, Yoon YH, Carlson J, Kramer J, Yansi K, Kelly M, et al (2019) The future of emerging IoT paradigms: Architectures and technologies. https://doi.org/10.20944/preprints201912.0276.v1
67. Salam A, Vuran MC, Dong X, Argyropoulos C, Irmak S (2019) A theoretical model of underground dipole antennas for communications in internet of underground things. IEEE Trans Antennas Propag 67(6):3996–4009
68. Salam A, Vuran MC, Irmak S (2019) Di-sense: in situ real-time permittivity estimation and soil moisture sensing using wireless underground communications. Comput. Netw. 151:31–41
69. Santi P, Blough DM (2003) The critical transmitting range for connectivity in sparse wireless ad hoc networks. IEEE Trans. Mob. Comput. 2(1):25–39
70. Seifi E, Atamanesh M, Khandani AK (2015) Media-based MIMO: a new frontier in wireless communications. CoRR, abs/1507.07516
71. Silva AR, Vuran MC (2009) Empirical evaluation of wireless underground-to-underground communication in wireless underground sensor networks. In: Proceedings of IEEE DCOSS '09, Marina Del Rey
72. Silva AR, Vuran MC (2010) Development of a testbed for wireless underground sensor networks. EURASIP J Wirel Commun Netw. Article number: 620307
73. Silva AR, Vuran MC (2010) Development of a testbed for wireless underground sensor networks. EURASIP J Wirel Commun Netw 2010:1–14
74. SoilNet - A Zigbee Based Soil Moisture Sensor Network. http://www.fz-juelich.de/icg/icg-4/index.php?index=739. Project Group – Institute of Chemistry and Dynamics of the Geosphere (ICG) – Agrosphere Institute – ICG 4

75. Sun Z, Akyildiz IF (2010) Connectivity in wireless underground sensor networks. In: Proceedings of IEEE communications society conference on sensor mesh and ad hoc communications and networks (SECON '10), Boston
76. Temel S, Vuran MC, Lunar MMR, Zhao Z, Salam A, Faller RK, Stolle C (2018) Vehicle-to-barrier communication during real-world vehicle crash tests. Comput Commun 127:172–186
77. Tiusanen MJ (2008) Wireless Soil Scout prototype radio signal reception compared to the attenuation model. Precis Agric 10(5):372–381
78. Vuran MC, Akyildiz IF (2008) Packet size optimization for wireless terrestrial, underwater, and underground sensor networks. In: Proceedings of IEEE INFOCOM '08, Phoeniz
79. Vuran MC, Akyildiz IF (2008) Cross-layer packet size optimization for wireless terrestrial, underwater, and underground sensor networks. In: INFOCOM 2008. The 27th conference on computer communications. IEEE, Piscataway, pp 226–230
80. Vuran MC, Salam A, Wong R, Irmak S (2018) Internet of underground things in precision agriculture: architecture and technology aspects. Ad Hoc Netw 81:160–173
81. Vuran MC, Salam A, Wong R, Irmak S (2018) Internet of underground things: sensing and communications on the field for precision agriculture. In: 2018 IEEE 4th world forum on internet of things (WF-IoT) (WF-IoT 2018), Singapore
82. Xhafa AE, Tonguz OK, Cepni AG, Stancil DD, Nikitin PV, Brodtkorb D (2005) On the capacity limits of HVAC duct channel for high-speed internet access. IEEE Trans Commun 53(2):335–342

Chapter 5
Underground Wireless Channel Bandwidth and Capacity

5.1 Introduction

The analysis of the capacity in wireless underground channel is challenging due to many factors: first, when an electromagnetic wave is incident into soil, the wavelength changes because of the higher permittivity of soil compared to that of air [7]. Soil permittivity depends on soil properties, such as bulk density, soil texture, soil moisture (volumetric water content), salinity, and temperature. Second, when an antenna is buried underground, its return loss property changes due to the high permittivity of the soil [40]. Moreover, with the variation in soil moisture and hence soil permittivity, the return loss of the antenna varies too. Figure 5.1 shows the organizational structure of the chapter. The wireless underground channel capacity is discussed next.

5.2 Wireless Underground Channel Capacity

The single- and multi-user capacity of wireless underground channel is discussed in the following sections.

5.3 Single User Capacity

The channel capacity, in a wireless communication, is calculated as :

$$C = B\log_2\left(1 + \frac{S}{N_0 B}\right), \tag{5.1}$$

A. Salam, U. Raza, *Signals in the Soil*,
https://doi.org/10.1007/978-3-030-50861-6_5

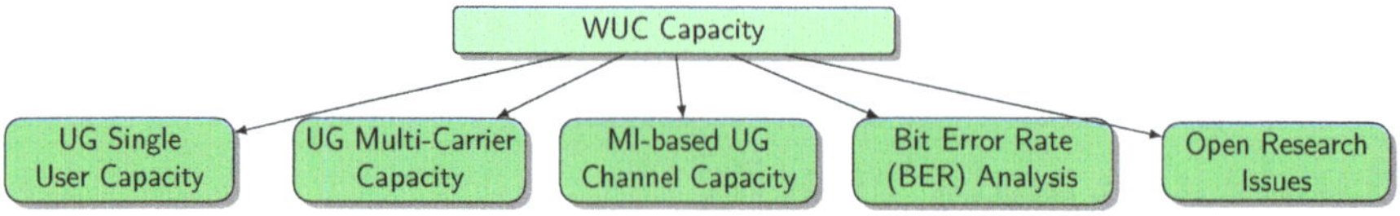

Fig. 5.1 Organization of the chapter

where B denotes the system bandwidth, S denotes the received signal strength at receiver, and N_0 denotes the noise power density. The analysis considers the maximum bandwidth for antenna design. Received signal strength of the signal is affected by the antenna return loss and is given as:

$$S_{\text{dB}} = P_t + 10\log_{10}\left(1 - 10^{-\frac{RL_{\text{dB}}}{10}}\right) - L, \tag{5.2}$$

where RL_{dB} is the antenna return loss, L is the path loss. The interference in wireless communication is not that big. It is because of less number of devices; hence, noise is mainly thermal noise and can be considered constant [21, 42].

5.3.1 Numerical Analysis

This section discusses the experiments conducted to understand the soil impact on underground wireless communication. In the upcoming sections, the discussion on how propagation loss, return loss, and bandwidth of antenna are affected by soil is presented. Moreover, it also analyzes how channel capacity responses to changing frequencies and soil moisture level. The default parameters for the experiments, unless a change is specified explicitly, are as follows: soil type is clay soil (having 31% clay and 29% sand), depth of underground device is 0.4 m and that of aboveground is 2.5 m, and length of antenna is 60 mm with diameter of 2 mm.

To assess the performance of UG channel we have analyzed the UG channel capacity. Our analysis shows that higher soil moisture affects the path loss and capacity of UG channel. Change in the path loss and capacity with soil moisture in silt loam testbed for 200, 500, and 700 MHz frequencies is shown in Fig. 5.2. We can see from Fig. 5.2a that when soil moisture is decreased, there is 5–8 dB decrease in path loss at 200 MHz frequency, 25 dB decrease at 500 MHz, and 15 dB decrease in path loss is observed at 700 MHz frequency. This is caused by higher permittivity of the soil at higher soil moisture, higher attenuation of waves causes higher path loss.

Similarly in Fig. 5.2b we can see that capacity is increased with decrease in soil moisture for all three frequencies. Due to the fact that in most situations, sandy soil has lower path loss, it has a higher capacity than the silt loam soil in most situations. Each soil has an optimal operation frequency region where the communication achieves the maximum capacity. However, unlike over-the-air communications, the

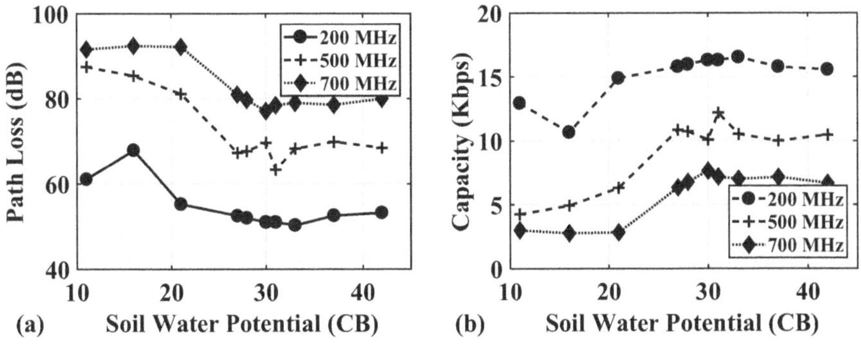

Fig. 5.2 Effects of change in soil moisture on (**a**) Path Loss. (**b**) Channel capacity

optimal frequency where the maximum capacity is achieved is not the same as the resonant frequency of the antenna. Moreover, since changes in soil moisture affect the path loss, return loss, and bandwidth of antenna, hence, capacity achieving frequency spectrum changes as well. This is because lateral wave and reflected wave [7] experience more attenuation as the burial depth of the antenna is increased. Therefore, in UG communications, the effects of the antenna and the soil need to be considered together to find the optimal operation frequency.

5.3.2 Soil Impact on Return Loss, Bandwidth, and Path Loss

Figure 5.3a shows the negative of return loss (S_{11}) at frequency range of 100 MHz to 1 GHz. Volumetric water content (VWC) is the major way to indicate soil moisture and is calculated as a ratio of water in a soil–water mixture. It analyzes the effect of VWC at 20, 25, 30, 35, and 40%. VWC inversely impacts the resonant frequency. Increasing the VWC values significantly decreases the resonant frequency, e.g., as shown, the frequency decreased approximately 33%, i.e., from 649 to 432 MHz, with a VWC increase of 20–40%.

Figure 5.3b plots the antenna bandwidth with operation frequency at changing soil moisture levels and return loss threshold δ of -10 dB. Bandwidth is given as a range of frequency where the return loss is less than the threshold. Therefore, system achieves the highest bandwidth at resonant frequency and it suddenly start to decrease out of the resonant frequency. For example, for VWC = 40% the bandwidth decreases approximately 16%, i.e., from 62 MHz (at resonant frequency of 434 MHz) to 52 MHz (operational frequency of 433 MHz). Similarly, the bandwidth is inversely affected by VWC at a given resonant frequency. It can be observed that, at resonant frequency, the bandwidth is 94 MHz at 20% VWC and drops to 74 and 62 MHz with a VWC increase of 20 and 40%, respectively.

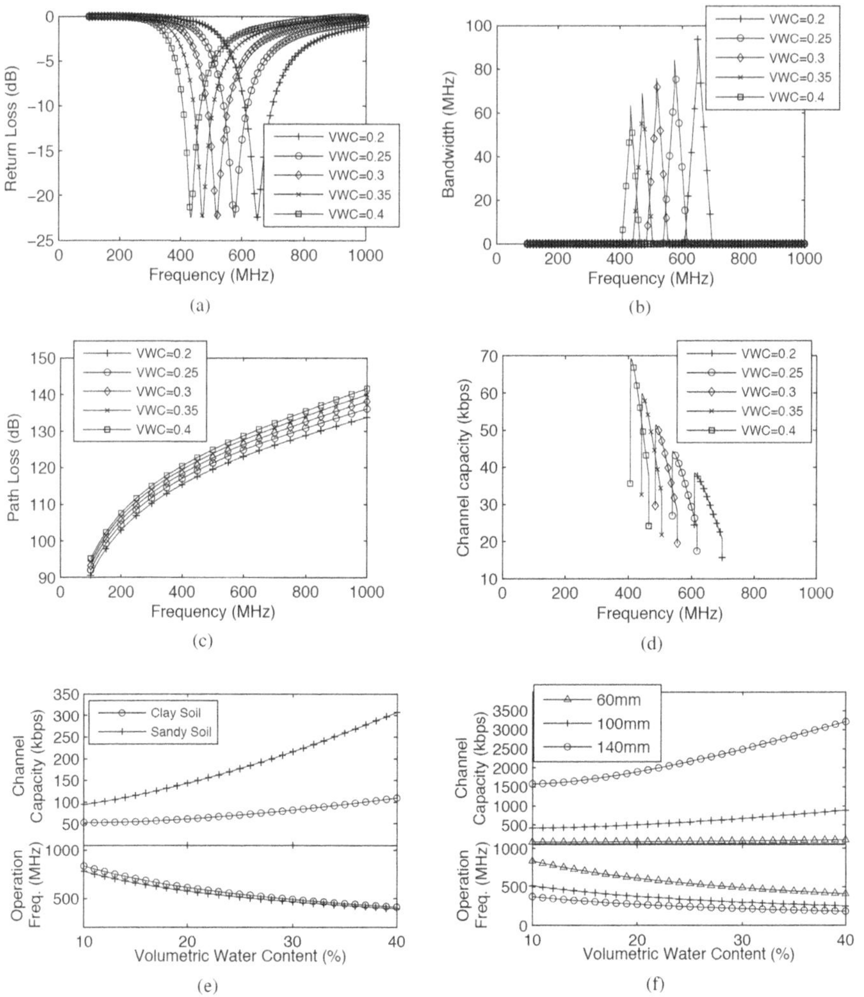

Fig. 5.3 Numerical results from experiments: (**a**) Return loss. (**b**) Bandwidth. (**c**) Path loss in UG2AG channel. (**d**) Capacity in UG2AG link. (**e**) Capacity at different frequencies and soil moisture values. (**f**) Capacity at different frequencies and antenna sizes [8]

Figure 5.3c shows the path loss for UG2AG channel, with frequency at changing soil moisture [22, 25]. It can be observed that path loss is directly proportional to VWC and frequency. Soil moisture affects the path loss and this effect is significantly higher at higher frequencies. For example, as shown in Fig. 5.3c, path loss is 107.6 dB for a VWC of 40% and frequency of 200 MHz. With the same frequency, it changes 102.9 dB at VWC of 20%. It shows that the path loss increases by 4.7 dB while doubling the soil moisture. However, at higher frequency of

900 MHz, path loss changes from 138.6–131.4 dB at VWC 40 and 20%, receptively. Hence, the difference increases to 7.2 dB.

5.3.3 Capacity Analysis

This section analyzes the underground communication on the basis of channel capacity and measures the impact of soil moisture. The parameters of analysis are as follows: transmit power of 10 dBm, noise power density is 1.5625×10^{-16} W Hz^{-1} [23, 42], maximum bandwidth of the system is same. Although new specific modulation schemes are needed to use the said maximum bandwidth, however, it is out of the scope of this discussion.

Figure 5.3d plots channel capacity, calculated by Eq. (5.1), with operational frequency. It shows the optimal frequency for each soil moisture level where the capacity is maximum. The channel capacity ranges from 38 to 70 kbps for VWC ranging from 20 to 40%. As in the case of antenna return loss, the operational frequency decreases with the increase in soil moisture. The interesting thing to note is that the optimal frequency, at which channel capacity is maximum, is much lower than the resonant frequency, e.g., at 20% VWC the optimal frequency (611 MHz) is 38 MHz lower than the resonant frequency (649 MHz). This is mainly because the system bandwidth is highest at resonant frequency but noise power also increases because of power density N0 being a constant. Moreover, as shown earlier in Fig. 5.3c, path loss decreases with decrease in frequency [27, 30].

Figure 5.3e plots channel capacity with optimal frequency as a function of VWC along with different soil types (clay and sandy). It can be observed that soil moisture and optimal frequency are inversely proportional where, for clay soil, a decrease in soil moisture from 40 to 10% causes optimal operational frequency to increase from 409 to 833 MHz and for sandy soil, the optimal frequency is between the range 393.1 and 778.6 MHz. It shows that frequency moves in a wide spectrum with change in soil moisture; therefore, the transceivers employed in WUSNs must be able to operate in wide and lower spectrum range depending on soil moisture values.

Channel capacity shows irregular behaviors in response to change in soil moisture especially in clay soil. For clay soil, there is a slight decrease in channel capacity for VWC increase of 10–11% and then increases to 109.3 kbps for VWC = 40% because of impact of soil moisture and operation frequency on path loss. From Fig. 5.3c, path loss is inversely proportional to the soil moisture at the same operational frequency and increases monotonically with increasing frequency. Even though the path loss increases with soil moisture, it also causes a decrease in optimal operational frequency because of wavelength shortening and low frequency in response to low path loss. Hence, the overall path loss of the system may not decrease with the increasing soil moisture. This holds true for the sandy soil as well where capacity is the direct function of soil moisture because sandy soil has relatively lower attenuation from clay soil especially for high soil moisture levels.

Sandy soil has much higher channel capacity than clay soil for all of the soil moisture levels. The channel capacity of sandy soil (94.22 kbps) is 78.2% higher than clay soil (52.86 kbps) at VWC = 10%. Similarly, at VWC = 40%, the channel capacity of sandy soil (307.8 kbps) is 181.6% higher than clay soil (109.3 kbps). This behavior is due to lower path loss in sandy soil [32]. In practical applications, soil type cannot be changed; however, antenna size can be changed. Figure 5.3f plots the operational frequency and corresponding capacity at different sized antennas. The lengths of antennas are kept at 60, 100, and 140 mm in a clay soil. It was observed that as the antenna size increased (longer antenna), the capacity also increased. This is because for longer antennas, the optimal operation frequency also decreased, hence, experiencing low path loss at lower frequency. At VWC = 15%, the operational frequency and corresponding capacity, in pair, were given as (703.4 MHz, 54.42 kbps) for 60 mm antenna, (443.4 MHz, 455.2 kbps) for 100 mm antenna, and (314.6 MHz, 1680 kbps) for 140 mm antenna. It can be seen that frequency is decreasing and capacity is increasing with increase in antenna size. The difference becomes greater with increase in soil moisture. For a soil moisture of VWC = 40%, the capacity increases from 109.3 to 3221 kbps when antenna size increases from 60 to 140 mm. Therefore, the analysis shows that long antennas are suitable for achieving lower path loss. However, the size depends upon the device and availability of the spectrum. It is also hard to employ longer antennas in underground environment [28].

Figure 5.4 shows the comparison of the fixed-frequency systems with cognitive radio system in clay soil. Four fixed-frequency systems are operating at 550, 600, 650, and 700 MHz and cognitive radios dynamically adjust its frequency with the soil moisture levels. For a fixed soil moisture level, the performance of both systems, in terms of channel capacity, is same, e.g., 600 MHz system has same capacity as of

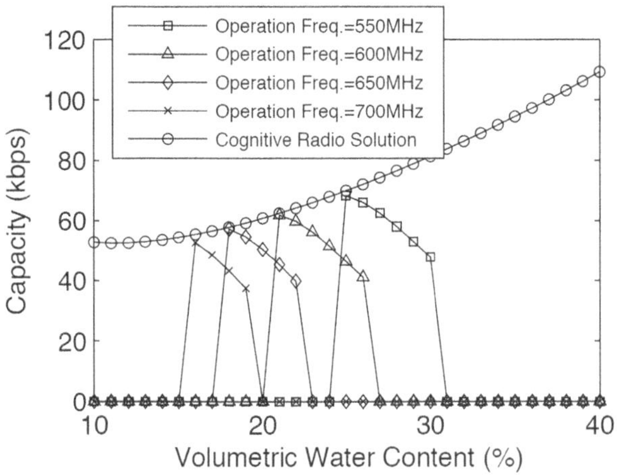

Fig. 5.4 Comparison of channel capacity from non-cognitive (fixed-frequency system) and cognitive radio system [8]

cognitive radio at VWC = 21%. However, it is not true when the soil moisture level varies, in which case, cognitive radio have better performance as compared to fixed-frequency system. Another important advantage of cognitive radios is that it can sustain capacity higher than the 50 kbps within a wide range of soil moisture level which is not possible with the fixed-frequency systems. For example, with fixed operational frequency of 550 MHz, system is operating with VWC in the range of 25–30% and this VWC range is 16–19% for fixed operational frequency of 700 MHz [24, 33].

5.4 Underground Multi-Carrier Capacity

A communication system is evaluated using probability of bit error rate as a metric for SNR values and data rate. However, UG nodes have very low transmission power to achieve a prolonged operating period. Therefore, the achievable data rate of wireless underground channel must be estimated for a fixed BER considering the low transmission power of UG nodes. Coherence bandwidth is used as sub-carrier bandwidth for ISI avoidance. It is known that channel capacity changes with the bandwidth [18, 26]; hence, it is highly recommended to use bandwidth of sender and receiver antenna pair and channel transfer function of the UG channel while evaluating the capacity of UG channel. Bandwidth of the channel is calculated from the return loss of the antenna. The effect of soil moisture and soil type on wireless UG channel is studied using multi-carrier modulation. The coherence bandwidth changes are adapted as per soil moisture changes, hence, adapting sub-carrier bandwidth and the system accordingly [29, 34].

The number of sub-carrier in a multi-carrier modulation is a minimum number required for inter-symbol interference (ISI) avoidance. If B_s denotes the system bandwidth, B_{cb} denotes the UG channel coherence bandwidth, then the total number of sub-carriers are given as $N_c = \lceil B_s/B_{cb} \rceil$. B_s is dependent upon the antenna bandwidth. Antenna bandwidth is calculated from its own return loss below a threshold value δ (e.g., [2] uses $\delta = -10\,\text{dB}$).

Figure 4.2a shows the experiment results where return loss of a dipole antenna operating at 433 MHz and saturated condition, i.e., soil matric potential = 0. Three different types of soils are used for this experiment. Bandwidth variation can also be seen whereas B_{cb} relies on channel characteristics. Empirical coherence bandwidth values are used for the analysis. These values are obtained from time-domain impulse response measurements and will be measured in testbed and field experiments [34].

The modulation scheme considered to determine the UG channel capacity is M-ary quadrature amplitude modulation (MQAM). This modulation scheme is used for each carrier of multi-carrier transmission system due to its high spectral efficiency[17]. The total UG channel bit rate is given as $R_{ug} = \sum_{i=1}^{N_c} r_i B_{cb}$, where N_c is the total number of sub-carriers, B_{cb} is the bandwidth of each sub-carrier, and r_i represents the number of bits per symbol for each carrier [38].

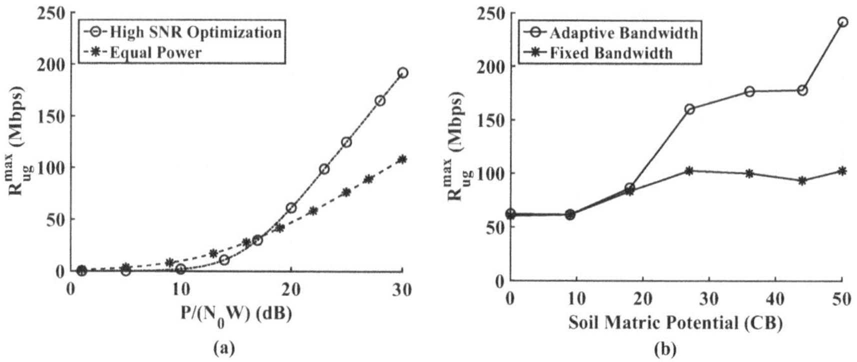

Fig. 5.5 (**a**) High-SNR optimization in the silty clay loam soil and (**b**) System bandwidth approaches

For maximum UG channel bit rate R_{ug}, optimized power allocation can be done between all sub-carriers on the basis of (a) fixed power constraint P and (b) probability of symbol error of each sub-carrier, $P^*_{sc} > P_{sc_i}$, $\forall i$.

The overall bit rate is given as [13]:

$$R_{ug} = \sum_{i=1}^{N_c} B_{cb} \log_2 \left\{ 1 + \frac{\frac{3\gamma_i P}{(N_0 B_{cb})|H_i(f)|^2}}{\left[Q^{-1}\left\{\frac{P^*_{sc}}{K_{r_i}}\right\}\right]^2} \right\}, \tag{5.3}$$

where P is transmit power constraint, N_c represents the number of sub-carriers, and γ_i's are selected such that $\sum_{i=1}^{N_c} \gamma_i \cdot P = P, \quad \gamma_i > 0$.

As discussed, the power between each sub-carrier is optimized to achieve maximum R_{ug}. This optimization problem [19, 48] of achieving optimum power allocation γ_i^* can be solved as water filling problem [4, 5] was solved by using a Lagrangian multiplier technique. Figure 5.5a shows the initial comparison of this approach with that of simple power allocation. This experiment is performed for silty clay loam soil. It can be seen that when equal power distribution is utilized, it results in higher data rate for low $P/(N_0 W)$ values of less than 17 dB. Therefore, this technique can be used as default in scenarios with mobile data collection when data collector is moving near the UG radio. The allocation can then be improved according to the improvement in channel state. It will give enough time to gauge the state of UG channel without any performance degradation [36, 37, 39].

Next, multi-carrier modulation capacity is analyzed through detailed antenna return loss and empirical channel transfer function. The experiments are performed in different soil types to study the impact of soil types with varying soil moisture levels.

5.4.1 Soil-Based UG Channel Estimation

One of the important part of multi-carrier modulation schemes is channel estimation. UG channel is estimated by using two techniques: (a) using statistics of soil moisture only, (b) extending approach of dynamic channel sensing. The first approach uses this channel transfer function to create a configuration database of channel states and impulse responses. The soil moisture measurements are taken at different instants and a channel state is selected based on those measurements. Next, a comparative analysis between theoretical (from (5.3)) and experimental multi-carrier communication is used for results for validation.

The second approach is an improvement of first one and gives more accurate channel state. It incorporates real-time channel sensing by adding extra hardware for dynamic channel estimation. However, it will also add an extra overhead. UG devices are being improved using advanced technologies making the dynamic channel sensing a feasible solution for UG communication systems [35, 40]. The channel state is not expected to change within close vicinity of a measured soil moisture level area. Hence, a powerful UG node can be used as a master node to perform dynamic channel estimation. This master node can disseminate the channel state to other low power UG nodes. Moreover, dynamic channel sensing is expected to consume more power. Unlike traditional sensing nodes and handheld devices, UG nodes do not have tight size limitations because they are buried. Hence, power consumption problem can be minimized by burying large batteries with UG nodes for prolonged operation.

5.4.2 Adaptive Sub-carrier and System Bandwidth

The initial results show that the capacity of UG channel with multi-carrier modulation is affected by the soil moisture, soil type, and distance between the transmitter and receiver. Similarly, transmission parameters (e.g., sub-carrier bandwidth) can be adjusted to improve performance.

Given these findings, the effect of adaptive sub-carrier bandwidth (ASB) on UG channel capacity was analyzed. In [3, 20], improvements were reported with the use of adaptive channel width, and [6, 43] reported improvements with adaptive sub-carrier bandwidth (ASB). As an initial work, a comparative analysis was performed between the theoretical data rate of a fixed sub-carrier (411 kHz) and system bandwidth (20 MHz) with an adaptive sub-carrier and system bandwidth approach. Figure 5.5b shows the results for different soil moisture levels. For a soil moisture level of 27 CB, fixed approach gives the channel capacity of 102 Mbps and adaptive gives 159 Mbps, i.e., approx. 56% higher channel capacity. Similar trend is experienced when the soil moisture is increased to 50 CB with adaptive approach achieving the channel capacity of 241 Mbps, i.e., approx. 136% higher than fixed approach.

5.5 Magnetic Induction-Based Wireless Channel Capacity

EM waves can only be used for small communication ranges in underground environment because it suffers from severe impairments due to rocks, soil, and sand. Therefore, a feasible alternative is to use magnetic induction for underground communication. First MI-based WUSN was used by Akyildiz and Stuntebeck [1]. In MI-based WUSNs, magnetic coils are used as communication antenna combined using a waveguide structure. Many relay nodes are used between the end nodes to facilitate the communication between two ends [45]. MI-based WUSN, similar to conventional wireless network, takes the advantage of lower path loss extending the transmission range as compared to its EM-based counterpart. Another advantage of MI-based relaying nodes is that they do not need to be charged; however, end transceiver nodes need energy and can be charged using mobile aboveground devices [14]. WUSNs are not demanding in terms of data rate; however, channel characterization of MI-based is important so that it can provide at least sufficient performance. In this section, a specific MI-based channel modeling scheme is discussed in detail.

5.5.1 *System Model*

Topology used to model system includes a one sender, one receiver, and $k - 1$ passive relays. All nodes are connected in a waveguide structure. The voltage source of sender is U_t and load impedance of receiver is Z_l. Relaying nodes are placed equidistantly between both end nodes [45]. Each node circuit consists of an antenna (magnetic), a capacitor C, and a resistor R. The parasitic effect of a circuit due to high frequencies is ignored. To that end, such frequency ranges are adopted for which the effect is negligible. Multi-layer air core coil is used as an antenna. The inductivity L and copper resistance R of such antenna coil are given as:

$$L = \frac{21\mu N^2 a}{4\pi}\left(\frac{a}{l+h}\right)^{0.5}, \tag{5.4}$$

$$R = \rho \cdot \frac{l_w}{A_w} = \rho \cdot \frac{2aN}{r_w^2}, \tag{5.5}$$

where N denotes the total coil windings, radius of the coil is a, length of the coil is $l = 0.5a$, h and μ denote the height and permeability of the coil. The capacitance $C = \frac{1}{(2\pi f_0)^2 L}$ is chosen such that the circuit operates at resonant frequency f_0. Moreover, in Eq. (5.5), $\rho \approx 1.678 \times 10^{-2}$ mm^2/m and represents the copper resistivity, l_w, A_w and r_W are the total length, cross-section area, and radius of the wire, respectively. The induced voltage is calculated through mutual inductance M [46] as follows:

$$M = \mu\pi N^2 \frac{a^4}{4r^3}(2\sin\theta_t \sin\theta_r + \cos\theta_t \cos\theta_r) \cdot G, \tag{5.6}$$

where θ_t is the angle between radial direction of the coil, θ_r is the angle between the line connecting the center of two coils, r gives distance between two coils. An additional loss factor which occurs due to eddy currents is given by G.

Loss in Soil Magnetic waves suffer from extra attenuation due to eddy current [16]. This effect is considered in this model by using a scaling factor $G = e^{-r/\delta}$, where the attenuation factor δ is given as:

$$\delta = \frac{1}{2\pi f \sqrt{\frac{\mu\epsilon}{2}\left(\sqrt{1 + \frac{\sigma^2}{(2\pi f)^2 \epsilon^2}} - 1\right)}}, \tag{5.7}$$

where ϵ is the permittivity and σ is the conductivity of the soil, f is the frequency.

Path Loss The path loss the system is calculated as [15]:

$$L_p(f) = \left|\frac{P_t(f)}{P_r(f)}\right| = \frac{|S(x, x_L, k) \cdot S(x, x_L, k+1)|}{|\text{Im}\{x_L\}|}, \tag{5.8}$$

where $P_t(f)$ is the required transmit power given as:

$$P_t(f) = \frac{1}{2}|U_t \cdot I_0| = \frac{1}{2}\frac{|U_t|^2}{|j2\pi f M|}\frac{|S(x, x_L, k)|}{|S(x, x_L, k+1)|} \tag{5.9}$$

and $P_r(f)$ is the active received power given as:

$$P_r(f) = \frac{1}{2}|I_k|^2 \text{Re}\{Z_L\} = \frac{|U_t|^2 \cdot \text{Re}\{Z_L\}}{2|j2\pi f M|^2 |S(x, x_L, k+1)|^2}, \tag{5.10}$$

where the induced voltage of coil $n-1$ is $U_n = j2\pi f M$ and

$$S(x, x_L, n) = F(x, n) + x_L \cdot F(x, n-1),$$

$$F(x, n) = \frac{\left(\frac{(x+\sqrt{x^2-4})}{2}\right)^{n+1} - \left(\frac{(x-\sqrt{x^2-4})}{2}\right)^{n+1}}{\sqrt{x^2-4}}.$$

Load Impedance The element of the circuits is susceptible to noise and can degrade the performance of the system; hence, the simplest possible load impedance is given as [15]:

$$Z_L = Z_{L,R} = \text{Re}\left\{j2\pi f_0 M \cdot \frac{F(x_0, k+1)}{F(x_0, k)}\right\}, \tag{5.11}$$

where $x_0 = \frac{R}{j2\pi f_0 J M}$ because of the very limited bandwidth of MI-based systems [45].

Noise Modeling There can be many random noise sources in the environment. The noise is basically EM waves 10 cm and given the high EM attenuation in the soil, the effect of these waves can be minimized to a large extent. Therefore, only thermal noise, from resistors of relay and end nodes, is considered as a dominant source. Noise power at load impedance is given as [15]:

$$P_{N,R}(f) = \frac{1}{2}\frac{Z_{L,R}}{|j2\pi f M|^2} \times \left|\sum_{m=0}^{k}\left(\sum_{n=0}^{m} U_{R,k-n}\frac{S(x,x_L,n)}{S(x,x_L,m)S(x,x_L,m+1)}\right)\right|^2. \tag{5.12}$$

If all the noise sources are independent with Johnson-noise [10] power spectral densities $\mathrm{E}\{|U_{R,n}(f)|^2\} = 4KTR,\ \forall\, n$ (E{·} = expected value), then average noise power due to resistor is given as:

$$\mathrm{E}\{P_{N,R}(f)\} = \frac{1}{2}\frac{4KTRZ_{L,R}}{|j2\pi f M|^2}$$
$$\times \sum_{n=0}^{k}\left(|S(x,x_L,n)|^2\left|\sum_{m=n}^{k}\frac{1}{S(x,x_L,m)S(x,x_L,m+1)}\right|^2\right),$$

where Boltzmann constant $K \approx 1.38 \times 10^{-23}$ J/K, temperature in Kelvin is $T =$ 290 K, and R is the copper resistance from Eq. (5.2).

Similarly, noise from the resistor with power density $\mathrm{E}\{|U_{Z_{L,R}}(f)|^2\} = 4KTZ_{L,R}$ is given as:

$$\mathrm{E}\{P_{N,Z_{LR}}(f)\} = \frac{1}{2}\frac{4KTZ_{L,R}^2}{|j2\pi f M|^2}$$
$$\times \left|\sum_{m=0}^{k}\frac{1}{S(x,x_L,m)S(x,x_L,m+1)}\right|^2.$$

Finally, the total power spectral density of noise for load resistor is given as:

$$\mathrm{E}\{P_{\text{noise}}(f)\} = \mathrm{E}\{P_{N,R}(f)\} + \mathrm{E}\{P_N, Z_{L,R}(f)\}. \tag{5.13}$$

5.5.2 *Channel Capacity*

To maximize the channel capacity of the system it is important to find the optimal waveguide parameters. To that end, three parameters are chosen: the number of coil windings, transmit power density spectrum, and the carrier frequency. Other parameters are either not possible to change or dependent upon these parameters. The optimization problem is given as follows [15] :

$$\underset{\forall f_0, N, P_t(f)}{\arg\max} \; C_{ch} = \int_{-\infty}^{+\infty} \log_2\left(1 + \frac{P_t(f)}{L_p(f)\mathrm{E}\{P_{\text{noise}}(f)\}}\right) \mathrm{d}f$$

$$\text{s.t.}: \quad (1) \int_{-\infty}^{+\infty} P_t(f)\mathrm{d}f = P, \qquad (2) N \in [N_{\min}, N_{\max}], \tag{5.14}$$

$$(3) f_0 \in [f_{0,\min}, f_{0,\max}], \quad (4) \frac{1}{(2\pi f_0)^2 L} \geq C_0,$$

where channel capacity is given as C_{ch} [15], and total power transmitted is P. The transmit power spectral density is denoted by $P_t(f)$. $P_t(f)$ can be adjusted by $U_t.L_p(f)$ (from Eq. (5.8)) and $\mathrm{E}\{P_{\text{noise}}(f)\}$ (from Eq. (5.11)). Some of the limitations, imposed on the systems, are: (1) limited number of coil windings are used in the range of $[N_{\min}, N_{\max}]$ as size of the relay nodes can cause an issue while deployment, (2) lower frequency ranges ($f_{0,\min}$ =1 kHz, $f_{0,\max}$ =300 MHz) are used to avoid parasitic effect. However, higher carrier frequencies with larger number of windings result in the low capacitance increasing the parasitic resistance significantly [44]. Hence, a lower bound on the penitence C_0 is applied.

Optimization Algorithm for Channel Capacity A 2-dimensional grid is spanned over extreme values of f_0 and N. Capacity is determined for each grid point over multiple iterations X_I. The grid point having the largest capacity is selected as a region of interest for each iteration of the algorithm. The values of f_0 and N are selected as per the constraints given in Eq. (5.14). A total of 40 grid points are used for the f_0 and 10 for N. The f_0 points are exponentially separated instead of being equidistant because of large difference in minimum and maximum values of f_0. Moreover, if a point is not in accordance with the constraint 4 of Eq. (5.14), channel capacity is set to zero. For other points, an optimal transmit power spectral density $P_t(f)$ is evaluated using water filling [47] as follows:

$$P_t(f) = \max\left(\left(\frac{1}{\lambda} - L_p(f)\mathrm{E}\{P_{\text{noise}}(f)\}\right), 0\right), \tag{5.15}$$

where λ is adapted to satisfy the constraint 4. Equation (5.15) is integrated in Eq. (5.14) which gives the channel capacity for a particular (f_0, N).

5.5.3 *Numerical Results*

The system is simulated for different arrangements of waveguide. Simulation parameters are given in Table 5.1. Two different deployment schemes, horizontal

Table 5.1 Parameters

Parameters	Description	Values
P	Total transmit power	10 mW
r_w	Wire radius	0.5 mm
a	Coil radius	4
σ	Coil conductivity	0.01 S/m for dry soil; 0.077 S/m for wet soil
ϵ_0	Electric constant	8.854×10^{-12} F/m
ϵ	Soil permittivity	$\epsilon = 7\epsilon_0$ for dry soil; $\epsilon = 29\epsilon_0$ for wet soil
$\mu_0 = \mu$	Magnetic constant	$4\pi \cdot 10^{-7}$ H/m

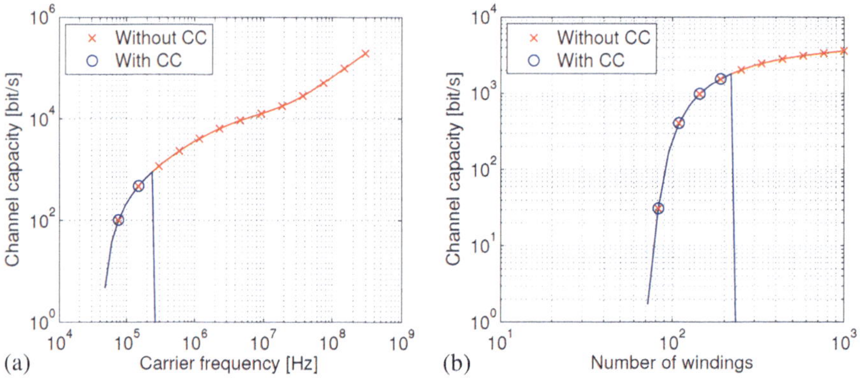

Fig. 5.6 (**a**) Effect of carrier frequency on channel capacity [15]; (**b**) Effect of number of windings on channel capacity at constant frequency [15]

and vertical, were analyzed [11]. For vertical deployment, θ_t and θ_r were kept equal, whereas for horizontal deployment $\theta_t = \pi/2$ and $\theta_r = -\pi/2$ were used. Vertical deployment gives the omnidirectional communication range, however, do not maximize the mutual inductance, hence, exhibiting a very high path loss for vertical deployment as given in the following equation:

$$L_{p,\text{vertical}}(f) \approx 2^{2k} \cdot L_{p,\text{horizontal}}(f). \tag{5.16}$$

In Fig. 5.6a, channel capacity is plotted with the carrier frequency with N=1000, 15 relays, and total distance of 50 m in a dry soil. The constrained capacitance is $C_0 = 1$ pF. Figure 5.6a shows the results for both constrained and unconstrained capacitance. Low capacity is observed for the constrained channel capacitance and, for unconstrained capacitance, channel capacity is high at large values of the frequency. Similar kind of behavior is observed in Fig. 5.6b where capacity is plotted with total number of windings at a constant frequency of $f_0 = 1$ MHz. Capacity is low for the constrained scenario and high for unconstrained scenario. Hence, both f_0 and N need to be optimized together.

Figure 5.7 gives the capacity results for optimum values of N and f_0. It plots the optimum capacity with minimum capacitance C_0 for $X_I = 4$ iterations. For large values of C_0, the channel capacity decreases more in dry soil than wet soil. Capacitor constraints do not affect the performance in wet soil because the constraints do not let maximum carrier frequency go significantly lower than optimal carrier frequency.

Figure 5.8 plots the channel capacity with the inter-relay distance. Parameters for the experiments are as follows: $C_0 = 1\,\text{pF}$, $B = 100\,\text{MHz}$, and $f_0 = 300\,\text{MHz}$. It is also compared with EM-based communication where noise power of EM-based transmission is $P_{\text{noise,EM}} = KTB$ [14]. Four different constellations were used for the MI-based communication: Direct MI (written as only MI in the figure), MI-

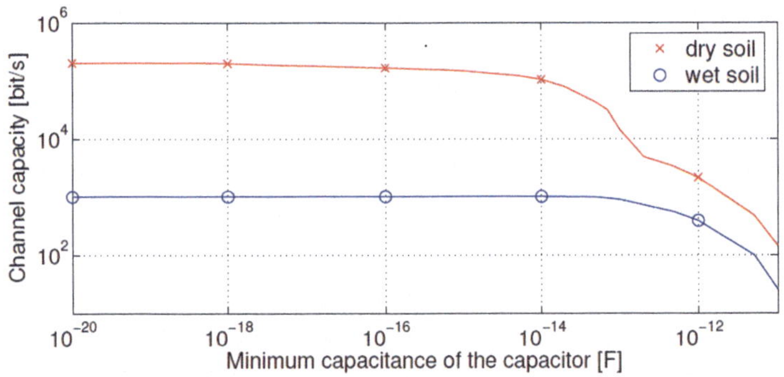

Fig. 5.7 Effect of capacitance on channel capacity [15]

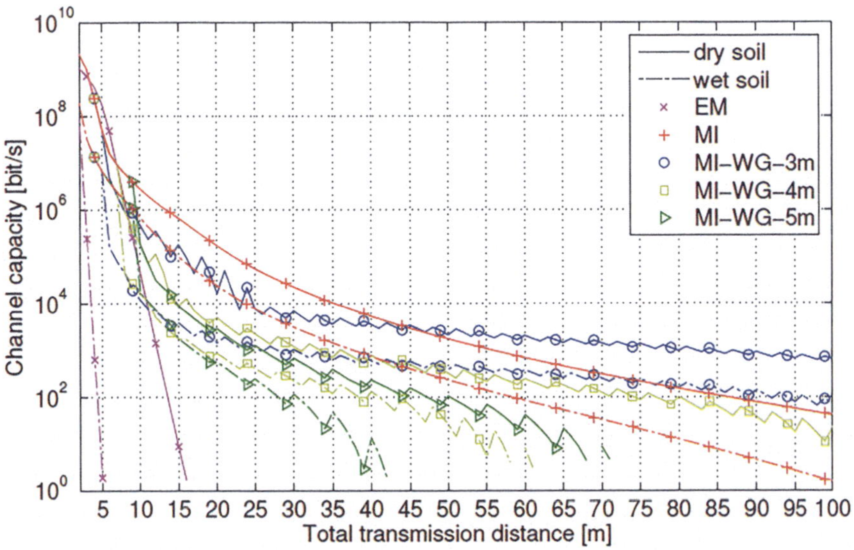

Fig. 5.8 Channel capacity for EM-based and different MI-based WUCs [15]

waveguides for distances of 3, 4, and 5 meters represented by MI-WG-3m, MI-WG-4m, and MI-WG-5m, respectively, in the Fig. 5.8. Here, minimum distance refers to the highest number of relay nodes. It was observed that EM-based transmission is effective for only $d < 7$ in dry soil. For wet soil, EM-based communication performance is always less than the MI-based. Under constrained capacitance the performance of the MI-waveguides suffers badly even less than the direct MI in case of 4 and 5 m. MI-waveguides with 3 m intercoil distance are performing well for the $d > 45$ m. Although MI-waveguide with high relay density, i.e., 3 m intercoil distance, increases the channel capacity (600 bits/s) and communication distance (100 m); however, deployment effort (deploying nodes and adjusting their angles) should also be considered for this case. Therefore, it is a trade-off between the relay efficiency and deployment effort.

5.6 Bit Error Rate (BER) Analysis

Experimental analysis of the BER in wireless underground channel is challenging due to the factors. First, when an electromagnetic wave is incident into soil, the wavelength changes because of the higher permittivity of soil compared to that of air. Soil permittivity depends on soil properties, such as bulk density, soil texture, soil moisture (volumetric water content), salinity, and temperature. Second, when an antenna is buried underground, its return loss property changes due to the high permittivity of the soil. Moreover, with the variation in soil moisture and, hence, soil permittivity, the return loss of the antenna varies too.

The BER statistics performance of channel is used to design error control schemes. For these experiments, the error statistics from [12] and [41] are used. These statistics are important to ascertain the channel behavior under different conditions and are useful in designing error coding schemes. A relation between BER and channel capacity has been established that can obtain capacity of channel in terms of BER obtained by experiments without the knowledge of statistical characterizations. (such as PDF, cumulative distribution function (CDF), moment generating function (MGF), moments, Laguerre moments, etc.)

The following bit error statistics are implemented:

- Bit error rate
- Probability of burst length b, P(b)
- Burst count of each burst length n
- Offset to start symbol
- Positions of bits in error of all symbol
- Number of correctly received symbols
- No of bits in error per symbol
- Average no of bits in error

The experiments are conducted for frequency range of 100–600 MHz using GNU Radio [11] and USRPs [9]. The 500K and 200K channel bandwidth are employed

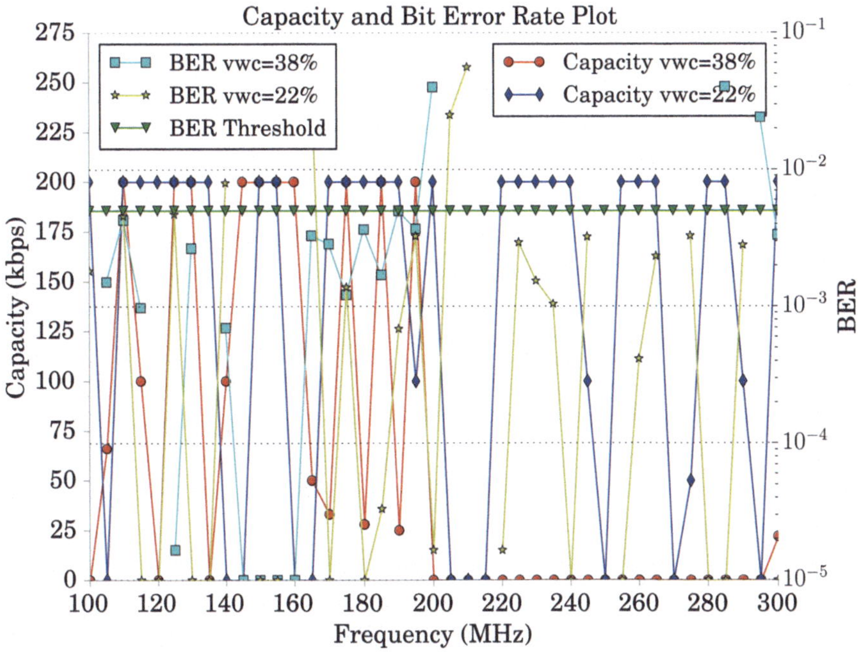

Fig. 5.9 Capacity and BER Plot—200k—different soil moisture values

for these experiments. The dipole antennas are used in these experiments. A series of sequences of 1000 bits long are sent from transmitter and at the receiver side error statistics of the channel are obtained by comparing the output with input. For each frequency, the bit error rate is calculated by adding the bits in error of each correct symbol and then dividing this sum by total number of bits in all correctly decoded symbols.

Figure 5.9 is of two experiments with volumetric water content 38 and 22% respectively. These experimental results verifies two theoretical capacity analysis findings and the strengthen the cognitive radio argument that to achieve capacity the transmitters receivers in underground channel should be able to work in wide range of spectrum:

First, Fig. 5.9 shows that with increase in soil moisture (VWC 38%) optimal frequency is between 100 and 200 MHz and with decrease in soil moisture (VWC = 22%) optimal frequency is between 100 and 300 MHz. As the soil moisture increases, the optimal operation frequency shifts to the lower spectrum. It also verifies that optimal operation frequency is a monotonically decreasing function of soil moisture.

Second, we can also see that for between 200 and 300 MHz, for a fixed frequency capacity decreases with increase in soil moisture.

Third, for the lower spectrum, i.e., operation frequency less than 200 MHz, channel capacity does not change with increase in soil moisture for most of the

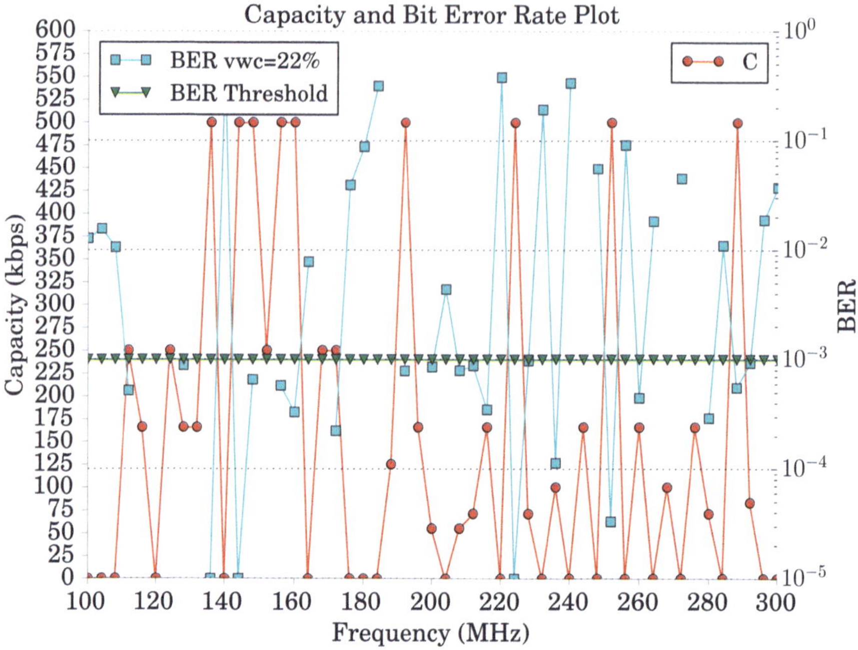

Fig. 5.10 Capacity and BER plot—500k

frequencies, whereas for some frequencies it is less than capacity of low soil moisture case.

Analysis of these plots reveals that higher bandwidth suffers more attenuation in soil and hence results in higher error rates. In 500K experiments for all four depths, bit error rates are very high. For frequencies where symbols were decoded correctly, bit error rate is around 10^{-1} to 10^{-2} (shown by red dots on the graph), for rest of the frequencies where symbols were decoded in error it is shown as 10^0.

Second, increasing depth also increases error rate. For example, for 10 cm depth we decoded symbols in more wider span of frequencies as compared to 40 cm depth where we decoded symbols only in fewer frequencies range. For 10 cm depth fewer symbols were in error than 40 cm depth. Going deep from 10 cm to 40 cm also decreases correctly received bits at higher frequencies.

Here in frequency range of 150–190 MHz error rate is less than 10^{-3}. Frequency range of symbols decoded with less errors is aligned with antenna return loss 120–230 MHz and 490–540 MHz. This underscores the importance of lower bandwidths and wideband planer antennas in wireless underground sensor networks.

Figure 5.10 shows the capacity for 500k rate experiments. It shows that capacity for the range of 200–300 MHz is less than 200k case this is because low BER threshold was used to analyze 500k experiment. It also shows that rate for 140–160 MHz is higher in 500k experiment as compared to 200k case.

Figures 5.11 and 5.12 show the rate for two 250k rate experiments for VWC value of 38%. This matches with the results of 200k experiments.

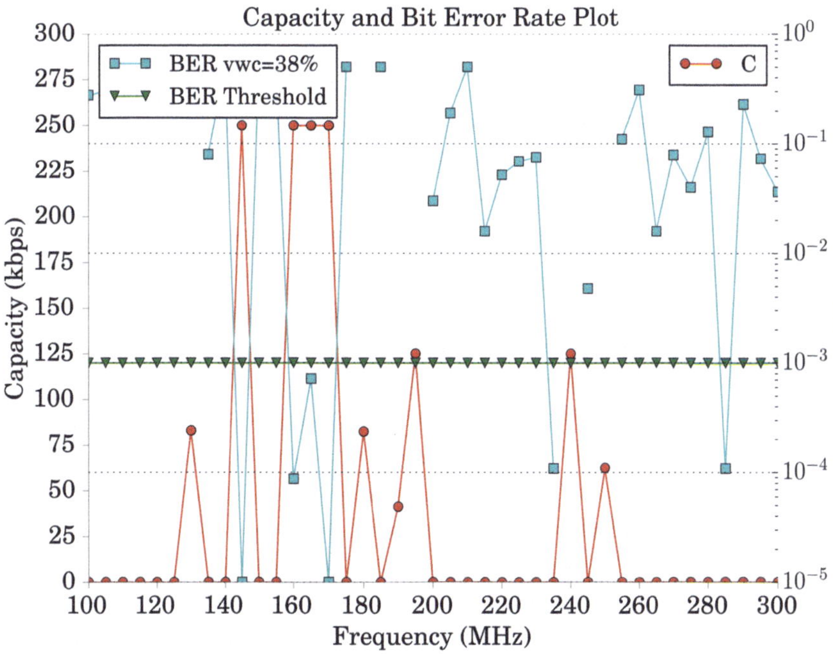

Fig. 5.11 Capacity and BER plot—250k

5.7 Open Research Issues

The BER analysis investigations confirm that lower frequencies are more suitable for wireless underground sensor networks (WUSNs) communication, as higher frequencies exhibit more error rate as compared to lower spectrum. Our BER results also underground channel capacity may be limited up to few hundred kilobits per second. Our BER results also show that the underground channel capacity may be limited up to few hundred kilobits per second.

For future work, we plan to conduct more experiments to continue validating theoretical model with channel sounding and multipath experiments. We also plan to use lower rate and bandwidth to ascertain channel capacity [31, 38]. Our findings also show that antenna return loss also affects underground channel communications. We plan to conduct these experiments with planer antennas. We also plan to characterize the effects of soil moisture, soil properties, and antennas by developing a highly robust, flexible, and configurable testbed in which these parameters, i.e., soil layers, soil moisture, etc., can be controlled efficiently and expediently [37, 39].

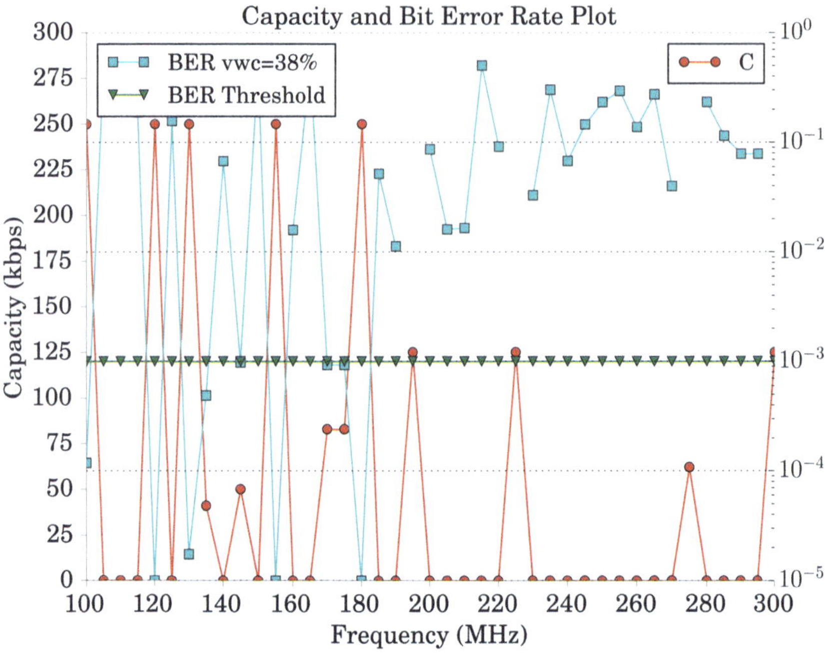

Fig. 5.12 Capacity and BER plot for 250k bandwidth

References

1. Akyildiz IF, Stuntebeck EP (2006) Wireless underground sensor networks: research challenges. Ad Hoc Netw 4(6):669–686
2. Bird TS (2009) Definition and misuse of return loss [Report of the transactions Editor-in-Chief]. IEEE Antennas Propag Mag 51(2):166–167. https://doi.org/10.1109/MAP.2009.5162049
3. Chandra R, Mahajan R, Moscibroda T, Raghavendra R, Bahl P (2008) A case for adapting channel width in wireless networks. SIGCOMM Comput Commun Rev 38(4):135–146. https://doi.org/10.1145/1402946.1402975
4. Cheng R, Verdu S (1993) Gaussian multiaccess channels with ISI: capacity region and multiuser water-filling. IEEE Trans Inf Theory 39(3):773–785. https://doi.org/10.1109/18.256487
5. Cover TM, Thomas JA (2006) Elements of information theory, 2nd edn. Wiley, Hoboken
6. Das S, de Carvalho E, Prasad R (2008) Performance analysis of OFDM systems with adaptive sub carrier bandwidth. IEEE Trans Wireless Commun 7(4):1117–1122. https://doi.org/10.1109/TWC.2008.060761
7. Dong X, Vuran MC (2011) A channel model for wireless underground sensor networks using lateral waves. In: Proceedings of IEEE Globecom'11, Houston
8. Dong X, Vuran MC (2013) Impacts of soil moisture on cognitive radio underground networks. In: 2013 first international black sea conference on communications and networking (BlackSeaCom). IEEE, Piscataway, pp 222–227
9. Ettus Research Website (2016) http://www.ettus.com
10. Friis HT (1944) Noise figures of radio receivers. Proc IRE 32(7):419–422

11. GNU Radio Website (2020) http://www.gnuradio.org
12. Jagath-Kumara KDR, Bebbington M (2005) Error content in frames transmitted over burst-error channels. IEEE Trans Wireless Commun 4(5):2533–2539
13. Kalet I (1989) The multitone channel. IEEE Trans Commun 37(2):119–124. https://doi.org/10.1109/26.20079
14. Karalis A, Joannopoulos JD, Soljačić M (2008) Efficient wireless non-radiative mid-range energy transfer. Ann Phys 323(1):34–48
15. Kisseleff S, Gerstacker W, Schober R, Sun Z, Akyildiz IF (2013) Channel capacity of magnetic induction based wireless underground sensor networks under practical constraints. In: 2013 IEEE wireless communications and networking conference (WCNC). IEEE, Piscataway, pp 2603–2608
16. Kriezis E, Tsiboukis TD, Panas SM, Tegopoulos JA (1992) Eddy currents: theory and applications. Proc IEEE 80(10):1559–1589
17. Ma R, Zhang W (2015) Adaptive MQAM for energy harvesting wireless communications with 1-bit channel feedback. IEEE Trans Wireless Commun 14:6459–6470. https://doi.org/10.1109/TWC.2015.2455494
18. Proakis J, Salehi M (2007) Digital communications, 5th edn. McGraw-Hill, New York
19. Rhee W, Cioffi J (2000) Increase in capacity of multiuser OFDM system using dynamic subchannel allocation. In: Proceedings of IEEE 51st VTC 2000-Spring, Tokyo. https://doi.org/10.1109/VETECS.2000.851292
20. Salam A (2019) A comparison of path loss variations in soil using planar and dipole antennas. In: 2019 IEEE international symposium on antennas and propagation. IEEE, Piscataway
21. Salam A (2019) Design of subsurface phased array antennas for digital agriculture applications. In: Proceedings of 2019 IEEE international symposium on phased array systems and technology (IEEE Array 2019), Waltham
22. Salam A (2019) Subsurface mimo: a beamforming design in internet of underground things for digital agriculture applications. J Sensor Actuator Netw 8(3). https://doi.org/10.3390/jsan8030041. https://www.mdpi.com/2224-2708/8/3/41
23. Salam A (2020) Internet of things for environmental sustainability and climate change. Springer, Cham, pp 33–69. https://doi.org/10.1007/978-3-030-35291-2_2
24. Salam A (2020) Internet of things for sustainability: perspectives in privacy, cybersecurity, and future trends. Springer, Cham, pp 299–327. https://doi.org/10.1007/978-3-030-35291-2_10
25. Salam A (2020) Internet of things for sustainable community development, 1st edn. Springer, Berlin. https://doi.org/10.1007/978-3-030-35291-2
26. Salam A (2020) Internet of things for sustainable community development: introduction and overview. Springer, Cham, pp 1–31. https://doi.org/10.1007/978-3-030-35291-2_1
27. Salam A (2020) Internet of things for sustainable forestry. Springer, Cham, pp 147–181. https://doi.org/10.1007/978-3-030-35291-2_5
28. Salam A (2020) Internet of things for sustainable human health. Springer, Cham, pp 217–242. https://doi.org/10.1007/978-3-030-35291-2_7
29. Salam A (2020) Internet of things for sustainable mining. Springer, Cham, pp 243–271. https://doi.org/10.1007/978-3-030-35291-2_8
30. Salam A (2020) Internet of things for water sustainability. Springer, Cham, pp 113–145. https://doi.org/10.1007/978-3-030-35291-2_4
31. Salam A (2020) Internet of things in agricultural innovation and security. Springer, Cham, pp 71–112. https://doi.org/10.1007/978-3-030-35291-2_3
32. Salam A (2020) Internet of things in sustainable energy systems. Springer, Cham, pp 183–216. https://doi.org/10.1007/978-3-030-35291-2_6
33. Salam A (2020) Internet of things in water management and treatment. Springer, Cham, pp 273–298. https://doi.org/10.1007/978-3-030-35291-2_9
34. Salam A (2020) Wireless underground communications in sewer and stormwater overflow monitoring: radio waves through soil and asphalt medium. Information 11(2):98
35. Salam A, Shah S (2019) Internet of things in smart agriculture: enabling technologies. In: 2019 IEEE 5th world forum on internet of things (WF-IoT). IEEE, Piscataway, pp 692–695

36. Salam A, Vuran MC (2016) Impacts of soil type and moisture on the capacity of multi-carrier modulation in internet of underground things. In: Proceedings of the 25th ICCCN 2016, Waikoloa, Hawaii
37. Salam A, Vuran MC (2017) Smart underground antenna arrays: a soil moisture adaptive beamforming approach. In: Proceedings IEEE INFOCOM 2017, Atlanta
38. Salam A, Vuran MC, Irmak S (2016) Pulses in the sand: impulse response analysis of wireless underground channel. In: The 35th annual IEEE international conference on computer communications (INFOCOM 2016), San Francisco
39. Salam A, Vuran MC, Irmak S (2017) Towards internet of underground things in smart lighting: a statistical model of wireless underground channel. In: Proceedings of 14th IEEE international conference on networking, sensing and control (IEEE ICNSC), Calabria
40. Salam A, Vuran MC, Dong X, Argyropoulos C, Irmak S (2019) A theoretical model of underground dipole antennas for communications in internet of underground things. IEEE Trans Antennas Propag 67:3996–4009
41. Salih O, Wang C, Ai B, Mesleh R (2014) Adaptive generative models for digital wireless channels. IEEE Trans Wireless Commun 13(9):517–5182. https://doi.org/10.1109/TWC.2014.2325028
42. Silva AR, Vuran MC (2009) Empirical evaluation of wireless underground-to-underground communication in wireless underground sensor networks. In: International conference on distributed computing in sensor systems. Springer, Berlin, pp 231–244
43. Singh S, Shahbazi M, Pelechrinis K, Sundaresan K, Krishnamurthy S, Addepalli S (2015) Adaptive sub-carrier level power allocation in OFDMA networks. IEEE Trans Mobile Comput 14(1):28–41. https://doi.org/10.1109/TMC.2014.2312716
44. Spyker RL, Nelms RM (2000) Classical equivalent circuit parameters for a double-layer capacitor. IEEE Trans Aerosp Electron Syst 36(3):829–836
45. Sun Z, Akyildiz I (2010) Magnetic induction communications for wireless underground sensor networks. IEEE Trans Antennas Propag 58(7):2426–2435. https://doi.org/10.1109/TAP.2010.2048858
46. Sun Z, Akyildiz IF (2012) On capacity of magnetic induction-based wireless underground sensor networks. In: 2012 Proceedings IEEE INFOCOM. IEEE, Piscataway, pp 370–378
47. Tse D, Viswanath P (2005) Fundamentals of wireless communication. Cambridge University Press, Cambridge
48. Xhafa A, Tonguz O, Cepni A, Stancil D, Nikitin P, Brodtkorb D (2005) On the capacity limits of HVAC duct channel for high-speed internet access. IEEE Trans Commun 53(2):335–342. https://doi.org/10.1109/TCOMM.2004.841949

Chapter 6
Signals in the Soil: Underground Antennas

6.1 Introduction

Antenna is a major design component of Internet of Underground Things (IOUT) communication system. This chapter focuses on developing a theoretical model for understanding the impact of soil on antenna by conducting experiments in different soil types (silty clay loam, sandy, and silt loam soil) and indoor testbed. Figure 6.1 shows the organizational structure of the chapter. The purpose of the model is to predict UG antenna resonance for designing efficient communication system for IOUT. Based on the model a wideband planar antenna is designed considering soil dispersion and soil–air interface reflection effect which improves the communication range five times from the antennas designed only for the wavelength change in soil [62, 71].

IOUT is being used for implementing many applications [1, 2, 16, 44, 58, 60, 69]. In all these applications, major challenge is to establish a reliable communication. To that end, an underground-specific antenna design challenge is necessary to address. Three major factors that can impact the performance of a buried antenna are: (1) effect of high soil permittivity changes the wavelength of EM waves, (2) variations in soil moisture with time affecting the permittivity of the soil, and (3) difference in how EM waves propagate during aboveground (AG) and underground (UG) communications.

For the third challenge above, it to be noted that lateral waves [26] are dominant component in EM [8, 47, 54, 58] during UG2UG communication and suffers lowest attenuation as compared to other, direct and reflected, components. Therefore, antennas used for over-the-air (OTA) communication will not be suitable for UG communication because of impedance mismatch. The chapter also focuses on developing an impedance model to study the effect of changing wavelength in underground communication. Furthermore, it is discussed how soil–air interface and soil properties effect the return loss of dipole antenna.

A. Salam, U. Raza, *Signals in the Soil*,
https://doi.org/10.1007/978-3-030-50861-6_6

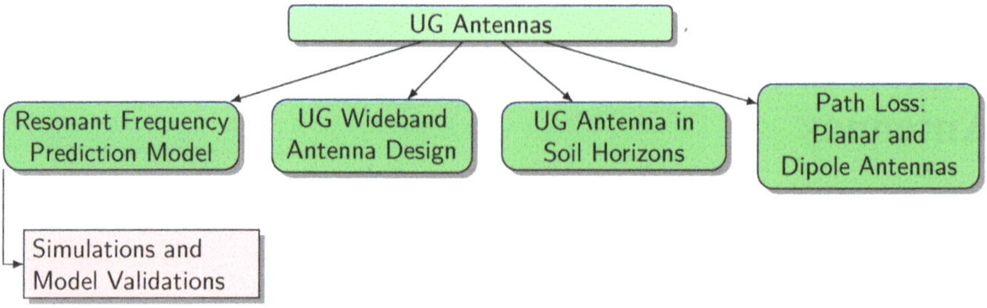

Fig. 6.1 Organization of the chapter

The use of antenna, in IOUT, differs from traditional communication in that it is buried in the soil. There has been lot of work being done to study electromagnetic wave propagation in subsurface stratified media [4–6, 15, 20, 28, 49, 65, 67, 72, 73]. These studies use fields of horizontal infinitesimal dipole of unit electric moment whereas, in practical applications, a finite size antenna is required. This section briefly sheds the light on work already done in the field.

In [28], authors calculate the depth attenuation and ground wave attenuation factor using two vector potentials for UG dipole without considering the impact of soil–air interface on current reflection. Currently, soil permittivity is calculated using soil dielectric model [3, 29, 51] which gives actual wavelength at a given frequency for elliptical planar antenna design in [48, 69]. The size of antenna in [69] is determined by wavelength comparison using the same frequency in air and soil. However, it does not provide the required impedance match. In [39, 50, 74], authors performed experiments for Impulse Radio Ultra-Wide Band (IR-UWB) IOUT without considering the impact of soil–air interface. In [9], circularly polarized patch antenna is analyzed without considering the interface effect. In another study [24, 40], communication between buried antennas is analyzed; however, the impact of orientation is not considered. Similarly, [18, 50] analyzes the performance of four buried antennas in refractory concrete without considering the concrete-air effect.

To the best of our knowledge, there is no study which considers the impact of soil properties while designing the underground antennas. Therefore, rest of the discussion in this chapter is focused on developing a resonant frequency model which is capable of predicting the resonance at different soil moisture levels, soil types, and depths. This information is useful in determining the transmission loss that may occur due to antenna mismatch in IOUT communications. The main focus of the model is to predict resonance, hence, impedance matching is ignored.

6.2 Resonant Frequency Prediction Model

6.2.1 Terminal Impedance and Soil Properties

Antenna Impedance Z_a is defined as a ratio of voltage and current at the input terminal of antenna. Complex power that is being radiated from the antenna can be calculated as by integrating Poynting's vector as given in [19, 47] as:

$$Z_a = \frac{1}{I^2} \int\int E \times H.da, \tag{6.1}$$

where I denotes the antenna current, da is perpendicular in the direction of surface of antenna, and $E \times H$ is energy per unit time. It can be assumed for perfectly conducting antenna that $\mathbf{E}(x, y, z) \equiv 0$, then impedance can be calculated as by integrating surface current density J_{se} and tangential electric field, and Eq. (6.1) becomes [19]:

$$Z_a = \frac{1}{I^2} \int\int E \times J_{se}.da, \tag{6.2}$$

By using the induced EMF method [17], Eq. (6.2) can be rewritten as:

$$Z_a = -\frac{1}{I(0)^2} \int_{-l}^{l} \mathbf{E_z}\, \mathbf{I}(\zeta)\, d\zeta \ , \tag{6.3}$$

The electric field $\mathbf{E_z}$ is used for calculating the self-impedance of UG dipole antenna. $\mathbf{E_z}$ is produced by an assumed current distribution $\mathbf{I(0)}$ and current and electric field are integrated over the antenna surface. Homogeneous soil is considered for the measuring impedance and return loss of the antenna. For a buried dipole antenna, current appears in simple sinusoidal waveform given as:

$$I_0(\zeta) = I_m \sin[k_s(l - |\zeta|)] \ , \tag{6.4}$$

where I_m is the current amplitude, k_s represents complex wave number of the soil, l is the half length of the antenna, and $k_s = \beta_s + i\alpha_s = \omega\sqrt{\mu_0 \hat{\epsilon}_s}$ is the wave number in soil. $\mathbf{E_z}$ is given as:

$$\mathbf{E_z} = -\int_{-l}^{l} \frac{1}{4\pi j\omega\epsilon_s} \frac{e^{-jk_s r}}{R} \left(\frac{\partial^2}{\partial\zeta^2} + k_s^2 \right) I(\zeta) d\zeta, \tag{6.5}$$

By substituting the $\mathbf{E_z}$ in Eq. (6.5) and $\mathbf{I(0)}$ from Eq. (6.4) in Eq. (6.2) we get [23, 51]:

$$Z_a \approx f_1(\beta l) - i\left(120\left(\ln\frac{2l}{d} - 1\right)\cot(\beta l) - f_2(\beta l)\right), \tag{6.6}$$

where

$$f_1(\beta_s l) = -0.4787 + 7.3246\beta_s l + 0.3963(\beta_s l)^2 + 15.6131(\beta_s l)^3 \tag{6.7}$$

$$f_2(\beta_s l) = -0.4456 + 17.0082\beta_s l - 8.6793(\beta_s l)^2 + 9.6031(\beta_s l)^3. \tag{6.8}$$

β_s is the real part of the wave number k_s, d is the diameter of the dipole, and l is half of the length of the dipole. βl is expressed as

$$\beta_s l = \frac{2\pi l}{\lambda_0}\mathrm{Re}\left\{\sqrt{\epsilon_s}\right\}, \tag{6.9}$$

where ϵ_s is the relative permittivity of soil and λ_0 is the wavelength in air. Since the ϵ_s is dependent on frequency, βl is not a linear function of l/λ_0. Therefore, when the medium is changed from soil to air, both, resonant frequency and impedance at the resonant frequency of the antenna, also change.

Practical IOUT has motes deployed at 0.3–1 m [12, 44] and there is high impact of soil–air interface at these depths, hence, environment cannot be modeled. Next, the environment is modeled to study the impact on antenna return loss and impedance due to reflection of waves by soil–air interface. Upon excitement of antenna, a current distribution of $I_0(\zeta)$ is generated and propagated wave is either reflected or refracted from soil–air interface. E_r and I_r are the reflected electric field and induced current, respectively, that reaches the antenna.

I_r and Z_r, resulting impedance are modeled due to field generated from imaginary dipole in homogeneous soil. As current distribution (6.4), E_r field reflected due to the soil–air interface at the antenna is [17, 45]:

$$E_r = -i30I_m\left(\frac{e^{-ik_s r_1}}{r_1} + \frac{e^{-ik_s r_2}}{r_2} - 2\cos k_s l\frac{e^{-ik_s r}}{r}\right) \times \Gamma, \tag{6.10}$$

where

$$r = [(2h)^2 + \zeta^2]^{1/2}, \tag{6.11}$$

$$r_1 = [(2h)^2 + (\zeta - l)^2]^{1/2}, \tag{6.12}$$

$$r_2 = [(2h)^2 + (\zeta + l)^2]^{1/2}, \tag{6.13}$$

h represents the burial depth of the antenna, and reflection coefficient at the soil–air interface Γ is measured as:

$$\Gamma = \frac{2}{1 + k_0/k_s} - 1 = \frac{2}{1 + \sqrt{\frac{1}{\epsilon_s}}} - 1 , \tag{6.14}$$

and k_0 is the wave number in air.

The antenna impedance is given as: $Z_a^u = Z_a.\frac{I_0}{I_r^2}$ and from this impedance values the return loss of the antenna is given as:

$$RL_{dB} = 20 \log_{10} \left| \frac{Z_s + Z_a^u}{Z_s - Z_a^u} \right| . \tag{6.15}$$

The reflection coefficient Γ is given as: $|\Gamma| = 10^{\frac{RL}{20}}$. Reflection coefficient is transformed to impedance by using: $Z_a^u = Z_s \frac{1+\Gamma}{1-\Gamma}$. Standing wave ratio (SWR) is expressed as: $SWR = \frac{1+|\Gamma|}{1-|\Gamma|}$

6.2.2 Defining Resonant Frequency

The frequency where the antenna's input impedance is pure resistance is known as resonant frequency f_r. i.e.,

$$Z_a^u|_{f=f_r} = Z_r = R_a \tag{6.16}$$

and where return loss is maximum such that:

$$f_r = \max(RL_{dB}). \tag{6.17}$$

A comparative performance analysis is done between this analytical model with resonant frequency of permittivity-based antenna by using: $f_r = f_0/\sqrt{\epsilon_s}$, where f_0 represents an OTA resonant frequency, and ϵ_s is the permittivity of the soil.

6.2.3 Bandwidth Expression

It is very difficult to find a closed-form bandwidth formula for the UG antenna because of involvement of many soil and antenna factors; however, a resonant frequency-based bandwidth expression (BW) can be calculated as [13]:

$$BW = \begin{cases} 0 & \text{if } -RL_{dB}(f) > \delta, \\ 2(f - f_m) & \text{if } -RL_{dB}(f) \leq \delta \text{ and } f < f_r, \\ 2(f_M - f) & \text{if } -RL_{dB}(f) \leq \delta \text{ and } f \geq f_r, \end{cases} \tag{6.18}$$

where f_r is the resonant frequency, f_m and f_M are the lowest and highest frequency at which $RL_{dB}(f) \leq \delta$. There is no fixed value of δ; however, value of 10 dB is generally used in the literature [7].

6.3 Simulations and Experiment Setup

Following simulation setup was used to analyze the performance of underground dipole antenna: CST Microwave Studio Suite (MWS) [11], an indoor testbed without changing the soil parameters. Simulations are conducted with antenna buried 20 cm inside the soil, and distances of 5–12 m from the first antenna. The results from this testbed were compared with outdoor testbed with dipole antenna in silty clay loam soil. Vector Network Analyzer (VNA) is used for measuring antenna S_{11} and channel responses to frequency.

6.4 Model Validations

6.4.1 Model, Simulation, and Empirical Results

Figures 6.2b, a, and c compares theoretical, measure, and simulated return loss at 20 cm of depth in silt loam, sandy, and silt clay soil type, respectively. It can be seen that all three results (measured, theoretical, and simulated) confirm each other with minor differences. For example, at resonant frequency, for silt loam soil, the measured and model return losses match whereas simulated return loss differs by 7% and this difference drops to 1% for sandy soil. This 1–7% difference is caused because of the uncertainties in the soil simulation software.

Figure 6.3 compares the resonant frequency and bandwidth from measured experiment and theoretical model for 20% VWC. The results are for sandy (Fig. 6.3a, b) and silt loam soil (Fig. 6.3c, d) at varying depths ranging from 10–40 cm. For sandy soil, both, measured resonant frequency and bandwidth, match the model value with minor difference of 0.01–1.93% in resonant frequency and 2.77–4 MHz in bandwidth. For silt loam soil, both, measured resonant frequency and bandwidth, matching trend between both models is same with minor difference of 1.01–3.53% in resonant frequency and 1–8 MHz in bandwidth. These differences between the models are because of change in return loss and resonant frequency at some particular depth which leads to difference in bandwidth. However, these variations do not effect the UG communication as antenna bandwidth is higher than these variations [46, 54, 68].

Other reasons for differences in model could be: (1) abrupt phase changes of waves while transition from one depth to other and due to soil–air interface impact, and (2) theoretical model do not consider the EM waves propagation effect in

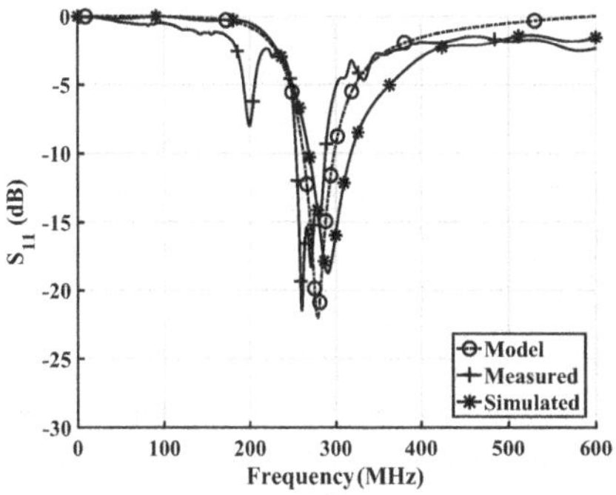

(a)

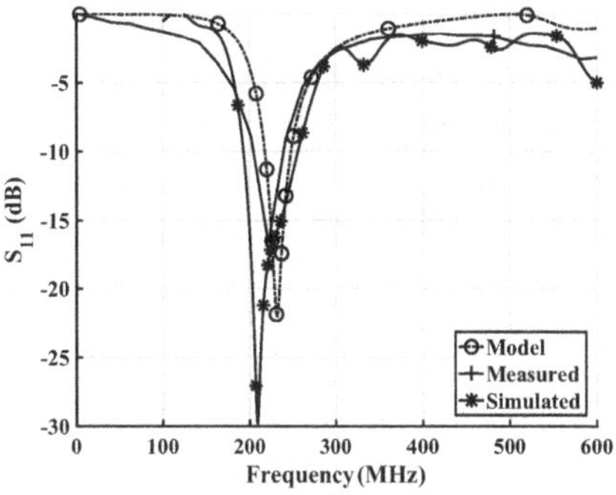

(b)

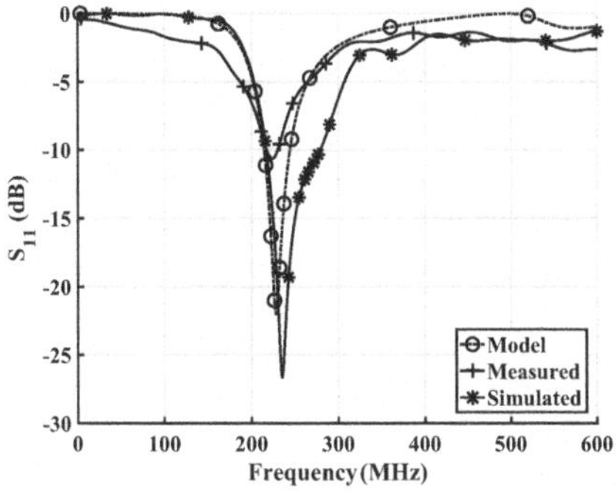

(c)

Fig. 6.2 Comparative analysis of return loss estimated from simulated, theoretical, measured experiments in [62]: (**a**)sandy soil. (**b**) Silt loam soil. (**c**) Silty clay loam soil

coaxial cable connected to antenna. Overall the resonant frequency matched with the model matched each other and comparing measurements with theoretical model makes it a powerful tool to analyze the underground antenna.

6.4.2 Analysis of Impact of Operation Frequency

Figure 6.5 plots the resonant frequency and return loss for 5–40% soil moisture level. The results are for sandy (Fig. 6.5a, b) and silt loam soil (Fig. 6.5c, d). Resonant frequency decreases 62% (from 369–137 MHz) for silt loam soil, and decreases 59% (from 357–146 MHz) for sandy soil [32, 34].

Resonant frequency, of a dipole antenna, in soil and OTA is represented by f_{rs} and f_{ro}, respectively. Figure 6.4 compares the ratio $\frac{f_{rs}}{f_{ro}}$ and antenna permittivity 433 and 915 MHz. The results are for sandy (Fig. 6.4b, d) and silt loam soil (Fig. 6.4a, c) at varying depths ranging from 10–40 cm. At different depths, change in resonant

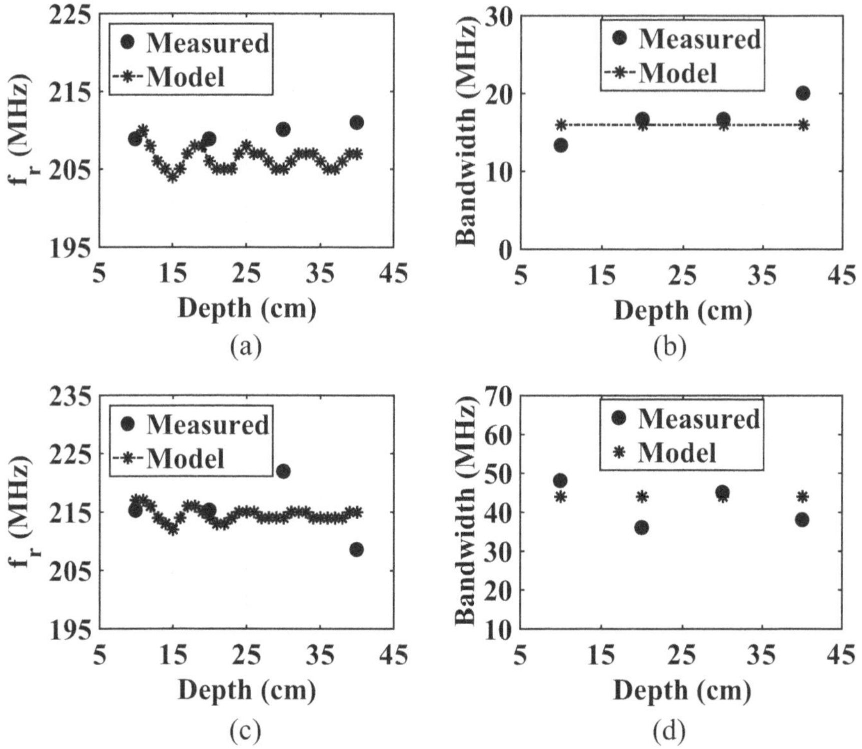

Fig. 6.3 Comparative analysis of theoretical and measured experiments at different depths for [62]: (**a**) Resonant frequency (sandy), (**b**) Bandwidth (sandy), (**c**) Resonant frequency (silt loam), and (**d**) Bandwidth (silt loam soil)

frequency difference is different , and ratio is also varying as compared to the OTA [42] (Fig. 6.5).

The difference is clear in Fig. 6.6 where difference in resonant frequency δ of theoretical model and antenna based on soil permittivity only, δ, is shown with varying soil moisture levels at 433 and 915 MHz. The results are for sandy (Fig. 6.6b, d) and silt loam soil (Fig. 6.6a, c) at varying depths ranging from 10–40 cm. It can be seen that δ is inversely proportional to soil moisture level. For example, δ increase by 10–15 MHz when frequency goes from 433–915 MHz which proves that only permittivity-based IOUT system suffers performance degradation and highlights the importance of considering impact of soil–air interface. Hence, consideration of burial depth is important for efficient IOUT communication system [41, 50] (Fig. 6.7).

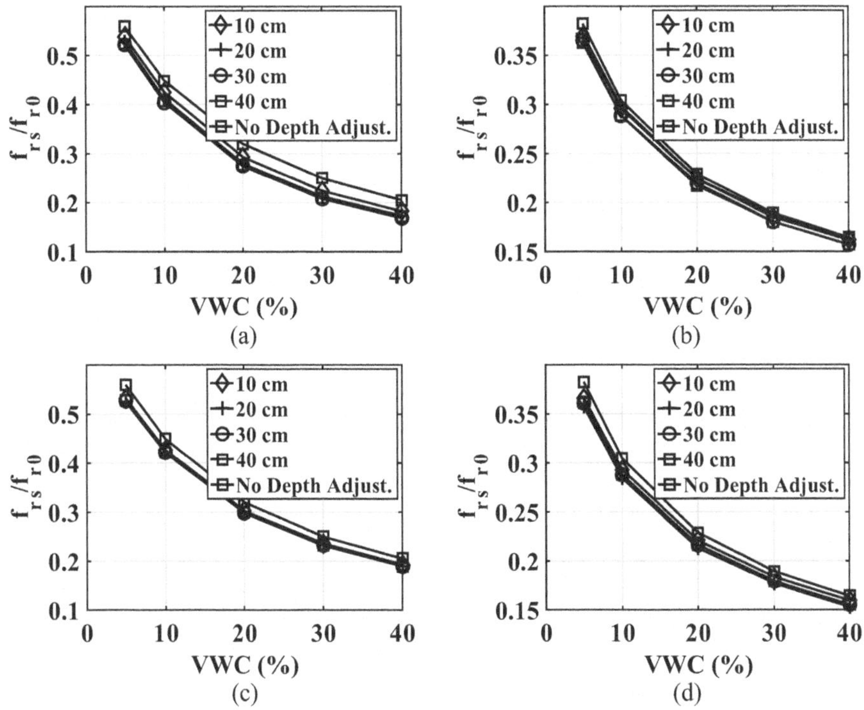

Fig. 6.4 Effect of VWC on ratio of resonant frequency in soil and OTA in [62]: (**a**) Silty Clay Loam Soil at 433 MHz, (**b**) Sandy soil at 433 MHz, (**c**) Silty Clay Loam Soil at 915 MHz, and (**d**) Sandy Soil at 915 MHz

6.5 Underground Wideband Antenna Design

To compensate for the shift of resonant frequency of UG dipole antenna, wideband antennas of different sizes and 433 MHz frequency are designed and fabricated for testing.

1. ***Radiation Pattern in UG Communications*:** The radiation pattern of the antenna is an added advantage for using this antenna. Out of three paths [26, 38] (direct wave, reflected wave, and lateral wave) in UG communication, lateral wave is the most dominant in far-field [12, 49, 52, 66]. Therefore, radiation pattern must have maximum lateral wave component. [26, 43, 66] shows that lateral wave only occurs when incident angle is at critical angle θ_c. θ_c changes with the varying soil moisture and is less than 15° in all soil moisture settings. The radiation pattern is unidirectional towards soil–air interface, thus, desirable radiation pattern can be achieved if antennas are placed parallel to soil–air interference.
2. ***The Return Loss*:** Figs. 6.8 and 6.9 shows the return loss and bandwidth at varying depths (0.13, 0.3, and 0.4 m) for three different soil moisture values (10,

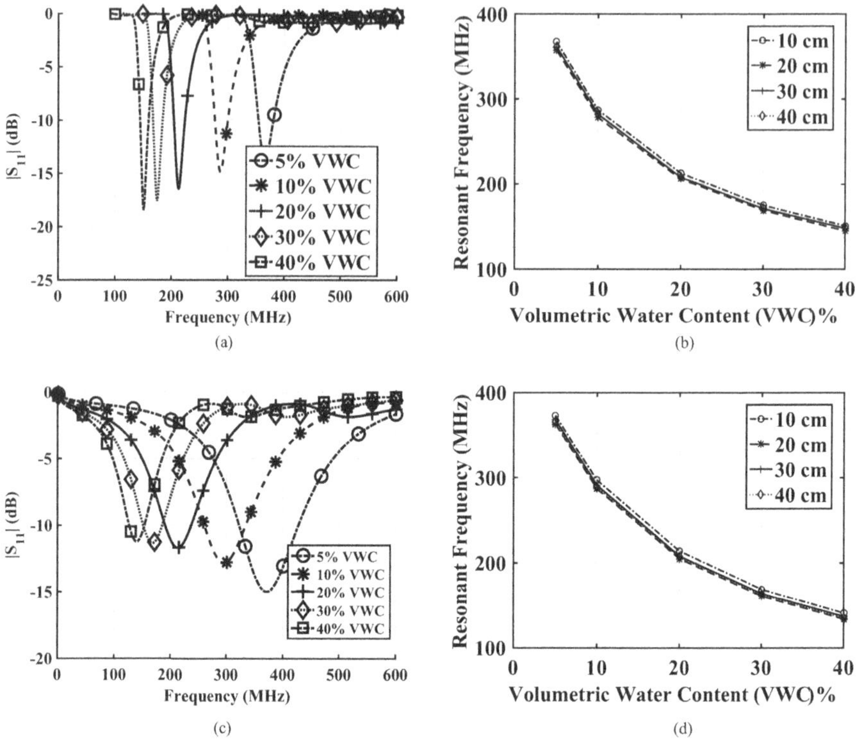

Fig. 6.5 Theoretical results [62]: (**a**) Return loss in sandy soil, (**b**) Resonant frequency in sandy soil, (**c**) Return loss in silt loam soil, and (**d**) Resonant frequency in silt loam soil

30, and 40%). The resonant frequency varies in all these scenarios, however, return loss remains below 10 dB for all depths and moisture levels [35].

3. ***Communication Results***: The designed planar antenna is compared with 25 mm wideband antenna and elliptical antenna in testbed to evaluate the performance for underground–aboveground communications. Two motes are used for UG and AG with planar and Yagi antenna, respectively, to accomplish UG2AG and AG2UG channel communication [37]. Figure 6.10 plots the received signal strength (RSS) with changing distance. It shows that although the communication range of 200m is achieved but practical multi-hop connectivity is still limited in underground communication. For UG2AG channel, the designed antenna increases the communication range by 587.5% as compared to elliptical antenna (from 8 to 55 m) and 223.5% as compared to circular antenna (from 17 to 55 m). For UG2AG channel, the designed antenna increases the communication range by 587.5% as compared to elliptical antenna (from 8 to 55 m) and 266.7% as compared to circular antenna (from 15 to 55 m) [36]. These results show that designing an antenna specific for UG environment is critical for IOUT system.

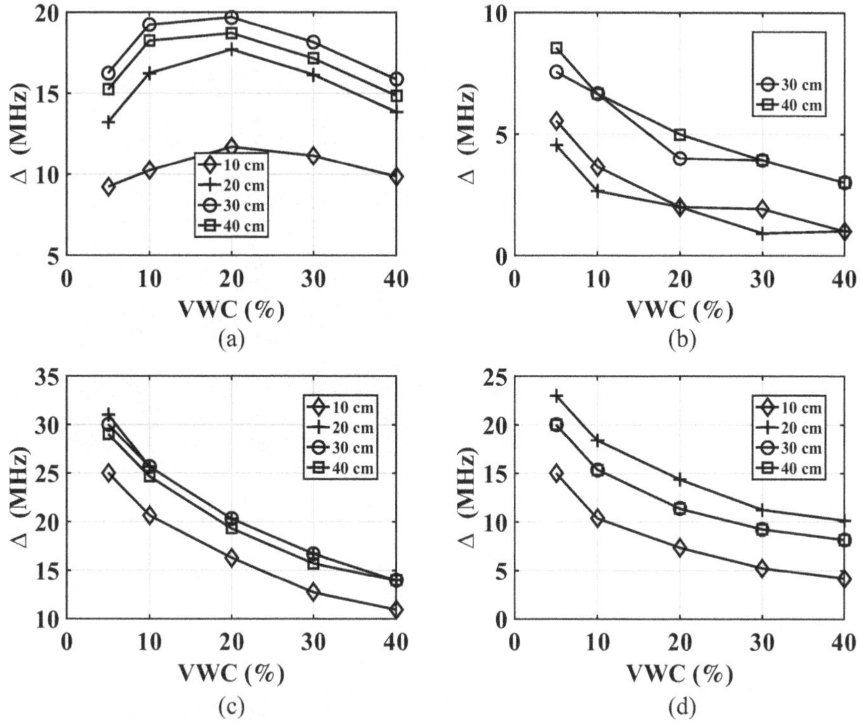

Fig. 6.6 Δ v/s VWC [62]: (**a**) Silty clay loam soil at 433 MHz. (**b**) Sandy soil at 433 MHz. (**c**) Silty clay loam soil at 915 MHz, and (**d**) Sandy soil at 915 MHz

6.6 Underground Antenna in Soil Horizons

Precision agriculture is the practice of accurately capturing the changing parameters of the soil including water infiltration and retention, nutrients supply, acidity, and other time changing phenomena by using the modern technologies. Using precision agriculture, fields can be irrigated more efficiently hence conserving water resources and increasing productivity. Wireless underground sensor networks (WUSN) are being used to monitor the soil for smart irrigation. Communication in wireless underground sensor networks is affected by soil characteristics such as soil texture, volumetric water content (VWC), and bulk density. These soil characteristics vary with soil type and soil horizons within the soil. In this section we have investigated the effects of these characteristics by considering Holdrege soil series and homogeneous soil. It is shown that consideration of soil characteristics of different soil horizons leads to (5–6 dB) improved communication in wireless underground sensor networks [33, 61].

Horizons are layers of soil which are formed by four soil processes and have unique chemical, physical, and visible characteristics. These soil processes are

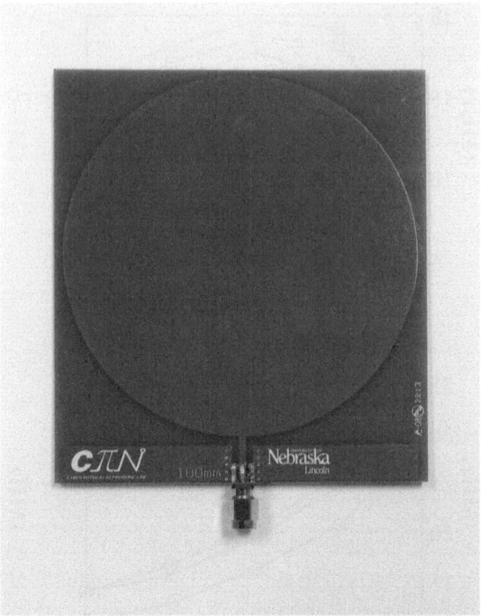

Fig. 6.7 UG wideband planar antenna [62]

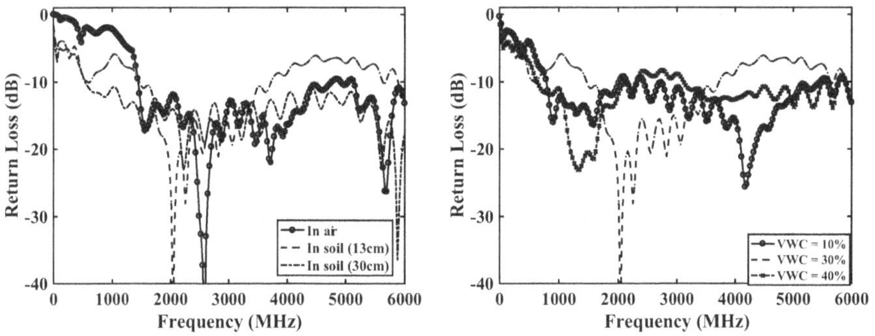

(a) (b)

Fig. 6.8 Return loss using wideband planar antenna [62]: (**a**) Depths, and (**b**) Volumetric water content

additions, losses, transformations, and translocation. There are five horizons: O, A, E, B, and C. In soil, these horizons can form in any order. Some soils do not contain all horizons and in some soils multiple horizons can repeat. Horizons A, B are of most interest because of their high impact on plant growth.

In wireless underground sensor networks sensor nodes are buried in soil. Establishment of wireless communication links is important for data communication. Each of the soil horizon have unique soil texture, bulk density and water holding capability. Also depth and width of each horizon differ in different type of soils. These factors have a significant influence on the performance of a buried antenna and communication. In [59], impact of these soil factors on underground communication is analyzed and given as follows:

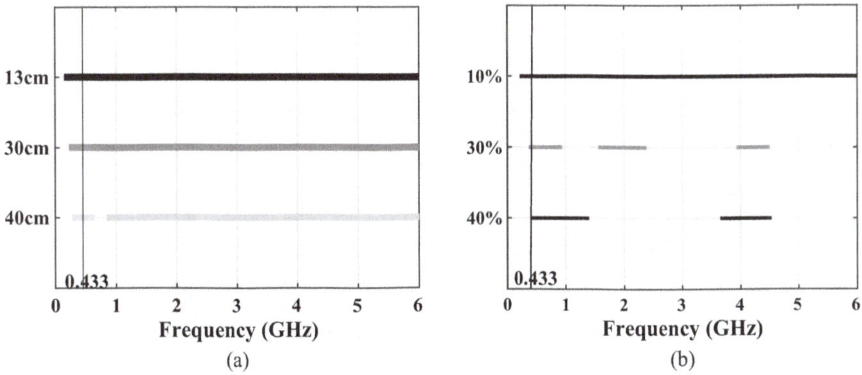

Fig. 6.9 Bandwidth of wideband planar antenna (100 mm) at (**a**) varying Depths and (**b**) varying soil moisture [62]

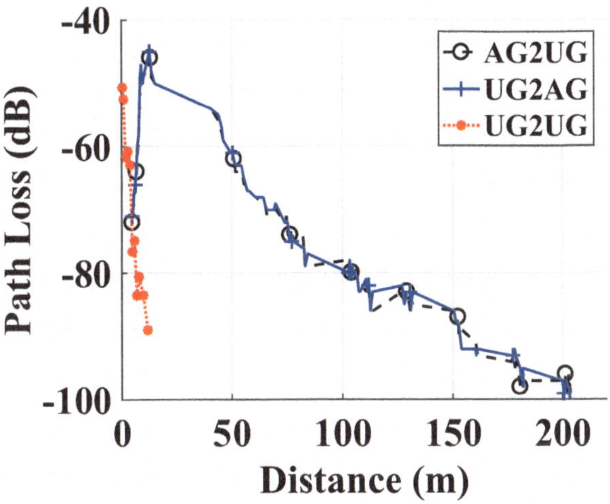

Fig. 6.10 Path loss with varying distance for different communication links [62]

Soil Moisture Soil moisture changes with time due to climate and irrigation, which influences the soil permittivity.

Soil permittivity Electromagnetic waves propagation in soil exhibit different characteristics in soil due to higher permittivity of soil.

Soil–Air Interface Impedance of underground antenna is changed because of current disturbance at antenna due to reflection from soil–air interface [30, 64, 69].

In this section, by using our model for underground to underground (UG2UG) communications [53], we have analyzed the performance of wireless underground channel by using Holdrege soil profile [10] and homogeneous soil. Moreover, we provide analytical results for path loss for three different scenarios including same soil moisture level across all horizons, water infiltration and water retention scenario.

Fig. 6.11 Holdredge soil profile

Based on the analysis it is shown that antennas buried into soil horizons by taking soil characteristics into account experience less path loss as compared to antenna buried in homogeneous soil and path loss is decreased from 5–6 dB. It is also shown that path loss varies with soil moisture and increase in soil moisture also increases the path loss for all type of soils. It is also evident that in underground wireless sensor networks path loss increases with frequency therefore low operation frequencies are suitable for wireless underground communication (Fig. 6.11).

Table 6.1 Holdrege soil—physical characteristics

Horizon	Depth in inches	Sand	Slit	Clay	Textual class
Ap	0–7	16.6	61.4	22.0	Silt loam
A	7–13	12.0	58.4	29.6	Silt clay loam
Bt1	13–16	13.3	55.3	31.4	Silt clay loam
Bt2	16–24	11.2	58.9	29.9	Silt clay loam

To get a wavelength in soil at a given frequency, soil permittivity is calculated using the dielectric model [3, 29, 56, 60]. This wavelength calculated by the dielectric gives insight into design of antenna [37, 69]. For underground communications antennas are buried in different depths in soil. Theoretical analysis EM field of antennas in infinite dissipative medium is presented in [25, 31, 39, 63]. Return loss of the antenna is not considered in this analysis. Measurements of dipole antenna in solution are presented in [22]. Because of the difference in permittivity of soil and permittivity solution this work is not applicable to wireless underground sensor networks. Current disturbance at antenna due to reflection from soil–air interface is mentioned in [21] but its impact is not analyzed. In [62] we have analyzed these impacts on underground antenna using homogeneous soil. We have also developed a three-wave channel model for wireless underground communications in [12, 14, 34].

6.6.1 Holdrege Soil Characteristics

We have used Holdrege soil and homogeneous soil for our analysis. Table 6.1 shows physical properties of Holdrege soil.

We have selected Holdrege series because it is one of the well-drained, highly productive, and most fertile soil in the Nebraska, United States. It is also official state soil of Nebraska and almost all the soil is under cultivation. As per United States Department of Agriculture [10]:

Prairie environment has contributed to formation of horizontal layers in profile of Holdrege series. Clay and lime particles have moved downward in profile due to drainage of water inside the profile. Due to interaction of these processes there is thick, dark color topsoil, a clay enriched subsoil, and a substratum containing free lime. Holdrege soil is very well irrigated and is an extensively cultivated soil. Corn and soy are the main crops.

6.6.2 Numerical Analysis

We have considered three cases for analytical evaluation. In first case we have compared the two soils under the same soil moisture case for all soil horizons and

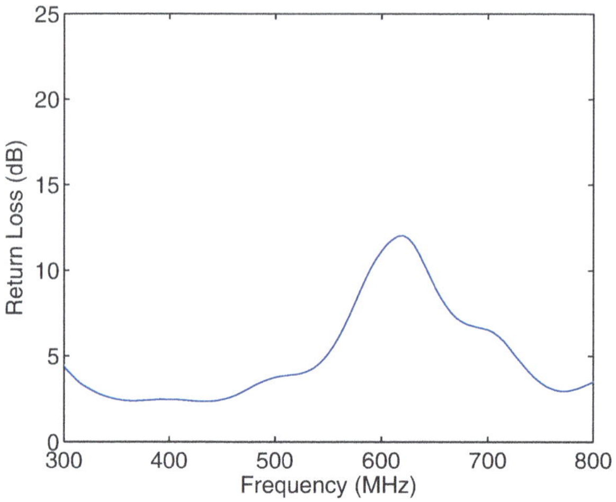

Fig. 6.12 Return loss of the antenna

depths. In second case we analyze the water infiltration scenario in which topsoil horizons have more water content than the subsoil horizons. Third case compares the water retention scenario in which subsoil is more saturated as compared to the topsoil. We have used frequency range of 300–800. Transmitted power is 15 dBm. Return Loss of the antenna used in the evaluation is shown in Fig. 6.12. Antennas are buried at four depths. Four antenna burial depth corresponds to four different horizons (Ap, A, Bt1, Bt2) of Holdrege soil as shown in Table 6.1. For homogeneous soil these are 10, 20, 30, and 40 cm. Horizontal distance between transmitter receiver is 50 cm. Bulk density is 1.5 g/cm^3 and particle density is 2.66 g/cm^3 (Fig. 6.12).

6.6.3 Same Soil Moisture Scenario

Figure 6.13 shows the path loss for two soil types for Volumetric Water Content (VWC) of value of 10%. For all depths and across all frequency range Path loss for homogeneous soil is 5–6 dB higher than as compared to Holdrege soil. Moreover between 550 and 650 MHz range path loss is low because of the low return loss of the antenna. It is also clear that path loss increases with frequency.

Figure 6.14 shows the path loss for two soil types for Volumetric Water Content (VWC) of value of 20%. For all depths and across all frequency range Path loss for homogeneous soil is 5–6 dB higher than as compared to Holdrege soil. Due to 10% increase in water content there is an increase of 8 dB for all horizons.

Figures 6.15 and 6.16 shows the path loss for two soil types for Volumetric Water Content (VWC) of value of 30 and 40%. For both soil moisture levels, for all depths and across all frequency range path loss for homogeneous soil is 5–6 dB increased

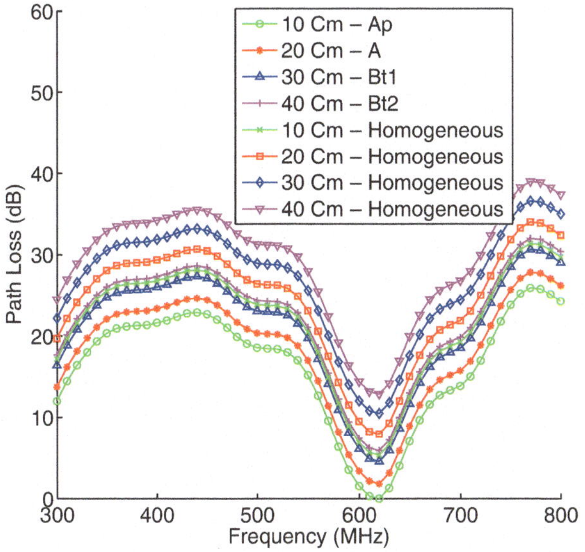

Fig. 6.13 Path loss vs. frequency—VWC 10%

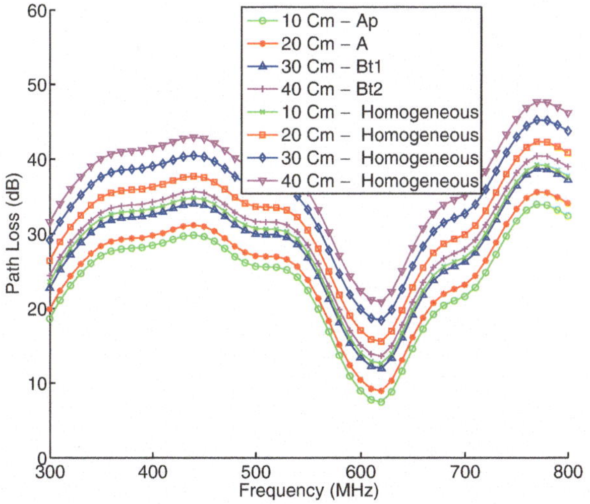

Fig. 6.14 Path loss vs. frequency—VWC 20%

as compared to Holdrege soil. Path loss for 30 and 40% is considerably higher than dry than the 10%.

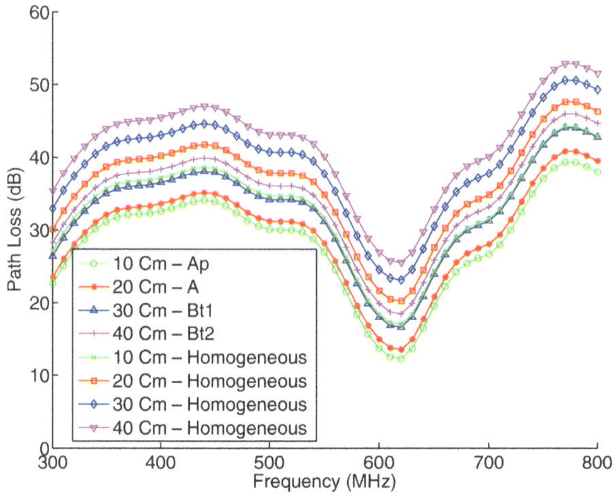

Fig. 6.15 Path loss vs. frequency—VWC 30%

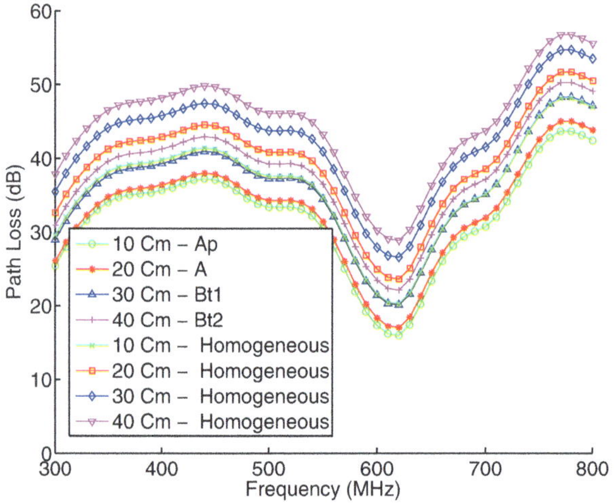

Fig. 6.16 Path loss vs. frequency—VWC 40%

6.6.4 Water Infiltration Scenario

In this case we consider the scenario in which higher horizons have more water content as compared to lower soil horizons. Figure 6.17 shows the path loss when Ap horizon have 40% VWC, A horizon have 30% VWC, Bt1 have 20% VWC, and Bt2 have 10% VWC. It is evident that communication performance is best at Bt2 horizon because of low water content.

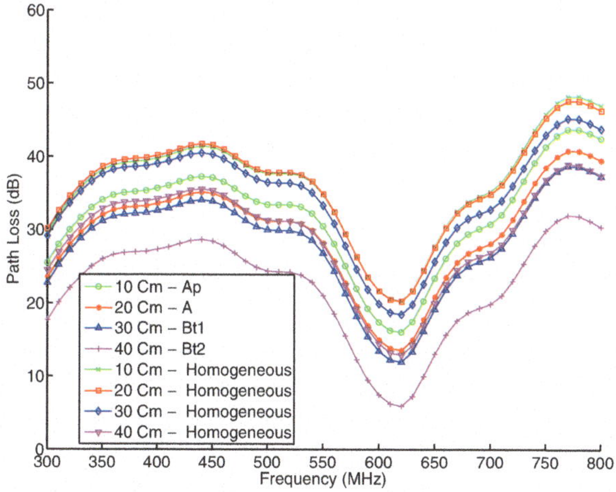

Fig. 6.17 Path loss vs. frequency—water infiltration scenario%

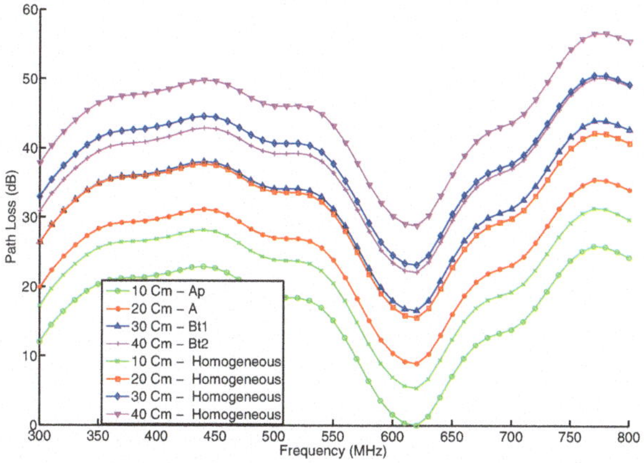

Fig. 6.18 Path loss vs. frequency—drainage scenario%

6.6.5 Water Retention Scenario

In this case we consider the scenario in which lower horizons have more water content as compared to higher soil horizons. Figure 6.18 shows the path loss when Ap horizon have 10% VWC, A horizon have 20% VWC, Bt1 have 30% VWC and Bt2 have 40% VWC. Antenna buried at the A horizon experiences lower path loss because of low attenuation due to lower VWC.

In this section, the impacts of soil texture, soil moisture on burial depth of antenna in different soil horizons, and on path loss are analyzed for underground wireless communications in Holdrege soil and homogeneous soil. It is shown that antennas buried into soil horizons by taking soil characteristics into account experience less path loss as compared to antenna buried in homogeneous soil. It is also shown that path loss varies with soil moisture and increase in soil moisture also increases the path loss for all type of soils. It is also evident that in underground wireless sensor networks path loss increases with frequency therefore low operation frequencies are suitable for wireless underground communication.

6.7 Path Loss Variations with Planar and Dipole Antennas

The digital agriculture [45, 55, 60, 63, 70] is the area in which technology is used to effectively manage agriculture by understanding the temporal and spatial changes in soil, crop, production, and management through innovative techniques. The analysis of the communication path loss is vital for an efficient communication system design in sensor-guided irrigation management system. To investigate propagation loss variations, the path loss experiments are conducted in sandy soil testbed, and greenhouse outdoor silty clay loam testbed using a wideband planar antenna [57, 62] and dipole antennas.

6.7.1 Experiment Setup

In a sandy soil testbed [59, 62], two planar antennas are buried at 20 cm depth at a distance of 1 m. The return loss and path loss measurements are taken. To analyze the effects of a planar in the middle of two planar, obstructing the communications, another planar antenna is buried in the middle at 50 cm distance and same depth (20 cm). Accordingly, the path loss and return loss measurements are taken again for 50 cm distance and 1 m distance [27, 55].

In the greenhouse, another testbed of planar antennas is commissioned in silty clay loam soil. To compare the results of the experiment with sandy soil testbed, same empirical parameters are used. First, the path loss and return loss measurements are taken for planar buried at 1 m distance at 20 cm depth. Afterward, another planar is installed at 50 cm distance and 20 cm depth, and return loss and path loss measurements are taken, again, first for 1 m distance and then for 50 cm distance [70, 71].

To compare the results of planar antennas with dipole antenna, a testbed of dipole antennas is developed outside of the greenhouse in silty clay loam soil. In this testbed, three dipole antennas are buried in soil at 50 cm distance each and burial depth is 20 cm. The physical properties of sandy soil and silty clay loam soil are shown in Table 6.2. The results of this empirical campaign are presented in Sect. 6.7.2. The return loss of dipole and planar antennas is shown in Fig. 6.19. The comparison of dipole and planar return loss in same soil is given in Fig. 6.20.

Table 6.2 Soil used in testbeds—physical characteristics

Textural class	Sand%	Silt%	Clay%
Silty clay loam	13	55	32
Sandy soil	86	11	3
Silt loam	33	51	16

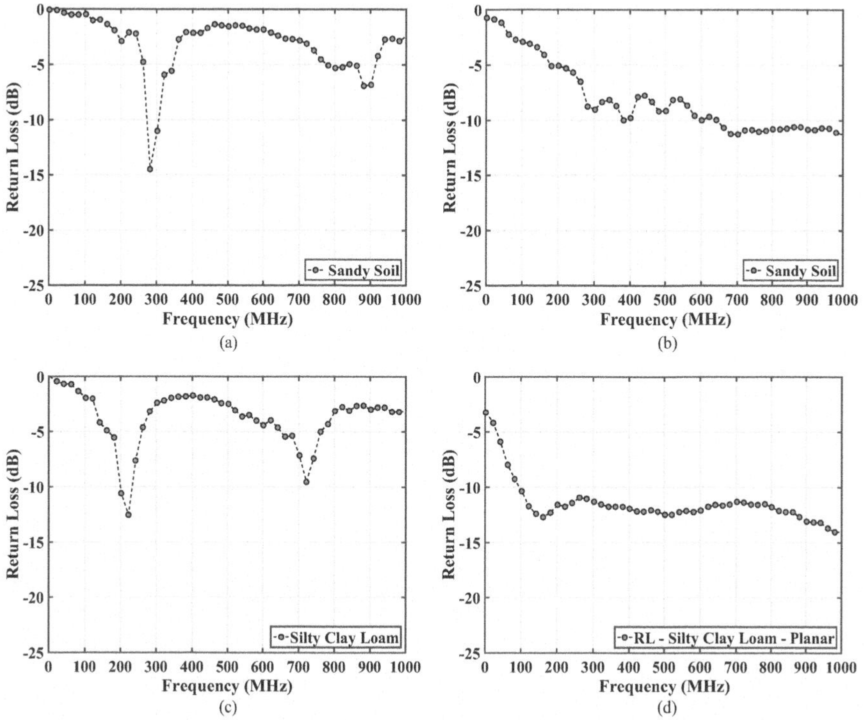

Fig. 6.19 Return loss: (**a**) sandy soil with dipole antenna. (**b**) sandy soil with planar antenna. (**c**) silty clay with dipole antenna. (**d**) silty clay with loam planar antenna

6.7.2 Results

The planar antenna path loss at 50 and 100 cm in sandy soil and silty clay loam testbed is shown in Fig. 6.21a, b, respectively. In sandy soil, there is 14 dB difference in path loss when communication distance is increased from 50 to 100 cm. Similarly, in silty clay loam soil, at frequencies higher than 500 MHz path loss is increased from 19 dB [32, 53].

In Fig. 6.20c, the path loss comparison of dipole and planar antenna is shown in sandy soil testbed at 50 cm. The variations in path loss with change in frequency, present in the case of dipole antenna, are not observed when measurements are taken using planar antenna. Similarly in Fig. 6.20d, the path loss comparison of dipole and planar antenna is shown in silty clay loam testbed at 50 cm. As observed in sandy

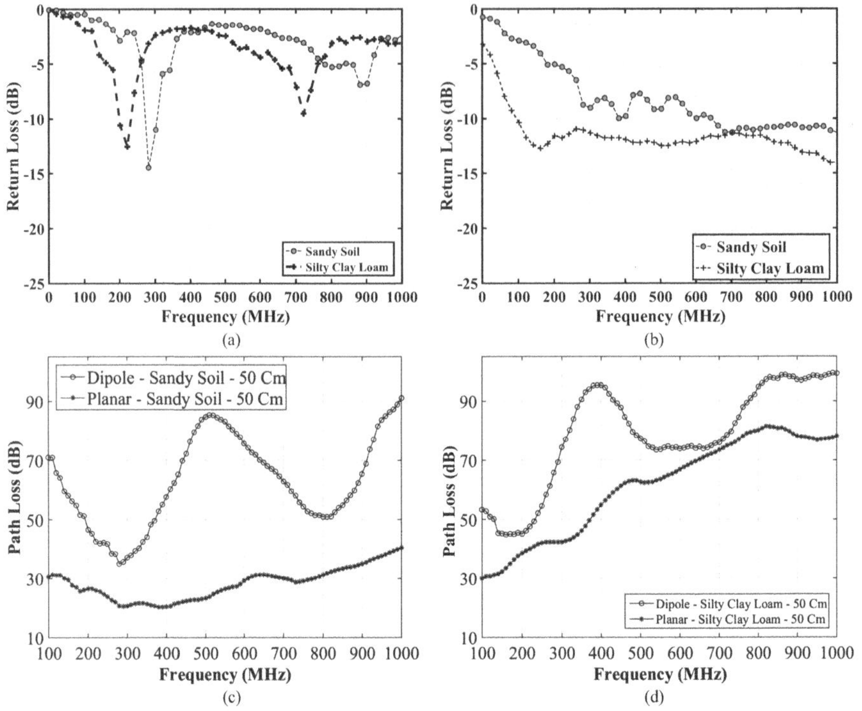

Fig. 6.20 Return loss comparison: (**a**) Dipole antennas only. (**b**) Planar antennas only. (**c**) using dipole and planar antenna in sandy soil. (**d**) Using dipole and planar antenna in silty clay loam soil

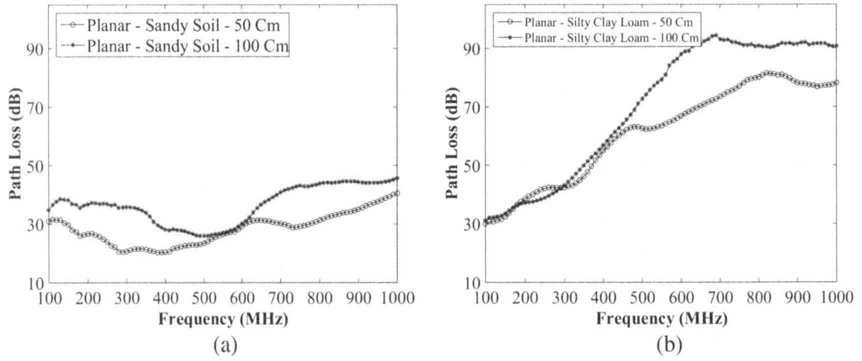

Fig. 6.21 Path loss comparison: (**a**) Planar antenna in sandy soil. (**b**) Planar antenna in silty clay loam

soil, the variations in path loss with frequency present in dipole antenna are not observed when using planar antenna [52, 54].

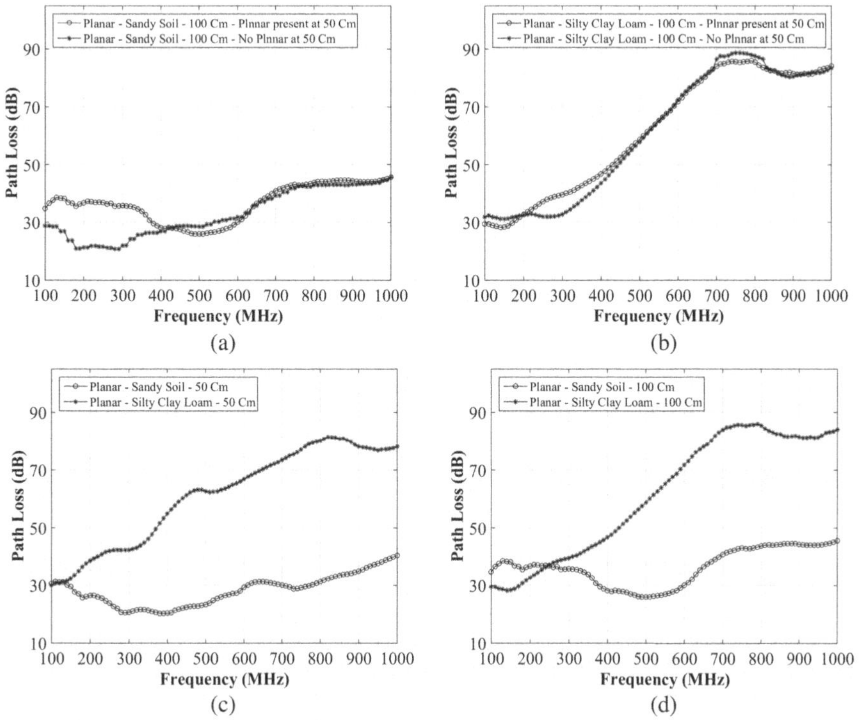

Fig. 6.22 (**a**) Planar antenna in sandy soil. (**b**) Planar antenna in silty clay loam. (**c**) Dipole and planar antenna in sandy soil. (**d**) Dipole and planar antenna in silty clay loam soil

The change in path loss when a planar is buried between planar antennas is shown in Fig. 6.22a for sandy soil and in Fig. 6.22b for silty clay loam. In sandy soil, difference of 8 dB is observed at frequencies less than 400 MHz, and in silty clay loam overall there is difference except 4–5 dB difference at 300 and 800 MHz [36, 38].

The path loss difference using same antenna at 50 and 100 cm distance in different soils is presented in Fig. 6.22c, d, respectively. A 28 dB lower path loss is observed in sandy soil when compared to silty clay loam both at 50 and 100 cm distance. This happens because the sandy soil holds less bounded water which is the major component in soil that absorbs electromagnetic waves [36]. A propagation path loss analysis has been presented using dipole and planar antennas in the sandy and silty clay loam. In the sandy soil, better radio wave propagation is observed. The results show that the planar antenna is more efficient for subsurface communications. The analysis is useful to determined inter-node distance in sensor-guided irrigation system.

References

1. Abrudan TE, Kypris O, Trigoni N, Markham A (2016) Impact of rocks and minerals on underground magneto-inductive communication and localization. IEEE Access 4:3999–4010. https://doi.org/10.1109/ACCESS.2016.2597641
2. Akyildiz IF, Stuntebeck EP (2006) Wireless underground sensor networks: research challenges. Ad Hoc Netw J 4(6):669–686
3. Akyildiz IF, Sun Z, Vuran MC (2009) Signal propagation techniques for wireless underground communication networks. Phys Commun J 2(3):167–183
4. Arnautovski-Toseva V, Grcev L (2016) On the image model of a buried horizontal wire. IEEE Trans Electromagn Compats 58(1):278–286
5. Banos A (1966) Dipole radiation in the presence of a conducting halfspace. Pergamon Press, Oxford
6. Biggs A (1968) Dipole antenna radiation fields in stratified Antarctic media. IEEE Trans Antennas Propag 16(4):445–448. https://doi.org/10.1109/TAP.1968.1139227
7. Bird TS (2009) Definition and misuse of return loss [Report of the transactions Editor-in-Chief]. IEEE Antennas Propag Mag 51(2):166–167. https://doi.org/10.1109/MAP.2009.5162049
8. Brekhovskikh LM (1980) Waves in layered media. Academic Press, New York
9. Castorina G, Donato LD, Morabito AF, Isernia T, Sorbello G (2016) Analysis and design of a concrete embedded antenna for wireless monitoring applications. IEEE Antennas Propag Mag 58(6):76–93
10. Census of Agriculture UD (2014). http://www.agcensus.usda.gov/
11. CST Microwave Studio (2020). www.cst.com/products/cstmws
12. Dong X, Vuran MC (2011) A channel model for wireless underground sensor networks using lateral waves. In: Proceedings of IEEE Globecom'11, Houston
13. Dong X, Vuran MC (2013) Impacts of soil moisture on cognitive radio underground networks. In: Proceedings of IEEE BlackSeaCom, Georgia
14. Dong X, Vuran MC, Irmak S (2012) Autonomous precision agriculture through integration of wireless underground sensor networks with center pivot irrigation systems. Ad Hoc Netw 11(7):1975–1987
15. Dong S, Yao A, Meng F (2015) Analysis of an underground horizontal electrically small wire antenna. J Electr Comput Eng 2851:9
16. Dong J, Shen F, Dong Y, Wang Y, Fu W, Li H, Ye D, Zhang B, Huangfu J, Qiao S, Sun Y, Li C, Ran L (2016) Noncontact measurement of complex permittivity of electrically small samples at microwave frequencies. IEEE Trans. Microwave Theory Tech. 64(9):2883–2893. https://doi.org/10.1109/TMTT.2016.2588487
17. Elliott RS (1981) Antenna theory and design. Prentice-Hall, Upper Saddle River
18. Fitzgerrell RG, Haidle LL (1972) Design and performance of four buried UHF antennas. IEEE Trans Antennas Propag 20(1):56–62
19. Galejs J (1969) Antennas in inhomogeneous media. Pergamon Press, Oxford
20. Hansen R (1963) Radiation and reception with buried and submerged antennas. IEEE Trans Antennas Propag 11(3):207–216
21. Hunt K, Niemeier J, Kruger A (2010) RF communications in underwater wireless sensor networks. In: IEEE international conference on electro/information technology (EIT)
22. Iizuka K (1964) An experimental investigation on the behavior of the dipole antenna near the interface between the conducting medium and free space. IEEE Trans Antennas Propag 12(1):27–35
23. Johnson RC (ed) (1993) Antenna engineering handbook, 3rd edn. McGraw-Hill, New York
24. Kesar AS, Weiss E (2013) Wave propagation between buried antennas. IEEE Trans Antennas Propag 61(12):6152–6156
25. King RWP, Smith G (1981) Antennas in matter. MIT Press, Cambridge
26. King RWP, Owens M, Wu TT (1992) Lateral electromagnetic waves. Springer, Berlin

27. Konda A, Rau A, Stoller MA, Taylor JM, Salam A, Pribil GA, Argyropoulos C, Morin SA (2018) Soft microreactors for the deposition of conductive metallic traces on planar, embossed, and curved surfaces. Adv Funct Mater 28(40):1803020. https://doi.org/10.1002/adfm.201803020
28. Moore RK, Blair WE (1961) Dipole radiation in conducting half space. J Res Natl Bur Stand. 65
29. Peplinski N, Ulaby F, Dobson M (1995) Dielectric properties of soil in the 0.3–1.3 Ghz range. IEEE Trans Geosci Remote Sens 33(3):803–807
30. Ritsema CJ, et al (2009) A new wireless underground network system for continuous monitoring of soil water contents. Water Resour Res J 45:1–9
31. Salam A (2018) Pulses in the sand: long range and high data rate communication techniques for next generation wireless underground networks. ETD collection for University of Nebraska - Lincoln (AAI10826112). http://digitalcommons.unl.edu/dissertations/AAI10826112
32. Salam A (2019) A comparison of path loss variations in soil using planar and dipole antennas. In: 2019 IEEE international symposium on antennas and propagation. IEEE, Piscataway
33. Salam A (2019) Design of subsurface phased array antennas for digital agriculture applications. In: Proceedings of 2019 IEEE international symposium on phased array systems and technology (IEEE Array 2019), Waltham
34. Salam A (2019) A path loss model for through the soil wireless communications in digital agriculture. In: 2019 IEEE international symposium on antennas and propagation. IEEE, Piscataway, pp 1–2
35. Salam A (2019) Sensor-free underground soil sensing. In: ASA, CSSA and SSSA international annual meetings (2019). ASA-CSSA-SSSA
36. Salam A (2019) Subsurface MIMO: a beamforming design in internet of underground things for digital agriculture applications. J Sensor Actuator Netw 8(3):41. https://doi.org/10.3390/jsan8030041. https://www.mdpi.com/2224-2708/8/3/41
37. Salam A (2019) Underground environment aware MIMO design using transmit and receive beamforming in internet of underground things. Springer, Cham, pp 1–15
38. Salam A (2019) An underground radio wave propagation prediction model for digital agriculture. Information 10(4). https://doi.org/10.3390/info10040147. http://www.mdpi.com/2078-2489/10/4/147
39. Salam A (2019) Underground soil sensing using subsurface radio wave propagation. In: 5th global workshop on proximal soil sensing, Columbia
40. Salam A (2020) Internet of things for environmental sustainability and climate change. Springer, Cham, pp 33–69. https://doi.org/10.1007/978-3-030-35291-2_2
41. Salam A (2020) Internet of things for sustainability: perspectives in privacy, cybersecurity, and future trends. Springer, Cham, pp 299–327. https://doi.org/10.1007/978-3-030-35291-2_10
42. Salam A (2020) Internet of things for sustainable community development, 1st edn. Springer, Berlin. https://doi.org/10.1007/978-3-030-35291-2
43. Salam A (2020) Internet of things for sustainable community development: introduction and overview. Springer, Cham, pp 1–31. https://doi.org/10.1007/978-3-030-35291-2_1
44. Salam A (2020) Internet of things for sustainable forestry. Springer, Cham, pp 147–181. https://doi.org/10.1007/978-3-030-35291-2_5
45. Salam A (2020) Internet of things for sustainable human health. Springer, Cham, pp 217–242. https://doi.org/10.1007/978-3-030-35291-2_7
46. Salam A (2020) Internet of things for sustainable mining. Springer, Cham, pp 243–271. https://doi.org/10.1007/978-3-030-35291-2_8
47. Salam A (2020) Internet of things for water sustainability. Springer, Cham, pp 113–145. https://doi.org/10.1007/978-3-030-35291-2_4
48. Salam A (2020) Internet of things in agricultural innovation and security. Springer, Cham, pp 71–112. https://doi.org/10.1007/978-3-030-35291-2_3
49. Salam A (2020) Internet of things in sustainable energy systems. Springer, Cham, pp 183–216. https://doi.org/10.1007/978-3-030-35291-2_6

50. Salam A (2020) Internet of things in water management and treatment. Springer, Cham, pp 273–298. https://doi.org/10.1007/978-3-030-35291-2_9
51. Salam A (2020) Wireless underground communications in sewer and stormwater overflow monitoring: radio waves through soil and asphalt medium. Information 11(2):98
52. Salam A, Karabiyik U (2019) A cooperative overlay approach at the physical layer of cognitive radio for digital agriculture. In: Third international Balkan conference on communications and networking 2019 (BalkanCom'19), Skopje
53. Salam A, Shah S (2019) Internet of things in smart agriculture: enabling technologies. In: 2019 IEEE 5th world forum on internet of things (WF-IoT). IEEE, Piscataway, pp 692–695
54. Salam A, Vuran MC (2016) Impacts of soil type and moisture on the capacity of multi-carrier modulation in internet of underground things. In: Proceedings of ICCCN 2016, Waikoloa
55. Salam A, Vuran MC (2017) EM-based wireless underground sensor networks. pp 247–285. https://doi.org/10.1016/B978-0-12-803139-1.00005-9
56. Salam A, Vuran MC (2017) Smart underground antenna arrays: a soil moisture adaptive beamforming approach. In: Proceedings of IEEE INFOCOM 2017, Atlanta
57. Salam A, Vuran MC (2017) Wireless underground channel diversity reception with multiple antennas for internet of underground things. In: Proceedings IEEE ICC 2017, Paris
58. Salam A, Vuran MC, Irmak S (2016) Pulses in the sand: impulse response analysis of wireless underground channel. In: Proceedings of IEEE INFOCOM 2016, San Francisco
59. Salam A, Vuran MC, Irmak S (2016) Pulses in the sand: impulse response analysis of wireless underground channel. In: The 35th annual IEEE international conference on computer communications (INFOCOM 2016), San Francisco
60. Salam A, Vuran MC, Irmak S (2017) Towards internet of underground things in smart lighting: a statistical model of wireless underground channel. In: Proceedings of 14th IEEE International Conference on Networking, Sensing and Control (IEEE ICNSC), Calabria
61. Salam A, Hoang AD, Meghna A, Martin DR, Guzman G, Yoon YH, Carlson J, Kramer J, Yansi K, Kelly M, et al (2019) The future of emerging IoT paradigms: architectures and technologies. https://doi.org/10.20944/preprints201912.0276.v1
62. Salam A, Vuran MC, Dong X, Argyropoulos C, Irmak S (2019) A theoretical model of underground dipole antennas for communications in internet of underground things. IEEE Trans Antennas Propag 67:3996–4009
63. Salam A, Vuran MC, Irmak S (2019) Di-sense: in situ real-time permittivity estimation and soil moisture sensing using wireless underground communications. Comput Netw 151:31–41. http://www.sciencedirect.com/science/article/pii/S1389128618303141
64. Silva AR, Vuran MC (2010) $(CPS)^2$: integration of center pivot systems with wireless underground sensor networks for autonomous precision agriculture. In: Proceedings of ACM/IEEE international conference on cyber-physical systems, Stockholm, Sweden, pp 79–88. http://doi.acm.org/10.1145/1795194.1795206
65. Sivaprasad K, King R (1963) A study of arrays of dipoles in a semi-infinite dissipative medium. IEEE Trans. Antennas Propag 11(3):240–256. https://doi.org/10.1109/TAP.1963.1138045
66. Staiman D, Tamir T (1966) Nature and optimisation of the ground (lateral) wave excited by submerged antennas. Electrical Engineers, Proceedings of the Institution of 113(8). https://doi.org/10.1049/piee.1966.0220
67. Tai CT, Collin RE (2000) Radiation of a Hertzian dipole immersed in a dissipative medium. IEEE Trans Antennas Propag 48(10):1501–1506
68. Temel S, Vuran MC, Lunar MM, Zhao Z, Salam A, Faller RK, Stolle C (2018) Vehicle-to-barrier communication during real-world vehicle crash tests. Comput Commun 127:172–186
69. Tiusanen MJ (2008) Wireless Soil Scout prototype radio signal reception compared to the attenuation model. Precis Agric 10(5):372–381
70. Vuran MC, Salam A, Wong R, Irmak S (2018) Internet of underground things in precision agriculture: architecture and technology aspects. Ad Hoc Netw https://doi.org/10.1016/j.adhoc.2018.07.017. http://www.sciencedirect.com/science/article/pii/S1570870518305067

71. Vuran MC, Salam A, Wong R, Irmak S (2018) Internet of underground things: sensing and communications on the field for precision agriculture. In: 2018 IEEE 4th world forum on internet of things (WF-IoT 2018), Singapore
72. Wait JR (1961) The electromagnetic fields of a horizontal dipole in the presence of a conducting half-space. Can J Phys 39(7):1017–1028
73. Wheeler HA (1961) Useful radiation from an underground antenna. J Res 65:89–91
74. Zemmour H, Baudoin G, Hamouda C, Diet A, Biancheri-Astier M (2015) Impact of soil on UWB buried antenna and communication link in IR-UWB WUSN applications. In: European radar conference (EuRAD), 2015, pp 353–356. https://doi.org/10.1109/EuRAD.2015.7346310

Chapter 7
Underground Phased Arrays and Beamforming Applications

7.1 Introduction

Underground (UG) communication is highly affected by the soil properties [57]. Limited communication range (8–12 m) due to high soil attenuation still remains the biggest challenge even after the developments in wireless communication [1, 52, 54–58, 73]. Therefore, an adaptive UG communication system, with integrated sensing and communication system, is needed which dynamically changes the communication parameters with change in environment. It is required to change the antenna parameters (operation frequency and bandwidth) for changing soil moisture [9]. The contribution of lateral waves is enhanced by directing the energy at a particular angle. The angle is determined by the nearby environment in which the UG antenna is buried [57]. Therefore, using the antenna without an ability to direct its beam direction causes the system performance to degrade; hence, a soil moisture adaptive beamforming technique is required for high throughput and gain.

It is shown in [57] that lateral waves are the dominant component of the signal in UG communication [1, 2, 64]. The lateral waves have the ability to carry the most energy and achieve high communication range, if antennas are aligned at a particular angle. The angle varies the variation in soil moisture, soil texture, and bulk density. On the other hand, the antenna also needs to be vertically aligned to prevent the refraction at the soil–air interface.

There are many factors which can affect the beamforming of antenna arrays. Some of the factors include: distance of antenna elements from soil–air interface, delay in speed beams causes refractive indices to change, soil moisture causing resonant frequency to change which in turn changes the bandwidth return loss and reflection coefficient. The beamforming of UG2UG and UG2AG differs such that the former requires lateral waves to travel along soil–air interface and the latter is focused on broadside. Due to different nature of propagation in both links, different angles are used for the links. Therefore, phase is adjusted at UG antenna elements for coherent addition to avoid errors in beam pointing direction. SMABF achieves

A. Salam, U. Raza, *Signals in the Soil*,
https://doi.org/10.1007/978-3-030-50861-6_7

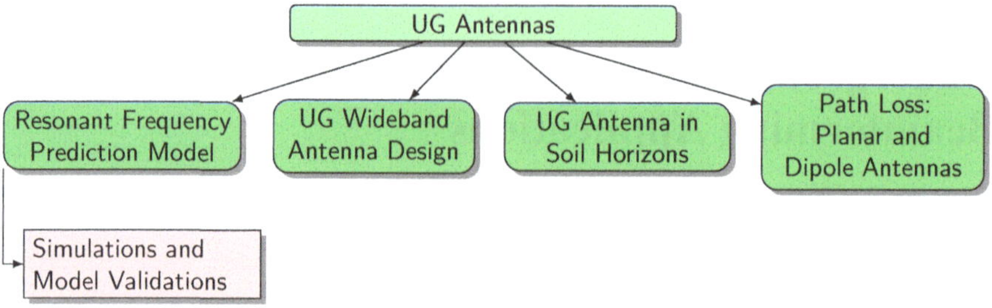

Fig. 7.1 Organization of the chapter

high performance by aligning the signal envelopes. It is important to understand the propagation in wireless UG communication to develop a reliable beamforming architecture and advanced mechanism is needed to extend the communication range. Therefore, optimum beamforming technique requires to determine the accurate physical characteristics of wireless UG channel propagation. Soil moisture adaptive beamforming (SMABF) uses an underground antenna array for communication through the soil. As per receiver position, the EM waves either go completely or partially through the soil. In partial case, some of it propagates through air. A beamforming method is developed using the optimal angle. An optimization algorithm is proposed which works on the feedback given by the soil moisture sensors (Fig. 7.1).

SAMBF is applied in various applications such as: ground penetrating radars (GPR), locating IEDs, aircraft communication from the runway, precision agriculture, hazardous object search, geology, and wireless underground sensor networks (WUSNs). Specifically, WUSNs use both electromagnetic (EM) [70] and magnetic induction-based propagation [67] and are being applied in landslide monitoring and pipeline monitoring [65], precision agriculture [1, 6, 10, 16, 26, 63, 69], border monitoring [2, 16, 63, 66].

A special beamforming antenna has been introduced and being used in wireless to achieve reduced interference and increased capacity. There has been effort in the literature for exploring OTA channel beamforming techniques [3–5, 11, 23, 24, 27, 28, 74] and MI-power transfer [21]. However, there have been no efforts made to realize the beamforming in UG communications. Lateral component [19] in UG communication can be improved further, via beamforming, to achieve long-range communication.

7.2 Channel Model for SMABF

It is important to completely understand the wave behavior in soil while designing the UG communication system. A UG channel model is given in [57] which identifies three major components, namely: direct waves, lateral waves, and reflected

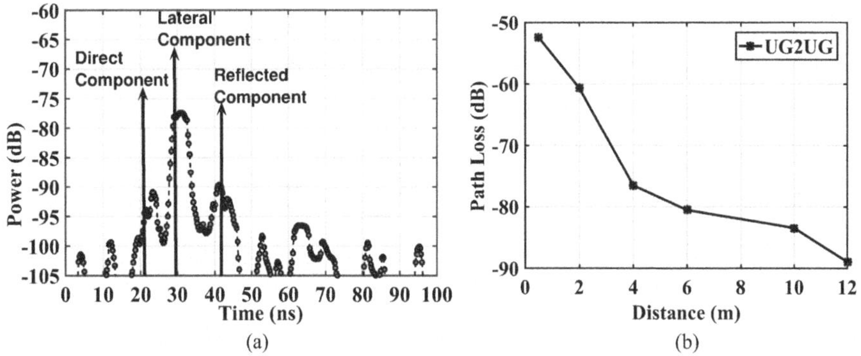

Fig. 7.2 (**a**) Power delay profile (PDP) [54], (**b**) Path loss in UG2UG channel [54]

components at the receiver. It can be seen in the power delay profile (Fig. 7.2a) that lateral wave is the strongest component as it travels most of the time through soil interface.

These factors contribute to the failure of using impedance matched antenna used for OTA to be used in soil motivating for developing new designs. Moreover, isotropic signal radiation will result in resource wastage because of buried deployment. Therefore, SMABF forms a focused narrow beam for communication between UG and AG devices to extend the communication range.

As given in [57], a UG channel can be given as sum of all wave components as follows:

$$h_{ug}(t) = \sum_{l=0}^{L-1} \alpha_l \delta(t - \tau_l) + \sum_{d=0}^{D-1} \alpha_d \delta(t - \tau_d) + \sum_{r=0}^{R-1} \alpha_r \delta(t - \tau_r) , \tag{7.1}$$

where subscripts L, D, and R refer to lateral, direct, and reflected components, α is the complex gain, and τ is the delays of each corresponding component.

7.2.1 Unique Features of UG Channel

Underground medium is considered unique and different from other medium because of combined interaction between soil, antennas, and channel. In soil wavelength of a signal is relatively lower than air because of relatively higher permittivity than air. Hence, for less attenuation, lower frequencies are used with small antenna in UG communication. This is beneficial because it permits the use of reasonably sized underground antenna array. Furthermore, in Fig. 7.2a, decreasing the delay spread can result in long-range communication with high data rates.

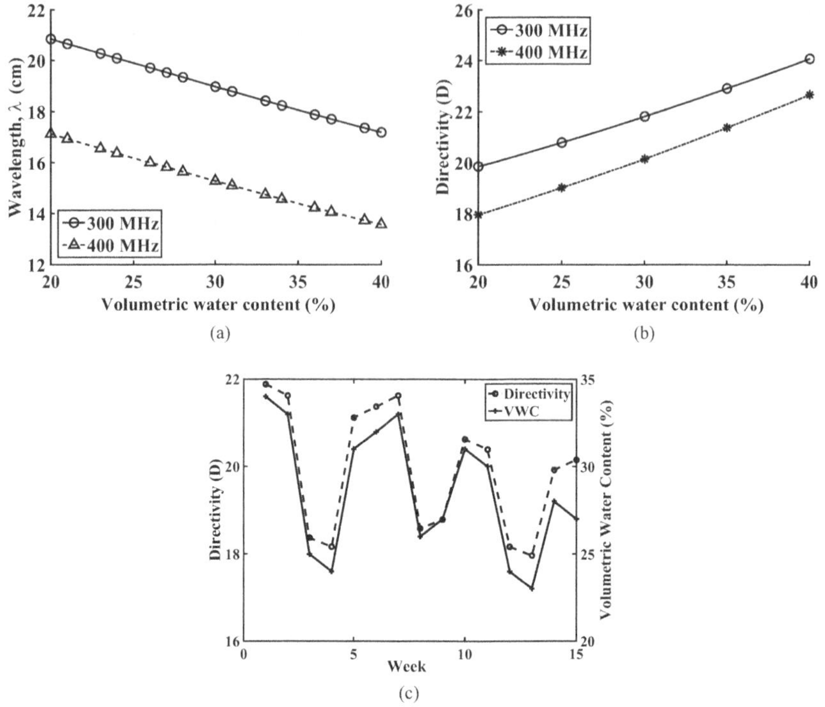

Fig. 7.3 (**a**) Wavelength as a function of soil moisture [54], (**b**) Array directivity as a function of soil moisture [54], (**c**) Antenna directivity for soybean field during growing season

UG communicationsuffers from (1) short-range of communication and (2) low data rate. Short range is due to the signal suffering from high attenuation in the soil as shown in Fig. 7.2b. In Fig. 7.2b, it is shown that path loss increases (30 dB increase) with increase in distance from 2 to 12 m. Moreover, in [9], a low data rate is observed in UG communication with commodity motes. These limitations are dealt by using SMABF.

7.2.2 *Underground Beamforming: Research Challenges*

This section lists the challenges observed in underground beamforming.

7.2.2.1 Analyzing the Effect of Soil Moisture on Signal Wavelength

Wavelength of a signal propagating in the soil is given as:

$$\lambda_s = \frac{2\pi}{k_s}, \tag{7.2}$$

where k_s gives the wavenumber in soil.

Figure 7.3a plots the wavelength with soil moisture. It can be seen that, at 300 MHz, wavelength decreases from 21 to 17 cm for a soil moisture decrease of 20 to 40%. Similarly, at 400 MHz, it decreases from 17 to 14 cm. Therefore, it is important to maintain the inter-element distance in such a way that wavelength change can be accommodated for sudden changes of soil moisture without affecting the beam pattern.

7.2.2.2 Analyzing the Effect of Soil Moisture on Directivity

Directivity is calculated by using the equation in [17]:

$$D \approx 2\frac{Nd}{\lambda_s}, \tag{7.3}$$

where N denotes the number of elements, d inter-element distance, and λ_s gives the wavelength in soil.

Figure 7.3b plots the directivity with soil moisture. This graph is for the inter-element distance of half-wavelength $\lambda_0/2$ antenna element. Similarly, Fig. 7.3c shows the measurement of directivity using 4-antenna element for 15 weeks of soil moisture data. The data was collected in 2015 from soybean field during summer. The figures show that directivity changes dynamically with the change in soil moisture. This effect requires mitigation through SMABF.

The antenna beam can be steered using the phased arrays that too without having any physical movement [13, 15, 18]. In UG communication, wavelength is continuously changing and needs an accurate phase control. Therefore, smart antenna with phase shifters suits the UG communication application. Wideband antenna arrays may face space and hardware issues while deployment in underground communication. Limited closed-form antenna models for UG environment make it difficult to understand the UG beamforming. In the upcoming sections, the soil effects on UG beamforming are studied to design an efficient SMABF technique with an aim to maximize the communication range and data rate.

7.2.3 Analysis of Single Array Element in Soil

Firstly, an empirical comparative analysis of a single soil antenna element with that of OTA antenna element is done in testbed shown in Fig. 7.4. This testbed deploys antenna using different depth and distance values and analyzes the effect of soil–air interface and impedance of antenna element (Table 7.1).

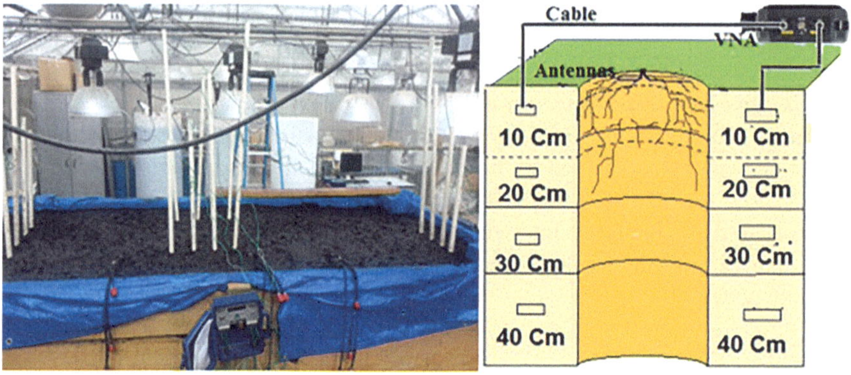

Fig. 7.4 Layout of the indoor testbed [57]

Table 7.1 Performance of 433 MHz OTA dipole antenna in soil

Property	Soil
Resonant frequency	Increases with the increase in soil moisture
	Increases with the increase in burial depth
Wavelength	Decreases with the increase in soil moisture
Return loss	Decreases with the increase in soil moisture

7.2.3.1 In-Soil v/s OTA Array Element

This section presents the comparison of UG dipole antenna element with antenna element in free space. The frequency is varied from 100 to 500 MHz. It can be seen that resonant frequency decreases because of shorter wavelength in the soil. The resonant frequency is 202, 209, and 278 MHz in silt loam soil, silty clay loam, and sandy soil, respectively. It is interesting to note that resonant frequency of sandy soil is higher than that of silt loam soil due to lower water holding capacity [7].

7.2.3.2 Effect of Soil Moisture on Individual Antenna Element Return Loss S_{11}

Figure 7.5a plots the return loss with frequency for varying soil moisture value from 0 to 255 CB. The experiment is performed using depth of 10 cm in silt loam soil. It can be seen that decrease in soil moisture causes the resonant frequency to increase from 278 to 305 MHz.

Figure 7.5b plots the resonant frequency with burial depth for varying soil moisture values from 0 to 255 CB. For a given depth value, the resonant frequency increases with change in soil moisture. For example, at the depths of 20 cm resonant frequency increases from what it is at 10 cm to 276 MHz when soil moisture changes from 0 to 255 CB. Similar trend is observed for all other depths where resonant

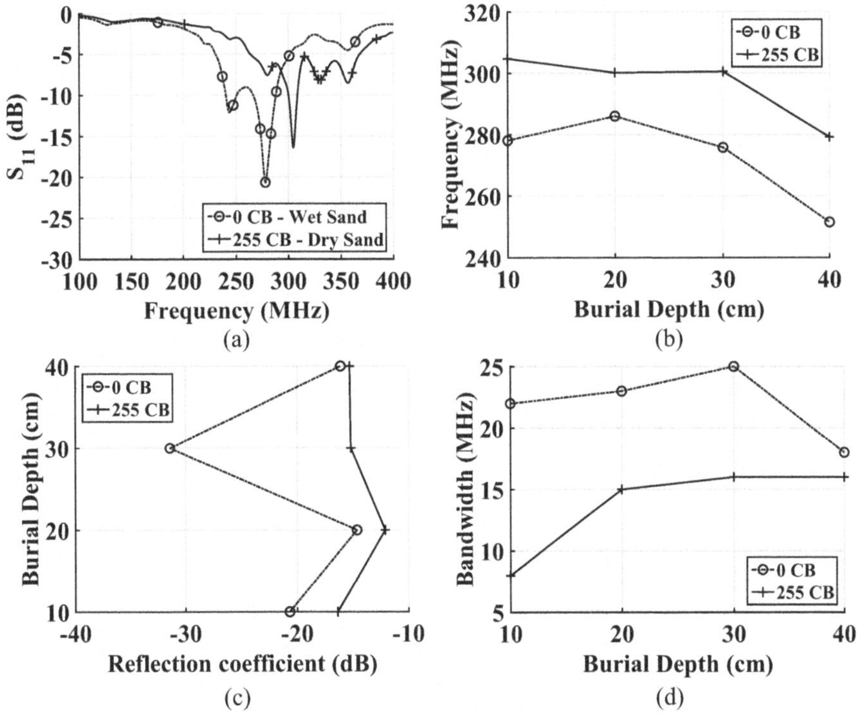

Fig. 7.5 (**a**) Effect of frequencies on return loss [54], (**b**) Effect of depth on resonant frequencies with varying soil moisture [54], (**c**) Effect of depth on reflection coefficient (dB) at different burial depths, (**d**) Effect of depth on bandwidth [54]

frequency increases from the previous depth when soil moisture is varied from 0 to 255 CB.

Figure 7.5a–d analyzes the return loss of an antenna for different depths and soil moisture values. It is observed that return loss and resonant frequency are affected by the change in soil moisture values. Furthermore, in contrast to OTA communication, optimal frequency is also different from the antenna resonant frequency [9].

7.2.3.3 Antenna Element Impedance in Soil

Input impedance Z_a of antenna element should be matched with output impedance of radio transceiver Z_s to achieve robust and efficient wireless communication. The matching is performed in such a way that maximum power is radiated and minimum power is transmitted back to the transmitter. In contrast to OTA medium, soil is not an infinite medium because of soil–air interface effect. Accordingly, return loss of an antenna RL is not just a spectrum shift but RL curve shape is also different.

7.2.3.4 Soil–Air Interface Impacts

On exciting the antenna, the current distribution of $I_0(\zeta)$ is generated from the antenna and travels towards soil–air interface where it is reflected and refracted. The reflected electric field E_r travels back to antenna where it induces an additional current I_r and affects the impedance of antenna [71]. I_r causes higher order of reflection effect; however, it can be ignored because of high attenuation in the soil, hence, only first-order effect is considered. Induced current I_r and impedance due to this induced current Z_r can be modeled considering an imaginary dipole placed in the soil environment. Similarly, Z_r is modeled as the mutual impedance from two dipole antennas [25] and reflection coefficient of soil–air interface. The mutual impedance Z_r along with self-impedance Z_a gives the total impedance of buried antenna in half-space [71].

Using the results from the analysis of individual antenna element, the design of multi-element SMABF array is presented in the upcoming sections.

7.3 Design of SMABF Array

The section will discuss the beamforming technique in the context of UG2UG and UG2AG communication. Basic concepts of beamforming are not discussed here and readers are encouraged to refer [14] for details on beamforming basic tutorial (Fig. 7.6).

7.3.1 Array Layout and Element Positioning

SMABF antenna array is expected to have the following characteristics:

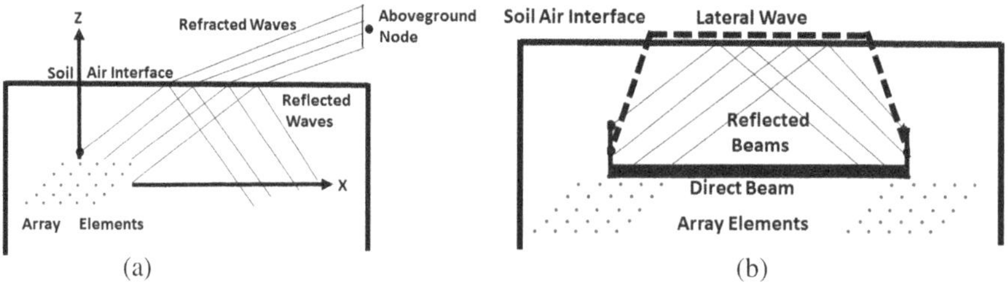

Fig. 7.6 Communications schematic for [54]: (**a**) UG2AG communications (**b**) UG2UG communications

Table 7.2 Steering angles

Communication link	θ	ϕ
UG2AG—No steering	0°	0°
UG2AG—Beam steering	0°–60°	0°
UG2UG—Lateral wave	VWC dependent see Sect. 7.3.3	0°
UG2UG—Direct wave—*X* orientation	90°	0°
UG2UG—Direct wave—*Y* orientation	90°	90°

1. Spacing between the antennas should consider the soil effect on the wavelength of the soil. It should be done in such a way that desired beam shape and directivity should not be lost because of soil effects.
2. It should be able to work with a wide range of frequencies.
3. Elements should be half-wavelength and support multiple inter-element spacing.
4. High number of elements provide high directivity. However, to determine the number of array elements, one should also consider the UG deployment conditions. Therefore, the total number must be easy to deploy and maintain high directivity.
5. Array patterns should support the steering angle. This applies to both UG2UG and UG2AG antenna arrays.
6. It should be adaptive to changes in soil moisture.

Multi-dimensional arrays structure, e.g., rectangular, planar, and circular has the advantage of providing simultaneous beams in more than one plane [13]. To fulfill the requirement of different array patterns, SMABF uses the planar array structure. The array lies on x–y plane and its z-plane points towards the soil–air interface. Elevation angle θ is measured for AG node and azimuthal angle ϕ is measured for the UG node. θ is measured from normal to soil–air interface, whereas ϕ measures steer beams to UG nodes. Table 7.2 lists the beam steering angles for UG2AG and UG2UG communication links. Planar antenna array is shown in Fig. 7.7.

Planar antenna array has the advantage of providing customized and desired radiation pattern. This customization is useful for communicating with AG nodes. Moreover, antenna elements are divided into multiple array groups on the basis of source strength. In this way source strength can be adjusted to make the lateral wave component stronger depending upon the depth and attenuation in vertical direction [19, 20].

7.3.1.1 Underground Communication Links

Figure 7.6 shows the communication schematic where UG nodes communicate with UG and AG nodes via UG2UG and UG2AG links, respectively. AG nodes can either be fixed or mobile. For AG communication, waves are reflected from soil–air interface and lateral waves are used in UG communication. The desired beam

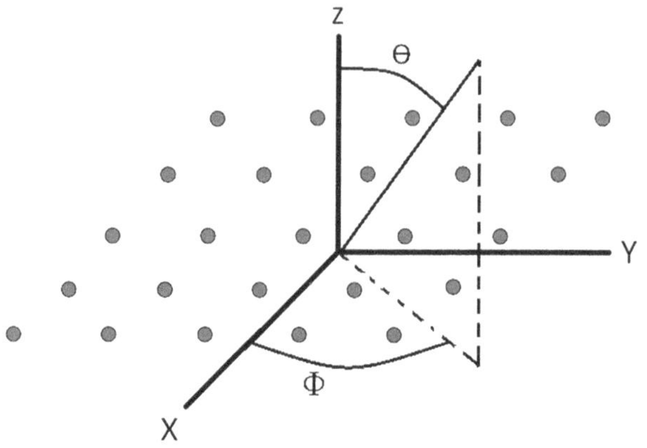

Fig. 7.7 Placements of antenna array elements in a 5×5 planar grid [54]

pattern is given in Fig.7.6. It is seen that for UG2AG communication, the EM waves reflected and refracted from soil–air interface affect the pattern towards AG node.

7.3.2 *Beam Pattern for Underground-to-Aboveground (UG2AG) Communication*

The difference in UG2AG and AG2UG communication links requires to determine the energy from UG antenna at different receiver angles. To that end, experiments were conducted for varying angles to assess the UG to AG communication [57]. The burial depth of sender UG node is kept at 20 cm and multiple distances of AG node, from the soil surface, were used. AG node is mounted at adjustable pole and distances were varied by changing the height. The following distances were used for the experiment: 2, 4, 5.5, and 7 m.

The following angles are used for the experiments: 0°, 30°, 45°, 60°, and 90°. Highest attenuation was observed at receiver angle 0° and lowest at 90°. The reason for the lowest attenuation is because of no refraction from soil–air interface at 90°. At 90°, due to lowest wave refraction, the energy is directed towards soil–air interface, thus, gives high gains and throughput in UG communication. There can be three cases in UG communication which are discussed below:

1. *UG communication with No Steering* For this case array factor for UG2AG pattern is given as follows [18]:

$$AF(\theta, \phi) = \sum_{i=1}^{N} w_i \exp\Big(-\big[jk_s r_i (x_i \sin\theta\cos\phi + y_i \sin\theta\sin\phi)\big]\Big), \tag{7.4}$$

where w_i represents the weight of the antenna element, N denotes the total number of antenna elements, and x_i, y_i gives the location coordinate of i-th element in the array.

2. *UG communication with Beam Steering* The no steering pattern can be used to create a broadside beam without considering the AG nodes location. However, with exact knowledge of AG node location, values for θ_{AG} and ϕ_{AG}, the beam can be steered by adding phase shifts δ_i at i-th element. The element then can steer the beam to (θ_{AG}, ϕ_{AG}) as follows:

$$AF(\theta, \phi) = \sum_{i=1}^{N} w_i \exp\big(- \big[jk_s r_i (x_i \sin\theta \cos\phi + y_i \sin\theta \sin\phi)\big] + \delta_i \big),$$

where δ_i can be calculated as:

$$\delta_i = -k_s (x_i \sin\theta_{AG} \cos\phi_{AG} + y_i \sin\theta_{AG} \sin\phi_{AG}).$$

3. *Refraction Adjustment* When beam steering is performed at angles different from the normal ones, refraction becomes significant and cannot be ignored. It occurs because of wave coupling with soil–air interface. This interface separates soil from air and difference in the properties of both medium causes refraction. Refraction not only deteriorates the UG beamforming performance but also causes the angle of arrival to change at AG nodes. Due to difference in the medium properties, refraction causes delay due to different speeds of the wave. This causes the spreading and decay of focused beam ultimately causing error in beam pointing direction and beam steering. This error due to refraction at soil–air interface is known as beam squint [18] resulting in signal dispersion.

SMABF deals with this issue by suing time delay beam steering [18] which aligns the signal envelope. This reduces the effects due to soil–air interface. Time delays and refraction angle are used for setting the beam direction. For an AG node location (θ_{AG}, ϕ_{AG}), time delay is given as:

$$\tau_{\mathrm{mn}} = \begin{pmatrix} \tau_{11} & \tau_{12} & \cdots & \tau_{1n} \\ \tau_{21} & \tau_{22} & \cdots & \tau_{2n} \\ \vdots & \vdots & \ddots & \vdots \\ \tau_{m1} & \tau_{m2} & \cdots & \tau_{mn} \end{pmatrix}, \tag{7.5}$$

where τ_{ij} is given as:

$$\tau_{ij} = \sin\theta_r [i \times d_i \cos\phi_r + j \times d_j \sin\phi_r]/S, \tag{7.6}$$

where S denotes the speed of signal wave in soil, d_i is the element spacing in x direction, and d_j in y direction. θ_r is estimated by Snell's law as follows:

$$\theta_r = \arcsin\left(\frac{\eta_s}{\eta_a}\sin\theta_{AG}\right), \tag{7.7}$$

where η_a and η_s are the refractive index of air and soil, respectively.

τ_{ij}, in Eq. (7.6), depends upon the burial depth and soil moisture. Higher refraction index decreases the speed of the wave, hence, increasing the delay as well. After estimating τ_{ij} and δ_i, array factor is expressed as:

$$AF(\theta,\phi) = \sum_{i=1}^{N} w_i \exp\Big(-\Big[jk_s r_i(x_i \sin\theta\cos\phi + y_i \sin\theta\sin\phi) + 2\pi f \sum_{j-1}^{M} \tau_{ij} + \delta_i\Big]\Big). \tag{7.8}$$

7.3.3 UG2UG Communication Beam Pattern

There can be two scenarios for the communication between underground devices, i.e., UG2UG communication. Both scenarios are given below:

1. *Estimation of Soil Moisture-Based Optimum Steering Angle to Maximize Lateral Wave* As discussed earlier, lateral waves travel along soil–air interface [8, 57]. These lateral waves can be maximized by adjusting the antenna radiation angle. The angle is based on the dielectric properties of soil and can be calculated as [62]:

$$\theta_{UG} = \frac{1}{2}\tan^{-1}\left(\frac{2\text{Re}(n^2-1)^{1/2}}{|n^2-1|-1}\right)\text{rad}, \tag{7.9}$$

   where n is the soil refractive index and calculated as:

$$n = \sqrt{\frac{\sqrt{\epsilon'^2+\epsilon''^2}+\epsilon'}{2}}. \tag{7.10}$$

   In the above equation, real and imaginary parts of soil permittivity are represented by ϵ' and ϵ'', respectively.
2. *Direct Wave* Direct waves are the relatively prominent and dominant component in short-range UG communication. Therefore, short-range UG communication is improved through direct UG beam going towards the receiver.

The values for steering angle used in both cases are given in Table 7.2 and Eq. (7.5) is for direct and lateral case.

Algorithm 2 SMABF Beam Steering Algorithm

Initialization :

Let A be the set of AG nodes
Let U be the set of the UG nodes
Let R be the receiver node
Sense the soil moisture level and determine the appropriate wavelength in soil
Select the array layout based on wavelength
Activate desired elements based on soil moisture and desired beam patterns
Produce the initial weights to achieve the desired beam pattern
Calculate the excitation and current distribution (root matching, pole-residue)
BEGIN
if R $\in$ A then **then**
  if θ_R is known **then**
    $AF(\theta, \phi) = \sum_{i=1}^{N} w_i \exp\big(-\big[jk_s r_i(x_i \sin\theta\cos\phi + y_i \sin\theta\sin\phi)\big] + \delta_i\big)$
  else if
   then
    Normal to the surface beam using
$AF(\theta, \phi) = \sum_{i=1}^{N} w_i \exp\big(-\big[jk_s r_i(x_i \sin\theta\cos\phi + y_i \sin\theta\sin\phi)\big]\big)$,
  end if
else if R $\in$ U **then**
  BEGIN
  Sense soil moisture
  Determine optimal angle using
$\theta_{UG} = \frac{1}{2}\tan^{-1}\left(\frac{2\mathrm{Re}(n^2-1)^{1/2}}{|n^2-1|-1}\right)$
  Output UG2UG Beam
  END
end if
Optimize to get low side lobe levels when wavelength changes
Optimize element positions
Activate virtual arrays
Adjust weights and excitation
Repeat this process to adjust these parameters when soil moisture changes
END

7.3.4 Directivity

Directivity of a SMABF array is calculated as:

$$D = \frac{4\pi |AF_{\max}|^2}{\int_0^{2\pi}\int_0^{\pi} |AF|^2 \sin\theta d\theta d\phi}, \tag{7.11}$$

where the main beam peak, i.e., maximum of the array factor $AF_{\max}$, is calculated as:

$$AF_{\max} = \sum_{i=1}^{N} w_i. \tag{7.12}$$

7.3.5 SMABF Steering Algorithm

Algorithm 2 gives multiple beam patterns and meets the requirements for both communications, i.e., UG2UG and UG2AG.

7.4 Results and Optimization

This section discusses the SMABF and compares the adaptive and non-adaptive beamforming technique.

7.4.1 Optimum UG Angle

The optimum angle is determined by using Eq. (7.9). At optimum angle, UG communication (with lateral waves) is improved adapting to soil properties. The analysis was performed on the lateral wave angle using frequency range of 100–1000 MHz, VWC between 0 and 40%, and soil types of sandy and silty clay loam. Table 7.3 gives the particle distribution for both soil types.

Figure 7.8 shows the optimal angle for varying VWC (0–40%) in two different soil types. The optimal angle in silty clay loam soil is higher than the sandy soil. The maximum optimal angle in silty clay loam soil is 16° and that in sandy soil

Table 7.3 Soil particle size distribution in testbeds [57]

Textural class	%Sand	%Silt	%Clay
Sandy soil	86	11	3
Silty clay loam	13	55	32

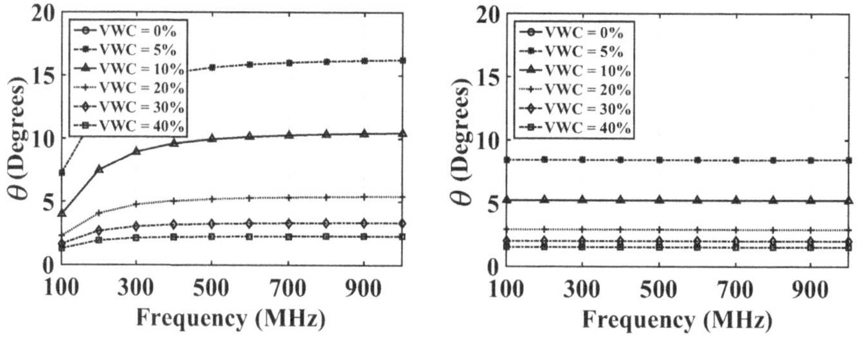

(a) (b)

Fig. 7.8 Optimal angle at different frequencies in: (**a**) Silty clay loam, (**b**) Sandy soil

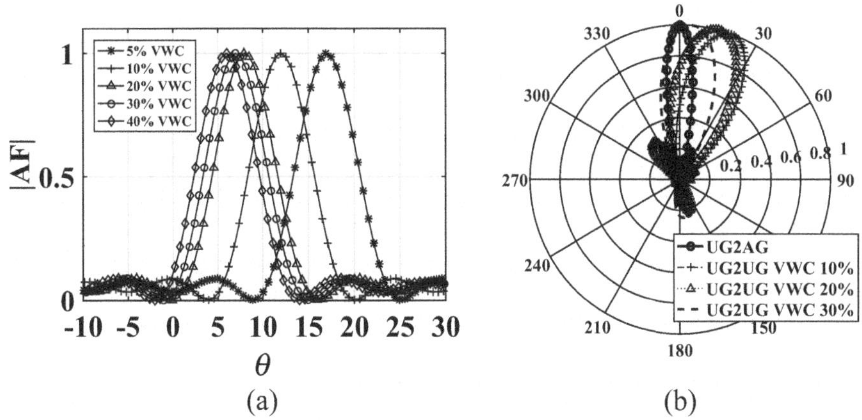

Fig. 7.9 (**a**) Array factor for varying soil moisture levels in UG2UG communications [54], (**b**) Underground-to-Aboveground (UG2AG) communications [54]

is 9°. This is because of silty clay loam soil having higher dielectric constant than sandy soil. A decreasing trend is observed between the optimal angle and VWC, i.e., optimal angle is decreasing with increase in VWC percentage, because of increase in soil permittivity with increasing soil moisture. Table 7.2 summarizes the steering angles for both UG2UG and UG2AG communication. Figure 7.9a and b gives the UG beam pattern with soil moisture.

7.4.2 Optimum UG Angle to Enhance Lateral Wave

For validity of lateral wave communication, different experiments were conducted in indoor and outdoor testbed. Different soils are used in both testbeds: Sandy for indoor and silty clay for outdoor. Burial depth of directional antenna is 20 cm and Keysights' Fieldfox Vector Network Analyzer (VNA) N9923A is used for taking measurements. Two experiments are performed for measuring the channel gain and channel transfer function: first without changing the orientations and the other one after finding an optimum angle. For sandy soil, optimal angle is 4° at 37% VWC and it is 16° at 0% VWC for silty clay loam soil.

Figure 7.10 shows the channel gain results for the two transmitter–receiver (T-R) separation distances: 50 cm and 1 m. For a distance of 50 cm, sandy soil type, and a frequency of 500 MHz, a 4 dB of gain is observed at 4° as compared to no orientation scenario (Fig.7.10a). Gain is increasing with the frequency when the energy is focused at optimum angle. This is because of the reason that soil path is affected by soil permittivity. Similarly, in Fig. 7.10b, for a distance of 1 m, a higher gain is observed as compared to what was observed at 50 m. This is because

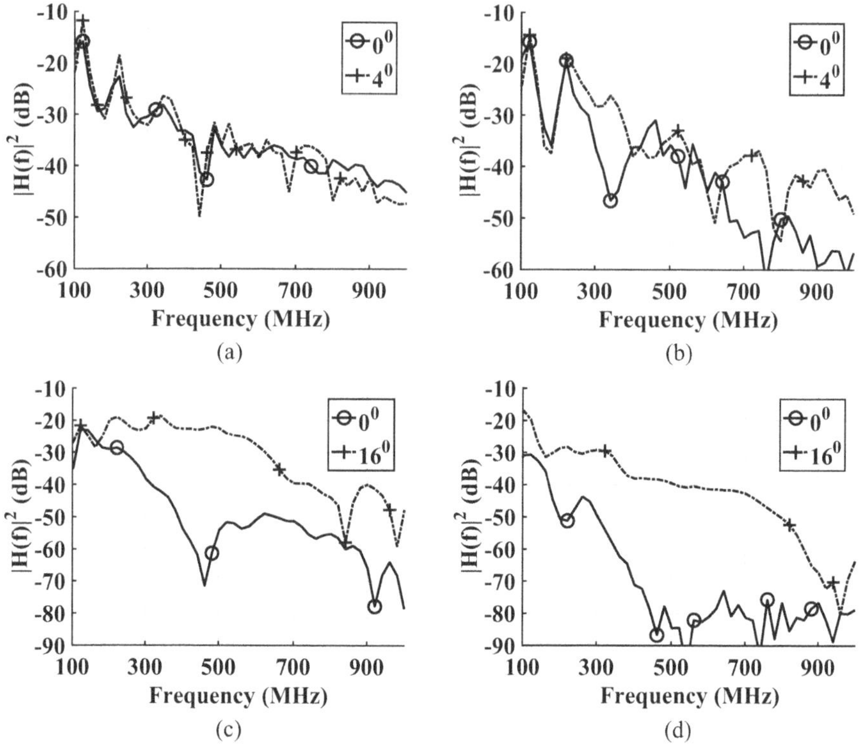

Fig. 7.10 Optimum angle in UG communication [54]: (**a**) Sandy soil (50 cm T-R separation), (**b**) Sandy soil (1 m T-R separation), (**c**) Silty clay loam soil (50 cm T-R separation), and (**d**) Silty clay loam soil (1 m T-R separation)

for large communication distances, the direct wave becomes negligible and lateral wave is significant.

Channel gain is much higher in silty clay loam soil for both distances, i.e., 50 cm (Fig. 7.10c) and 1 m (Fig. 7.10d). The reason is that silty clay loam soil has high permittivity value as compared to sandy soil, hence, exhibiting higher losses and observing high channel gains.

7.4.3 Comparing SMABF with Non-adaptive Beamforming

In this section the performance of adaptive (SMABF) and non-adaptive beamforming techniques is compared. A 5×5 planar antenna array is used for the experiment and change in array factor and directivity of the antenna, due to soil moisture change, is studied in detail. Figure 7.11 shows the elements weight of the planar antenna array at 40% VWC and frequency of 433 MHz.

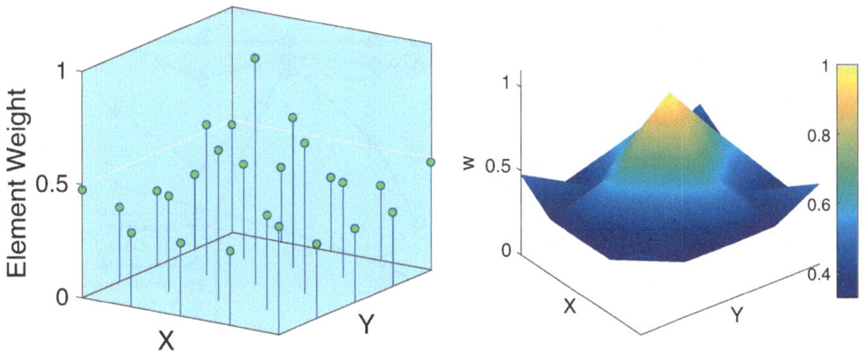

(a) (b)

Fig. 7.11 Weights of a grid element (433 MHz and 40% Soil moisture) [54]: (**a**) Stem plot, (**b**) Surf plot

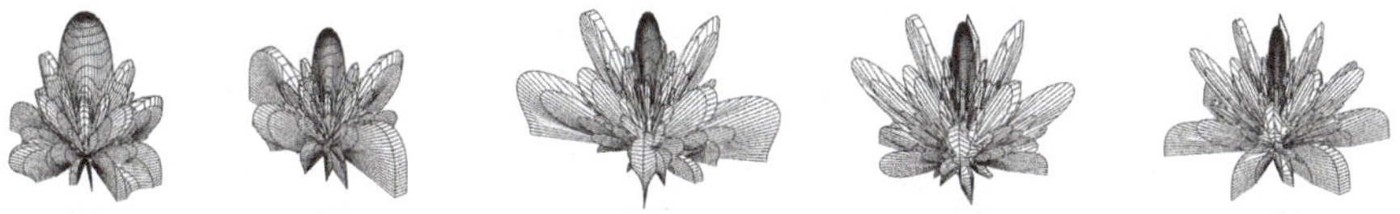

(a) (b) (c) (d) (e)

Fig. 7.12 Non-adaptive beamforming in silty clay loam soil: array factor deterioration with changing soil moisture [54]. (**a**) 5%. (**b**) 10%. (**c**) 20%. (**d**) 30%. (**e**) 40%

(a) 5% (b) 10% (c) 20% (d) 30% (e) 40%

Fig. 7.13 Non-adaptive beamforming in sandy soil: array factor deterioration with changing soil moisture [54]. (**a**) 5%. (**b**) 10%. (**c**) 20%. (**d**) 30%. (**e**) 40%

Figure 7.12 shows the array factor deterioration for silt clay loam soil and Fig. 7.13 for sandy soil. Both results are obtained using a non-adaptive beamforming technique with VWC ranging from 5 to 40%. In both soils, the side lobes are growing higher as the soil moisture is increasing. High soil moisture values significantly impact the wavelength of sandy soil as compared to silty clay loam soil; therefore, the effect of high soil moisture will be severe in sandy soil as compared to silty clay loam soil.

Figure 7.14 compares the directivity of adaptive SMABF with non-adaptive beamforming for both sandy and silty clay loam soils. SMABF maintains the

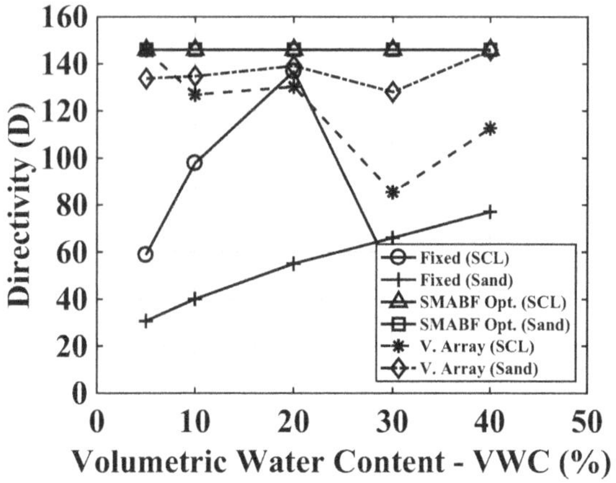

Fig. 7.14 Effect of soil moisture on directivity in sandy and SCL soil [54]

directivity along with the soil moisture change, whereas, for non-adaptive approach, the drastic change in directivity is observed. For 5% VWC, in silty clay loam soil, non-adaptive approach has lower directivity than SMABF; for 10 and 30% VWC, in sandy soil, non-adaptive approach has the same directivity as of SMABF and even shows higher directivity at 20% VWC. It is important to note that SMABF is not maximizing the directivity instead adapting to wavelength with change in soil moisture; however, it can be maximized by using optimal inter-element spacing.

7.4.4 Antenna Element Thinning Through Virtual Arrays

In UG array thinning, a small subset of antenna elements is selected from the full planar array using the optimization approach and a virtual array is constructed. Physical antenna elements are triggered on/off using the virtual array. This approach is used to discover the elements with optimal configuration which can be used to form current soil moisture level. Antenna element weights, w_i are toggled (on/off) as follows:

$$w_i = \begin{cases} 1 & \text{if } d = \lambda_s/2, \\ 0 & \text{otherwise}, \end{cases} \tag{7.13}$$

where λ_s is the wavelength in soil, d is the distance between the i-th antenna element and the previous element. d is chosen in such a way that half-wavelength inter-element spacing is not disturbed due to change in soil moisture and wavelength. Table 7.4 shows the change in SMABF half-wavelength inter-element spacing for a frequency of 433 MHz and VWC in the range of 10–40%. Firstly, array uses the initial configuration for the operation and then virtual adaptive thinning is applied based on soil moisture values. Virtual SMABF maintains the side lobes and

Table 7.4 SMABF half-wavelength inter-element spacing (cm) for different VWC percentage in soil

Soil type	Volumetric water content (VWC)			
	10%	20%	30%	40%
Silt loam	30.79	23.72	20.25	18.03
Sandy	46.83	39.28	34.62	31.28
Silty clay loam	27.86	20.53	17.12	15.01

directivity and no distortion in side lobes is observed as in the case of non-adaptive beamforming.

7.4.5 SMABF Directivity Maximization

Virtual arrays have fixed directivity. This issue can be improved using directivity maximization. Goal is to optimize the inter-element spacing and avoid grating lobes. The optimization problem is given as follows:

$$\wp : \quad \max D = \frac{4\pi |AF_{\max}|^2}{\int_0^{2\pi} \int_0^{\pi} |AF|^2 \sin\theta d\theta d\phi}, \tag{7.14}$$

$$s.t. \quad \frac{d}{\lambda_s} < \frac{1}{1+\sin\theta}, \tag{7.15}$$

where θ is the steering angle, inter-element spacing is given by d, and D represents the directivity.

Genetic algorithms [18] are used where inter-element position can be chosen randomly or specified beforehand. A priori position can be selected without considering the soil moisture level. Inter-element spacing is then determined using a cost function.

7.4.6 Feedback Control

Feedback signals are used to dynamically adjust the element weights through array gain feedback loops. This feedback is in addition to the weight adaption on the basis of soil moisture sensing. A pilot signal is used for the maximization of array gain. SMABF transmitting array, operating in receiver mode, receives the pilot signal and based on this signal changes its parameter for the transmission. The transmitting array determines the channel state and adjusts the parameters on the basis of best signal-to-noise (SNR) ratio. Figure 7.15 shows the SMABF feedback control mechanism (Table 7.5).

Table 7.5 Comparison of SMABF and UG phased array

Property	SMABF	UG phased array
Inter-element spacing	Element thinning through virtual arrays	Depends upon soil moisture
Frequency	433 MHz	–
Directivity	Adapts to the soil moisture $\frac{4\pi \lvert AF_{\max}\rvert^2}{\int_0^{2\pi}\int_0^{\pi} \lvert AF\rvert^2 \sin\theta d\theta d\phi}$	Increases with the decrease in wavelength and inter-element spacing
Steering angle	Silt clay loam = 16° Sandy Soil = 9°	–
Element weighting	Soil moisture adaptive weights $w^i_{sm,l+1} = w^i_{sm,l} + s(-\nabla_l)$	Soil moisture adaptive weights $w^i_{sm,l+1} = w^i_{sm,l} + s(-\nabla_l)$
Element excitation	Initial distribution (w_i^0): Root matching method [12] Subsequent weights: $w_i = w_i^0 \pm \delta w_i$	Initial distribution (w_i^0): Root matching method [12] Subsequent weights: $w_i = w_i^0 \pm \delta w_i$
Effective isotropic radiated power (EIRP)	–	$P_r = 10\log_{10}\left(10^{\frac{P_r^d}{10}} + 10^{\frac{P_r^r}{10}} + 10^{\frac{P_r^L}{10}}\right)$
Implementation	Planar antenna	SDR-based implementation

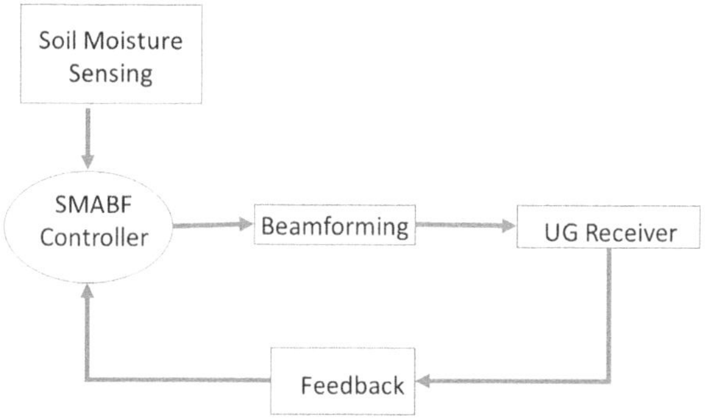

Fig. 7.15 SMABF with feedback [54]

7.4.7 *Adaptive SMABF Element Weighting*

Amplitude weighting [74] can be used to get a desired beam pattern from antenna elements in the array. SMABF uses the soil moisture values as an input to determine the element weight which is the reason for improved SNR in SMABF. Weights are adjusted dynamically with soil moisture for both UG2UG and UG2AG communication. These adaptive weights are given as:

$$\boldsymbol{w} = \{w_0, w_1, w_2 \ldots w_{n-1}\}^T. \tag{7.16}$$

Change in soil moisture causes the soil permittivity to change which in turn changes the wavelength. Weight factor γ_s is given as:

$$\gamma_s = \frac{1}{\lambda_s} \times d \times \pi \times \sin\theta_0. \tag{7.17}$$

Using this weight factor, adaptive weight of the i-th antenna is given as:

$$w_{sm}^{i} = \alpha_i \exp(-j\gamma_s(2i - n - 1)), \tag{7.18}$$

where α_i represents the element coefficient. Beam patterns and required side lobe levels are obtained using these element coefficients via positioning and element thinning technique. After determining the adaptive weight vector, desired beam pattern is obtained as:

$$F = \boldsymbol{X} w_{sm}^{i}, \tag{7.19}$$

where $\boldsymbol{X}$ represents an intended signal. The same process is repeated using the gradient method in which adaptive weights are adjusted for upcoming iteration $(l + 1)$ as follows [74]:

$$w_{sm,l+1}^{i} = w_{sm,l}^{i} + s(-\nabla_l), \tag{7.20}$$

where s is the stability and gradient estimate vector is given by ∇_l. Adaptive antenna array does not perform well with faster adaption rate [74]. As change in soil moisture is a slow process, this approach has minimum noise and high tolerance for performance degradation.

7.4.8 SMABF Element Excitation

Antenna element excitation also needs to be optimized as change in soil moisture can cause the performance degradation, if traditional excitation distribution methods (e.g., Hansen, Bayliss, Uniform, Dolph/Chebyshev, Binomial, Taylor, radial taper, etc.) are used. In SMABF excitation method, an initial current distribution w_i^0 is determined using the root matching method [12]. New patterns are determined for each soil moisture change. Due to the slow pace of soil moisture change, patterns differ from each other in small angle steps. The changes in w_i are given by δw_i:

$$w_i = w_i^0 \pm \delta w_i. \tag{7.21}$$

With the knowledge of desired and current patterns, matrix inversion method is used to solve pattern difference equation for getting δw_i. This δw_i is used for calculating new distribution for new beam pattern. The process is repeated multiple times until the desired beam pattern is obtained.

7.4.9 Simulations

A simulation program, CST Microwave Studio (MWS), is used to simulate SMABF design. A 5×5 phased array in sandy soil is used for the simulation. The operating frequency of the array lies between 0.2 and 0.6 GHz with beam steering support for different communication types and angles given in Table 7.2.

Operating bandwidth, S-parameters, and radiation pattern of an array depend upon the individual element; therefore, to analyze these parameters, a single dipole antenna is simulated initially. The simulation with dipole antenna is performed using resonant frequency of 433 MHz and feed impedance of 50 Ω. Elements are simulated as PEC cylindrical material and these elements are excited using the gap between the spacing of elements. Simulated S-parameter values are compared with measured S-parameter values for validation. Both results are discussed in Sect. 7.2.3 and confirm each other. Simulated results show that change in soil moisture affects the resonant frequency, bandwidth, and reflection coefficient. More details on this effect are given in Sect. 7.2.3.

After verifying the design of individual element, full array is simulated for the validation. Figure 7.16a shows the structure of simulated array. Distribution matrix is estimated after giving the beam pattern specification for UG2UG and UG2AG

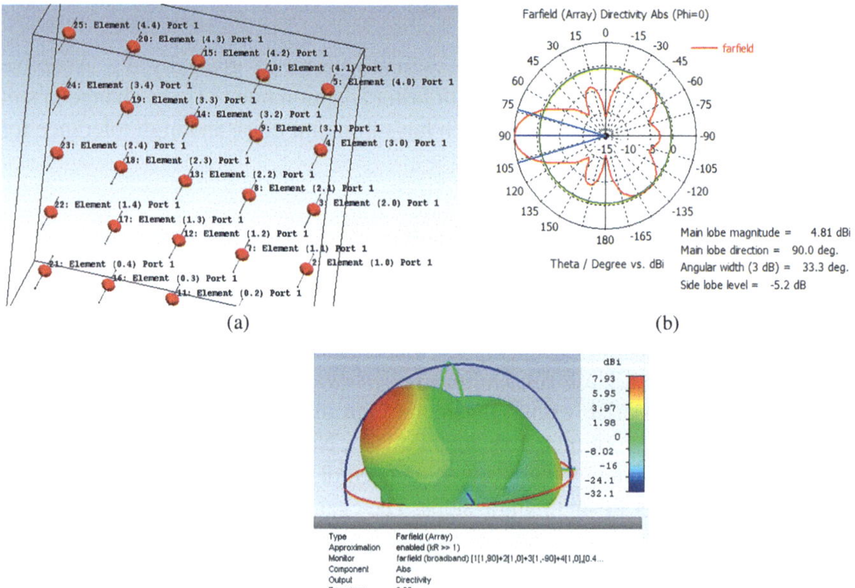

Fig. 7.16 (**a**) Design of 5×5 array in CST MWS [54], (**b**) UG2AG far-field simulated in CST MWS [54], (**c**) 3D view of UG2AG Beam [54]

communication. The matrix is used for exciting the element and obtaining the desired beam pattern. The distribution matrix is recalculated for every variation in soil moisture. Accordingly, beam steering angle is adjusted and the process is repeated multiple times to adapt to soil moisture changes. Figure 7.16b shows the UG2AG results for the simulation with beam parameters and Fig. 7.16c shows the same in 3D. It is observed that simulation results confirm the beam synthesis analysis from Sect. 7.3.

7.5 Underground Phased Array Antennas

Underground transmitter, with phased array antennas, uses UG transmit beamforming to maximize the contribution of lateral wave by using a particular angle to transmit energy [52, 54, 57]. This saves energy loss by using narrow-width beam instead of transmitting signal in isotropic direction. The primary goals of wireless underground communications are: (a) increase the received signal strength and (b) minimize the interference at receiver end [54, 57].

Soil uses weights which adapts to the soil moisture values of the soil. Moreover, UG phased antenna arrays use array gain feedback loops as feedback signals to adjust weights. This is done to maximize the array gain with pilot signals. This method uses the phased array antenna transmitter for receiving the pilot signal (as a receiver) and adjusts parameter for transmission of the signal as a transmitter. While transmitter is working in a receiver mode, channel state is estimated by changing the scan angles. The best SNIR values are selected and soil moisture values are changed to adjust the parameters. Figure 7.15 shows soil moisture-adaptive feedback controller with the feedback.

7.5.1 Adaptive Element Weighting

The array elements signals of beamforming antennas can be controlled to produce the desired beam by phase and amplitude weighting [72, 74]. In UG phased arrays, the current environment and the soil moisture information is used to weight the elements which leads to improvements in received SNR. The adaptive weight adjustment is done to keep the desired UG2UG and UG2AG characteristics based on the soil moisture variations.

Soil moisture adaptive weights are expressed as:

$$\boldsymbol{w} = \{w_0, w_1, w_2 \ldots w_{n-1}\}^T. \tag{7.22}$$

The permittivity of the soil changes with the change in soil moisture and hence the wavelength. The weight factor γ_s is defined as:

$$\gamma_s = \frac{1}{\lambda_s} \times d \times \pi \times \sin\theta_0, \tag{7.23}$$

where d is the inter-element distance. Accordingly, with this weight factor, ith soil moisture adaptive weight w^i_{sm} becomes

$$w^i_{sm} = \alpha_i \exp(-j\gamma_s(2i - n - 1)), \tag{7.24}$$

where α_i is the element coefficient. These element coefficients are optimized to obtain desired side lobe levels and beam patterns through element thinning and positioning approach. Once the beamforming vector is populated with the adaptive weights, the desired beam pattern is produced as follows:

$$F = \boldsymbol{X} w^i_{sm} \tag{7.25}$$

where $\boldsymbol{X}$ is the intended signal. To repeat this process with soil moisture change, gradient method is used. In this method, soil moisture adaptive weights are adjusted for next $(l + 1)$ iteration as [55, 74]:

$$w^i_{sm,l+1} = w^i_{sm,l} + s(-\nabla_l), \tag{7.26}$$

where s ensures stability and convergence and ∇_l is the gradient estimate vector. It is well known that performance of an adaptive antenna array system degrades with faster adaption [32, 74]. Since the soil moisture is a slowly varying process, this simple-to-implement approach exhibits minimum noise and high tolerance to performance degradation that is caused by faster adoption and limited sampling.

7.5.2 Phased Array Element Excitation

In addition to wavelength and element position optimization, current excitation also need to be optimized. Due to the soil moisture variations, the use of conventional excitation distributions (e.g., Uniform, Dolph/Chebyshev, Binomial, Taylor, Hansen, Tseng, Bayliss, separable, radial taper, and radial taper squared) lead to degraded beam patterns with soil moisture variations [39, 49, 74]. Therefore, in this section we develop a soil moisture adaptive phased array excitation method to maintain the desired pattern. This method works as follows [22, 46].

First, an initial current distribution w^0_i is found for the desired pattern by using the root matching method [12, 30, 35]. Once the soil moisture changes, the new pattern is determined. The soil moisture does not vary drastically with time in the field. Therefore, with change in soil moisture, the new pattern varies gradually from the old one in small angle steps. Hence, the w_i is expressed in minor variations δw_i to the w^0_i:

$$w_i = w_i^0 \pm \delta w_i. \tag{7.27}$$

Since the current and desirable patterns are known, the pattern difference equation is solved using the matrix inversion method to obtain the δw_i. Accordingly, it is used to get the new current which defines new pattern. This method is repeated iteratively until the desired beam pattern is achieved. To further improve the pattern, a received power based feedback method can be employed which will enhance performance and converge faster. The received power and the directivity are defined in the next section.

7.5.3 Effective Isotropic Radiated Power (EIRP)

The phased array effective isotropic radiated power (EIRP) is expressed as the product of the transmitted power and antenna gain:

$$P_{\text{rad}} = G_t P_t, \tag{7.28}$$

where P_t is the transmitted power and G_t is the array gain.

The far-field power density P_{av} is expressed as [8]:

$$P_{av} = P_{av}^D + P_{av}^R + P_{av}^L, \tag{7.29}$$

where D (direct), R (reflected), L (lateral) denote the power densities of corresponding wave component [53, 57]. The received power is the product of far-field power density P_{av} and antenna aperture $(\lambda_s^2/4\pi)$. The received power is calculated as [8, 61]:

$$\begin{aligned}
P_r^d &= P_t + 20\log_{10}\lambda_s - 20\log_{10} r_1 - 8.69\alpha_s r_1 \\
&\quad -22 + 10\log_{10} D_{rl}, \\
P_r^r &= P_t + 20\log_{10}\lambda_s - 20\log_{10} r_2 - 8.69\alpha_s r_2 \\
&\quad +20\log_{10}\Gamma - 22 + 10\log_{10} D_{rl}, \\
P_r^L &= P_t + 20\log_{10}\lambda_s - 40\log_{10} d - 8.69\alpha_s (h_t + h_r) \\
&\quad +20\log_{10} T - 22 + 10\log_{10} D_{rl},
\end{aligned} \tag{7.30}$$

where Γ and T are reflection and transmission coefficients [8, 33], and λ_s is the wavelength in soil. The received power, for an isotropic antenna, is expressed as [8, 51, 60]:

$$P_r = 10\log_{10}\left(10^{\frac{P_r^d}{10}} + 10^{\frac{P_r^r}{10}} + 10^{\frac{P_r^L}{10}}\right). \tag{7.31}$$

The directivity of the underground phased array antenna is defined as [54, 73]:

$$D = \frac{4\pi |AF_{\max}|^2}{\int_0^{2\pi} \int_0^{\pi} |AF|^2 \sin\theta d\theta d\phi}, \tag{7.32}$$

where the $AF_{\max}$ is the maximum of the array factor [31, 40, 54].

7.6 Beyond Optimal: Enhancement of Array Dimensions

This section discusses the performance of UG phased array through directivity settings [54, 58]. Figure 7.17 plots the change in directivity with deployment at 30% moisture level. It also observes the effect of extending array dimension by increasing the number of elements. It concludes that inter-element spacing must increase with low soil moisture levels. Increasing the inter-element spacing also increases the overall array dimension. Moreover, if the soil moisture is decreased for virtual array scenarios with fixed (e.g., 5 × 5) configuration, then array size also decreases despite maintaining optimum value of directivity. However, at the

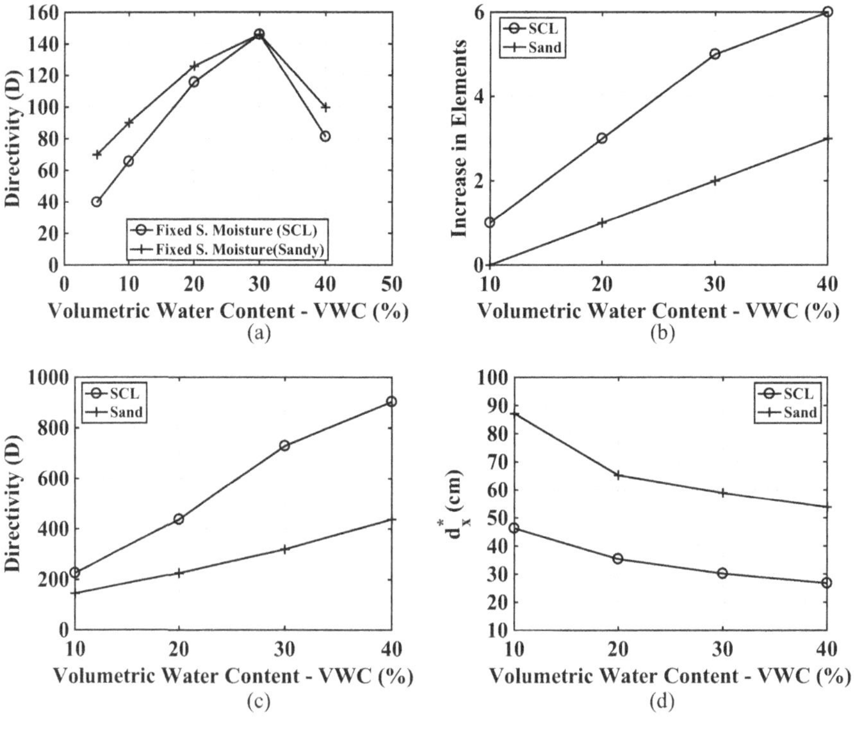

Fig. 7.17 (**a**) Directivity at 30% soil moisture level for optimized deployment, enhanced array dimension: (**b**) Number of elements, (**c**) Directivity, (**d**) d_x^*

same time, fixed configuration (5 × 5) at high soil moisture levels results in wastage of resources, i.e., elements over fixed dimensions. For high soil moisture levels, UG phased array performance is improved by using more elements, hence, getting beyond optimum directivity [37, 50].

An ideal case is considered first where an array (5 × 5) is designed with optimum inter-element spacing d_x^* at 5% soil moisture levels. This configuration is used for both types of soils. This design gives the maximum possible physical dimensions. After that, if the soil moisture level is increased, the number of elements are determined and deployed within these fixed physical dimensions. It is also important to note that no limitation is placed on practical inter-element spacing d_x. The increased number of elements, within fixed maximum dimension, are also computed using d_x^* of the current soil moisture level. For the second case, practical limitations are applied on d_x, and the performance of the system is studied [29, 36].

Figure 7.17b plots the effect of increased element in 5 × 5 array along with increasing soil moisture level. The experiment is performed in both types of soil. In silt clay loam soil, at all soil moisture levels array dimension is enhanced to 6 × 6, 8 × 8, 10 × 10, and 11 × 11 for addition of 1, 3, 5, and 6 elements at soil moisture level of 10%, 20%, 30%, and 40%, respectively. In sandy soil, the array dimension is not effected at soil moisture level of 10%; however, it is enhanced to 6 × 6, 7 × 7, and 8 × 8 at soil moisture level of 20%, 30%, and 40%, respectively.

Figure 7.17c plots the directivity of enhanced configuration. For silt clay loam soil, at higher soil moisture level, the directivity is increased significantly with enhanced configuration. This happens because of two reasons: (a) decrease in wavelength when soil moisture levels are increased and (b) smaller inter-element spacing d_x^* in silt clay loam soil (see Fig. 7.17d) results in increased number of elements in array dimensions which in turn increase its directivity in silt clay loam soil. Hence, enhanced array deployment can be advantageous for the soil with higher clay content and high soil moisture level [34, 68].

7.7 Antenna Array Implementation

This section discusses the UG phased array in the context of software and hardware implementation.

7.7.1 Software-Defined Implementation

Evolution in SDR-based technology has enabled the robust UG phased arrays implementations. Software-defined array elements allow the development of steering solutions with the ability to process complex algorithms. These steering solutions can be used by antenna elements to communicate with static and mobile AG nodes. SDR-based UG beamforming [28] faces various challenges for its implementation. Among all challenges, different phase shifts of antenna element is the major one. A desired beam pattern requires the same phase shift for all antenna element in a particular direction. This can be done by doing proper calibration of phase shifter, real-time synchronization, and phase correction for all the elements.

Soil moisture-based digital beamforming method uses planar antenna array. It consists of its own phase shifter and is programmed on predefined communication parameters for communicating with the UG and AG arrays. Multiple beams can be stitched in such a way that the total number of beam patterns can be calculated after analyzing the UG and AG devices and can be stored in a database for later use [48].

SDR-based phase shifting approach dynamically adapts to the wavelength with the change in soil moisture values. The processing is done in the software-defined radios. The major benefit of this technique is that it adapts to the uncertainties of UG environment without changing the arrangement of the antenna array, hence, consuming less energy as compared to traditional mechanical techniques [13, 38].

7.7.2 Hardware Components

Hardware-based solutions for UG phased array include using dipole antenna elements and printed circuit antenna elements. Moreover, other microwave components, e.g., amplifiers, phase shifters, dividers, and hybrids can be used as printed circuits [13, 41, 59]. To achieve good wideband properties, a stripline configuration includes using a closely spaced antenna elements with large diameter. A prototype system can be designed and tested using EM-based simulations. The performance of an antenna array suffers at the edges. This can be avoided using resistive elements as a dummy element at the edge. Using the results from the simulation-based design, an initial array layout can be configured. This layout can be optimized using vector network analyzer [42, 45].

7.8 Future Directions

As a future research direction, the effect of soil–air interface on communication performance of UG phased array should be investigated in detail. Reflection from the interface can be studied by adjusting the array factor. Moreover, the receiving array should also be analyzed in the context of mutual coupling between array elements, antenna reciprocity principle, impedance matching, and simultaneous sending and receiving functionality [43, 44, 47].

It is important to realize UG phased array with soil moisture sensing capabilities. It is a challenging task incurring more complexity in the system. UG phased antenna array is an enabling technology for the next-generation wireless UG systems due to its decreased cost, lower hardware complexity, extended communication range, and high data rate. It can be considered as a serious contender to replace traditional OTA solutions.

References

1. Akyildiz IF, Stuntebeck EP (2006) Wireless underground sensor networks: research challenges. Ad Hoc Netw J 4:669–686
2. Akyildiz IF, Sun Z, Vuran MC (2009) Signal propagation techniques for wireless underground communication networks. Phys Commun J 2(3):167–183
3. Alexandropoulos GC, Ferrand P, Gorce JM, Papadias CB (2016) Advanced coordinated beamforming for the downlink of future LTE cellular networks. IEEE Commun Mag. https://doi.org/10.1109/MCOM.2016.7509379
4. Anand N, Lee SJ, Knightly EW (2012) Strobe: actively securing wireless communications using zero-forcing beamforming. In: INFOCOM, 2012 proceedings. IEEE, Piscataway, pp 720–728. https://doi.org/10.1109/INFCOM.2012.6195817
5. Aryafar E et al (2013) Adam: an adaptive beamforming system for multicasting in wireless lans. IEEE/ACM Trans Netw. https://doi.org/10.1109/TNET.2012.2228501
6. Bogena HR, Herbst M, Huisman JA, Rosenbaum U, Weuthen A, Vereecken H (2010) Potential of wireless sensor networks for measuring soil water content variability. Vadose Zone J 9(4):1002–1013
7. Dobson M et al (1985) Microwave dielectric behavior of wet soil—Part II: dielectric mixing models. IEEE Trans Geosci Remote Sens GE-23(1):35–46. https://doi.org/10.1109/TGRS.1985.289498
8. Dong X, Vuran MC (2011) A channel model for wireless underground sensor networks using lateral waves. In: Proc. of IEEE Globecom '11, Houston, TX
9. Dong X, Vuran MC (2013) Impacts of soil moisture on cognitive radio underground networks. In: Proc. IEEE BlackSeaCom, Georgia
10. Dong X, Vuran MC, Irmak S (2012) Autonomous precision agriculture through integration of wireless underground sensor networks with center pivot irrigation systems. Ad Hoc Netw 11(7):1975–1987
11. Du Y et al (2014) iBeam: intelligent client-side multi-user beamforming in wireless networks. In: IEEE INFOCOM 2014. https://doi.org/10.1109/INFOCOM.2014.6848009
12. Elliott RS (1981) Antenna theory and design. Prentice-Hall, Englewood Cliffs
13. Fenn A, Hurst P (2015) Ultrawideband phased array antenna technology for sensing and communications systems. MIT Press, London
14. Godara LC (1997) Application of antenna arrays to mobile communications. II. Beam-forming and direction-of-arrival considerations. Proc IEEE 85(8):1195–1245. https://doi.org/10.1109/5.622504
15. Gross F (2015) Smart antennas with MATLAB. McGraw-Hill, New York
16. Guo H, Sun Z (2014) Channel and energy modeling for self-contained wireless sensor networks in oil reservoirs. IEEE Trans Wirel Commun 13(4):2258–2269. https://doi.org/10.1109/TWC.2013.031314.130835
17. Hansen R (2009) Phased array antennas. Wiley, New York
18. Haupt R (2015) Timed arrays. Wiley, New York
19. King RWP, Smith G (1981) Antennas in matter. MIT Press, Cambridge
20. King RWP, Owens M, Wu TT (1992) Lateral electromagnetic waves. Springer, New York
21. Kisseleff S, Akyildiz IF, Gerstacker W (2015) Beamforming for magnetic induction based wireless power transfer systems with multiple receivers. In: 2015 IEEE GLOBECOM. https://doi.org/10.1109/GLOCOM.2015.7417006
22. Konda A, Rau A, Stoller MA, Taylor JM, Salam A, Pribil GA, Argyropoulos C, Morin SA (2018) Soft microreactors for the deposition of conductive metallic traces on planar, embossed, and curved surfaces. Adv Funct Mater 28(40):1803020. https://doi.org/10.1002/adfm.201803020
23. Kontaxis D et al (2016) Optimality of transmit beamforming in spatially correlated MIMO Rician fading channels. Wirel Pers Commun 88:371–384

24. Lakshmanan S, Sundaresan K, Kokku R, Khojestepour A, Rangarajan S (2009) Towards adaptive beamforming in indoor wireless networks: an experimental approach. In: INFOCOM 2009. IEEE, Piscataway. https://doi.org/10.1109/INFCOM.2009.5062199
25. Mailloux R (2005) Phased array antenna handbook. Artech House, London
26. Markham A, Trigoni N (2012) Magneto-inductive networked rescue system (MINERS): taking sensor networks underground. In: Proceedings of the 11th ICPS, IPSN '12. ACM, New York, pp 317–328. https://doi.org/10.1145/2185677.2185746
27. Nitsche T et al (2015) Steering with eyes closed: Mm-wave beam steering without in-band measurement. In: IEEE INFOCOM. https://doi.org/10.1109/INFOCOM.2015.7218630
28. Quitin F et al (2013) A scalable architecture for distributed transmit beamforming with commodity radios: design and proof of concept. IEEE Trans Wirel Commun. https://doi.org/10.1109/TWC.2013.012513.121029
29. Salam A (2018) Pulses in the sand: long range and high data rate communication techniques for next generation wireless underground networks. ETD collection for University of Nebraska - Lincoln (AAI10826112). http://digitalcommons.unl.edu/dissertations/AAI10826112
30. Salam A (2019) A comparison of path loss variations in soil using planar and dipole antennas. In: 2019 IEEE international symposium on antennas and propagation. IEEE, Piscataway
31. Salam A (2019) Design of subsurface phased array antennas for digital agriculture applications. In: Proc. 2019 IEEE international symposium on phased array systems and technology (IEEE Array 2019), Waltham, MA
32. Salam A (2019) A path loss model for through the soil wireless communications in digital agriculture. In: 2019 IEEE international symposium on antennas and propagation. IEEE, Piscataway, pp 1–2
33. Salam A (2019) Sensor-free underground soil sensing. In: ASA, CSSA and SSSA international annual meetings (2019), ASA-CSSA-SSSA
34. Salam A (2019) Subsurface MIMO: a beamforming design in internet of underground things for digital agriculture applications. J. Sens. Actuator Netw. 8(3). https://doi.org/10.3390/jsan8030041, https://www.mdpi.com/2224-2708/8/3/41
35. Salam A (2019) Underground environment aware MIMO design using transmit and receive beamforming in internet of underground things. Springer International Publishing, Cham, pp 1–15
36. Salam A (2019) An underground radio wave propagation prediction model for digital agriculture. Information 10(4):147
37. Salam A (2019) Underground soil sensing using subsurface radio wave propagation. In: 5th global workshop on proximal soil sensing, Columbia, MO
38. Salam A (2020) Internet of things for environmental sustainability and climate change. Springer International Publishing, Cham, pp 33–69. https://doi.org/10.1007/978-3-030-35291-2_2
39. Salam A (2020) Internet of things for sustainability: perspectives in privacy, cybersecurity, and future trends. Springer International Publishing, Cham, pp 299–327. https://doi.org/10.1007/978-3-030-35291-2_10
40. Salam A (2020) Internet of things for sustainable community development, 1st edn. Springer Nature, Cham. https://doi.org/10.1007/978-3-030-35291-2
41. Salam A (2020) Internet of things for sustainable community development: introduction and overview. Springer International Publishing, Cham, pp 1–31. https://doi.org/10.1007/978-3-030-35291-2_1
42. Salam A (2020) Internet of things for sustainable forestry. Springer International Publishing, Cham, pp 147–181. https://doi.org/10.1007/978-3-030-35291-2_5
43. Salam A (2020) Internet of things for sustainable human health. Springer International Publishing, Cham, pp 217–242. https://doi.org/10.1007/978-3-030-35291-2_7
44. Salam A (2020) Internet of things for sustainable mining. Springer International Publishing, Cham, pp 243–271. https://doi.org/10.1007/978-3-030-35291-2_8
45. Salam A (2020) Internet of things for water sustainability. Springer International Publishing, Cham, pp 113–145. https://doi.org/10.1007/978-3-030-35291-2_4

46. Salam A (2020) Internet of things in agricultural innovation and security. Springer International Publishing, Cham, pp 71–112. https://doi.org/10.1007/978-3-030-35291-2_3
47. Salam A (2020) Internet of things in sustainable energy systems. Springer International Publishing, Cham, pp 183–216. https://doi.org/10.1007/978-3-030-35291-2_6
48. Salam A (2020) Internet of things in water management and treatment. Springer International Publishing, Cham, pp 273–298. https://doi.org/10.1007/978-3-030-35291-2_9
49. Salam A (2020) Wireless underground communications in sewer and stormwater overflow monitoring: radio waves through soil and asphalt medium. Information 11(2), 98
50. Salam A, Karabiyik U (2019) A cooperative overlay approach at the physical layer of cognitive radio for digital agriculture. In: Proceedings of the 3rd international Balkan conference on communications and networking (2019 BalkanCom)
51. Salam A, Shah S (2019) Internet of things in smart agriculture: enabling technologies. In: 2019 IEEE 5th world forum on internet of things (WF-IoT). IEEE, Piscataway, pp 692–695
52. Salam A, Vuran MC (2016) Impacts of soil type and moisture on the capacity of multi-carrier modulation in internet of underground things. In: Proc. of the 25th ICCCN 2016, Waikoloa, Hawaii
53. Salam A, Vuran MC (2017) Em-based wireless underground sensor networks, pp 247–285. https://doi.org/10.1016/B978-0-12-803139-1.00005-9
54. Salam A, Vuran MC (2017) Smart underground antenna arrays: a soil moisture adaptive beamforming approach. In: Proc. 36th IEEE INFOCOM 2017, Atlanta
55. Salam A, Vuran MC (2017) Wireless underground channel diversity reception with multiple antennas for internet of underground things. In: Proc. IEEE ICC 2017, Paris
56. Salam A, Vuran MC (2018) EM-based wireless underground sensor networks. In: Pamukcu S, Cheng L (eds) Underground sensing. Academic, London, pp 247–285. https://doi.org/10.1016/B978-0-12-803139-1.00005-9
57. Salam A, Vuran MC, Irmak S (2016) Pulses in the sand: impulse response analysis of wireless underground channel. In: The 35th annual IEEE international conference on computer communications (INFOCOM 2016), San Francisco
58. Salam A, Vuran MC, Irmak S (2017) Towards internet of underground things in smart lighting: a statistical model of wireless underground channel. In: Proc. 14th IEEE international conference on networking, sensing and control (IEEE ICNSC), Calabria
59. Salam A, Hoang AD, Meghna A, Martin DR, Guzman G, Yoon YH, Carlson J, Kramer J, Yansi K, Kelly M et al (2019) The future of emerging IoT paradigms: architectures and technologies. https://doi.org/10.20944/preprints201912.0276.v1
60. Salam A, Vuran MC, Dong X, Argyropoulos C, Irmak S (2019) A theoretical model of underground dipole antennas for communications in internet of underground things. IEEE Trans Antennas Propag 67(6):3996–4009
61. Salam A, Vuran MC, Irmak S (2019) Di-sense: in situ real-time permittivity estimation and soil moisture sensing using wireless underground communications. Comput Netw 151:31–41. https://doi.org/10.1016/j.comnet.2019.01.001, http://www.sciencedirect.com/science/article/pii/S1389128618303141
62. Staiman D, Tamir T (1966) Nature and optimisation of the ground (lateral) wave excited by submerged antennas. Proc Inst Electr Eng 113(8). https://doi.org/10.1049/piee.1966.0220
63. Sun Z, Akyildiz I (2010) Channel modeling and analysis for wireless networks in underground mines and road tunnels. IEEE Trans Commun. https://doi.org/10.1109/TCOMM.2010.06.080353
64. Sun Z, Akyildiz I (2010) Magnetic induction communications for wireless underground sensor networks. IEEE Trans Antennas Propag 58(7):2426–2435. https://doi.org/10.1109/TAP.2010.2048858
65. Sun Z et al (2011) MISE-PIPE: MI based wireless sensor networks for underground pipeline monitoring. Ad Hoc Netw 9(3):218–227
66. Sun Z, Wang P, Vuran MC, Al-Rodhaan MA, Al-Dhelaan AM, Akyildiz IF (2011) Border patrol through advanced wireless sensor networks. Ad Hoc Netw 9(3):468–477

67. Tan X, Sun Z, Akyildiz IF (2015) Wireless underground sensor networks: MI-based communication systems for underground applications. IEEE Antennas Propag Mag 57(4). https://doi.org/10.1109/MAP.2015.2453917
68. Temel S, Vuran MC, Lunar MM, Zhao Z, Salam A, Faller RK, Stolle C (2018) Vehicle-to-barrier communication during real-world vehicle crash tests. Comput Commun 127:172–186
69. Tiusanen MJ (2006) Wideband antenna for underground Soil Scout transmission. IEEE Antennas Wirel Propag Lett 5(1):517–519
70. Vuran MC, Akyildiz IF (2010) Channel model and analysis for wireless underground sensor networks in soil medium. Phys Commun 3(4):245–254. https://doi.org/10.1016/j.phycom.2010.07.001
71. Vuran MC, Dong X, Anthony D (2015) Antenna for wireless underground communication. US Patent App. 14/415,455. http://www.google.com/patents/US20150181315
72. Vuran MC, Salam A, Wong R, Irmak S (2018) Internet of underground things in precision agriculture: architecture and technology aspects. Ad Hoc Netw. https://doi.org/10.1016/j.adhoc.2018.07.017, http://www.sciencedirect.com/science/article/pii/S1570870518305067
73. Vuran MC, Salam A, Wong R, Irmak S (2018) Internet of underground things: sensing and communications on the field for precision agriculture. In: 2018 IEEE 4th world forum on internet of things (WF-IoT) (WF-IoT 2018)
74. Widrow B et al (1967) Adaptive antenna systems. Proc IEEE. https://doi.org/10.1109/PROC.1967.6092

Part III
Novel Soil Sensing Techniques

Chapter 8
Signals in the Soil: Subsurface Sensing

8.1 Introduction

Soil is a primary resource in agriculture. Soil fertility is ability of a soil to sustain the plant growth. Several physical (particle size, structure, water, etc.), chemical (mineralogy, organic matter, and acidity, etc.), and biological properties (beneficial organism) are used by the scientist to describe soil fertility (see Table 8.1). It is important to know these properties of soil fertility in order to optimize the plant production. However, complex nature of the soil makes it difficult to assess the soil fertility. Soil properties may vary on micro or macro level. Micro-variability is due to granularity of the soil. Macro-variability is due to the climate, parent material, time, and how human treats the soil [93]. Soil properties show various spatial and temporal variations. Observing soil in different areas within the field or between multiple fields to study its spatial variations is known as soil survey. Studying the temporal variations refers to the soil monitoring [41, 166].

Traditional approach to investigate soil fertility involves manual collection of soil samples from the field and analyzing them in laboratories. This method specially applies to study the chemical properties and soil texture. However, manual sampling of soil is time intensive, requires a lot of labor work, and highly expensive. Due to these drawbacks, studying spatial and temporal properties of soil, using traditional approach, does not make digital agriculture a viable farming method. The implementation and widespread of digital agriculture rely on use of fast and cost-effective methods [169, 190].

Another approach to study soil properties is to use sensor-based technologies. Soil sensors collect the data for sensing chemical, biological, and physical properties directly from the field. However, due to the complex nature of the soil, only few sensor systems have been successful for use in agriculture. Complex nature of soil makes it difficult to separate the sensor stimuli. The minerals components in the soil cause the mechanical stress and wear in the sensor. Soil absorbs or attenuates most of the EM spectrum due to which remote sensors receive signal from the surface

A. Salam, U. Raza, *Signals in the Soil*,
https://doi.org/10.1007/978-3-030-50861-6_8

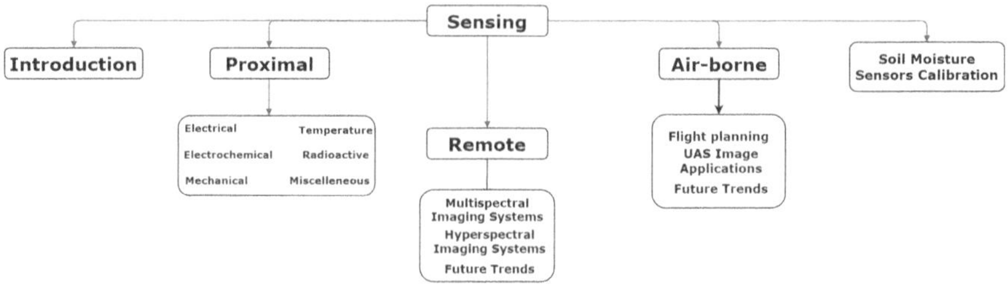

Fig. 8.1 Organization of the chapter

of the soil only. Hence, the success of mapping soil through remote sensing is very limited. Sensing has always been a key technology in the farming industry. In earlier days, farmers used to evaluate the crop properties manually by assessing each plant individually. They used to estimate yield potential, identify stresses from specific symptoms, differentiate water deficiency from nutrient deficiency, and identify diseases from insect infestation. Modern farmers are no different than their ancestors, however, scale of the crops is a major concern nowadays. Farmers, in old days, used to manage small portion of the land, e.g., some fraction of the hectare, however, modern farmers may have to manage hundreds even thousands of hectares of fields. Large field size makes it impossible to manually analyze and manage them. Hence, farmers are becoming increasingly reliant on sensors in their day-to-day farming operations [164, 210] (Fig. 8.1).

In this section we are going to discuss different types of sensor technologies that can be used in monitoring of soil and crop conditions.

8.2 Current Challenges in Sensor Development

EM-based crop canopy sensors have high potential to improve N-fertilizer management, however, some factors may influence its accuracy. These factors may include sensors' operating characteristics (wavelengths and VIs), seasonal variations, genotype effects, and stresses [167, 173]. N-fertilizer management, using proximal crop canopy sensors, assumes that the effect of other stresses, e.g., water, insects, and nutrient, is either absent or equally present in reference and target area, thus, canceling the effect of each other. Even in the irrigated environments, water stress can confound N management during the critical times of the growing season. Some VIs are found to be successful differentiating between N stress and water stress [26]. These indices include three wavebands in the equation (eq. 1) namely the canopy chlorophyll content index [57], the DATT index [39, 141], and the MERIS terrestrial chlorophyll index [38, 157]. However, these three indices are currently not available in any of the commercially available crop canopy sensor system.

Most of the research with proximal crop sensors is being done in N-fertilizer management. However, there is a huge potential of application of proximal crop sensor in identifying other stresses. Few commercial systems are available which identify plant stresses other than the N stress. WeedSeeker spray system [87] uses the same sensors as GreenSeeker system. It successfully finds the weed in the crop and applies herbicides on the go. WeedSeeker system successfully solves the weed infestation problem in cotton [170, 195]. Salam [151] and Slaughter et al. [187] studies machine vision-based weed detection and control systems using reflectance with the crop sensor for real-time weed control. Crop canopy sensor can also be used in disease and insect detection. High-resolution and multispectral imagery is used to detect disease onset in wheat. Feng et al. [63] use visible to red-edge bands (580–710 nm) to inspect powdery mildew infestation during growth season. Hyperspectral imaging is used to detect damages in sugar beet due to nematodes and fungal pathogens [79]. In Bravo et al. [24], authors used multispectral fluorescence images for the detection of foliar diseases in winter wheat. Huanglongbing disease in citrus canopy is detected by using the ratio of yellow fluorescence to simple fluorescence.

Another trending current of area of research is the concept of sensor fusion. Sensor fusion refers to using multiple proximal crop sensors over single sensor. Combining various sensing techniques can give more accurate measurements promoting the widespread adoption of sensor-based technologies from crop management [5, 150]. Veris Technologies combines multiple sensors for soil electrical conductivity (EC), pH level of the soil, and soil organic matter to a single platform [209]. Salam and Vuran [164] and Shiratsuchi et al. [186] use three different types of sensors: optical for canopy reflectance, thermal for temperature, and ultrasonic distance sensor for height, to assess the N status in the crop. The authors in [213] combine four different sensors to study the herbicide dosage control. Sensor fusion in [108, 197] documents the information about crop and straw yield and grain protein content for stress evaluation in harvesting wheat. Salam [142] and Lamb et al. [103] use the combination of active and passive sensors to map a cotton field. Sensor fusion is also used to study high throughput phenotyping. A sensor fusion system for field phenotyping in [28, 148] integrates a color camera, Light Detection and Ranging (LIDAR), time-of-flight cameras, light curtain imaging systems, and a hyperspectral imaging sensors. The system could measure plant moisture, lodging biomass yield and tiller density. High throughput phenotyping platforms [10, 154, 210] use multispectral active crop canopy sensors, LIDAR or ultrasonic sensors, and thermal sensors. Crop Circle Phenom [85, 210] incorporates sensors for canopy temperature, humidity atmospheric pressure, air temperature, and canopy reflectance wavelengths of 670, 730, and 780 nm.

8.2.1 *Proximal Sensing*

The term proximal indicates the close proximity of the sensors to the crops. They are deployed close to the crops; thus, different from remote sensors. Due to advancement in unmanned aerial systems (UASs), which used to be a remote sensing technique, is now considered as a proximal sensing technique. We consider the proximal sensors as the sensors which determine and evaluate the crops' properties ranging from physical contact with canopy to a few meters above the canopy [165, 172].

Crop producers are interested in the properties which may influence yield and quality, and issues which can be detected and managed during crop producing seasons. Some of the properties that may interest a farmer include biomass accumulation, water status, nutrient deficiency (particularly nitrogen), disease onset, and insect infestation. Monitoring these properties is crucial during some specific periods of growing seasons. Remote sensing, e.g., aerial platforms, is easily affected by weather and clouds condition limiting their use. However, proximal sensing is not affected by these making it an appropriate choice during growing season [165, 172].

- **Contact or in-situ sensors:** Some sensors are either directly attached or placed among the plants. Sap flow sensors, attached to the stem of the plant, are used to estimate transpiration [149, 171, 191]. Ground-based sensors are used to measure leaf area index (LAI) [25, 144]. However, measuring sap flow and LAI with static instrument is a time-consuming process. Crop-meter [56, 170] is a simple biomass sensor for the cereals which works on the pendulum principle. Crop-meter is mounted on vehicle driven through the crops. If crops' biomass is higher, crop-meter is highly deflected.
- **Ranging Sensors**: Range-finding or distance measurement is another approach to find biomass and height of the crop. Salam [148] and Dworak et al. [55] measure the biomass, characteristics of the canopy, and crop stand using acoustic and electromagnetic (EM) wave ranging. Geometric characteristics of the citrus trees can be quantified by laser scanner. It is used for yield prediction, measurement of water consumption, health, and long-term productivity monitoring of the crop [104]. Water stress in maize can be detected by integrating acoustic ranging sensors with multispectral and thermal sensors [162, 186].
- **Electromagnetic (EM) Sensors:** Most of the crop canopy sensors use EM spectrum. EM sensors can be classified as active-with internal energy source, or passive-using external emitted energy source. The four regions of interest in EM spectrum include visible, near-infrared, mid-infrared, and thermal infrared ranging from 400 nm (visible) to 14,000 nm (thermal infrared) [100, 143]. The properties of the plants, e.g., plant pigment concentrations, cell structure, water content in canopy and leaves, are determined by reflection or transmission in these spectrum ranges. Reflectance does not give much information about plant stress. Vegetation index (VI) allows to infer the plant stress and specific properties of plants by using the relationship between or reflectance in multiple spectral regions. Among 150+ VIs, the most famous VI is normalized difference VI (NDVI) [138, 197]. NDVI is calculated using following equation:

$$\mathrm{NDVI} = \frac{(\mathrm{NIR} - \mathrm{RED})}{(\mathrm{NIR} + \mathrm{RED})} \tag{8.1}$$

Chlorophyll meter, frequently used instrument in the plant research, is used to measure the chlorophyll content in a plant. It is based on the reflectance meter presented by Wallihan [212]. Konica Minolta SPAD 502 Plus [118] is a commonly used and mostly available chlorophyll meter today. The meter is used in determining chlorophyll concentration [113] at 650 and 940 nm of wavelengths. The performance of this meter has been studied [113] for determining the chlorophyll content and ability to detect N stress and, assuming the high correlation between chlorophyll and nitrogen content in a plant, consequently predict the demand for N fertilization [22, 27, 152, 179].

The Dualex Scientific [51, 154] is another handheld senor which measures the chlorophyll and polyphenol contents in ultraviolet, visible, and NIR wavelengths. Polyphenols can be used to determine plant stress factors such as N availability [157, 173]. When used together, Dualex and SPAD meter are more sensitive in calculating the corn N status rather than being used alone. Force-A also manufactures the handheld sensors used to determining abiotic stresses in the plant especially in detecting disease [145, 151, 161, 175].

Chlorophyll meters are useful in estimating the properties of a given plant, however, it becomes laborious to measure the properties of a group of a plants or fields. It is a useful proximal sensing tool for a plot of small size. However, the labor challenges associated with the chlorophyll meter prevent it to be used by the farmers for N-fertilizer management.

- **Mobile EM Sensors:** Yara N-Sensor [150, 223] is a passive spectrometer system that can be mounted on a tractor. It consists of two spectrometers which are used to scan crop canopy and real-time correction of the reflected signal in a wavelength selected between 450 and 900 nm [159, 173, 226]. NDVI and various other VIs are calculated through reflectance. The system then adjusts the application rate of N fertilizers for scanned region in real-time. N-Sensor is majorly used for spatial N management of wheat [21], however, its applications are also found in corn [156, 202] and potato [155, 226].

  Passive sensors face the challenges of clouds, angle of the sun, and time of the day. Active sensors have been developed to address the challenges of passive sensors. Active sensors, as discussed earlier, use internal light source [42, 160]. Optical sensors, a type of proximal active sensors, use radiometric principle. Holland et al. [84] discuss how this principle, especially inverse square law of optics, affects the functionality and use of the sensor. Salam [153] and Raun et al. [132] showed that, as compared to the traditional methods, sensor-based N application methods can increase the nitrogen use efficiency (NUE) by 15%. This system was developed by Oklahoma State University and was named GreenSeeker. GreenSeeker system [87, 168] is an active crop canopy sensor system commercially available for the N-fertilizer management. It uses an in-field reference for the calibration of the sensor for specific field conditions. It measures reflectance at the wavebands of 671 ± 6 and 780 ± 6 nm. The system generates the response index (RI) by dividing the in-field reference NDVI by targets' NDVI. RI and in-season estimate of yield (INSEY) are used in various crop and locale specific algorithms to generate the real-time fertilizer N rate.

This system can be used for many crops, however, majorly used for corn [2, 169] and winter wheat [147, 196]. Handheld version of this sensor has also been developed by Trimble for the exploration purposes. It comes with an integrated power supply datalogger and GPS.

Crop circle suite of sensors, initially developed by Holland Scientific as ACS-210, is set of active crop canopy sensor. ACS-210 is two-band active sensor which measures reflectance at 590 ± 5.5 and 880 ± 10 nm. It is not used commercially anymore and is replaced by various other three-band sensors such as ACS-430 and ACS-470. They also have handheld sensor, RapidScan CS-45, equipped with a datalogger, power supply, and a GPS. It has the optics similar to ACS-430.

The set of sensors described above are primarily used for the research purposes. Holland Scientific has integrated the technology into OptRx®crop sensor system [152, 192] for the commercial use. OptRX uses the ACS-430 and measures the reflectance at wavebands of 670, 730, and 780 nm. OptRX, like GreenSeeker system, uses in-field calibration. OptRX is designed to be used with a universal N recommendation algorithm which can either be adjusted by the user [82, 162] or virtual reference approach [83, 144]. In virtual reference approach, healthiest area of the field is selected as a reference N status of the crop and the area is scanned in real-time. This prevents the grower to establish a nitrogen rich area as a reference strip in the field. Yara N-Sensor ALS (active light system) uses xenon flash lamp as a light source. CropSpec system uses laser as a light source and measures reflectance at 730–740 nm and 800–810 nm bands. The operating characteristics of both, CropSpec and N-Sensor ALS, are similar to those of passive N-Sensor.

Sensors sense analog physical or chemical stimuli, e.g., temperature, heat, etc., and convert them to a digital signal for further analyses. Data analysis methods are used to get information from this data and finally incorporated in decision making to take appropriate agricultural decisions. As discussed in previous section, complex nature of soil requires sensors to be placed in the soil at very short distance from the soil. Proximal sensing is highly effective in creating high-resolution soil maps. Salam and Shah [163], Salam et al. [172], and Rossel et al. [137] define proximal sensing as the application and development of sensor that operates close by or inside the soil. A proximal soil sensor system ranges from simple stationary systems with one or more sensors and data recording and transmission unit to complex automated and mobile systems. Complex proximal system may be equipped with a platform to carry sensors, sampling unit, sample heads, sample preparation equipment, and GNSS.

8.2.1.1 Future Research Directions

Proximal sensing is used to achieve sustainability and reducing the environmental impact that may occur due to crops. Proximal sensing enables spatial management and manages temporal variations for the site-specific management in precision

Table 8.1 Soil properties relevant for plant production as indicated by the notion of soil fertility [69]

Physical	Chemical	Biological
Particle size: gravel, sand, loam, clay	Mineralogy: quartz, clay type	Mineralization fixation
Water: water content, water potential	Organic matter: total content, fractions (labile, stable)	Beneficial organisms
Structure: bulk density, porosity	Nutrients: total content, plant available (e.g., N, P, K)	Bioturbation by animals and plants
Thermal properties: thermal diffusivity, heat capacity, heat conductivity, specific heat	Acidity pH	
Relief: slope, exposure	Redox potential: O_2	
	Toxic substances	

agriculture. Proximal sensors allow an efficient use of resources by detecting and taking preemptive measures in a timely manner.

There is a huge research potential in developing active proximal sensors for use with UAV. Apart from N stress, it can be used to explore water and nutrient stress and many other diseases. There is potential to study different stresses using multispectral fluorescence, mid-infrared, and thermal region of EM spectrum. Non-EM sensors, e.g., pheromone and spore detectors can be investigated for disease or insects.

8.2.2 *Electrical Soil Sensing*

Some sensors use electric circuits to measure the electrical conductivity (EC) of a soil, capacity of the soil to become polarized, or form magnetic fields. Such sensors are known as electrical sensors [105, 143]. Electrical sensors are assessed in the frequency range of 0 (direct current) to 300 (radar) GHz. Electrical sensors are the most common proximal soil sensors. In the next section, we discuss relevant electrical properties of soil [99, 178].

8.2.2.1 Electrical Soil Properties

Electrical conductivity (EC) of a soil refers to the amount of salt in the soil (salinity). It indicates the health of the soil. EC affects the crop growth, crop suitability, and plant nutrient availability (RSEC). EC in soil can be due to movement of free electrons, movement of ions in dissolved water, and surface conductivity [99]. EC due to all three mechanisms is known as bulk electrical conductivity (ECa). The ECa is mainly associated with the properties of soil such as: water content, hydraulic permeability, temperature, bulk density, and surface changes. In non-saline soils, spatial variations of soil ECs within the field are due to the soil texture. Coarse sands have limited contact and low capacity of holding moisture, hence are poor conductors. Heavy clays have high capacity of holding moisture and high particle contact, hence are good conductors.

Tendency of a medium to become polarized upon passing electricity is its dielectric permittivity [159, 178]. Dielectric permittivity increases with the decrease in frequency. Soil sensors measure the permittivity at frequencies between 1 MHz and 1 GHz. It is used to measure the soil–water content. Dielectric permittivity of free water is 80 [155, 178]. The electromagnetic sensors cannot measure the dielectric permittivity of the soil directly. They use the travel time or frequencies to derive the values. As a result, other interfering soil properties (ECs, temperature, magnetic permeability, and signal frequency) are also taken into account while measuring these properties [61, 119, 158].

The ability of the soil to form a magnetic field is known as magnetic permeability. Ferromagnetic compounds such as iron oxide and super paramagnetic minerals can be found in the soil due to atmospheric deposition and human activities [75, 100, 119]. If present, the magnetic permeability is proportional to the volume of these compounds. Many proximal sensors measure soil susceptibility which is the ratio of mediums' permeability to permeability of free space minus one. Magnetometers or EM sensors are used to measure the magnetic properties of the soil. Magnetic properties of soil, both susceptibility and permeability, are mainly used in the field of environmental pollution and archeology [75, 165]. Very limited work has been published on the use of magnetic properties in soil mapping from agricultural context [15, 171]. All of the above three properties are highly dependent upon the frequency of applied EM field. Lower frequency methods are more related to EC and high frequency methods, e.g., TDR and radar, are more associated with dielectric and magnetic permeability.

8.2.2.2 Electrical Soil Sensors

Galvanic coupled resistivity (GCR) measures the bulk electrical resistivity (ERa) in Ω-m under the low frequencies of less than 50 Hz. High frequencies can be used to analyze polarization effect. It uses two electrodes, in direct contact with the soil, and an ohmmeter to measure the electricity resistance by the soil. Wheatstone bridge uses four electrodes, in pair of two, to measure more accurate readings. One pair is used to inject electric current into the soil and the other is used to measure the potential difference. GCR is a relatively cheap, robust, and low power method of measuring electrical resistivity. Contrary to other electrical methods, GCR is less sensitive to electromagnetic sources. Main disadvantages of GCR are that it requires good contact with the soil and the invasive nature of galvanic coupling. The performance of electromagnetic induction (EMI) methods is much better in frozen, stony, or dry soils as compared to GCR methods [149, 211].

Electromagnetic induction (EMI) sensors are greatly impacting the digital agriculture [37, 77]. It uses two electrical coils (solenoids) and operates at the frequencies of 100–60 kHz (0.4–40 kHz, [37, 146]). On applying alternating current (AC) to the transmitter coil, an EM field termed as primary field is generated. This primary field induces an eddy current through the soil, which results in a secondary field. The secondary field differs from primary field in terms of amplitude and phase [116, 153]. The phase and amplitude difference between primary and secondary field depends upon various soil properties [12, 145], spacing between the transmitting and receiving coil, distance between and coil and soil surface, and orientation of the coils. ECa is calculated by using amplitude and phase differences between primary and secondary fields, and inter-coil spacing. Tuller and Islam [203] study the magnetic susceptibility of the soil. However, few studies in agriculture use it.

Capacitive methods use the capacitor principle to analyze the soil properties. Electrical oscillator is connected to electrode at the frequency of 0.1–0.25 GHz (38–150 MHz) to create an electric field penetrating the dielectric medium (soil). The dielectric permittivity of the soil can be determined by estimating the charging time of the capacitor with that medium [121, 156]. FDR and capacitance probes are mostly used for measuring the water content in the soil. Both are sensitive to clay and temperature variation of the soil. However, they are cheaper with a flexible electrode geometry. Mobile mapping of ERa is also one of the applications of capacitance principle. Capacitive coupled resistivity (CCR) is based upon classical GCR method with only difference of using capacitive plates/antennae instead of galvanic electrode [102, 160].

Time-domain reflectometry (TDR) is used to determine the water content of the soil by measuring the travel time of electromagnetic waves through the soil under high frequencies. The travel time is used to measure the dielectric permittivity of the soil which in turn is used to measure the soil–water content. TDR sends the electromagnetic signals via two electrodes buried in the soil. It measures the propagation velocity of a step voltage with a bandwidth around 20 kHz to 1.5 GHz [135, 161]. TDR is less affected by the interference due to EC because they operate at frequency >0.5 Hz.

Other variants of TDR and FDR are also used as electrical measurement methods. Some of them discussed briefly in [121] includes amplitude domain reflectometry (ADR), phase transition, and time-domain transition (TDT).

8.2.2.3 Stationary Electrical Sensors

Stationary soil sensors are very useful in monitoring soil–water properties. It has advantage over traditional methods, e.g., tensiometers, in having high range of tension (up to 1500 Pa), ability to be placed for long time, and low cost. However, slow reaction and intervention of soil salinity are some of the disadvantages [54, 168].

Stationary soil sensors must be carefully calibrated to get the correct measures [61]. Even a small scale variation can cause several problems. Durner et al. [54], for example, perform an experiment with 107 sensors on 60 m^2 test site. Sensors readings were varied significantly over the two months due to small scale variation (growth of algae) in the soil surface. It is difficult to distinguish the effect of water on electromagnetic properties from other factors such as salinity, bulk density, etc. [90]. It is recommended to use standard procedure, multiple sensors, and rain gauge records for detection of error and troubleshooting [54, 146, 197]. Salam et al. [172] and Muñoz-Carpena et al. [121] present a detailed survey of various commercially available soil sensors. Some sensors are briefly described in coming section.

- **Stationary TDR and capacitance sensors** can be used for stationary measurements. TDR uses metal rods and capacitance sensors are flexible and cheap. TDR gives more accurate reading under high frequencies, however, comparative studies suggest mixed results [61, 148, 162]. Durner et al. [54] studied 15 different types of sensors (FDR, TDR, capacitance) and posited that most sensors respond to temporal changes reasonably well, however, with a considerable difference in absolute values.
- **Gypsum Block method** is another method to assess the soil–water content. Apparatus contains a porous block with embedded electrodes. In an ideal scenario, the soil–water content and the bock water should be uniform. Fiber glass and gypsum can be used as a porous block. Gypsum is a cheaper option but it needs calibration and it decomposes with time. Fiberglass is durable and gives more stable calibrations. The simplest implementation consists of a gypsum block with two wires connecting to resistivity meter.
- **Electrical resistivity tomography** is a method which is used to study the depth profiles and spatio-temporal properties of soil–water content in high resolution. It extends GCR method with array of multiple electrodes (20–100) placed equidistantly in a transect. Four of the electrodes placed at different locations and spacing are switched on to study the depth variations[94, 210, 211].

8.2.2.4 Mobile Electrical Sensors

Mobile electrical sensors were the first sensors used to measure the soil spatial variability in digital agriculture [151, 161, 174]. EMI and GCR based sensor systems are still the most commonly used systems in digital agriculture. Salam [150, 158] and Gebbers et al. [70] did a comparative study on GCR, EMI, and capacitive coupled sensors which is discussed in the coming section.

- **Galvanic coupled resistivity** Mobile GCR sensors, use four-point method, consist of four wheels acting as electrode. Four-point arrangement is extendable by adding more pair of electrodes to get the readings at different depth variations. Depth sensitivity is controlled by the spacing and position of electrodes. The most common arrangement of the electrode is Wenner array equally spaced electrode in a straight line. Veris 3100 uses injected current of 100 mA and frequency of 150 Hz and works with the two depths [193]. Six rolling electrodes are arranged in Wenner array. Spacing between the electrode is 24 cm for the shallow and 72 cm for the deep measurement. Automatic resistivity profiling (ARP) is made by Geocarta company in France. It operates at frequency of 225 Hz and arranges the rolling electrodes in trapezoid pattern [174]. The spacing between the current electrode is 1 m and that of between voltage electrode is 0.5, 1, and 2 m. GEOPHILLUS is an advance system from Germany operating at the frequency of up to 1 kHz [111, 142]. It uses 6 pairs of galvanic coupled electrodes arranged in an equatorial dipole–dipole array. First pair measures the electric current while other 5 pairs measure the voltage drop simultaneously.

- **Electromagnetic induction (EMI)** EM38 instrument from Geonic Ltd., is one of the most popular mobile electromagnetic instruments in digital agriculture. Salam and Vuran [166] and Heil and Schmidhalter [77] recently gave the review of EM38 applications. Orientation and spacing of the coil, and frequency often affect the DOI characteristics. EM38 uses the inter-coil spacing of 1 m and provides the readings for magnetic susceptibility and electrical conductivity simultaneously. EM38-DD with the same inter-coil spacing uses two EM38. EM38-MK2 operates at frequency of 14.5 kHz and uses three coils with two receiver coils separated by 0.5 and 1 m from transmitter coil. Topsoil Mapper (Geo-prospectors, Traiskirchen, Austria) is the first EMI instrument which provides various interesting features specially for the digital agriculture. It can be mounted to the front of the tractor because of its ability to suppress interference from the metal. It can be used for the estimation of bulk density, water content, texture, real-time tillage control, and seed rate.
- **Capacitance and CCR sensors** [3, 167] reviews the mobile sensors that use capacitive principle and galvanic coupling. Mobile capacitive coupled sensors (CCR) have been in the market for 10 years, however, their use in agriculture has been very limited. They work better than the galvanic coupled sensors in places where EC is very low and a mechanical contact is difficult. However, places where EC is high, receiver dipoles' voltage becomes too small and its measurements are unreliable [7, 147]. Coaxial cable or metallic conductors are used as a capacitor plates. One pair is used to generate current in the ground and other is used to measure the potential distribution at the surface of ground [102, 165].

  OhmMapper (Geometrics Inc., San Jose, USA (http://www.geometrics.com/)) is a CCR system with capacitive coupling. It can continuously collect the data even at the short time interval of 0.5 s. A coaxial cable is divided into transmitter and receiver sections and both are of 5 m in length. An alternating current is generated by the transmitter at 16 kHz. It consists of a power supply, a datalogger, and rope separating two dipoles from each other. Distance and length of dipoles are used to measure the DOI. Distances can also be set arbitrarily.
- **Mobile TDR and GPR sensors** GPR systems are available commercially in the market, however, only prototype of TDR systems can be found in the literature. Salam [141] and Thomsen et al. [198] study the modified version of stationary TDR with longer probes designed for stop-and-go measurements. For continuous mobile measurement, geometry of the traditional TDR probes needs to be modified along with the consideration of other aspects such as contact and

heating. Due to time taking data analysis, use of GPR in agriculture is for research purpose only. However, it is commonly used in geophysical and archeology.

8.2.3 Soil Temperature Sensors

- **Thermal sensors** One of the oldest sensor systems in agriculture are thermal sensor systems. Electrical and infrared (IR) thermometers are used to measure the temperature. Electrical thermometers need to be in physically contact with soil, whereas IR ones can be used for stand-off readings. Thermistor and thermocouple are the examples of electrical thermometers. Thermistor relies on change in resistance and thermocouple uses the thermoelectric effect. Electrical thermometers are cheaper and are integrated with various sensors systems such as TDR and FDR. An IR thermometer uses a lens to focus on thermal radiations emitted from an object onto a detector. The lens is sensitive in the range of 0.7–1.4 μm. Thermal properties such as volumetric heat capacity, thermal conductivity, and thermal diffusivity can be utilized for the mineralization, germination, and other growth related processes.
- **Heat-pulse sensors** Heat-pulse sensors [29, 163] are primarily used for measuring the volumetric water content. Sensor consists of at least two probes: heater probe and a temperature probe (thermocouple). A heat pulse is applied at the heater probe and the temperature of the soil is measured at the temperature probe. This approach is dependent on the fact that specific heat of the water is higher than the other constituents of the soil. After applying the heat impulse, temperature depends upon the volumetric heat of the soil medium. More water content results in low temperature rise and vice versa. In commercial systems, heater probe and thermocouple are insulated in a porous matrix. Water potential of the soil should be same as of the matrix. By this assumption, the water content of the matrix becomes the indirect measure of the water content of the soil [133, 152]. Heat-pulse sensors have slow response and are sensitive to soil contact. Due to these reasons they are mainly used for the stationary measurements. A very less information is available on the continuous heat-pulse mobile sensors in the current literature. Authors in [9, 159] describe an approach of IR thermometer measuring the temperature variations in soil because of warming up by solar irradiation.

8.2.4 Electrochemical Soil Sensors

Electrochemical sensors are used to measure the chemical properties of the soil, e.g., nutrient content and pH level of soil. Electrochemical sensor can be categorized into potentiometric, amperometric, and electromechanical biosensors. The working principle is based on chemical interaction between the sensor and chemical component of the liquid. The two most popular methods used for electrochemical sensors are ion-selective electrode (ISE) and ion-selective field-effect transistor (ISFET). The electrochemical sensors for pH use the combination of ISE and ISFET and fall into the category of potentiometer sensors. Salam [157] and Kim et al. [97] extensively review the potentiometer sensors. Their main advantage is that they directly measure the liquids' ions concentration.

The ISE system consists of a sensing membrane (glass, polyvinyl chloride, or metal) and a reference electrode both assembled in a single probe. The potential difference between both electrodes is measured and concentration of selective ions is estimated using Nernst equation (reference for Nernst). ISFET uses field-effect transistor technology along with the selectivity of ISE. The ion-selective membrane acts as a gate electrode and controls the current between the two semiconductor electrodes. ISFET differs from ISE in that it does not have internal solution and the ion-selective membrane is firmly attached on the gate. The pH ISE with antimony membrane is being used in on-the-go commercial systems for pH mapping [5, 100].

8.2.5 Soil Radioactive Radiation Sensors

- **Gamma ray sensors** Gamma rays are produced from decayed nuclei of radioactive elements and have highest energy and lowest wavelength within the EM spectrum [59, 170]. There are naturally occurring nuclides in soil which emit gamma rays in the range of 0.4–2.81 MeV. Large amount of gamma rays is harmful that is why active gamma sensors are avoided due to high energy and ability of penetration into the material. Gamma ray detectors convert the incoming radiation into light photons which are further amplified by photomultiplier and detected by photodetector. Passive sensors detect the gamma photons released from radioactive decay present in the soil. Passive gamma sensing is very good for quick soil mapping because of already established technology, strong theoretical background, and robust instrument that can quickly collect data.

  In proximal sensing, gamma sensors are used as ex-situ systems in both continuous and stop-and-go mode with 90% of the radiations coming from upper 0.3 to 0.5 m of the soil. Bulk density of the soil can affect the readings [112]. Many radionuclides occur naturally in the soil, however, only potassium (^{40}K), uranium (^{238}U), and thorium ^{232}Th are the ones producing gamma rays with sufficient energy and intensity. Radiation not coming from the earth are known as background and mainly come from radon (^{222}Rn) [112].

  Clay (soil texture) is usually correlated with the gamma count because K, U, and Th are incorporated by the major fraction of the clay, i.e., clay minerals [76]. Gamma count has also been formulated through indirect correlation to pH, organic carbon, gravel, and moisture [76, 154]. These indirect relationships are highly dependent upon relation of soil to other properties such as the total K in the soil. These relationships showed high spatial variability which mandated the separate calibrations for each site. Heggemann et al. [76] proposed a general model for predicting soil texture fractions (clay, sand, silt) from non-linear support vector machine (SVM). Most of the studies counts only three regions (K, Th, Ur) and total gamma counts [137], however, some researchers also recommend to study full spectrum [206]. Though not much performance gains in prediction model are shown when the full spectrum is considered [137, 145].

- **Neutron sensors** Neutron sensors are used to measure the soil moisture. It can be classified as active and passive neutron sensors. Neutron probes (active sensors) use the waters' moderator properties for neutrons. The high-energy neutrons from the sensors collide with the H nuclei of water. The number of neutrons scattered back at the device is directly proportional to the water content in the soil. More water content results in more neutrons scattered at device and vice versa. Neutron probes though accurate, however, are very expensive and their operation is strictly regulated by the law because of radioactive neutron source. Hence, they are rarely in use today.
  Cosmic-ray probes are the passive neutron sensors. They measure the low energy neutron (1 keV) generated within the soil by moderation of cosmic-ray neutrons [32, 171]. This moderation is primarily controlled by the soil–water content. This method provides a continuous and aboveground (without contact) method of monitoring of water content. Cosmic-ray neutron sensor can be used to measure the soil–water content over footprint 600 m and depth varies from 0.76 m (dry soils) to 0.12 m (wet soils). The depth of exploration is highly dependent on soil moisture. Soil moisture is calculated from neutron intensity using a universal function which is indifferent to changes in soil chemistry [32, 155]. Mobile probes, mountable on vehicles, are also designed for in-depth spatial investigation [32].

8.2.6 Mechanical Soil Sensors

- **Cone penetrometers** Vertical cone penetrometer is a device that is used to measure the soil resistance to penetration, i.e., soil strength, as they are inserted to the soil. In agriculture, they have been used for a long time to detect the soil compaction [190]. Some of the soil properties that may affect the index of the cone includes bulk density, soil type, soil moisture, and structure of the soil [48, 164]. In 2000, Veris technologies were the first to design the stop-and-go automated cone penetrometers (Profiler 3000) for soil mapping [50, 144]. However, high variation of penetration resistance makes the soil mapping expensive [48, 149]. Therefore, systems with continuous mapping were more favored.

  There were three main approaches to use: (1) horizontal penetrometers to measure horizontal penetrations [8, 156], (2) use of draft force sensors and vertical force sensors between tractor and tillage [78], and (3) measuring fuel consumption of tractor during tillage [23]. In a review of penetrometer and draft/vertical force sensors, [78] emphasizes that if soil texture and moisture is known then variation in penetration resistance can be understood in terms of bulk density. Mapping of bulk density has been done using the integrated multi-sensors system in [184].

- **Tensiometers** Tensiometer is the device used to measure the soil matric potential, i.e., soil moisture tension. A matric potential is found in the water when the water in soil is under tension. The name, tensiometer, owes to its ability of measuring tension. It consists of a porous cup connected to manometer through a water-filled tube, vacuum gauge, pressure transducer, or any other pressure measuring device. The plant needs to overcome the tension to pull water from the soil. Tensiometer is advantageous in that they provide direct and easy to interpret measurements. However, its maintenance requirement is high and range of measurement is limited. They need to be protected in frost and embedded properly to establish good contact between porous cup and soil. They are not suitable for operation under -85 kPa and it takes several minutes to establish equilibrium between porous cup and soil matrix. Due to these reason, tensiometer should only be used as stationary sensors [143].

8.2.7 Other Sensors

- **Gas sensors** Gas sensors are becoming popular for detection of acetylene. Plants emit acetylene in unsuitable conditions, e.g., drought or fungal infections. CO_2 emission is analyzed to study the biological activities in plants. Non-dispersive infrared (NDIR) CO_2 sensors are used to detect simpler gaseous molecules such as CO_2, SO_2, and NO_2. NDIR is affected by the cross-sensitivity of these gases in low concentration [95]. Electronic noses are used for assessing the complex molecules [160, 216].
- **Capillary electrophoresis** Capillary electrophoresis (CA) is used to separate solute ions from the liquid soil extract after applying electric field. Soil is put in a capillary tube and ions are identified from their time of travel inside the capillary. The tuning parameter for the setup contains selection of electrolyte solution, capillary's' length, and applied voltage to electrolyte. CA has been a commonly applied method in the labs, however, recent development in making portable systems is also being reported [153, 189]. The iMETOS Mobilab was recently released by Pessl Instruments GmbH. It is based on small and inexpensive microfluidic chip CA and measures to NO_3^-, PO_4^-, and K^+ (Weiz, Austria).

8.2.8 Integration of Proximal Sensors in Digital Agriculture

- **Sample preparation for the soil:** Ideally, sampling is not required in proximal sensing. Due to the impact of several properties of soil on sensor, selectivity of a sensor is degraded. Unfavorable soil conditions, e.g., very dry soil, presence of plant residue, are also one of the obstacles in taking measurements. Hence, sample preparation can solve these issues [107, 168]
- **Calibration and measurement errors in soil sensing:** As discussed, sensors are affected by various soil properties. EC of a soil is affected by water, clay content, and salt. In field, measurements can vary beyond control due to soil properties and other environmental factors. Data must be interpreted very carefully. Hence, a traditional lab analysis of the soil is advised for calibration of sensor readings.
- **Sampling design for calibration samples for soil sensing:** Several studies have highly emphasized on the sample calibration for successfully translating the sensor measurement to soil fertility properties. Determining sampling site for the stationary sensors is easy whereas it is difficult for mobile sensors. Adamchuk et al. [4] provide an algorithmic solution and give three criteria for optimal sampling: (1) accounting for spatial separation to avoid readings from auto-correlated samples, (2) increasing spread of value for stability, (3) local spatial homogeneity so that sample fully represents the sampling site instead of being an outlier. Adamchuk et al. [4] and Salam and Vuran [165] also transformed these criteria into mathematical model.
- **Robustness, safety, ease of handling and economic efficiency:** Due to deployment in rough agricultural environment robustness and safety (protected cables and watertight plugs) of proximal sensors is a pre-requisite for daily usage. Some sensors (XRF systems and GCR) can cause serious injuries and operators need to be trained properly as a safety measure. It contributes towards the efficiency and safety of the users.
- **Integrating to decision-making algorithm:** Sensor data alone does not make any sense. It gives information only when integrated into some decision-making algorithm. In digital agriculture, decision support system links the input data (temperature, moisture) to output (crop yield, profitability). These systems can help in decreasing effect of agronomic measures and can be used to do cost-benefit analysis.

Future Trends

Currently, commercially available proximal sensors capture very limited number of soil properties and that too with insufficient accuracy. New proximal sensors systems covering wide range of soil fertility properties and more accuracy are needed. These sensor systems must be affordable and manageable in order for the farmer to be comfortable to use them. Low-cost handheld sensors can be used to introduce farmers to the benefits of using digital agriculture.

A combination of multiple sensors is used to capture different properties of the soil. There is no single platform that combines multiple sensors capabilities. There is a need of establishment of reliable calibration database and protocols for evaluating the sensor data. It also allows to compare the sensors readings. Only few countries offer a scientific sound and robust decision-making algorithms to their farmer for the management purpose. Only ad-hoc approaches are being used in digital agriculture. These approaches have not been validated over a different range of conditions. Authorities have to support and establish guidelines for the promotion of sensor-based farming.

Most of the mobile soil sensor data is evaluated offline. Due to the requirement of quick response in crop management, farmer demands a real-time management. There is a lot of uncertainty about the proper soil sensor usage due to which immediate rate of investment cannot be guaranteed to the farmers. However, digital agriculture must not be evaluated for only its economic potential. A lot of environmental benefits (avoiding pollution and erosion) associated to the adoption of digital agriculture. These benefits must be quantified to support farmers. Farmers and agriculture advisors do not want to waste their time in data calibrating, cleaning, and integration to decision support systems. They need smooth transition of raw data into information for decision making that too with robust and user-friendly systems. These systems must be flexible enough to work with various data formats. Decision support algorithms, e.g., for fertilizer recommendations are based on simple model and require few input parameters. Research in proximal sensing should start out by matching these simple models. However, proximal sensing can provide information which is neglected by the standard algorithms for best management. Advanced soil-crop model can integrate this information for further improvements.

8.3 Remote Sensing

Science of collecting data from the surface of the earth without direct contact is known as remote sensing. Remote sensors are the instruments that collect this information by detecting and measuring the reflected electromagnetic radiations. Platforms used to carry remote sensors include manned aircraft, satellites, and unmanned aircraft systems (UAS). Some sensors can be mounted on ground-based vehicle or may be integrated into handheld systems. Few factors which can be considered for selection of appropriate platform are: size of the area to be imaged, complexity of crop types, time, and cost. Manned aircraft-based sensors are also known as airborne sensors and satellite-based sensors are also known as spaceborne sensors. Different platforms should be evaluated for their suitability of and efficacy for digital agriculture applications [144, 147].

This section provides an overview of airborne imaging system and spaceborne remote sensors being used in digital agriculture. Remote sensors can broadly be categorized into two types: (1) imaging (cameras) and (2) non-imaging (spectroradiometers). Imaging sensors give the vertical (nadir) view of the target area. As with the proximal sensing, remote sensing is also passive (electro-optical sensors) and active (imaging radar) depending on the source of energy. However, active sensors can take measurement regardless of time of the day and season. Electro-optical sensors are the imaging sensors that detect and convert the reflected radiation to the electrical signals which then can be viewed as images on the computer. Electro-optical can detect the radiations of wavelength ranging from 0.3 to 15 μm. Most of the airborne and satellite remote sensors in digital agriculture are electro-optical sensors. In the coming sections, we discuss the advances in airborne and satellite-based remote sensors along with the examples, advantages, and disadvantages in digital agriculture [145].

8.3.1 Multispectral and Hyperspectral Imaging

8.3.1.1 Multispectral Imaging Systems Using Industrial Cameras

With overall advancement in imaging sensor technology, cameras used in multi-spectral imaging systems have also been evolved. Many commercial and customized multispectral systems have been developed and used for the different remote sensing applications such as cropland assessment, digital agriculture, and pest management, etc. [120]. Most systems provide 8–16 bit images with 3–12 narrow spectral bands in the visible to near infrared (NIR) region of EM spectrum [60, 218, 222].

Multispectral imaging systems are based on different approaches. One approach employs monochrome charged couple device (CCD) industrial cameras. Each camera in multispectral system uses different band pass filter. It gives advantage of individual adjustment of camera for focus and aperture settings. However, one disadvantage is that multiple band images need to be properly aligned. Another approach uses beam-splitting prism and integrates multiple sensors to achieve the effect of multispectral imagery. One example of such system is CS-MS1920 multispectral 3-CCD camera (Teledyne Optech, Inc., Vaughan, Ontario, Canada). It uses 3 CCD sensors to produce images in 3–5 spectral ranges with EM spectrum range of 400–100 nm. In this approach, band images are aligned mechanically as well as optically.

The four-camera imaging system developed by U.S. Department of Agriculture, Agricultural Research Service (USDA-ARS) consists of four monochrome CCD cameras, PC with frame grabber and image acquisition software [162, 218]. The camera uses spectral range of 400–100 nm and 12-bit data depth. Spectra-View by Airborne Data Systems, Inc. (Redwood Falls, Minnesota USA) can accommodate up to eight different cameras. These cameras can vary in size, format, and wavelength and contain global positioning system/inertia navigation system (GPS/INS) for precise geo-registration of the images. Spectra-View 5WT captures 12-bit images in six (Blue, Green, Red, NIR, MWIR, LIR) spectral bands. A cheap alternative, Agri-View, can be used to capture same green, NIR, and red band as captured by Spectra 5WT.

Tetracam's multispectral imaging systems (Tetracam, Inc., Chatsworth, California, USA) come in two product families: (1) Agricultural Digital Camera (ADC) and (2) Multiple Camera Array. ADC family is equipped with single camera along with fixed filters at red, green, and NIR wavelength. Some ADC systems (ADC Micro, ADC Lite and ADC) captures 8/10- bit images with 2048 × 1536 pixels while other (ADC Snap) captures 1280 × 1024 pixels. MCA family contains 4, 6, and 12 cameras synchronized to take 8/10-bit images with 1280 × 1024 pixels in visible to NIR wavelengths. The family comes is two versions: standard Micro-MCA and Micro-MCA Snap versions. RGB+3 system by Tetracam has four cameras: one RGB and three monochrome cameras. RGB senses three broad visible bands and monochrome senses three narrow bands which are 680, 780, and 800 nm, respectively. Mcaw (Multiple camera array wireless), comes with a Linux computer system, a storage device for computations and six 1280 × 1024 snap shutter sensors. Each Tetracam's systems has PixelWrench2 software which allows the editing of the images [164, 169].

Teledyne Optech produces RGB color cameras and thermal cameras both as standalone or integrated in the LIDAR (light detection and ranging) system. RGB CS-10000 and CS-LW640 are the examples of Teledyne's RGB and thermal camera, respectively. CS-10000 has a resolution of 10,320 by 7760 pixels and CS-LW640 comes with resolution of 640 × 480 pixels. Integrated with Orion C LIDAR system, CS-LW640 is a very powerful tool for 3D mapping of thermal features.

FLIR and ITRES Research Ltd. also offers some thermal cameras. FLIR's T600 series cameras come with the resolution of 640 × 480 pixels and SC8000-series have the resolution of 1024 × 1024 pixels. TABI-1800 by ITRES research can differentiate temperature difference of 1/10th of a degree.

8.3.1.2 Multispectral Imaging Systems Using Consumer Grade Cameras

Low cost, compact size, data storage, and user-friendliness are some of the advantages that make a consumer grade camera an attractive choice for remote sensing. Consumer grade cameras are mostly equipped with CCD or CMOS sensors, and Bayer color filter mosaic for arranging the RGB color pixel [16, 80, 167]. Various mosaicing algorithms are used to interpolate complete RGB values for each pixel which aligns the three ban images perfectly. Therefore, these cameras have been used frequently by researcher for the agricultural purposes [6, 140, 166].

In remote sensing, images in visible and NIR bands are commonly used especially in vegetation monitoring. NDVI uses spectral information in NIR and red bands. Consumer grade camera uses filters to block UV and infrared light. These filters can be replaced by long-pass infrared filter to convert the consumer grade RGB camera to NIR camera and obtain NIR images. Some companies, e.g., Life Pixel, provide services for conversion of camera. Long-pass filter of 720- and 830-nm are used to replace NIR blocking filter. All three channels can be used to record NIR radiation and any of the three can be used as NIR channel, however, red channel is mostly preferred because of best sensitivity. These NIR-converted cameras are proved to be simple and cheaper tools for plant monitoring, crop monitoring, and stress detection [123, 217].

A cheaper and user-friendly imaging systems are required to capture geotagged images at varying altitudes on any traditional or agricultural aircrafts at normal airplane speed [222]. Agricultural aircrafts are readily available platform for airborne remote sensing. If equipped with imaging system, they can be used to get aerial imagery for various applications such as monitoring, detecting stress, and analyses of aerial applications usefulness. Aerial imaging and aerial spraying must not be done simultaneously to avoid contamination of camera.

A USDA-ARS single-camera uses Nikon D90 camera (Nikon Inc., Melville, New York, USA) to capture color images of up to 4288 × 2848, geotag the image and is equipped with LCD monitor to view live image. For dual-camera system, it uses another same D90 camera but modified to NIR camera. Giga T Pro II wireless timer remote receiver (Hahnel Industries Ltd) and a transmitter is attached to the camera to start and stop the image capturing. Both cameras can be mounted on the aircraft with little or no modification [210, 219]. USDA-ARS also produces two other systems: one consists of two Canon (Canon USA Inc.) EOS 5D Mark III cameras to capture images up to 5760 × 3840 pixels; other system by USDA-ARS consists of two Nikon D810 cameras to capture images up to 5760 × 3840 pixels. In both of the systems, one camera is used to capture RGB color image and the other is converted NIR camera with 830-nm long-pass filter. Both systems use the same sensor size (36 × 24 mm) and focal length (20 nm). Cost of each system is around 6500 USD [172, 217, 222].

Due to increasing demand of light-weight and cheaper imaging systems, many consumer grade camera systems are being converted to capture B-G-NIR or G-R-NIR images using single sensor. LDP LLC, for example, provides either of the modified cameras or services for modification. Both cameras can also be used simultaneously and images can be aligned to create five band images. However, unlike NIR-converted cameras, these cameras may suffer from light contamination depending on filters and algorithms used for the band separation. Salam [156] and Rabatel et al. [130] converted a standard RGB camera by replacing NIR blocking filter with long-pass filter to obtain NIR and R bands. However, spatial resolution of the image is reduced due to smoothing effect in the process.

8.3.1.3 Hyperspectral Imaging Systems

Hyperspectral imaging system can capture images in tens to hundreds of narrow and spectral bands from visible to thermal spectral regions. The airborne visible/infrared imaging spectrometer (AVIRIS) was the first hyperspectral imaging system developed by Jet Propulsion Laboratory and proposed to NASA in 1983. It consists of a flight system, ground data system, and a calibration system. It captures images in 224 continuous spectral bands under solar spectral region of EM spectrum. Different detectors are used which are separated by four panels of wavelength ranging from 400 to 2500 nm. It provides 12-bit spectral data (AVARIS). The AVARIS system has been extensively studied and improved over the time to meet the requirements of the scientists for the research and application purposes. Green et al. [73] provide the overview of AVARIS sensor along with its various scientific applications [143, 160].

HyMap is another popular hyperspectral imaging sensor developed by Integrated Spectronics Pty Ltd. (Sydney, Australia). Initially, it was used for mineral exploration with 96 channels in the 550–2500 nm spectral range [35, 151]. Current HyMap uses 128 bands in the spectral range of 450–2500 nm. The system is mounted with 3-axis gyro stabilized platform for reduced image distortion. Sensor can capture up to 512-pixel images with spatial resolution of 3–10 m for agricultural applications [43, 68, 211]. The compact airborne spectrographic image (CASI) was the first programmable hyperspectral sensor first introduced in 1989 by ITRES Research Ltd. It allowed user to collect the data in specific band and bandwidth by programming the sensor. CASI-1500H, a lighter and smarter design, captures 14-bit images at 288 bands in the spectral range of 380–1500 nm. SASI-1000A captures 1000-pixel images at 200 bands in spectral range of 950–2450 nm. MASI-600, first commercial system, is available with 600 pixels and 64 bands in the spectral range of 3–5 μm. TASI-600 is a hyperspectral thermal sensor which captures 600 spatial pixel images and 32 bands in the spectral range of 8–11.5 nm.

Commercial hyperspectral imaging systems have become advanced with improved spatial and spectral resolutions and improve GPS units for position accuracy. Specim's AISA hyperspectral systems (Spectral Imaging Ltd., Oulu, Finland) is available with spectral ranges from 380 nm to 12.3 μm. It covers VNIR, SWIR, and thermal LWIR spectral ranges. Specims ASIA family of hyperspectral systems include: AisaKESTREL 10 (spectral range of 400–1000 nm), AisaKESTREL 16 (spectral range 600–1640 nm), AisaFENIX and AisaFENIX 1K (spectral range of 380–2500 nm), and AisaOWL (spectral range of 7.7–12.3 μm). All systems come with a GPS system for accurate positioning [152].

Headwall Photonics, Inc. (Fitchburg, Massachusetts, USA) manufactures hyperspectral imaging sensors for UV to visible, VNIR, NIR, and SWIR in the spectral range of 250–2500 nm. It also produces VNIR-SWIR sensor with spectral coverage of 400–2500 nm. It can capture 1600-pixel swath image at hundreds of bands. USDA-ARS in College Station, Texas, uses hyperspectral imaging system with VNIR E-Series imaging spectrometer, a GPS/INS unit, and hyperspectral data processing unit. It can capture 16-bit images within 923 spectral bands, 1600-pixels of swath in the wavelength of 380–1000 nm [161].

8.3.2 Future Trends

This section discusses some of the challenges and future directions for remote sensing. First challenge is that growers do not know about the availability of imagery in the fast changing market; they do not know what type of imagery to select and how to order imagery from the archived data for their particular application. Image providers and different vendors must develop easy instructions for customers and growers to select and order appropriate imagery. Timely acquisition and delivery of the images is also one of the challenges.

Numerous literature exists on processing and conversion of imagery into useful information and map, however, there exists no standard procedure for converting imagery to vegetation index maps, classification and prescription maps. There are many image processing software with different capabilities, complexities, and prices. However, grower may have difficulty in choosing a particular software.

There is a dire need of practical guidelines for the growers and other end users for the conversion of images to appropriate agricultural maps. Researchers need to focus on this area. Users having some familiarity with GIS and image processing must be able to select and use appropriate software with the help of documentation and tutorial. If a grower cannot learn these skills, they can use commercial services to process their images and create the relevant maps. Some dealers do provide services of image acquisition, prescription map creation, and variable rate application.

Environmental changes in agriculture may result in large variations in soil moisture, plant nutrition, crop growth and yields. Fast crop canopy changes need continuous crop monitoring [180]. Remote imagery and satellite imagery have been successfully used in crop prediction. However, coarse spatial and temporal resolution makes their application in agriculture very limited. Airborne multispectral (e.g., [220, 221] and hyperspectral [45] have been used to monitor crop condition and yield.

The recent development in small unmanned aerial systems (UASs) makes agriculture sector a largest commercial market for its use and it will see an increase of 80–90% in market share (Association of Unmanned Vehicle Systems International, 2013). Cost-efficient, ultra-high spatial imagery, and easy image acquisition makes UAS an ideal option for crop monitoring. It is also known by various different names: unmanned aerial vehicles (UAVs), drones, unmanned aircraft systems, and remotely piloted vehicle [122].

Applications of UAS in agriculture include: monitoring of physiological characteristics of crops, leaf area index (LAI), disease and crop stress, monitoring of crop growth [19], yield, removal of rainwater [228], spraying fungicide, herbicide, and pesticide [125, 161], air broadcasting of seeds [106], and measuring of crop and environmental parameters (temperature, humidity, etc., [136]). There remains a security and privacy concern regarding the civilian use of UAS, however, its usefulness has already been established among the public specially in the agricultural sector [66].

8.4 Soil Sensing from the Air

Airborne imaging systems are relatively cheaper, provide high spatial resolution, and have the ability to obtain data in visible to shortwave infrared (SWIR) region of EM spectrum [98, 115, 141]. There are two type of imaging systems: (1)

multispectral and (2) hyperspectral. They differ in spectral bands and bandwidth. Multispectral uses 3–12 and hyperspectral uses tens to hundreds spectral bands to measure the reflected energy. Imagery from hyperspectral has great details of spectral bands and multispectral is great in detecting subtle difference among similar objects. We will discuss in detail the airborne multispectral and hyperspectral imaging systems in coming sections.

UAVs are comprised of two main components: UAVs and sensors. UAVs act as a platform for the sensors. UASs can be classified into two categories: (1) fixed-wing, (2) rotatory-wing [171, 176]. Fixed-wing has long range and is faster, however, ideal takeoff and finding landing spot for them are a challenging task. On contrary, rotatory-wing UASs have short range and flight duration but improved maneuverability. In addition to the sensors, platform may be equipped with the global navigation and satellite system (GNSS) and an inertial management unit (IMU). GNSS provides information about the position of the platform and IMU provides information about the altitude of the platform. This information is integrated with the autopilot system to adjust the course of the flight.

One of the challenges in operating UAS is the restriction put by a country on their operation. Canada, for example, does not allow to fly a UAS more the 90 m of height which gives the images with small footprints. This makes it difficult to map a large crop field especially when the average crop field size was 315 ha in Canada in 2010.

Figure 8.2 shows variety of sensors and cameras are available for UAS. The performance of both, multispectral and commercial, cameras is very good [74, 149]. Although, data quality is a concern for the commercial camera [214], however, its low cost for data acquisition makes it an appealing option for agriculture. Simple RGB camera is not only cost-efficient but also a powerful tool for monitoring plant condition and plant phenology [71]. Consumer grade cameras are sensitive to illumination, hence they either must be used under stable lights or adjustment should be made with variable illumination [131, 142].

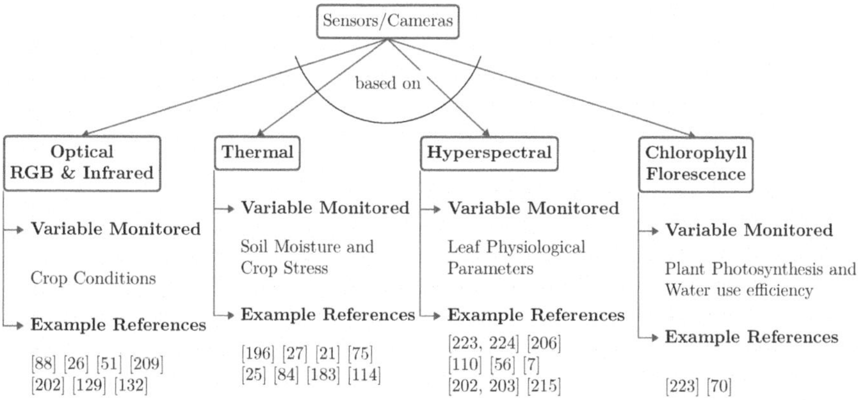

Fig. 8.2 Sensor types and crop monitoring variables [33]

8.4.1 Flight Planning

8.4.1.1 Image Acquisition

Most UASs are using autopilot flight planning for image acquisition. A flight plan includes: area, height course, and speed of the flight, camera setting, forward and side lap; and it is generated using a planning software. Manual control is also possible, however, may cause issues with image post-processing. High-resolution images or high-scale map is shown as a background for the planning and it is stored prior to field trip. GNSS and IMU data are also recorded which assist to determine image center position and camera orientation estimations [163, 188, 214]. All data are initially stored in a storage device in UAS and must be downloaded to the computer for further usage and processing.

The altitude of the flight determines the spatial resolution of images; lower altitude, e.g., 100 and 120 m are common, gives images with the high spatial resolution. For some tasks, e.g., weed mapping, a much lower altitudes (30 m) are also used which results in much higher spatial resolution. However, altitude is also limited by aviation regulations. Flight altitude does not affect VIs, however, it greatly influences the image segmentation with mixed pixels in images with higher altitudes [131, 148]. Four pixels are necessary to identify and find ground object.

Image overlapping is also an important factor to consider for flight planning. Salam [158] and Colomina and Molina [36] suggest to set the minimum value of forward and side overlap to 80% and 60%, respectively. High image overlap is recommended to avoid mismatch between estimated and actual ground image [170, 214], assist in identifying common points in image pair, and minimize the impact of bidirectional reflectance by allowing image processing software to extract the central points of the image for image mosaic [89, 146].

Prior to image acquisition, researchers have used ground control points (GCP), as an artificial target, for spectral calibration. It is recommended to use minimum of three GCPs evenly distributed covering the whole study area [188]. Position of these GCPs should be measured using a total station or differential GNSS to guarantee positional accuracy of image mosaic [47, 157].

8.4.1.2 Image Processing

Various photogrammetry algorithms are used to rectify and mosaic images. After downloading images and logs, initial position and orientation estimates are determined using log files. Ortho-rectified mosaic is generated using various photogrammetric software (e.g., GeoLink [181] and MicMac [13]). Structure from Motion (SfM) photogrammetry has recently become popular for many UAS applications. It uses bundle adjustment algorithms for establishing the structure of the scene, internal and the external orientations [1, 154, 188]. SfM has advantages of having a simple processing workflow, ability to calculate the camera position, orientation, and scene of the geometry from overlapping images only, not requiring camera calibration parameters and lastly, getting height from two-dimensional images [188]. Pix4Dmapper [131, 155], Agisoft PhotoScan Pro [19], and Automatic Photogrammetric Software are some of the commercially available SfM software. Freely available SfM Web services include Autodesk 123D Catch and Microsoft Photosynth. However, both services were discontinued in 2016 and 2017. Bundler, VisualSFM, Multi-View Stereo (PMVS2), and Ecosynth fall under the category of open-source SfM packages. Although they have been criticized for their computational time and reliability, their performance is sometimes found to be at par with some of the commercially available options [100, 188].

8.4.1.3 UAS Image Applications in Digital Agriculture

Researchers have been using UAS imagery for extracting glut of agricultural data. UAS imagery data includes: plant height, crop biological parameters, and plant stress.

Plant height is a key indicator for predicting many parameters such as: crop biomass and yield potential, growth, treatments and stress monitoring, underlying biophysical, ecological, and hydrological processes [19, 71, 150]. Plant height can be derived from LiDAR or SfM-based photogrammetry. Terrestrial laser scanning (TLS) accurately measures plant height for modeling of crop surface and growth monitoring [185, 197, 199].

Although Direct DEM product of SfM produces a digital surface model, vegetation canopy points can be filtered from point cloud to obtain ground surface elevation point (DTM). Separating ground from non-ground points helps in estimating biomass and other relevant parameters [1, 71, 188]. However, separation of vegetation information has been only partially successful [64, 159]. DTM can also be extracted using UAS before and after growing season [19]. SfM-based heights are found to be more accurate than the TLS-based heights. Moreover, lower altitudes (e.g., 40 m) give more accurate crop heights [86, 153].

Many studies have shown the application of UAS in monitoring crop biological parameters. Information from UAS imaging has been used for evaluation of plant growth, biomass, physiological changes, stresses, and many other crop biological properties. LiDAR, thermal, and hyperspectral sensors are used in this type of research. The biological parameters are highly effected if crop is stressed (water, diseases, and infection from pests, etc.). 20% of irrigated land of the world has high salt concentration [117]. It can cause stomatal closure, decreased photosynthesis, increased leaf temperature [177]. Crop stress can be monitored by the data from periodic thermal and visible to NIR UAS imaging during the growth season.

Thermal UAS imaging can be used to measure temperature for calculating crop water stress index for leaf anomalies [18, 81]. Thermal remote sensing is also used to measure the soil moisture and texture, crop residue cover, field drainage tiles, and yield [96]. Apart from remote sensing, optical sensing can also be used to identify plant conditions. Nutrient deficiencies in plants make them more susceptible to herbivores pests [168, 177]. Optical UAS images have been used to detect symptoms for plant nutrient deficiencies [34, 183].

Weeds compete with the crops for the natural resources, e.g., solar radiation, nutrition, and water. Site-specific weed management can decrease the cost of crop production and lower the environmental impact. Ultra-high resolution UAS images help in early and late season detection of weed's species, density, and patches [127, 128]. Due to spectral similarities between weeds and crops, hyperspectral sensors are more useful than multispectral sensors [109]. UAS imaging have been used in assessing disease development, atmospheric pathogen development and monitoring, and precision spraying [11, 100, 165]. However, UAS-based sensing for assessment of diseases is still not in the mature stage [144, 147, 177].

8.4.1.4 Image Analysis

Prior to analysis of UAS images, some categorical information, e.g., crop type, lodged crop, stressed crop ,etc., needs to be extracted. Many researchers adopt classification methods while doing qualitative analysis for assigning different class labels. These classification methods can be supervised or unsupervised methods. Some machine learning algorithms are also developed for segmentation of vegetation and bare soil [53].

For quantitative analysis, VIs are computed to find relation between spectral and field biological data [88]. Digital numbers, Vis, and reflectance can be linked to crop biological and environmental parameter. Most common VI, NDVI, is used to measure biomass, nitrogen content, chlorophyll content, and other biological parameters. Soil adjusted VI is used to amplify spectral difference between vegetation and soil [185]. RGB camera-based VIs (Green NDVI, Excess Green Index (ExG), Green Ratio VI) proved to be a good indicator of leaf area, pigment content, and canopy structure [131]. Some VIs are also used to detect vineyard water stress [14, 141, 163]. Studies have indicated that VIs derived from UAS are comparable to those derived from other remote sensing method (e.g., satellite and manned aircraft) and also provide more spatial details.

8.4.2 Future Trends and Conclusion

UAS has become very popular option for monitoring of crops' biological and environmental parameters, however, they are relatively expensive option. There exists no straightforward procedure for image interpretation, processing, and analyses and it requires considerable amount of expertise and skilled technicians to use this option [13]. A straightforward and an automated tool should be developed for wider adoption of UAS in agriculture [185]. UAS may not be able to cover large enough area which can cause technical difficulties. Furthermore, advancement and development of UAS technology also rely on loosening of aviation regulations.

Local producers must be educated on what cost-effective UAS options are available to them. Due to involvement of different equipment's and requiring high technical expertise from different area, a group of researcher can be formed to raise awareness among farmers on using these systems [185, 227]. Farmers, either as a group or individually, can also hire the consultancy service to receive proper guidance [227].

8.4.3 NASA Soil Moisture Active Passive Mission

NASA sent a Soil Moisture Active Passive (SMAP) satellite for research purpose carrying an L-band radiometer (resolution = 40 km) on January 31, 2015. This satellite was capable of detecting passive microwave radiation emitted by the Earth. It was also equipped with the radar (resolution $\approx$3 km) for sending microwave to the surface of the earth and detecting the reflection. The mission was to measure soil moisture and detect water freeze/thaw states. Backscatters from radars were affected by the vegetation, bodies of water, and surface irregularities. Therefore, signals from radiometers and radars were combined to get soil moisture reading within 9 km resolution [58, 197]. The radar amplifier was failed on September 2, 2015, because of which it stopped working. However, NASA was successful in getting images of 20 km resolution by using different methods such as interpolating and de-convolution and also the oversampling of radiometer. Another approach can read image of 1 km resolution by applying active-pass algorithm to the images from other radars [126, 162]. The mission's data plays an important role in developing weather, soil moisture model, and carbon cycle but SMAP accuracy is major concern.

8.4.3.1 Cosmic-Ray Neutron Probes

A neutron probe is an amalgamation of beryllium and a radioactive material which releases high- and low-speed neutrons. High-speed neutron collides with light hydrogen atom to produce a high amount of energy, whereas low-speed neutron forms a cloud whose density is directly proportional to the soil–water content, e.g., dry soil will create a less dense cloud, fewer low-speed neutron, and less amount of water [134]. Cosmic-ray neutron probes (CRNP) work on the same principle. High-speed neutrons are produced by cosmic rays coming from space and can reach several hundred meters. Soil moisture is inversely proportional to the effective depth. However, while calibrating the instruments, it is very important to consider the other hydrogen sources, e.g., decomposing soil organic component and humus, and need average in-situ soil moisture measurement and neutron intensity [124]. It is also important to properly follow security and safety infrastructure to avoid any radiation hazard while CRNP [17, 30].

Franz et al. [65] did a comparative investigation for soil–water content (SWC) readings taken by in-situ time-domain transmissivity (TDT) sensors and CRNP. The mean absolute error of 0.0286 m^3/m^3 was measured. However, while taking SWC measurement from TDT sensors, the sensors were not permanently installed in the field due to their sparse distribution. Similarly, readings from CRNP do not include vegetation type.

8.4.3.2 GPS Interferometric Reflectometry

Global positioning system (GPS) also uses L-band microwaves. GPS interferometric reflectometry can detect conditions like snow depth, soil moisture, and vegetation water content. To does so by detecting the temporal changes in SNR of line-of-sight (direct wave) and multipath reflected component [31, 172]. Soil permittivity affects the reflected wave and can change frequency, phase, and amplitude of the signal. These signal snapshots are referred to as interferograms. Geodetic-quality GPS antennas can detect soil moisture from a very small distance (2–5 cm) from the surface. It is fixed on airborne devices. However, one disadvantage of the scheme is that it highly dependent on constellation of GPS satellites. These satellites move around the earth few times a day, hence limiting the number of estimations [124, 168].

8.4.3.3 Wireless Sensor Networks

Precision irrigation applications require high spatio-temporal resolution for proper working which is provided by the sensor networks. Wireless communication helps in providing remote information access. This information is provided in real-time so that manual manipulation can be avoided to get an idea about the field conditions. Dong and Vuran [49] used small amount of sensors for measuring soil moisture because of precipitation duration and rainfall cell radius magnitude.

Vereecken et al. [207] discuss various techniques for estimating exact location of the sensor nodes. In general, examining different soil properties with varying soil moisture level can give an average soil moisture value for the field. Delgado and Martínez [44] calculate location-specific solar radiation intensity values to estimate the evaporation rate corresponding to that certain location. A mobile application is used for this approach. The calculations matched the reading from agriculture station which helped in generalizing the result to all those locations which have same radiation conditions. Sensor cluster is then placed in chosen area with following the recommendations given in [49, 164, 169].

The primary task in the WUSN is to determine number of sample readings. It is important because sensing and communicating data consumes large amount of power [49, 166]. Shallow roots and high porosity cause speedy water infiltration and evaporation in soil. Therefore, large sampling rate is needed to overcome this highly fluctuating effect.

8.5 Soil Moisture Sensors Calibration

Extensive research is being done in designing efficient and accurate soil moisture sensors for the irrigation applications [40, 62, 91, 101, 129, 139, 142, 229]. The aim is to avoid financial losses because of over- and under-irrigation. The studies help in (1) selecting accurate soil moisture sensor based on the soil texture, (2) error rate of the sensor, and (3) using calibration method that can help in accurate decision making. The adoption rate of sensing technologies has been very slow not only in America but globally as well. For example, in the United States, Nebraska leads the sensors adoption rate by 30.5% in 2013 [204]. Nebraska has taken a firm position in agricultural water management. The organizations like Nebraska Agricultural Water management (NAWMN) [92] spread the farm-level technologies among 1500 collaborators to increase the adoption rate among stakeholders, i.e., producers, advisors, and crop consultants, etc. However, there is still need to improve the agricultural use and management of water given the fact that on 11.2% of the United States farms use soil moisture based sensor devices [171, 204].

About 90% of farms in the United States do not adopt science-based irrigation water management technologies which raises the questions about the effectiveness of ongoing research. The major challenge in soil moisture adoption rate is the lack of well-defined guidelines to train users to understand and what to expect of technology. When one has to choose from multitude of available sensors, it can cause uncertainty due to different soil textures, therefore, suitability is an important aspect to look for while increasing the adoption rate. In addition to accuracy, operational feasibility, e.g., financial, ease of operation, durability and logistic features of a sensor are also very important. Although, accuracy is the primary concern for scientific users, however, operational feasibility plays an important role in changing preferences of commercial in selecting sensors. For example, a field with high spatial variability will require many sensors in to cover multiple sites and labor to deploy those sensors. Therefore, operational feasibility parameters (time, cost, and labor) will be important inputs to consider, in addition to accuracy, while implementing the system [150, 160].

In coming sections, a framework is provided to evaluate the sensors on the basis of accuracy and operational feasibility. A total of nine commercially available sensors are used along with the two different soil types: Silt loam and sandy soil. Two different orientations of sensing equipment, i.e., horizontal and vertical to ground surface, are used. Finally a decision-making guide is presented to help selecting the sensors on the basis of accuracy and operational feasibility [144, 149].

8.5.1 Materials and Methods

- ***Experimental Sites***: Outdoor field experiments were conducted in two sites with each having different soil types (sandy and silt loam) in Nebraska during growing

Table 8.2 Experiment site details [101]

Site features	Site 1	Site 2
Location	University of Nebraska-Lincoln South Central Agricultural Laboratory (SCAL), near clay center, Nebraska	Central City approximately 10 km north of the Platte River, Nebraska
Average annual precipitation	730 mm	732 mm
Average growing season precipitation	437 mm	464 mm
Soil type	Well-drained Hastings silt loam soil	Deep, moderately drained, and moderately permeable loamy sand
Soil field capacity (FC)	0.34 $m^3\ m^{-3}$	0.19 $m^3\ m^{-3}$
Soil permanent wilting point (PWP)	0.14 $m^3\ m^{-3}$	0.05 $m^3\ m^{-3}$
Crops grown	Field maize and Soybean	Buffalo grass and tall fescue
Size of the field	16.5 ha	70 ha

season of 2017 and 2018. Rest of the discussion will refer these sites as per the soil type, i.e., sandy and silt loam. These two soil types provide an opportunity for sensor evaluation in irrigated and rainfed agricultural systems. Table 8.2 lists the important details pertaining to both experimental sites.

- ***Sensors***: As discussed earlier, a total of nine sensors. These nine sensors were evaluated in two sets of each sensor: horizontal orientation and vertical orientation. The only exceptions were JD multi-sensor probe and TDR315L (Acclima) because that can only be used vertically and horizontally, respectively. Following sensors were used for the experimentation [20, 46, 52, 67, 72, 110, 114, 182, 194, 200, 201, 205, 208, 215, 224, 225]:

  - TrueTDR-315L (Acclima, Inc., Meridian, ID)
  - CS616 (Campbell Scientific, Inc., Logan, UT)
  - CS655 (Campbell Scientific, Inc., Logan, UT)
  - 5TE (Meter Group, Pullman, WA)
  - SM150 (Delta-T Devices, Cambridge, U.K.)
  - 10HS (Meter Group, Pullman, WA)
  - John Deere (JD) Field Connect (John Deere Water, San Marcos, Cal.)
  - EC-5 (Meter Group, Pullman, WA)
  - Dielectric Water Potential-based SensorTEROS 21 (MPS-6) (Meter Group, Pullman, WA)

  The first three are time-domain reflectometry (TDR)-based sensors and others are capacitance-based sensors. The sensors measure volumetric water content (θ_v) ($m^3\ m^{-3}$%vol) except TEROS 21 (MPS-6), which gives soil matrix potential (Ψ_m) (kPa) and is converted to θ_v by soil-specific soil–water release curves given by.

- ***Reference (True) Moisture Measurement*:** Reference soil moisture values θ_v (θ_{vref}) have been measured using Troxler Model 4302 neutron probe (NP) soil moisture gauge (Troxler Electronic Laboratories, Inc., Research Triangle Park, N.C.). All other sensor values are calibrated on the basis of NP measurements. Factory calibrated NP measurement is correlated with gravimetric-sample determined θ_v to develop site-specific calibration using following equations:

$$y = 0.9061x + 0.0354 \tag{8.2}$$

$$y = 1.0848x - 0.0246, \tag{8.3}$$

  where y is θ_{vref}, x is θ_v and Eq. (8.2) and (8.3) are measures calibration for silt loam and sandy soil, respectively.
- ***Installation Specifications*:** Sensor output accuracy is also dependent upon how they are installed in the field. Four soil type-orientations were referred as: silt loam H, silt loam V, loamy sand H, and loamy sand V, where H and V represent the vertical and horizontal orientation. For horizontal orientation (silt loam H, loamy sand H): soil pits were dug on both sides of sensors in furrow (silt loam soil) and grassed area (loamy sand soil). Sensors were installed parallel and perpendicular for horizontal and vertical orientation, respectively. The distance between the sensors is kept such that one sensor may not affect the readings from other sensor. For vertical orientation, distance between the horizontal plane and the ground surface is kept 30 and 50 cm for silt loam and loamy sand soil, respectively. The JD probe uses different orientation because of its distinguishing characteristic of being a multi-depth probe and is compared with NP probe at multiple depths (10, 20, 30, 50, and 100 cm).
- ***Soil Moisture Data Measurement and Retrieval*:** Dataloggers are used to collect data from the sensors about soil moisture every minute and hourly averages. Only in the case of JD probes, telemetry was used to collect data. NP measurements were taken every week throughout growing season.
- ***Statistical Analysis*:** For performance evaluation of sensors, root mean squared error (RMSE, $m^3\,m^{-3}$) is computed as follows:

$$\text{RMSE} = \sqrt{\frac{\sum_{i=1}^{n}(E_i - M_i)^2}{n}}, \tag{8.4}$$

  where n is total number of observations, M_i is sensor values, and E_i are corresponding NP-probe measure values. RMSE value is used as absolute value of error for a particular sensor.
- ***Evaluation Metrics*:** The two metrics used for the each sensor evaluation are: Operational feasibility and performance accuracy. The next two sections briefly explain how sensors are evaluated for these metrics.

8.5.2 Operational Feasibility

Operational feasibility of a sensors can be expressed by the following four characteristics:

- *Telemetry.* Telemetry (TM) refers to real-time access of soil moisture data from the site on mobile or web platforms which is transmitted using some terrestrial or radio system. It prevents users from labor and time investment of going physically to the site and monitoring water profile of the soil [157]. The TM information is quantified by following equation and referred to as $Score_1$:

$$Score_1 = \begin{cases} 0 & \text{No TM} \\ 100 & \text{With TM} \end{cases} \tag{8.5}$$

- *Sensor Cost.* Sensor cost plays a very important role in selection of sensor. The absolute cost of sensor (Table 8.3) is quantified and rescaled to scale of 0–100 to be consistent with other factors scores and is represented by $Score_2$ as follows:

$$Score_2 = 100 - \left(\frac{Max_{scaled} - Min_{scaled}}{Max_{cost} - Min_{cost}} \right) \times (Sensorcost - Min_{cost}) + Min_{scaled}, \tag{8.6}$$

where Max_{scaled} and Min_{scaled} are extremes of score metric (0 and 100), and Max_{cost} and Min_{cost} are extremes of absolute cost of sensors in USD. Cheapest sensor will have the score of 100 and expensive sensor will have score of 0.
- *Cost of Sensing and DataLogging.* Accurate sensing and datalogging after fixed intervals (e.g., 30–60 min) is an important part of sensor operations. It gives daily status as well as the historical soil moisture data for decision making and scrutiny of data for quality purposes. Therefore, Eq. (8.7) and (8.8) give the total cost of sensing and datalogging with and without TM, respectively:

$$\text{Total Cost}_{NoTM} = \text{Sensor Cost + DL cost} \tag{8.7}$$

$$\text{Total Cost}_{TM} = \text{Per sensor cost + DL cost + TM cost} \tag{8.8}$$

The total cost of the sensor (Table 8.3) is quantified and rescaled as $Score_3$ using method similar to used in $Score_2$ computation as follows:

$$Score_3 = 100 - \left(\frac{Max_{scaled} - Min_{scaled}}{Max_{Totalcost} - Min_{Totalcost}} \right) \times (Totalcost - Min_{Totalcost}) + Min_{scaled}, \tag{8.9}$$

Table 8.3 Prices (in USD) of sensors and corresponding dataloggers as on March 2019 [101]

Sensor	Sensor cost ($)	Sensing and logging cost ($) (No TM)	Sensing and logging cost ($) (TM)
CS655	228	1928	2378
CS616	148	1848	2298
SM150	230	1590	3590
10HS	128	624	1274
EC-5	120	616	1266
5TE	225	721	1371
TEROS 21 (MPS-6)	225	721	1371
JD Probe	193	1193	2193
TDR315L (Acclima)	295	670	N/A (not available from manufacturer)

where Max_{scaled} and Min_{scaled} are extremes of score metric (0 and 100), and $Max_{Totalcost}$ and $Min_{Totalcost}$ are the extremes of absolute total cost of sensors in USD (Table 8.3). Cheapest sensor will have the score of 100 and expensive sensor will have score of 0.

- *Ease of Operation.* Ease of operation can be assessed by measuring ease of interaction with users at various stages. The stages can be categorized to setting up datalogger, collecting data from datalogger and post-processing the data. The following $Score_4$ for this factor is quantified based on the factor if in a sensor datalogger comes with a graphical user interface (GUI) and if post-processing of data is needed for the sensor:

$$Score_4 = \begin{cases} 0 & \text{No GUI} \\ 100 & \text{GUI, no data post-processing} \\ 50 & \text{GUI, require data post-processing} \\ 1 & \end{cases} \tag{8.10}$$

All scores apply to any orientation and soil type and remain unchanged if site-specific calibrations (S.S.C.) or factory calibrations (F.C.) are used.

8.5.3 Performance Accuracy

The performance accuracy shows the ability of sensor to accurately sense the data. It is dependent upon soil type and orientation, therefore, for all soil type and orientation combinations, root mean squared error (RMSE), for each sensor's θ_v, is measured against reference values from neutron probes. To be consistent with operational feasibility, RMSE values are scaled to have score of 0–100 referred as a performance accuracy (P.A.) score. P.A. score is computed as follows:

Table 8.4 Root mean squared error RMSE ($m^3\ m^{-3}$) for each sensor using multiple orientations, soil types, and calibration types [101]

		RMSE ($m^3\ m^{-3}$)			
		Vertical		Horizontal	
Soil type	Sensor	F.C.	S.S.C.	F.C.	S.S.C.
Silt loam	CS655	0.05	0.03	0.15	0.05
	CS616	0.06	0.03	0.40	0.05
	SM150	0.0	7 0.02	0.06	0.04
	10HS	0.10	0.03	0.07	0.06
	EC-5	0.15	0.03	0.15	N/A
	5TE	0.05	0.02	0.06	0.04
	TEROS 21 (MPS-6)	0.08	0.03	0.11	0.05
	JD Probe	0.05	0.06	N/A	N/A
	TDR315L (Acclima)	N/A	N/A	0.06	0.04
Loamy sand	CS655	0.03	0.03	0.01	0.02
	CS616	0.03	0.02	0.03	0.025
	SM150	0.04	0.04	0.02	0.02
	10HS	0.04	0.02	0.14	0.02
	EC-5	0.05	0.04	0.09	0.02
	5TE	0.04	0.04	0.03	0.01
	TEROS 21 (MPS-6)	0.21	0.03	0.22	N/A
	JD Probe	0.01	0.02	N/A	N/A
	TDR315L (Acclima)	N/A	N/A	0.02	0.02

$$P.A.score = 100 - \left(\frac{Max_{scaled} - Min_{scaled}}{Max_{RMSE} - Min_{RMSE}}\right) \times (RMSE - Min_{RMSE}) + Min_{scaled}, \tag{8.11}$$

where Max_{scaled} and Min_{scaled} are extremes of score metric (0 and 100), and Max_{RMSE} and Min_{RMSE} are the extremes of sensor's RMSE value (Table 8.4). Most accurate sensor will have the score of 100 and least accurate sensor will have score of 0.

8.5.4 *Results and Discussion*

Scores from Eqs. (8.5), (8.6), (8.9), and (8.10) were used to assess operational feasibility of a sensor (see Table 8.5). All sensors, except TDR-315L (Acclima), were scored 100 for $Score_1$. TDR-315L (Acclima) was scored 0 because of non-availability of TM. $Score_2$ shows more variability than $Score_1$ with EC-5 having lowest score of 0 (most expensive) and TDR-315L (Acclima) sensors having the maximum score of 100 (cheapest). Rest of the sensors did not had much difference

Table 8.5 Total score calculated for each sensor for operational feasibility (O.F.) and performance accuracy (P.A.)

Operational feasibility (O.F.) score					
Column ID	1	2	3	4	5
Sensor	Score 1	Score 2	Score 3 (non-TM)	Score 3 (TM)	Score 4
CS655	100	38	0	52	0
CS616	100	84	6	56	0
SM150	100	37	26	0	100
10HS	100	95	99	100	100
EC-5	100	100	100	100	100
5TE	100	40	92	95	100
TEROS 21 (MPS-6)	100	40	92	95	50
JD Probe	100	58	N/A	60	100
TDR315L (Acclima)	0	0	96	N/A	100

Performance accuracy (P.A.) score							
6							
Silt loam V		Silt loam H		Loamy sand V		Loamy sand H	
F.C.	S.S.C.	F.C.	S.S.C.	F.C.	S.S.C.	F.C.	S.S.C.
100	76	73	47	90	66	100	53
87	80	0	31	94	74	95	79
74	100	100	80	85	0	98	0
44	84	95	2	87	100	41	8
0	75	73	N/A	84	17	65	84
97	94	100	100	87	11	91	100
67	80	85	72	0	23	0	N/A
99	0	N/A	N/A	100	81	N/A	N/A
N/A	N/A	98	83	N/A	N/A	97	8

P.A. scores may differ from parameters of experiments, e.g., soil types, orientations, and calibration whereas scores for O.F. are universal [101]

(\$225–\$230) with the scores ranging between 37 and 40. For $Score_3$, there can be two cost cases: with or without TM. In category of sensors without TM sensing and datalogging, EC-5 and CS655 were the cheapest and the most expensive sensors, respectively. In category of sensors with TM sensing and datalogging, EC-5 and SM150 were the cheapest and the most expensive sensors, respectively. TM options had a significant impact on the total cost of the sensor, hence, also on the selection of sensor. Finally, all sensors, except CS616, CS655, and TEROS 21 (MPS-6), score 100 in $Score_4$. Because of need of programming for setting up the datalogger for CS616 and CS655, they were scored 0 for $Score_4$. Similarly, TEROS 21 required data post-processing and was scored 50 for $Score_4$ [149, 152].

P.A. scores were calculated for four soil type and orientation combinations: silt loam H, silt loam V, loamy sand H, and loamy sand V, using Eq. (8.11) (see

Table 8.5). P.A. scores are studied from the three perspective: on the basis of site, calibration type, and orientation. For silt loam V, CS655 with highest P.A., and EC-5 with lowest P.A., had extreme P.A. score values under F.C., and SM150 with highest P.A., and JD Probe with lowest P.A., had extreme P.A. score values under S.S.C. It is interesting to note that changing the calibration method completely alters the P.A. scores. For silt loam H, 5TE SM150 with highest P.A., and CS616 with lowest P.A. had extreme P.A. score values under F.C., and 5TE with highest P.A., and 10 HS with lowest P.A., had extreme P.A. score values under S.S.C. It is interesting to note that 5TE performed well in all conditions and can be a suitable choice for irrigation applications. For the soil type, following changes were observed for P.A. scores: 10 HS (loamy sand V under S.S.C.), JD probe (for loamy sand V under F.C.), CS655 (loamy sand H under F.C.), and 5TE (loamy sand H under S.S.C.). P.A scores are singular for the sensors and do not have constituent as in operational feasibility scores and it is significantly affected by changing the orientation, soil type, and calibration type [100, 165, 210, 211].

8.6 A Guide for Sensor Selection

Figure 8.3 gives a step-by-step sensor selection framework to help choosing appropriate sensors for given conditions. The steps of this framework are as follows:

- Choose appropriate factors among operational feasibility components and P.A which are most relevant to the users condition to recognize the sensors with characteristics more closer to user demand, e.g., for a highly skilled research ease of operation (Score_4) can be ignored.
- Each factor is assigned a weight on the basis of importance to user, e.g., for research purpose P.A. scores are assigned high weight.
- The assigned weight is multiplied with the corresponding score of the factor. As an example, an equation is shown in Fig. 8.3-step 3 where various individual

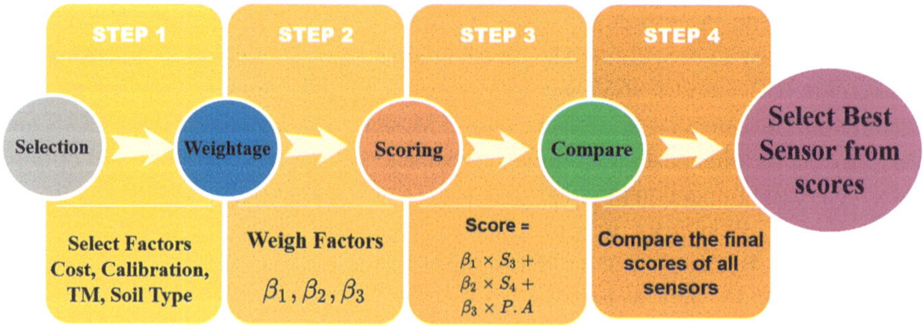

Fig. 8.3 Selection procedure of soil moisture sensors

factors scores for ease of operation, sensing and datalogging cost, and P.A. scores are multiplied by corresponding weights (β_2, β_1, β_3), respectively, and a final score is computed.

- All sensors are compared for the final scores for evaluation on the basis of degree of success and operational feasibility that can be achieved by a particular sensor. The sensor with the highest final score will be most suited for the application.

References

1. Aasen H, Burkart A, Bolten A, Bareth G (2015) Generating 3d hyperspectral information with lightweight UAV snapshot cameras for vegetation monitoring: from camera calibration to quality assurance. ISPRS J Photogramm Remote Sens 108:245–259
2. Abendroth L, Elmore R, Boyer M, Marlay S (2011) Corn growth and development. pmr 1009
3. Adamchuk VI, Hempleman CR, Jahraus DG (2005) On-the-go capacitance sensing of soil water content. In: Mid-central conference. American Society of Agricultural and Biological Engineers, St. Joseph, p 1
4. Adamchuk VI, Rossel RAV, Marx DB, Samal AK (2011) Using targeted sampling to process multivariate soil sensing data. Geoderma 163(1-2):63–73
5. Adamchuk VI, Rossel RV, Sudduth KA, Lammers PS (2011) Sensor fusion for precision agriculture. Sensor fusion-foundation and applications. InTech, Rijeka, pp 27–40
6. Akkaynak D, Treibitz T, Xiao B, Gürkan UA, Allen JJ, Demirci U, Hanlon RT (2014) Use of commercial off-the-shelf digital cameras for scientific data acquisition and scene-specific color calibration. JOSA A 31(2):312–321
7. Allred B, Daniels JJ, Ehsani MR (2008) Handbook of agricultural geophysics. CRC Press, Boca Raton
8. Andrade-Sánchez P, Upadhyaya SK, Jenkins BM (2007) Development, construction, and field evaluation of a soil compaction profile sensor. Trans ASABE 50(3):719–725
9. Antonucci F, Pallottino F, Costa C, Rimatori V, Giorgi S, Papetti P, Menesatti P (2011) Development of a rapid soil water content detection technique using active infrared thermal methods for in-field applications. Sensors 11(11):10114–10128
10. Araus JL, Cairns JE (2014) Field high-throughput phenotyping: the new crop breeding frontier. Trends Plant Sci 19(1):52–61
11. Aylor DE, Schmale DG III, Shields EJ, Newcomb M, Nappo CJ (2011) Tracking the potato late blight pathogen in the atmosphere using unmanned aerial vehicles and Lagrangian modeling. Agric For Meteorol 151(2):251–260
12. Babaeian E, Sadeghi M, Jones SB, Montzka C, Vereecken H, Tuller M (2019) Ground, proximal, and satellite remote sensing of soil moisture. Rev Geophys 57(2):530–616
13. Ballesteros R, Ortega J, Hernández D, Moreno M (2014) Applications of georeferenced high-resolution images obtained with unmanned aerial vehicles. part I: description of image acquisition and processing. Precis Agric 15(6):579–592
14. Baluja J, Diago MP, Balda P, Zorer R, Meggio F, Morales F, Tardaguila J (2012) Assessment of vineyard water status variability by thermal and multispectral imagery using an unmanned aerial vehicle (UAV). Irrig Sci 30(6):511–522
15. Barrios MDR, Marques J Jr, Matias SSR, Panosso AR, Siqueira DS, Scala N Jr (2017) Magnetic susceptibility as indicator of soil quality in sugarcane fields. Revista Caatinga 30(2):287–295
16. Bayer BE (1976) Color imaging array. US Patent 3,971,065
17. Bell J (1987) Neutron probe practice. Institute of Hydrology, Wallingford
18. Bellvert J, Zarco-Tejada PJ, Girona J, Fereres E (2014) Mapping crop water stress index in a 'pinot-noir'vineyard: comparing ground measurements with thermal remote sensing imagery from an unmanned aerial vehicle. Precis Agric 15(4):361–376

19. Bendig J, Bolten A, Bennertz S, Broscheit J, Eichfuss S, Bareth G (2014) Estimating biomass of barley using crop surface models (CSMS) derived from UAV-based RGB imaging. Remote Sens 6(11):10395–10412
20. Berni J, Zarco-Tejada P, Suárez L, González-Dugo V, Fereres E (2009) Remote sensing of vegetation from UAV platforms using lightweight multispectral and thermal imaging sensors. Int Arch Photogramm Remote Sens Spatial Inform Sci 38(6):6
21. Berntsen J, Thomsen A, Schelde K, Hansen O, Knudsen L, Broge N, Hougaard H, Hørfarter R (2006) Algorithms for sensor-based redistribution of nitrogen fertilizer in winter wheat. Precis Agric 7(2):65–83
22. Blackmer T, Schepers J (1995) Use of a chlorophyll meter to monitor nitrogen status and schedule fertigation for corn. J Prod Agric 8(1):56–60
23. Boon N, Yahya A, Kheiralla A, Wee B, Gew S (2005) A tractor-mounted, automated soil penetrometer–shearometer unit for mapping soil mechanical properties. Biosyst Eng 90(4):381–396
24. Bravo C, Moshou D, Orberti R, West J, McCartney A, Bodria L, Ramon H (2004) Foliar disease detection in the field using optical sensor fusion. Agric Eng Int CIGR J, 1–14
25. Breda NJ (2003) Ground-based measurements of leaf area index: a review of methods, instruments and current controversies. J Exp Bot 54(392):2403–2417
26. Bronson KF, White JW, Conley MM, Hunsaker DJ, Thorp KR, French AN, Mackey BE, Holland KH (2017) Active optical sensors in irrigated durum wheat: nitrogen and water effects. Agron J 109(3):1060–1071
27. Bundy LG, Andraski TW (2004) Diagnostic tests for site-specific nitrogen recommendations for winter wheat. Agron J 96(3):608–614
28. Busemeyer L, Mentrup D, Möller K, Wunder E, Alheit K, Hahn V, Maurer HP, Reif JC, Würschum T, Müller J et al (2013) BreedVision—a multi-sensor platform for non-destructive field-based phenotyping in plant breeding. Sensors 13(3):2830–2847
29. Campbell G, Calissendorff C, Williams J (1991) Probe for measuring soil specific heat using a heat-pulse method. Soil Sci Soc Am J 55(1):291–293
30. Chanasyk D, Naeth MA (1996) Field measurement of soil moisture using neutron probes. Can J Soil Sci 76(3):317–323
31. Chew C, Small EE, Larson KM (2016) An algorithm for soil moisture estimation using GPS-interferometric reflectometry for bare and vegetated soil. GPS Solut 20(3):525–537
32. Chrisman BB, Zreda M (2013) Quantifying mesoscale soil moisture with the cosmic-ray rover. Hydrol Earth Syst Sci 17:5097–5108
33. Chunhua Zhang DW, Kovacs JM (2019) The use of unmanned aerial systems (UASs) in precision agriculture, chap 4. Burleigh Dodds Science Publishing, Cambridge, pp 107–128
34. Cilia C, Panigada C, Rossini M, Meroni M, Busetto L, Amaducci S, Boschetti M, Picchi V, Colombo R (2014) Nitrogen status assessment for variable rate fertilization in maize through hyperspectral imagery. Remote Sens 6(7):6549–6565
35. Cocks T, Jenssen R, Stewart A, Wilson I, Shields T (1998) The HyMapTM airborne hyperspectral sensor: the system, calibration and performance. In: Proceedings of the 1st EARSeL workshop on imaging spectroscopy, EARSeL, pp 37–42
36. Colomina I, Molina P (2014) Unmanned aerial systems for photogrammetry and remote sensing: a review. ISPRS J Photogramm Remote Sens 92:79–97
37. Corwin DL (2008) Past, present, and future trends in soil electrical conductivity measurements using geophysical methods. Handbook of agricultural geophysics. CRC Press, Taylor & Francis, Boca Raton, pp 17–44
38. Dash J, Curran P (2004) The MERIS terrestrial chlorophyll index. Int J Rem Sens 25(23):257
39. Datt B (1999) A new reflectance index for remote sensing of chlorophyll content in higher plants: tests using eucalyptus leaves. J Plant Physiol 154(1):30–36
40. Datta S, Taghvaeian S, Ochsner TE, Moriasi D, Gowda P, Steiner JL (2018) Performance assessment of five different soil moisture sensors under irrigated field conditions in Oklahoma. Sensors 18(11):3786

41. De Gruijter J, Brus DJ, Bierkens MF, Knotters M (2006) Sampling for natural resource monitoring. Springer Science & Business Media, Berlin
42. de Souza EG, Scharf PC, Sudduth KA (2010) Sun position and cloud effects on reflectance and vegetation indices of corn. Agron J 102(2):734–744
43. Dehaan R, Taylor G (2002) Field-derived spectra of salinized soils and vegetation as indicators of irrigation-induced soil salinization. Remote Sens Environ 80(3):406–417
44. Delgado B, Martínez M (2015) Software application for calculating solar radiation and equivalent evaporation in mobile devices. Agric Water Manag 151:30–36
45. DeTar WR, Chesson JH, Penner JV, Ojala JC (2008) Detection of soil properties with airborne hyperspectral measurements of bare fields. Trans ASABE 51(2):463–470
46. Diaz-Varela RA, Zarco-Tejada PJ, Angileri V, Loudjani P (2014) Automatic identification of agricultural terraces through object-oriented analysis of very high resolution DSMs and multispectral imagery obtained from an unmanned aerial vehicle. J Environ Manag 134:117–126
47. d'Oleire Oltmanns S, Marzolff I, Peter KD, Ries JB (2012) Unmanned aerial vehicle (UAV) for monitoring soil erosion in Morocco. Remote Sens 4(11):3390–3416
48. Domsch H, Ehlert D, Giebel A, Witzke K, Boess J (2006) Evaluation of the soil penetration resistance along a transect to determine the loosening depth. Precis Agric 7(5):309–326
49. Dong X, Vuran MC (2010) Spatio-temporal soil moisture measurement with wireless underground sensor networks. In: 2010 The 9th IFIP Annual Mediterranean ad hoc networking workshop (Med-Hoc-Net). IEEE, Piscataway, pp 1–8
50. Drummond P, Christy C, Lund E et al (2000) Using an automated penetrometer and soil EC probe to characterize the rooting zone. In: Proceedings of the 5th international conference on precision agriculture, Bloomington, Minnesota, 16–19 July 2000. American Society of Agronomy, Madison, pp 1–9
51. Dualex (2020). https://www.force-a.com/fr/produits/dualex
52. Duan Z, Zhou Q (2015) CRLB-weighted intersection method for target localization using AOA measurements. In: 2015 IEEE international conference on computational intelligence and virtual environments for measurement systems and applications (CIVEMSA). IEEE, Piscataway, pp 1–6
53. Duan T, Zheng B, Guo W, Ninomiya S, Guo Y, Chapman SC (2017) Comparison of ground cover estimates from experiment plots in cotton, sorghum and sugarcane based on images and ortho-mosaics captured by UAV. Funct Plant Biol 44(1):169–183
54. Durner W, Germer K, Jackisch C, Andrä I, Schulz K, Schiedung M, Haller-Jans J, Schneider J, Jaquemotte J, Helmer P et al (2020) Feldstudie zur bodenfeuchtesensorik
55. Dworak V, Selbeck J, Ehlert D (2011) Ranging sensors for vehicle-based measurement of crop stand and orchard parameters: a review. Trans ASABE 54(4):1497–1510
56. Ehlert D, Dammer KH (2006) Widescale testing of the crop-meter for site-specific farming. Precis Agric 7(2):101–115
57. El-Shikha DM, Barnes EM, Clarke TR, Hunsaker DJ, Haberland JA, Pinter P Jr, Waller PM, Thompson TL (2008) Remote sensing of cotton nitrogen status using the canopy chlorophyll content index (CCCI). Trans ASABE 51(1):73–82
58. Entekhabi D, Njoku EG, O'Neill PE, Kellogg KH, Crow WT, Edelstein WN, Entin JK, Goodman SD, Jackson TJ, Johnson J et al (2010) The soil moisture active passive (SMAP) mission. Proc IEEE 98(5):704–716. https://doi.org/10.1109/jproc.2010.2043918
59. Erdi-Krausz G, Matolin M, Minty B, Nicolet J, Reford W, Schetselaar E (2003) Guidelines for radioelement mapping using gamma ray spectrometry data: also as open access e-book. International Atomic Energy Agency (IAEA), Vienna
60. Everitt J, Escobar D, Cavazos I, Noriega J, Davis M (1995) A three-camera multispectral digital video imaging system. Remote Sens Environ 54(3):333–337
61. Evett S, Parkin G (2005) Advances in soil water content sensing. Vadose Zone J 4(4):986–991
62. Evett S, Steiner J (1995) Precision of neutron scattering and capacitance type soil water content gauges from field calibration. Soil Sci Soc Am J 59(4):961–968

63. Feng W, Shen W, He L, Duan J, Guo B, Li Y, Wang C, Guo T (2016) Improved remote sensing detection of wheat powdery mildew using dual-green vegetation indices. Precis Agric 17(5):608–627
64. Fernández T, Pérez JL, Cardenal J, Gómez JM, Colomo C, Delgado J (2016) Analysis of landslide evolution affecting olive groves using UAV and photogrammetric techniques. Remote Sens 8(10):837
65. Franz TE, Wahbi A, Vreugdenhil M, Weltin G, Heng L, Oismueller M, Strauss P, Dercon G, Desilets D (2016) Using cosmic-ray neutron probes to monitor landscape scale soil water content in mixed land use agricultural systems. Appl Environ Soil Sci 2016
66. Freeman PK, Freeland RS (2016) Media framing the reception of unmanned aerial vehicles in the United States of America. Technol Soc 44:23–29
67. Gago J, Douthe C, Coopman R, Gallego P, Ribas-Carbo M, Flexas J, Escalona J, Medrano H (2015) UAVs challenge to assess water stress for sustainable agriculture. Agric Water Manag 153:9–19
68. Galvao L, Ponzoni F, Epiphanio J, Rudorff B, Formaggio A (2004) Sun and view angle effects on NDVI determination of land cover types in the Brazilian Amazon region with hyperspectral data. Int J Remote Sens 25(10):1861–1879
69. Gebbers R (2019) proximal soil surveying and monitoring techniques, chap 2. Burleigh Dodds Science Publishing, Cambridge, pp 29–77
70. Gebbers R, Lück E, Dabas M, Domsch H (2009) Comparison of instruments for geoelectrical soil mapping at the field scale. Near Surface Geophysics 7(3):179–190
71. Geipel J, Link J, Claupein W (2014) Combined spectral and spatial modeling of corn yield based on aerial images and crop surface models acquired with an unmanned aircraft system. Remote Sens 6(11):10335–10355
72. Gonzalez-Dugo V, Zarco-Tejada P, Nicolás E, Nortes PA, Alarcón J, Intrigliolo DS, Fereres E (2013) Using high resolution UAV thermal imagery to assess the variability in the water status of five fruit tree species within a commercial orchard. Precis Agric 14(6):660–678
73. Green RO, Eastwood ML, Sarture CM, Chrien TG, Aronsson M, Chippendale BJ, Faust JA, Pavri BE, Chovit CJ, Solis M et al (1998) Imaging spectroscopy and the airborne visible/infrared imaging spectrometer (AVIRIS). Remote Sens Environ 65(3):227–248
74. Haghighattalab A, Pérez LG, Mondal S, Singh D, Schinstock D, Rutkoski J, Ortiz-Monasterio I, Singh RP, Goodin D, Poland J (2016) Application of unmanned aerial systems for high throughput phenotyping of large wheat breeding nurseries. Plant Methods 12(1):35
75. Hannam JA, Van Dam RL, Harmon RS (2020) Emerging applications and new frontiers: report
76. Heggemann T, Welp G, Amelung W, Angst G, Franz SO, Koszinski S, Schmidt K, Pätzold S (2017) Proximal gamma-ray spectrometry for site-independent in situ prediction of soil texture on ten heterogeneous fields in Germany using support vector machines. Soil Tillage Res 168:99–109
77. Heil K, Schmidhalter U (2017) The application of EM38: determination of soil parameters, selection of soil sampling points and use in agriculture and archaeology. Sensors 17(11):2540
78. Hemmat A, Adamchuk V (2008) Sensor systems for measuring spatial variation in soil compaction. Comput Electron Agric 63(2):89–103
79. Hillnhütter C, Mahlein AK, Sikora RA, Oerke EC (2012) Use of imaging spectroscopy to discriminate symptoms caused by Heterodera schachtii and Rhizoctonia solani on sugar beet. Precis Agric 13(1):17–32
80. Hirakawa K, Wolfe PJ (2018) Spatio-spectral sampling and color filter array design. In: Single-sensor imaging. CRC Press, Boca Raton, pp 157–172
81. Hoffmann H, Jensen R, Thomsen A, Nieto H, Rasmussen J, Friborg T (2016) Crop water stress maps for an entire growing season from visible and thermal UAV imagery. Biogeosciences 13, 6545–6563
82. Holland K, Schepers J (2010) Derivation of a variable rate nitrogen application model for in-season fertilization of corn. Agron J 102(5):1415–1424

83. Holland KH, Schepers JS (2013) Use of a virtual-reference concept to interpret active crop canopy sensor data. Precis Agric 14(1):71–85
84. Holland KH, Lamb DW, Schepers JS (2012) Radiometry of proximal active optical sensors (AOS) for agricultural sensing. IEEE J Sel Top Appl Earth Obs Remote Sens 5(6):1793–1802
85. Holland Scientific - NVDI Sensors (2020). https://hollandscientific.com/
86. Holman FH, Riche AB, Michalski A, Castle M, Wooster MJ, Hawkesford MJ (2016) High throughput field phenotyping of wheat plant height and growth rate in field plot trials using UAV based remote sensing. Remote Sens 8(12):1031
87. Home (2020). https://agriculture.trimble.com/
88. Hunt ER, Cavigelli M, Daughtry CS, Mcmurtrey JE, Walthall CL (2005) Evaluation of digital photography from model aircraft for remote sensing of crop biomass and nitrogen status. Precis Agric 6(4):359–378
89. Hunt ER, Hively WD, Fujikawa SJ, Linden DS, Daughtry CS, McCarty GW (2010) Acquisition of NIR-green-blue digital photographs from unmanned aircraft for crop monitoring. Remote Sens 2(1):290–305
90. Inoue M, Ahmed BO, Saito T, Irshad M (2008) Comparison of twelve dielectric moisture probes for soil water measurement under saline conditions. Am J Environ Sci 4(4):367–372
91. Irmak S, Irmak A (2005) Performance of frequency-domain reflectometer, capacitance, and psuedo-transit time-based soil water content probes in four coarse-textured soils. Appl Eng Agric 21(6):999–1008
92. Irmak S, Rees JM, Zoubek GL, van DeWalle BS, Rathje WR, DeBuhr R, Leininger D, Siekman DD, Schneider JW, Christiansen AP (2010) Nebraska agricultural water management demonstration network (NAWMDN): integrating research and extension/outreach. Appl Eng Agric 26(4):599–613
93. Jenny H (2012) The soil resource: origin and behavior, vol 37. Springer Science & Business Media, Berlin
94. Käthner J, Ben-Gal A, Gebbers R, Peeters A, Herppich WB, Zude-Sasse M (2017) Evaluating spatially resolved influence of soil and tree water status on quality of European plum grown in semi-humid climate. Front Plant Sci 8:1053
95. Kaur J, Adamchuk VI, Whalen JK, Ismail AA (2015) Development of an NDIR CO2 sensor-based system for assessing soil toxicity using substrate-induced respiration. Sensors 15(3):4734–4748
96. Khanal S, Fulton J, Shearer S (2017) An overview of current and potential applications of thermal remote sensing in precision agriculture. Comput Electron Agric 139:22–32
97. Kim HJ, Sudduth KA, Hummel JW (2009) Soil macronutrient sensing for precision agriculture. J Environ Monit 11(10):1810–1824
98. King DJ (1995) Airborne multispectral digital camera and video sensors: a critical review of system designs and applications. Can J Remote Sens 21(3):245–273
99. Knödel K, Krummel H, Lange G (2013) Handbuch zur Erkundung des Untergrundes von Deponien und Altlasten: Band 3: Geophysik. Springer, Berlin
100. Konda A, Rau A, Stoller MA, Taylor JM, Salam A, Pribil GA, Argyropoulos C, Morin SA (2018) Soft microreactors for the deposition of conductive metallic traces on planar, embossed, and curved surfaces. Adv Funct Mater 28(40):1803020. https://doi.org/10.1002/adfm.201803020
101. Kukal MS, Irmak S, Sharma K et al (2019) Development and application of a performance and operational feasibility guide to facilitate adoption of soil moisture sensors. Sustainability 12(1):1–19
102. Kuras O, Beamish D, Meldrum PI, Ogilvy RD (2006) Fundamentals of the capacitive resistivity technique. Geophysics 71(3):G135–G152
103. Lamb D, Schneider D, Stanley J (2014) Combination active optical and passive thermal infrared sensor for low-level airborne crop sensing. Precis Agric 15(5):523–531
104. Lee K, Ehsani R (2009) A laser scanner based measurement system for quantification of citrus tree geometric characteristics. Appl Eng Agric 25(5):777–788

105. Lesmes DP, Friedman SP (2005) Relationships between the electrical and hydrogeological properties of rocks and soils. In: Hydrogeophysics. Springer, Berlin, pp 87–128
106. Li Z, Isler V (2016) Large scale image mosaic construction for agricultural applications. IEEE Robot Autom Lett 1(1):295–302
107. Li Y, Zhang M, Zheng J, Pan L, Kong P, Lei Z (2017) Design and experiment of prototype soil pretreatment device for ISE-based soil nitrate-nitrogen detection. Trans Chin Soc Agric Eng 33(1):120–125
108. Long DS, McCallum JD (2015) On-combine, multi-sensor data collection for post-harvest assessment of environmental stress in wheat. Precis Agric 16(5):492–504
109. López-Granados F (2011) Weed detection for site-specific weed management: mapping and real-time approaches. Weed Res 51(1):1–11
110. Lucieer A, Malenovskỳ Z, Veness T, Wallace L (2014) HyperUAS—imaging spectroscopy from a multirotor unmanned aircraft system. J Field Robot 31(4):571–590
111. Lueck E, Rühlmann J (2013) Resistivity mapping with geophilus electricus—information about lateral and vertical soil heterogeneity. Geoderma 199:2–11
112. Mahmood HS, Hoogmoed WB, Van Henten EJ (2013) Proximal gamma-ray spectroscopy to predict soil properties using windows and full-spectrum analysis methods. Sensors 13(12):16263–16280
113. Markwell J, Osterman JC, Mitchell JL (1995) Calibration of the Minolta SPAD-502 leaf chlorophyll meter. Photosynth Res 46(3):467–472
114. Martins CH, Alshehri AA, Akyildiz IF (2017) Novel MI-based (FracBot) sensor hardware design for monitoring hydraulic fractures and oil reservoirs. In: 2017 IEEE 8th annual ubiquitous computing, electronics and mobile communication conference (UEMCON). IEEE, Piscataway, pp 434–441
115. Mausel P, Everitt J, Escobar D, King D (1992) Airborne videography: current status and future perspectives. Photogramm Eng Remote Sensing 58(8):1189–1195
116. McNeill J (1980) Electromagnetic terrain conductivity measurement at low induction numbers. Geonics Ltd., Mississauga
117. Metternicht G, Zinck J (2003) Remote sensing of soil salinity: potentials and constraints. Remote Sens Environ 85(1):1–20
118. Minolta K (2009) Chlorophyll meter SPAD-502plus. Konica Minolta
119. Mohamed A (2008) Impact of soil magnetic permeability on water content prediction using TDR. In: The 12th international conference of international association for computer methods and advances in geomechanics (IACMAG), Citeseer
120. Moran MS, Inoue Y, Barnes E (1997) Opportunities and limitations for image-based remote sensing in precision crop management. Remote Sens Environ 61(3):319–346
121. Muñoz-Carpena R, Shukla S, Morgan K (2004) Field devices for monitoring soil water content, vol 343. University of Florida Cooperative Extension Service, Institute of Food and Agricultural Sciences
122. Nex F, Remondino F (2014) UAV for 3d mapping applications: a review. Appl Geomat 6(1):1–15
123. Nijland W, De Jong R, De Jong SM, Wulder MA, Bater CW, Coops NC (2014) Monitoring plant condition and phenology using infrared sensitive consumer grade digital cameras. Agric For Meteorol 184:98–106
124. Ochsner TE, Cosh MH, Cuenca RH, Dorigo WA, Draper CS, Hagimoto Y, Kerr YH, Njoku EG, Small EE, Zreda M (2013) State of the art in large-scale soil moisture monitoring. Soil Sci Soc Am J 77(6):1888–1919
125. Pan Z, Lie D, Qiang L, Shaolan H, Shilai Y, Yande L, Yongxu Y, Haiyang P (2016) Effects of citrus tree-shape and spraying height of small unmanned aerial vehicle on droplet distribution. Int J Agric Biol Eng 9(4):45–52
126. Panciera R, Walker JP, Jackson TJ, Gray DA, Tanase MA, Ryu D, Monerris A, Yardley H, Rudiger C, Wu X et al (2014) The soil moisture active passive experiments (SMAPEx): toward soil moisture retrieval from the SMAP mission. IEEE Trans Geosci Remote Sens 52(1):490–507

127. Peña JM, Torres-Sánchez J, de Castro AI, Kelly M, López-Granados F (2013) Weed mapping in early-season maize fields using object-based analysis of unmanned aerial vehicle (UAV) images. PLoS ONE 8(10), e77151
128. Pérez-Ortiz M, Peña J, Gutiérrez PA, Torres-Sánchez J, Hervás-Martínez C, López-Granados F (2015) A semi-supervised system for weed mapping in sunflower crops using unmanned aerial vehicles and a crop row detection method. Appl Soft Comput 37:533–544
129. Quinones H, Ruelle P, Nemeth I (2003) Comparison of three calibration procedures for TDR soil moisture sensors. Irrig Drain 52(3):203–217
130. Rabatel G, Gorretta N, Labbe S (2014) Getting simultaneous red and near-infrared band data from a single digital camera for plant monitoring applications: theoretical and practical study. Biosyst Eng 117:2–14
131. Rasmussen J, Ntakos G, Nielsen J, Svensgaard J, Poulsen RN, Christensen S (2016) Are vegetation indices derived from consumer-grade cameras mounted on UAVs sufficiently reliable for assessing experimental plots? Eur J Agron 74:75–92
132. Raun WR, Solie JB, Johnson GV, Stone ML, Mullen RW, Freeman KW, Thomason WE, Lukina EV (2002) Improving nitrogen use efficiency in cereal grain production with optical sensing and variable rate application. Agron J 94(4):815–820
133. Reece CF (1996) Evaluation of a line heat dissipation sensor for measuring soil matric potential. Soil Sci Soc Am J 60(4):1022–1028
134. Risinger M, Carver K (2020) Neutron moisture meters. http://sanangelo.tamu.edu/extension/agronomy/agronomy-publications/grain-sorghum-production-in-west-central-texas/how-to-estimate-soil-moisture-by-feel/soil-moisture-measuring-devices/neutron-moisture-meters/
135. Robinson D, Jones SB, Wraith J, Or D, Friedman S (2003) A review of advances in dielectric and electrical conductivity measurement in soils using time domain reflectometry. Vadose Zone J 2(4):444–475
136. Roldán JJ, Joossen G, Sanz D, Del Cerro J, Barrientos A (2015) Mini-UAV based sensory system for measuring environmental variables in greenhouses. Sensors 15(2):3334–3350
137. Rossel RV, Adamchuk V, Sudduth K, McKenzie N, Lobsey C (2011) Proximal soil sensing: an effective approach for soil measurements in space and time. In: Advances in agronomy, vol 113. Elsevier, Amsterdam, pp 243–291
138. Rouse J, Haas R, Schell J, Deering D (1974) Monitoring vegetation systems in the great plains with ERTS. NASA Spec Publ 351:309
139. Rudnick DR, Djaman K, Irmak S (2015) Performance analysis of capacitance and electrical resistance-type soil moisture sensors in a silt loam soil. Trans ASABE 58(3):649–665
140. Sakamoto T, Gitelson AA, Nguy-Robertson AL, Arkebauer TJ, Wardlow BD, Suyker AE, Verma SB, Shibayama M (2012) An alternative method using digital cameras for continuous monitoring of crop status. Agric For Meteorol 154:113–126
141. Salam A (2018) Pulses in the sand: long range and high data rate communication techniques for next generation wireless underground networks. ETD collection for University of Nebraska - Lincoln (AAI10826112), http://digitalcommons.unl.edu/dissertations/AAI10826112
142. Salam A (2019) A comparison of path loss variations in soil using planar and dipole antennas. In: 2019 IEEE international symposium on antennas and propagation. IEEE, Piscataway
143. Salam A (2019) Design of subsurface phased array antennas for digital agriculture applications. In: Proc. 2019 IEEE international symposium on phased array systems and technology (IEEE Array 2019), Waltham, MA
144. Salam A (2019) A path loss model for through the soil wireless communications in digital agriculture. In: 2019 IEEE international symposium on antennas and propagation. IEEE, Piscataway, pp 1–2
145. Salam A (2019) Sensor-free underground soil sensing. In: ASA, CSSA and SSSA international annual meetings, ASA-CSSA-SSSA
146. Salam A (2019) Subsurface MIMO: a beamforming design in internet of underground things for digital agriculture applications. J. Sens. Actuator Netw. 8(3). https://doi.org/10.3390/jsan8030041, https://www.mdpi.com/2224-2708/8/3/41

147. Salam A (2019) Underground environment aware MIMO design using transmit and receive beamforming in internet of underground things. Springer International Publishing, Cham, pp 1–15
148. Salam A (2019) An underground radio wave propagation prediction model for digital agriculture. Information 10(4):147
149. Salam A (2019) Underground soil sensing using subsurface radio wave propagation. In: 5th global workshop on proximal soil sensing, Columbia, MO
150. Salam A (2020) Internet of things for environmental sustainability and climate change. Springer International Publishing, Cham, pp 33–69. https://doi.org/10.1007/978-3-030-35291-2_2
151. Salam A (2020) Internet of things for sustainability: perspectives in privacy, cybersecurity, and future trends. Springer International Publishing, Cham, pp 299–327. https://doi.org/10.1007/978-3-030-35291-2_10
152. Salam A (2020) Internet of things for sustainable community development, 1st edn. Springer Nature, Berlin. https://doi.org/10.1007/978-3-030-35291-2
153. Salam A (2020) Internet of things for sustainable community development: introduction and overview. Springer International Publishing, Cham, pp 1–31. https://doi.org/10.1007/978-3-030-35291-2_1
154. Salam A (2020) Internet of things for sustainable forestry. Springer International Publishing, Cham, pp 147–181. https://doi.org/10.1007/978-3-030-35291-2_5
155. Salam A (2020) Internet of things for sustainable human health. Springer International Publishing, Cham, pp 217–242. https://doi.org/10.1007/978-3-030-35291-2_7
156. Salam A (2020) Internet of things for sustainable mining. Springer International Publishing, Cham, pp 243–271. https://doi.org/10.1007/978-3-030-35291-2_8
157. Salam A (2020) Internet of things for water sustainability. Springer International Publishing, Cham, pp 113–145. https://doi.org/10.1007/978-3-030-35291-2_4
158. Salam A (2020) Internet of things in agricultural innovation and security. Springer International Publishing, Cham, pp 71–112. https://doi.org/10.1007/978-3-030-35291-2_3
159. Salam A (2020) Internet of things in sustainable energy systems. Springer International Publishing, Cham, pp 183–216. https://doi.org/10.1007/978-3-030-35291-2_6
160. Salam A (2020) Internet of things in water management and treatment. Springer International Publishing, Cham, pp 273–298. https://doi.org/10.1007/978-3-030-35291-2_9
161. Salam A (2020) Wireless underground communications in sewer and stormwater overflow monitoring: radio waves through soil and asphalt medium. Information 11(2), 98
162. Salam A, Karabiyik U (2019) A cooperative overlay approach at the physical layer of cognitive radio for digital agriculture. In: Proceedings of the 3rd international Balkan conference on communications and networking (2019 BalkanCom)
163. Salam A, Shah S (2019) Internet of things in smart agriculture: enabling technologies. In: 2019 IEEE 5th world forum on internet of things (WF-IoT). IEEE, Piscataway, pp 692–695
164. Salam A, Vuran MC (2016) Impacts of soil type and moisture on the capacity of multi-carrier modulation in internet of underground things. In: Proc. of the 25th ICCCN 2016, Waikoloa, Hawaii
165. Salam A, Vuran MC (2017) EM-based wireless underground sensor networks, pp 247–285. https://doi.org/10.1016/B978-0-12-803139-1.00005-9
166. Salam A, Vuran MC (2017) Smart underground antenna arrays: a soil moisture adaptive beamforming approach. In: Proc. IEEE INFOCOM 2017, Atlanta
167. Salam A, Vuran MC (2017) Wireless underground channel diversity reception with multiple antennas for internet of underground things. In: Proc. IEEE ICC 2017, Paris
168. Salam A, Vuran MC, Irmak S (2016) Pulses in the sand: impulse response analysis of wireless underground channel. In: The 35th annual IEEE international conference on computer communications (INFOCOM 2016), San Francisco
169. Salam A, Vuran MC, Irmak S (2017) Towards internet of underground things in smart lighting: a statistical model of wireless underground channel. In: Proc. 14th IEEE international conference on networking, sensing and control (IEEE ICNSC), Calabria

170. Salam A, Hoang AD, Meghna A, Martin DR, Guzman G, Yoon YH, Carlson J, Kramer J, Yansi K, Kelly M et al (2019) The future of emerging IoT paradigms: architectures and technologies. https://doi.org/10.20944/preprints201912.0276.v1
171. Salam A, Vuran MC, Dong X, Argyropoulos C, Irmak S (2019) A theoretical model of underground dipole antennas for communications in internet of underground things. IEEE Trans Antennas Propag. https://doi.org/10.1109/TAP.2019.2902646
172. Salam A, Vuran MC, Irmak S (2019) Di-sense: in situ real-time permittivity estimation and soil moisture sensing using wireless underground communications. Comput Netw 151:31–41. https://doi.org/10.1016/j.comnet.2019.01.001, http://www.sciencedirect.com/science/article/pii/S1389128618303141
173. Samborski SM, Tremblay N, Fallon E (2009) Strategies to make use of plant sensors-based diagnostic information for nitrogen recommendations. Agron J 101(4):800–816
174. Samouëlian A, Cousin I, Tabbagh A, Bruand A, Richard G (2005) Electrical resistivity survey in soil science: a review. Soil Tillage Res 83(2):173–193
175. Sankaran S, Ehsani R (2012) Detection of huanglongbing disease in citrus using fluorescence spectroscopy. Trans ASABE 55(1):313–320
176. Sankaran S, Khot LR, Carter AH (2015) Field-based crop phenotyping: multispectral aerial imaging for evaluation of winter wheat emergence and spring stand. Comput Electron Agric 118:372–379
177. Sankaran S, Khot LR, Espinoza CZ, Jarolmasjed S, Sathuvalli VR, Vandemark GJ, Miklas PN, Carter AH, Pumphrey MO, Knowles NR et al (2015) Low-altitude, high-resolution aerial imaging systems for row and field crop phenotyping: a review. Eur J Agron 70:112–123
178. Santamarina JC, Rinaldi VA, Fratta D, Klein KA, Wang YH, Cho GC, Cascante G (2005) A survey of elastic and electromagnetic properties of near-surface soils. Near-Surface Geophys 1:71–87
179. Scharf PC, Brouder SM, Hoeft RG (2006) Chlorophyll meter readings can predict nitrogen need and yield response of corn in the North-Central USA. Agron J 98(3):655–665
180. Schirrmann M, Giebel A, Gleiniger F, Pflanz M, Lentschke J, Dammer KH (2016) Monitoring agronomic parameters of winter wheat crops with low-cost UAV imagery. Remote Sens 8(9):706
181. Selsam P, Schaeper W, Brinkmann K, Buerkert A (2017) Acquisition and automated rectification of high-resolution RGB and near-IR aerial photographs to estimate plant biomass and surface topography in arid agro-ecosystems. Exp Agric 53(1):144–157
182. Sepúlveda-Reyes D, Ingram B, Bardeen M, Zúñiga M, Ortega-Farías S, Poblete-Echeverría C (2016) Selecting canopy zones and thresholding approaches to assess grapevine water status by using aerial and ground-based thermal imaging. Remote Sens 8(10):822
183. Severtson D, Callow N, Flower K, Neuhaus A, Olejnik M, Nansen C (2016) Unmanned aerial vehicle canopy reflectance data detects potassium deficiency and green peach aphid susceptibility in canola. Precis Agric 17(6):659–677
184. Shamal S, Alhwaimel SA, Mouazen AM (2016) Application of an on-line sensor to map soil packing density for site specific cultivation. Soil Tillage Res 162:78–86
185. Shi Y, Thomasson JA, Murray SC, Pugh NA, Rooney WL, Shafian S, Rajan N, Rouze G, Morgan CL, Neely HL et al (2016) Unmanned aerial vehicles for high-throughput phenotyping and agronomic research. PloS ONE 11(7). https://doi.org/10.1371/journal.pone.0159781
186. Shiratsuchi L, Ferguson R, Shanahan J, Adamchuk V, Rundquist D, Marx D, Slater G (2011) Water and nitrogen effects on active canopy sensor vegetation indices. Agron J 103(6):1815–1826
187. Slaughter D, Giles D, Downey D (2008) Autonomous robotic weed control systems: a review. Comput Electron Agric 61(1):63–78
188. Smith M, Carrivick J, Quincey D (2016) Structure from motion photogrammetry in physical geography. Prog Phys Geogr 40(2):247–275
189. Smolka M, Puchberger-Enengl D, Bipoun M, Klasa A, Kiczkajlo M, Śmiechowski W, Sowiński P, Krutzler C, Keplinger F, Vellekoop M (2017) A mobile lab-on-a-chip device for on-site soil nutrient analysis. Precis Agric 18(2):152–168

190. Stafford J, Hendrick J (1988) Dynamic sensing of soil pans. Trans ASAE 31(1):9–0013
191. Steinberg S, van Bavel CH, McFarland MJ (1989) A gauge to measure mass flow rate of sap in stems and trunks of woody plants. J Am Soc Hortic Sci 114(3):466–472
192. Stiekema H (2012) The Ag leader OptRx crop sensor. Ag Leader Technology Inc, Ag Leader Europe bv
193. Sudduth KA, Kitchen N, Bollero G, Bullock D, Wiebold W (2003) Comparison of electromagnetic induction and direct sensing of soil electrical conductivity. Agron J 95(3):472–482
194. Sugiura R, Noguchi N, Ishii K (2007) Correction of low-altitude thermal images applied to estimating soil water status. Biosyst Eng 96(3):301–313
195. Sui R, Thomasson JA, Hanks J, Wooten J (2008) Ground-based sensing system for weed mapping in cotton. Comput Electron Agric 60(1):31–38
196. Teal R, Tubana B, Girma K, Freeman K, Arnall D, Walsh O, Raun W (2006) In-season prediction of corn grain yield potential using normalized difference vegetation index. Agron J 98(6):1488–1494
197. Temel S, Vuran MC, Lunar MM, Zhao Z, Salam A, Faller RK, Stolle C (2018) Vehicle-to-barrier communication during real-world vehicle crash tests. Comput Commun 127:172–186
198. Thomsen A, Schelde K, Drøscher P, Steffensen F (2007) Mobile TDR for geo-referenced measurement of soil water content and electrical conductivity. Precis Agric 8(4-5):213–223
199. Tilly N, Hoffmeister D, Cao Q, Huang S, Lenz-Wiedemann V, Miao Y, Bareth G (2014) Multitemporal crop surface models: accurate plant height measurement and biomass estimation with terrestrial laser scanning in paddy rice. J Appl Remote Sens 8(1):083671
200. Torres-Sánchez J, Peña JM, de Castro AI, López-Granados F (2014) Multi-temporal mapping of the vegetation fraction in early-season wheat fields using images from UAV. Comput Electron Agric 103:104–113
201. Torres-Sánchez J, Lopez-Granados F, Serrano N, Arquero O, Peña JM (2015) High-throughput 3-D monitoring of agricultural-tree plantations with unmanned aerial vehicle (UAV) technology. PloS ONE 10(6). https://doi.org/10.1371/journal.pone.0130479
202. Tremblay N, Wang Z, Ma BL, Belec C, Vigneault P (2009) A comparison of crop data measured by two commercial sensors for variable-rate nitrogen application. Precis Agric 10(2):145
203. Tuller M, Islam MR (2005) Field methods for monitoring solute transport. CRC Press, Boca Raton, FL
204. UC of Agriculture (2018) Irrigation and water management survey. https://www.nass.usda.gov/Publications/AgCensus/2017/Online_Resources/Farm_and_Ranch_Irrigation_Survey/fris.pdf
205. Uto K, Seki H, Saito G, Kosugi Y (2013) Characterization of rice paddies by a UAV-mounted miniature hyperspectral sensor system. IEEE J Sel Top Appl Earth Obs Remote Sens 6(2):851–860
206. Van Egmond F, Loonstra E, Limburg J (2010) Gamma ray sensor for topsoil mapping: the mole. In: Proximal soil sensing. Springer, Berlin, pp 323–332
207. Vereecken H, Huisman J, Pachepsky Y, Montzka C, Van Der Kruk J, Bogena H, Weihermüller L, Herbst M, Martinez G, Vanderborght J (2014) On the spatio-temporal dynamics of soil moisture at the field scale. J Hydrol 516:76–96
208. Verger A, Vigneau N, Chéron C, Gilliot JM, Comar A, Baret F (2014) Green area index from an unmanned aerial system over wheat and rapeseed crops. Remote Sens Environ 152:654–664
209. Veris Technology (2020). https://www.veristech.com/the-sensors/v3100
210. Vuran MC, Salam A, Wong R, Irmak S (2018) Internet of underground things in precision agriculture: architecture and technology aspects. Ad Hoc Netw. https://doi.org/10.1016/j.adhoc.2018.07.017, http://www.sciencedirect.com/science/article/pii/S1570870518305067
211. Vuran MC, Salam A, Wong R, Irmak S (2018) Internet of underground things: sensing and communications on the field for precision agriculture. In: 2018 IEEE 4th world forum on internet of things (WF-IoT) (WF-IoT 2018), Singapore

212. Wallihan E (1973) Portable reflectance meter for estimating chlorophyll concentrations in leaves 1. Agron J 65(4):659–662
213. Weis M, Andújar D, Peteinatos G, Gerhards R (2013) Improving the determination of plant characteristics by fusion of four different sensors. In: Precision agriculture'13. Springer, Berlin, pp 63–69
214. Whitehead K, Hugenholtz CH (2014) Remote sensing of the environment with small unmanned aircraft systems (UASS), part 1: a review of progress and challenges. J Unmanned Veh Syst 2(3):69–85
215. Willkomm M, Bolten A, Bareth G (2016) Non-destructive monitoring of rice by hyperspectral in-field spectrometry and UAV-based remote sensing: case study of field-grown rice in North Rhine-Westphalia, vol 41. International Archives of the Photogrammetry, Remote Sensing & Spatial Information Sciences
216. Wilson AD, Baietto M (2009) Applications and advances in electronic-nose technologies. Sensors 9(7):5099–5148
217. Wu M, Yang C, Song X, Hoffmann WC, Huang W, Niu Z, Wang C, Li W (2017) Evaluation of orthomosics and digital surface models derived from aerial imagery for crop type mapping. Remote Sens 9(3):239
218. Yang C (2012) A high-resolution airborne four-camera imaging system for agricultural remote sensing. Comput Electron Agric 88:13–24
219. Yang C, Hoffmann WC (2015) Low-cost single-camera imaging system for aerial applicators. J Appl Remote Sens 9(1):096064
220. Yang C, Everitt JH, Bradford JM, Murden D (2004) Airborne hyperspectral imagery and yield monitor data for mapping cotton yield variability. Precis Agric 5(5):445–461
221. Yang C, Everitt JH, Du Q, Luo B, Chanussot J (2012) Using high-resolution airborne and satellite imagery to assess crop growth and yield variability for precision agriculture. Proc IEEE 101(3):582–592
222. Yang C, Westbrook JK, Suh CPC, Martin DE, Hoffmann WC, Lan Y, Fritz BK, Goolsby JA (2014) An airborne multispectral imaging system based on two consumer-grade cameras for agricultural remote sensing. Remote Sens 6(6):5257–5278
223. Yara International (2018) Yara publishes 2017 Annual Report. https://www.yara.com/corporate-releases/yara-publishes-2017-annual-report/
224. Zarco-Tejada PJ, González-Dugo V, Berni JA (2012) Fluorescence, temperature and narrow-band indices acquired from a UAV platform for water stress detection using a micro-hyperspectral imager and a thermal camera. Remote Sens Environ 117:322–337
225. Zarco-Tejada PJ, Diaz-Varela R, Angileri V, Loudjani P (2014) Tree height quantification using very high resolution imagery acquired from an unmanned aerial vehicle (UAV) and automatic 3d photo-reconstruction methods. Eur J Agron 55:89–99
226. Zebarth B, Rees H, Tremblay N, Fournier P, Leblon B (2002) Mapping spatial variation in potato nitrogen status using the N sensor. In: XXVI international horticultural congress: toward ecologically sound fertilization strategies for field vegetable production 627, pp 267–273
227. Zhang C, Walters D, Kovacs JM (2014) Applications of low altitude remote sensing in agriculture upon farmers' requests–a case study in northeastern Ontario, Canada. PLoS ONE 9(11):e112894
228. Zhou J, Khot LR, Peters T, Whiting MD, Zhang Q, Granatstein D (2016) Efficacy of unmanned helicopter in rainwater removal from cherry canopies. Comput Electron Agric 124:161–167
229. Zhu Y, Irmak S, Jhala AJ, Vuran MC, Diotto A (2019) Time-domain and frequency-domain reflectometry type soil moisture sensor performance and soil temperature effects in fine-and coarse-textured soils. Appl Eng Agric 35(2):117–134

Chapter 9
Soil Moisture and Permittivity Estimation

9.1 Introduction

IOUT can be applied to many fields of precision agriculture [1, 2, 4, 8, 11, 22, 28, 49, 62]. It is being used to provide important information to the farmers. IOUT has an ability to estimate soil properties and monitor soil moisture. An important component of precision agriculture applications (e.g., making real-time agricultural decision, smart agriculture variable-rate irrigation (VRI), and water conservation) is continuous soil moisture sensing [17, 54]. Permittivity plays an important role in propagation analysis of electromagnetic (EM) waves on the basis of soil medium, depth, UG localization, and subsurface imaging. Therefore, efficient measurement techniques for soil in-situ properties are very important. Method to determine soil permittivity includes: ground penetrating radar (GPR) measurements [5, 13, 27], time-domain reflectometry (TDR) [25, 32, 55, 64], and remote sensing [18, 56, 65, 66]. Furthermore, disadvantage of laboratory method of estimating permittivity is offline measuring the soil sample, and that of remote sensing is limitations of depth up to 20 cm. In-situ methodologies can be used to measure soil properties for higher depths and that too with accuracy.

This chapter focus on developing an in-situ technique, Di-Sense, for measuring soil moisture and permittivity based on wireless underground communications (WUC) in IOUT. When EM waves propagate through the soil they are effected by the distance, depth, frequency, soil moisture [45, 50], and soil properties [6]. Path loss at receiver can be used to determine the properties change of propagating signal. The approach in this chapter uses path for measuring permittivity and moisture of the soil. Path loss is highly dependent upon depth, distance, and soil moisture. Di-Sense uses a buried transmitter antenna (fixed depth) for transmitting wideband signal using the frequency range of 100–500 MHz and received signal is analyzed for measuring path loss. Di-Sense allows IOUT system to communicate and estimate permittivity and soil moisture sensing simultaneously. A model is developed to measure soil moisture and permittivity using path loss and is validated

A. Salam, U. Raza, *Signals in the Soil*,
https://doi.org/10.1007/978-3-030-50861-6_9

in an indoor testbed and a software-defined radio (SDR) testbed. For validation, various soil parameters (depth, type, and moisture level) are changed to study the impact of those parameters.

There has been lot of work done previously for determining soil permittivity and moisture. The literature review in this chapter will try to cover only the most recent work and try to identify the similarities and differences between the works. The process of estimating permittivity and soil water can be classified into two main approaches. For soil water estimation, TDR, gravimetric method, capacitance probes, GPR, hygrometric techniques, remote sensing, electromagnetic induction, neutron thermalization, tensiometry, nuclear magnetic resonance, gamma ray attenuation, optical methods, and resistive sensors can be used. Some of these approaches are discussed below. First estimation technique is soil properties based. Authors in [12] derive EM parameters of soil on the basis of soil moisture, frequency, and soil density; however, the model has limitation of working with 20% soil moisture weight and also needs rigorous sample preparation. In [6, 38, 40], a probe-based equipment is developed for the laboratory use. The probe works with vector network analyzer (VNA) in the frequency range of 45–26.5 MHz. In [71], author proposes a model to calculate the dielectric soil permittivity on the basis of empirical evaluations. Similarly, a dielectric model for soil properties is developed by [7]. This model works for frequencies >1.44 Mhz. Peplinski extended the model for characterization of dielectric behavior of soil under the frequency range of 300 MHz to 1.3 GHz [26, 39, 41]. Curtis [6] extensively reviews the techniques for estimating soil permittivity. These methods involve taking sample for laboratory measurements and are labor extensive. The laboratory sample technique does not depict the in-situ soil conditions. Hence, there is a need of automated approaches for monitoring soil moisture (Fig. 9.1).

Another approach is given by [25] which uses TDR for measuring soil properties. It needs to measure refractive index and impedance of the soil. Authors in [64] uses cross-well radar (CWR) to determine EM properties of soil. The purpose of the methodology is to detect dense non-aqueous phase liquids (DNAPLs) hazardous materials. A wideband wave is transmitted under the frequency range of 0.5–1.5 GHz. Transmission simulations and reflection is used to measure soil permittivity in dry sand. References [30, 33, 67] extensively reviews such time-domain-based techniques for estimating soil permittivity. In TDR-based approaches, sensors are placed at each location where measurements are needed. However, it is

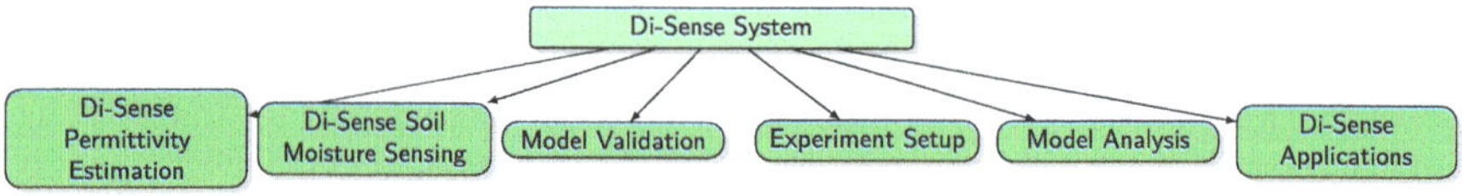

Fig. 9.1 Organization of the chapter

important to obtain real-time soil moisture sensing data so that effective decision making can be achieved in agricultural applications.

Third category of estimating soil properties falls under the antenna based approaches. References [57] and [58] propose a method of measuring electrical properties of the earth using buried antennas. However, length of antenna is adjusted so that the input reactance remains zero. Solimene et al. [59] uses Fresnel reflection coefficient to measure the soil permittivity on the basis of GPR measurements. However, the result in this study is without any empirical validation and also needs to be analyzed in time-domain. Authors in [3] measure soil dielectric properties for the frequency range of 0.1–1 GHz. They use wideband frequency domain method and need LCR meter for measuring impedance and VNA. Complex dielectric properties are measured in frequency domain by using probes [24, 37, 72].

In [13], GPR-based technique is utilized for permittivity estimation by correlating cross-talk of GPR signal and dielectric properties of soil. However, GPR methods are only applied for the implementations which require calibration and low depths (0–20 cm).

Measurements from remotes sensing methods have wider range [66], but are more susceptible to soil water content [18]. Remote sensing methods are classified into active and passive remote sensing [20]. Passive techniques have low spatial resolution (in the order of kilometers) and active has high spatial resolution (in the order of meters). However, with active methods measurements values are limited to few centimeters of topsoil layer and its accuracy is also effected by vegetation cover [56].

This review of measurement methods identifies the gap between large-scale and point-based measurements which can be covered by WUC. The chapter focuses on using WUC for permittivity and soil moisture estimation. It is known that soil properties and soil moisture impact the EM waves propagating through soil [46, 50], hence even a small amount of water can highly effect the IOUT wireless channel between transmitter and receiver. Analysis of path loss at receiver gives a more detailed idea about these changes. This method can use the field IOUT infrastructure without using soil moisture sensors. Therefore, WUC are successfully being used for soil sensing. Last decade has seen significant improvement in various aspects of IOUT communication such as UG communications, characterization of impact of soil type and moisture, and UG channel modeling [1, 4, 8, 11, 22, 34, 44, 48, 50, 60, 62, 68]. Authors in [29, 50] characterize the wireless UG channel in a very detailed manner. Similarly, [46] studies the impact of soil moisture and soil type on the capacity of multi-carrier modulations and also validates the results empirically. To the best of our knowledge, no study provides WUC-based in-situ estimation of these properties in real-time. Therefore, this will be the first work to estimate soil permittivity and soil moisture in response to channel path loss and velocity of EM wave in UG channel.

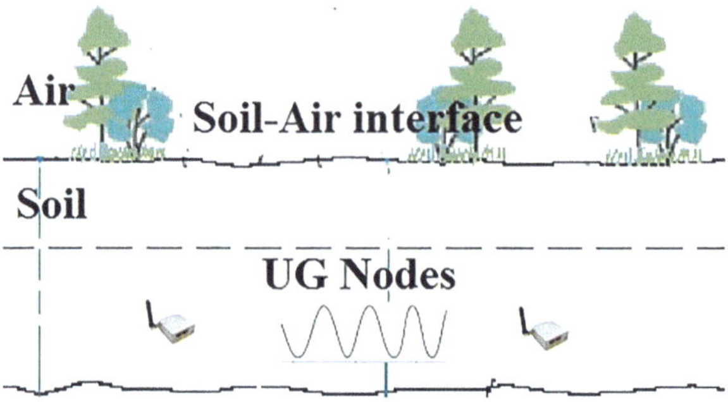

Fig. 9.2 WUC implementation for determining soil properties [54]

9.2 System Models

Expressions are derived for calculating soil permittivity and soil moisture at the distance of 1–15 m. These expressions connect both of the soil parameters to WUC. It is connected to propagation path loss and wave velocity for calculating permittivity. To that end, problem is formulated as follows: derive a function for calculating soil permittivity and soil moisture as an output for an input of path loss of the link to the system. Figure 9.2 shows the schema of WUC along with the sensing UG nodes. For a low electrical loss, there is no difference between the effective permittivity and complex permittivity. This chapter refers the permittivity as a relative permittivity. Section 9.2.1 discusses the Di-Sense permittivity estimation and Sect. 9.2.2 discusses the soil moisture model.

9.2.1 *Di-Sense Permittivity Estimation*

Propagation Path Loss Approach IOUT communication is mainly carried out through EM waves propagation in the soil. The propagation loss of the EM waves due to water in the soil is highly dependent upon the real effective permittivity, i.e., dielectric constant of the soil. Therefore, it is possible to use the propagation loss for calculating soil moisture (within the range of 100–500 MHz) and relative permittivity. For modeling the permittivity of the soil, a known signal is transmitted using narrow bandwidth. A lowest path loss (LPL) is calculated for the signal among all frequency ranges and frequency is increased sequentially in predefined intervals, Δf. The propagation loss of the received signal is measured at the receiver end. Path loss can be defined as the ratio of transmit signal power at sender P_t to the received signal at the receiver P_r. It is given as follows:

$$PL = P_t - P_r = 10.\log 10(P_t/P_r)\ , \tag{9.1}$$

where PL denotes system path loss, along with the effects of both antenna gains, i.e., transmitting G_t and receiving G_r. After measuring the path loss, the frequency of the lowest path loss is calculated as follows:

$$f_{min} = F(\min(PL(f))) , \tag{9.2}$$

where f_{min} represents the frequency of the minimum path loss. The distance between the transmitting and receiving antenna has no effect on f_{min} due to antennas gains. Therefore, PL already contains the antenna gains. Since narrowband is used for calculating PL measurements, noise effects in the signal are minimal. Next the soil factor, ϕ, is given as follows:

$$\phi_s = f_{min}/f_0 , \tag{9.3}$$

where f_0 denotes the resonant frequency of the antenna in the free space. After calculating the soil factor ϕ_s, the wavelength at frequency f_0 is calculated as follows:

$$\lambda_0 = c/f_0 , \tag{9.4}$$

where c is the speed of light. Relative permittivity of the soil is calculated as follows:

$$\epsilon_r = \frac{1}{(\phi_s \times \lambda_0)^2} . \tag{9.5}$$

Permittivity Estimation Through Velocity of Wave Propagation in Soil Soil permittivity varies greatly because of it being non-homogeneous characteristics which results in variation in phase velocity and wavelength of the signal as it propagates through the soil [53]. Therefore, the velocity of the signal can also be used to calculate the permittivity of the soil. Power delay profile (PDP) is known by the geometry layout of the testbed which in turn is used to measure the velocity of the signal. Direct wave component travels completely through the soil. Hence, after calculating the signal velocity, C_s, difference between the arrival and transmission time of direct wave component is used to measure the relative permittivity. Consequently, ϵ_r is calculated as follows:

$$\epsilon_r = \left[C_s \times \frac{(\tau_{dr} - \tau_{dt})}{l} \right] , \tag{9.6}$$

where the distance between receiver and transmitter antennas is given as l, travel time of the direct component in the soil is given as $\tau_{dr} - \tau_{dt}$, and the wave propagation velocity in the soil is denoted by C_s. Propagation velocity of wave is different in soil and air. Due to this difference the direct wave has lower attenuation as compared to lateral wave and travels along the soil–air interface through air.

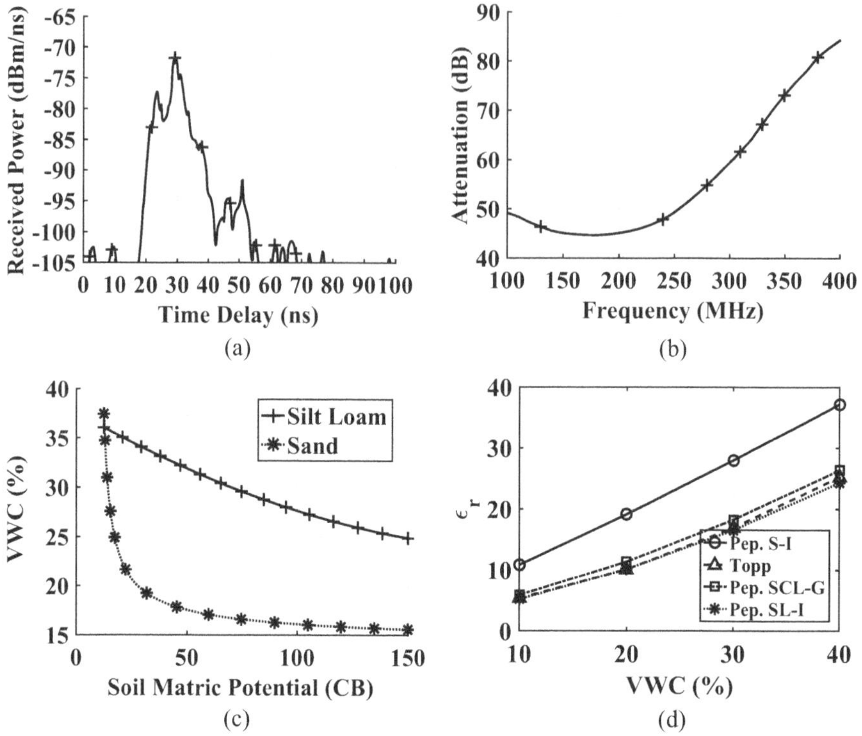

Fig. 9.3 (**a**) Power delay profile (PDP) (Silt Loam Soil and Indoor Testbed), (**b**) signal attenuation in soil with changing operational frequency, (**c**) conversion of soil matric potential to VWC [14, 15], (**d**) relation between VWC and soil permittivity as per Topp and Peplinski model [26, 54]

Figure 9.3, as an example, gives PDP in silt loam soil. It plots the soil attenuation with operational frequency.

9.2.2 Di-Sense Soil Moisture Sensing

Soil moisture-permittivity relation is not dependent upon bulk density, soil texture, and frequency[63]. It is possible to determine the soil water content from soil permittivity as soil moisture is only dependent upon the soil permittivity [19, 63]. [1]

[1] Soil moisture-permittivity relation is proved to work in coarse textured soils and fine textured soil, however, some errors were found in the relationship and this relation is weak in mineral soils [21, 54].

Relative permittivity of the dry soil is 3 and that of water is 80, therefore, it is calculated by Eqs. (9.5) and (9.6) and soil moisture is calculated as given in [19, 63]:

$$VWC\,(\%) = \frac{\epsilon_r - 3}{0.77} + 14.97\,. \tag{9.7}$$

9.3 Model Validation Techniques

Soil moisture sensors are used to measure the soil water content and validate the model. Two main methods used to represent the soil water content are: volumetric water content (VWC) and soil matric potential (SMP). Watermark sensors are used for measurement of SMP in centibars (CB)/kilopascals (kPa).[2] Soil–water retention curve in [14, 31] is used to convert soil matric potential to soil volumetric water content (VWC). Figure 9.3c shows the water retention curves for silt loam soil and sandy soil. For sandy soil, SMP is inversely proportional to VWC and a small increase in SMP causes the VWC to drop significantly because of large pore size[16, 42]. Therefore, developing soil texture-based water retention curves for different soil types with varying soil moisture levels is very important [15, 35].

In addition to these sensor measurements, Peplinski's dielectric and Topp's model is also used to validate the Di-Sense model. This validation is performed with different soil types and soil water content. Soil type has no effect on Topp model [35, 63], and it related soil water content to soil permittivity. Topp model is given as follows:

$$\theta = 4.3 \times 10^{-6}\epsilon^3 - 5.5 \times 10^{-4}\epsilon^2 + 2.92 \times 10^{-2}\epsilon - 5.3 \times 10^{-2}, \tag{9.8}$$

where θ represents the soil water content, and ϵ represents the dielectric constant of the soil.

Soil dielectric constant is measured by the Peplinski model [26]. Soil dielectric constant is given as $\gamma = \alpha + j\beta$, where

$$\alpha = \omega \sqrt{\frac{\mu\epsilon'}{2}\left[\sqrt{1 + \left(\frac{\epsilon''}{\epsilon'}\right)^2} - 1\right]}\,, \tag{9.9}$$

$$\beta = \omega \sqrt{\frac{\mu\epsilon'}{2}\left[\sqrt{1 + \left(\frac{\epsilon''}{\epsilon'}\right)^2} + 1\right]}\,, \tag{9.10}$$

[2]Higher matric potential values equal low soil moisture and, similarly, near saturation point is denoted by zero matric potential 1 CB = 1 kPa.

where μ is the magnetic permeability, $\omega = 2\pi f$ is the angular frequency, and ϵ' and ϵ'' denote the real and imaginary parts of the dielectric constant. Figure 9.3d plots the relation between VWC and permittivity given by Topp and Peplanski model. Peplanski model is used for three different soil types, whereas Topp model is independent of soil type.

9.4 Experiment Setup

This section discusses the experiment setup and methodology used for the validation of the model. Table 9.1 lists the values for burial depths, empirical VWC range, testbed classification, and particle size distribution. For rest of the chapter, soil's name abbreviation is also used as given in Table 9.1.

9.4.1 Experiment Methodology

SDR testbed uses USRPs [9] and GNU Radio [10] for experiments. A UG dipole antenna buried at 40 cm is used to transmit Gaussian signal of 2 MHz bandwidth via transmitter USRP. Receiver USRPs receive the signal which is connected to dipole antennas buried at different depths of 10, 20, 30, and 40 cm whereas the distance between the transmitter and receiver is fixed at 50 cm. Several experiments are performed for all depths and distances of 2 and 4 m [43, 61]. For a given frequency, signal is transmitted by the transmitter for just 1 s and receivers receive IQ data of four mega samples. Transmitter transmits for the next frequency only after receiving the acknowledgment from the receiver. This is done for all the frequency in the range

Table 9.1 Antenna depths, VWC percentage, and particle distribution for different types of soil used in a testbed

Textural class	Sand (%)	Silt (%)	Clay (%)	VWC range (%)	Depth
Silty Clay Loam—Greenhouse (SCL-G)	13	55	32	32–38	20 cm
Silt Loam—Field (SL-F)	17	55	28	22–38	10, 20, 30, and 40 cm
Sandy Soil—Indoor Testbed (S-I)	86	11	3	15–38	10, 20, 30, and 40 cm
Silt Loam—Indoor Testbed (SL-I)	33	51	16	30–37	10, 20, 30, and 40 cm

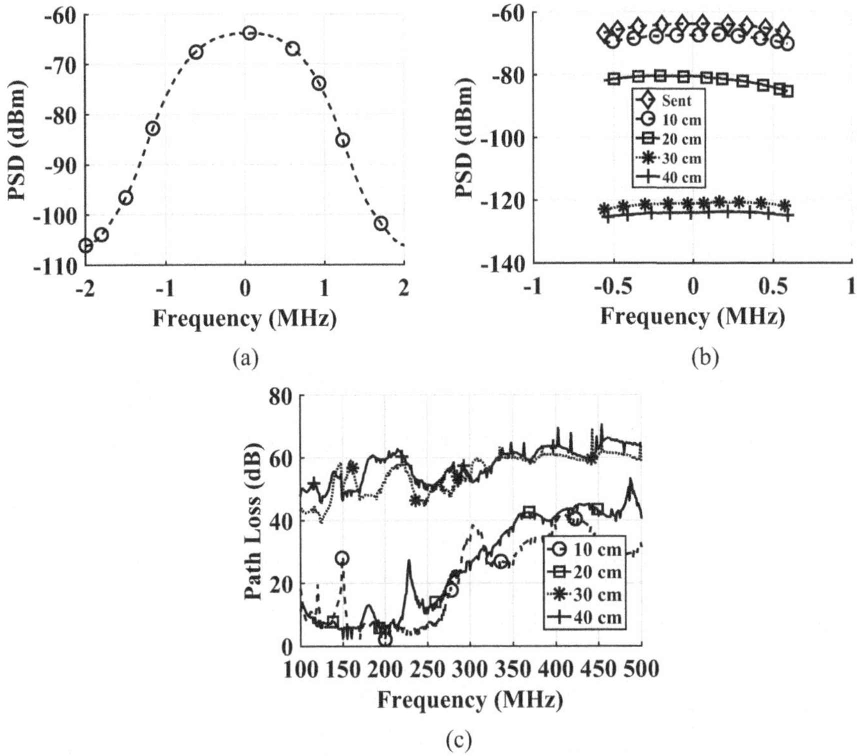

(a) (b) (c)

Fig. 9.4 Power spectral density [54]: (**a**) transmitted signal, (**b**) received signal; and (**c**) effect of frequency on path loss [54]

of 100–500 MHz at each depth and distances. Finally, three measurements are taken and Matlab [23] is used for the post-processing.

Welch's method [46, 73], an advanced form of periodogram analysis, is used for path loss analysis and spectral estimation. It uses discrete Fourier transform to divide the data into fixed blocks for calculation and modification of periodogram which are averaged out for estimating power spectrum. Figure 9.4a shows the periodogram of the transmitted signal.

Figure 9.4b shows the PSD for all depths at a distance of 50 cm and the burial depth of the transmitter is 40 cm, hence shows that PSD is inversely proportional to the burial depths.

Figure 9.4b shows the path loss for all depths. It can be observed that path loss is directly proportional to the burial depths, i.e., increase with the increase in frequency. Therefore, lower frequency, normally <500 MHz, works better in WUC channel.

9.4.2 Power Delay Profile Measurements

The purpose of measuring PDP is to estimate the velocity of propagating wave. To that end, Keysight Technologies N9923A FieldFox VNA is used for measuring PDP. Experimental setup of indoor testbed is used for the purpose with sandy and silt loam soils, varying burial depths (10, 20, 30, and 40 cm). For greenhouse testbed, silty clay loam soil and depth of 20 cm is used with OTA resonant frequency of 433 MHz for all types of soil. Finally, PDP and channel transfer function are measured for changing soil moisture levels [30, 54].

For PDP measurement, VNA transmits a sinusoidal signal with frequency increasing in incremental increment. This frequency domain data is used to obtain the time-domain equivalent impulse response, $h(t)$, through inverse Fourier transform (IFFT). To that end, 401 complex tones are stored in range of 10 MHz–4 GHz, and side lobes are suppressed by windowing the impulse response of the channel. The velocity is then calculated by the process in Sect. 9.2.

9.5 Model Analysis: Performance, Validation, and Error

This section analyzes the performance of the model and validates it. Section 9.5.1 discusses the impact of soil moisture, burial depth, and soil moisture on path loss calculated by using the methodology described in Sect. 9.2. Section 9.5.2 provides the model validation and Sect. 9.5.3 performs the error analysis for the model.

9.5.1 WUC Path Loss

Figure 9.5 shows the change in lowest path loss frequency of UG channel. The results are shown for three soil types: silt loam, silty clay loam, and sandy soil (Fig. 9.5a), varying soil moisture (Fig. 9.5b) and burial depth (Fig. 9.5c). It can be seen in Fig. 9.5a that for sandy soil, increase in soil moisture (15–36%) causes 56% decrease in lowest path loss frequency. For silt loam soil, increase in soil moisture (38–22%) causes 60% decrease in lowest path loss frequency and for silty clay loam soil this decrease is 15.62% for 32–38% increase in soil moisture. Generally, it can be concluded that lowest path loss is inversely proportional to soil moisture for these soil types. This is because of the fact that soil permittivity is greater than that of air causing it to increase with increasing soil moisture, hence decreasing lowest path loss [48, 49, 52].

In Fig. 9.5b change in frequency loss is plotted for all three soil types under different burial depths. For silt loam soil, the lowest path frequency is decreased 12.32% as depth increases from 10 to 40 cm. For silt loam soil, the lowest path frequency is decreased by 12.32% (from 211 to 185 MHz) as depth increases from

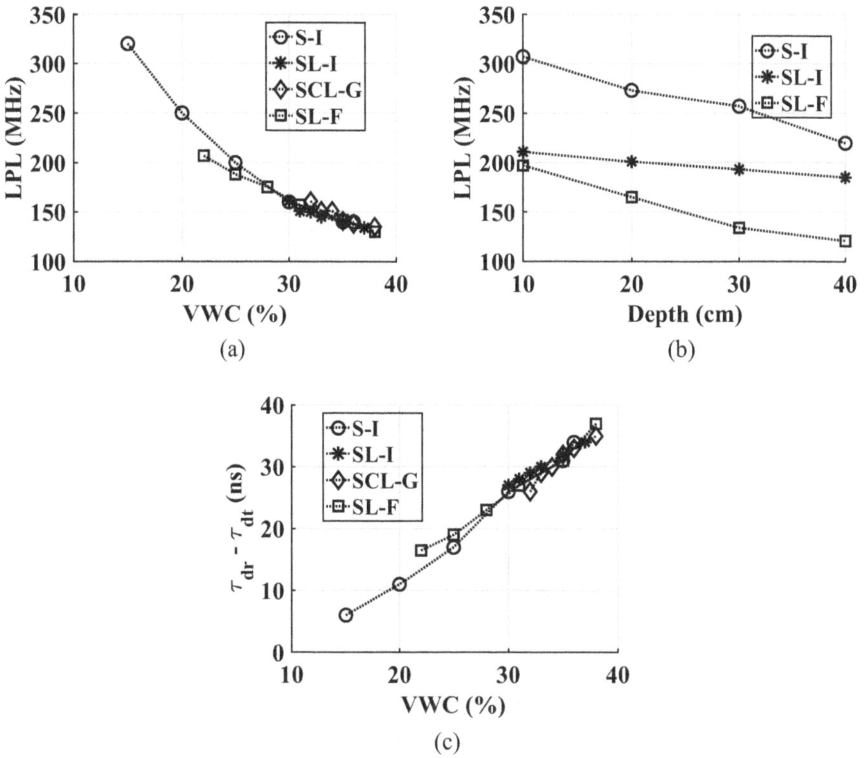

Fig. 9.5 VWC effect, in different soil types, on [54]: (**a**) lowest path loss frequency, (**b**) depth, and (**c**) wave velocity

10 to 40 cm, and that for silt loam soil (field) it decreases by 38% (from 197 to 121 MHz). Similarly, for sandy soil, the lowest path frequency is decreased by 28.24% (from 308 to 221 MHz) as depth increases from 10 to 40 cm. The reason for this behavior of lowest frequency is that wave reflected form the soil–air interface produces a current at antenna causing impedance to change, hence resulting in lowest path loss frequency to change. The distance between soil–air interface increases with the increase in depth; therefore, the reflected waves attenuates because of high absorption rate of the soil. As the relative permittivity is directly proportional to the water content [7, 44], therefore, the change in lowest path loss frequency is greater in sandy soil than silt loam soil because silt loam soil high capacity of holding water. Sandy soil has lower permittivity causing the lowest path frequency shift to higher spectrum.

Figure 9.5c plots the arrival time of direct wave component along with the changing soil moisture. It can be seen that velocity of the wave is inversely proportional to the soil moisture. The velocity decreases with the increasing soil moisture for all types of soil. Wave velocity decreases by five times for soil moisture increase of 15–36%, and for silt loam soil, it decreases three times for soil moisture increase of 22–38% [69, 70].

9.5.2 Validation

This section discusses the results for model validation. Equations (9.5), (9.6), and (9.7) are used to calculate the values for soil permittivity and soil moisture and are shown in Fig. 9.6. Figure 9.6a, b compares the VWC calculated from Topp model and ground truth measurements with Di-Sense VWC. Similarly, Fig. 9.6c compares Di-Sense permittivity with Peplinski model, and Fig. 9.6d compares the Di-Sense permittivity estimated by time-domain velocity of propagation method with Di-Sense path loss propagation permittivity method. The graphs confirm the result from Di-Sense model with ground truth measurements. There are some interesting points

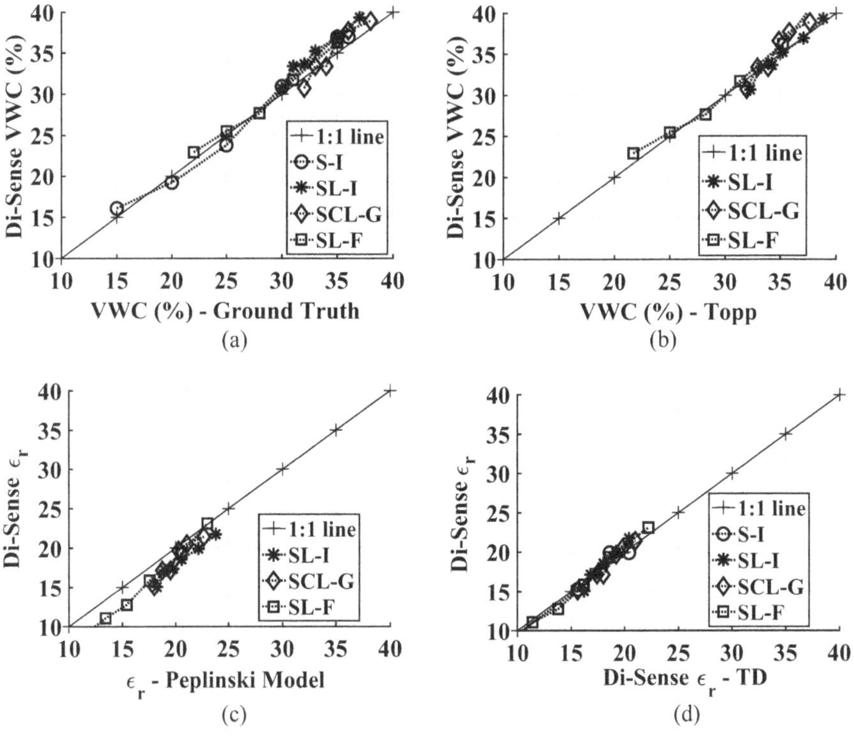

Fig. 9.6 Comparison of VWC of Di-Sense Model with [54]: (**a**) VWC measure from experiments, (**b**) VWC from Topp model; (**c**) comparison of permittivity of Di-Sense Model with permittivity from Peplinski model [54], (**d**) comparison of permittivity from Di-Sense Model from time-domain velocity of propagation method with Di-Sense permittivity from path loss propagation method [54]

to consider from Fig. 9.6, e.g., decrease in lowest path loss frequency causes increase in soil permittivity leading to increase in soil moisture.

9.5.3 Model Error Analysis

Figure 9.7 shows the result from the model error analysis. Di-Sense VWC estimation error is compared with ground truth soil moisture sensing in Fig. 9.7a. It can be observed that the error varies more in silt loam soil (1–8%) as compared to the sandy soil underscoring the importance of clay contents in soil. Overall estimation error remains less than 8%.

Figure 9.7b compares the Di-Sense soil moisture estimation error with that of from Topp model. As in the case of Di-Sense VWC estimation error, Di-Sense soil

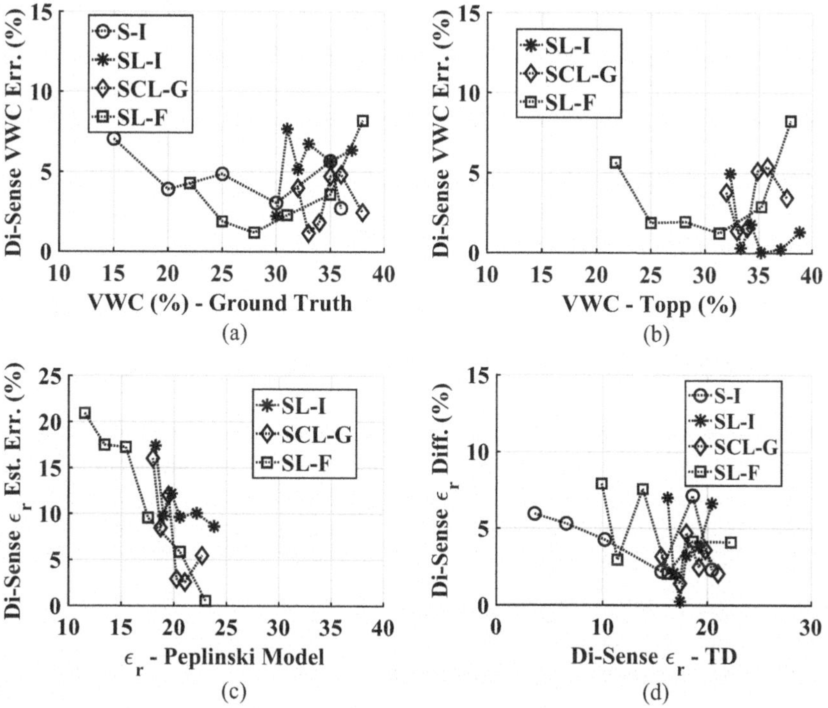

Fig. 9.7 Error Analysis of Di-Sense Model [54]: (**a**) comparison of VWC from Di-Sense Model with VWC measured from experiments, (**b**) comparison of VWC from Di-Sense Model with VWC from Topp model, (**c**) comparison of permittivity of Di-Sense Model with permittivity from Peplinski model, and (**d**) comparison of permittivity from Di-Sense Model from time-domain velocity of propagation method with Di-Sense permittivity from path loss propagation method

moisture estimation error also varies more in silt loam soil. Moreover, Di-Sense soil moisture estimation error is 7% less than the one measured from Topp model.

Figure 9.7c shows the permittivity estimation error from Di-Sense and Peplinski model. It can be seen that, in comparison with Peplinski model, Di-Sense estimation error is high (21%) for silt loam (field) than silt loam (error < 15%) and silty clay loam. It can be observed that error is inversely proportional to the soil moisture levels, i.e., low error is observed for the high soil moisture. Water permittivity depends upon numerous factors; therefore, soil dielectric constant becomes complicated under high soil moisture level. Moreover, soil permittivity also relies on many other factors such as soil temperature, soil type/texture, percentage of clay particles, bulk density, soil bulk density, porosity, and salinity. For Di-Sense soil moisture estimation model, the effect of these factors is very low. To summarize, modeled and measured values are in confirmation with each other and Di-Sense method can be considered as a feasible method for measuring soil permittivity and soil water content [29, 36].

An important thing to note is that Fig. 9.6c does not compare permittivity for sandy soil using Peplinski model. The reason is that Peplinski model does not work well with the sandy soil having sand content of 86% [26]. Figure 9.7d shows the difference between Di-Sense path loss propagation permittivity method and Di-Sense permittivity by time-domain velocity of propagation method and both methods are in confirmation with testbed soil with estimation difference of less than 8%. Hence, Di-Sense model for estimating soil permittivity and soil moisture is suitable for the soils having same particle size classification and distribution as of used in these experiments [47, 54].

9.5.4 *Transfer Functions of Di-Sense*

Although results are most relevant for determining soil permittivity and soil moisture, however, it is also suitable for designing IOUT communication system. Moreover, soil permittivity is less effected at higher depths because the intensity of reflected wave from soil–air interface is reduced at higher depths. Following procedure is used for estimation:

- First, lowest path loss frequency is measured.
- Soil permittivity is determined by using Eqs. (9.5) and (9.6).
- Soil moisture is estimated by using Eq. (9.7).

Figure 9.8 shows the Di-Sense transfer functions for soil moisture and soil permittivity. These graphs can be used for measuring the permittivity and soil moisture for a given IOUT propagation path loss. Di-Sense measurement method is very simple which requires no knowledge of IOUT deployment parameters, i.e., type of radios, antenna knowledge, and communication parameters. Only requirement is to accurately measure the propagation path loss. It can be used with different operation frequencies f_0 as Eqs. (9.3) and (9.4) scale with the f_0. Di-Sense has

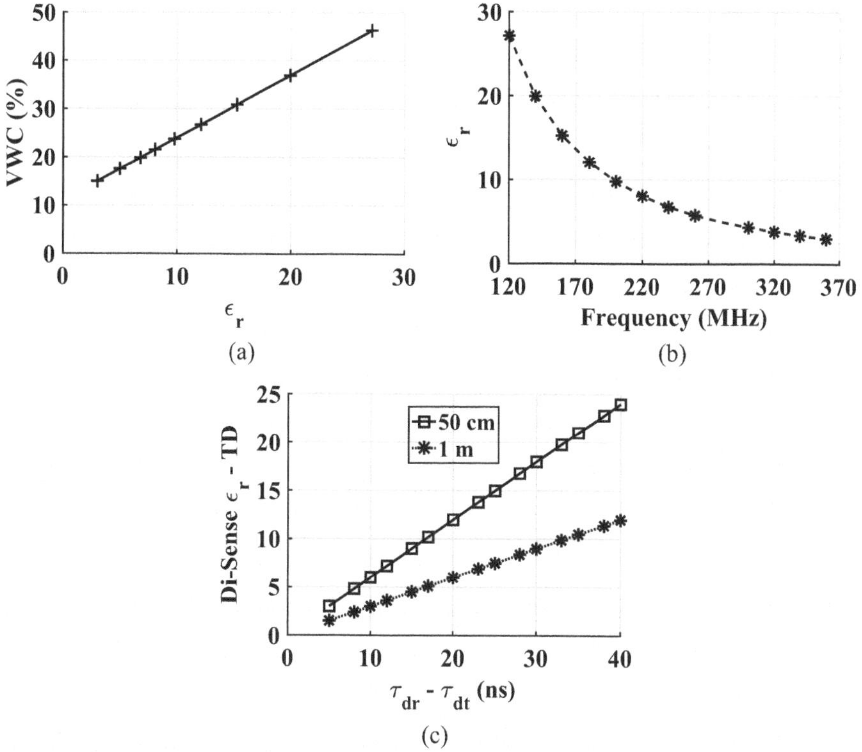

Fig. 9.8 Transfer function of Di-Sense model for [54]: (**a**) soil moisture, (**b**) soil permittivity (time-domain), (**c**) soil permittivity

some limitations as well: it requires accurate measurement of propagation path loss of soil under observation. For applications requiring higher accuracy, soil–water retention capability and specific soil properties can be represented by an empirical factor [51].

9.6 Di-Sense Applications

Di-Sense can be used for irrigation scheduling in IOUT-based agricultural systems [8]. In construction-based IOUT systems (building and bridges), Di-Sense can be used to examine the health of the building structure. In geophysical applications, Di-Sense can be used for estimating permittivity of ice and rocks. It can also be used in detection of soil contamination, and in the domains of meteorology, geophysics, and civil engineering.

References

1. Abrudan TE, Kypris O, Trigoni N, Markham A (2016) Impact of rocks and minerals on underground magneto-inductive communication and localization. IEEE Access 4:3999–4010. https://doi.org/10.1109/ACCESS.2016.2597641
2. Akyildiz IF, Stuntebeck EP (2006) Wireless underground sensor networks: research challenges. Ad Hoc Netw J 4(6):669–686
3. Bobrov P, Repin A, Rodionova O (2015) Wideband frequency domain method of soil dielectric property measurements. IEEE Trans Geosci Remote Sens 53(5):2366–2372. https://doi.org/10.1109/TGRS.2014.2359092
4. Bogena HR, Herbst M, Huisman JA, Rosenbaum U, Weuthen A, Vereecken H (2010) Potential of wireless sensor networks for measuring soil water content variability. Vadose Zone J 9(4):1002–1013
5. Comite D, Galli A, Lauro SE, Mattei E, Pettinelli E (2016) Analysis of GPR early-time signal features for the evaluation of soil permittivity through numerical and experimental surveys. IEEE J Select Top Appl Earth Obs Remote Sens 9(1):178–187. https://doi.org/10.1109/JSTARS.2015.2466174
6. Curtis JO (2001) A durable laboratory apparatus for the measurement of soil dielectric properties. IEEE Trans Instrum Meas 50(5):1364–1369. https://doi.org/10.1109/19.963211
7. Dobson M, etal (1985) Microwave dielectric behavior of wet soil—Part II: dielectric mixing models. IEEE Trans Geosci Remote Sens GE-23(1):35–46. https://doi.org/10.1109/TGRS.1985.289498
8. Dong X, Vuran MC, Irmak S (2012) Autonomous precision agriculture through integration of wireless underground sensor networks with center pivot irrigation systems. Ad Hoc Networks (Elsevier) 11(7):1975–1987
9. Ettus Research Website (2016). http://www.ettus.com
10. GNU Radio Website (2020) http://www.gnuradio.org
11. Guo H, Sun Z (2014) Channel and energy modeling for self-contained wireless sensor networks in oil reservoirs. IEEE Trans Wireless Commun 13(4):2258–2269. https://doi.org/10.1109/TWC.2013.031314.130835
12. Hipp JE (1974) Soil electromagnetic parameters as functions of frequency, soil density, and soil moisture. Proc IEEE 62(1):98–103. https://doi.org/10.1109/PROC.1974.9389
13. Hislop G (2015) Permittivity estimation using coupling of commercial ground penetrating radars. IEEE Trans Geosci Remote Sens 53(8):4157–4164. https://doi.org/10.1109/TGRS.2015.2392110
14. Irmak S, Haman D (2001) Performance of the watermark. Granular matrix sensor in sandy soils. Appl Eng Agric17(6):787
15. Irmak S, Irmak A (2005) Performance of frequency-domain reflectometer, capacitance, and pseudo-transit time-based soil water content probes in four coarse-textured soils. Appl Eng Agric 21(6):999–1008
16. Irmak S, Payero JO, Eisenhauer DE, Kranz WL, Martin D, Zoubek GL, Rees JM, VanDeWalle B, Christiansen AP, Leininger D (2006) Ec06-783 watermark granular matrix sensor to measure soil matric potential for irrigation management
17. Irmak S, etal (2012) Large scale on-farm implementation of soil moisture-based irrigation management strategies for increasing maize water productivity. Trans ASABE 55(3):881–894
18. Jonard F, Weihermüller L, Schwank M, Jadoon KZ, Vereecken H, Lambot S (2015) Estimation of hydraulic properties of a sandy soil using ground-based active and passive microwave remote sensing. IEEE Trans Geosci Remote Sens 53(6):3095–3109. https://doi.org/10.1109/TGRS.2014.2368831
19. Josephson B, Blomquist A (1958) The influence of moisture in the ground, temperature and terrain on ground wave propagation in the VHF-band. IRE Trans Anten Propag 6(2):169–172. https://doi.org/10.1109/TAP.1958.1144574

20. Kim S, Ouellette JD, van Zyl JJ, Johnson JT (2016) Detection of inland open water surfaces using dual polarization l-band radar for the soil moisture active passive mission. IEEE Trans Geosci Remote Sens 54(6):3388–3399. https://doi.org/10.1109/TGRS.2016.2517010
21. Ledieu J, Ridder PD, Clerck PD, Dautrebande S (1986) A method of measuring soil moisture by time-domain reflectometry. J Hydrol 88(3):319–328. http://dx.doi.org/10.1016/0022-1694(86)90097-1
22. Markham A, Trigoni N (2012) Magneto-inductive networked rescue system (miners): taking sensor networks underground. In: Proceedings of the 11th ICPS, ACM, IPSN '12, pp 317–328. https://doi.org/10.1145/2185677.2185746
23. MATLAB (2020) http://wwwmatlab.com
24. Nassar EM, Lee R, Young JD (1999) A probe antenna for in situ measurement of the complex dielectric constant of materials. IEEE Trans Anten Propag 47(6):1085–1093. https://doi.org/10.1109/8.777136
25. Nicolson AM, Ross GF (1970) Measurement of the intrinsic properties of materials by time-domain techniques. IEEE Trans Instrum Measurement 19(4):377–382. https://doi.org/10.1109/TIM.1970.4313932
26. Peplinski NR, Ulaby FT, Dobson MC (1995) Dielectric properties of soils in the 0.3–1.3-GHz range. IEEE Trans Geosci Remote Sens 33(3):803–807
27. Pettinelli E, Matteo AD, Mattei E, Crocco L, Soldovieri F, Redman JD, Annan AP (2009) GPR response from buried pipes: measurement on field site and tomographic reconstructions. IEEE Trans Geosci Remote Sens 47(8):2639–2645. https://doi.org/10.1109/TGRS.2009.2018301
28. Salam A (2018) Pulses in the sand: long range and high data rate communication techniques for next generation wireless underground networks. ETD collection for University of Nebraska - Lincoln (AAI10826112). http://digitalcommons.unl.edu/dissertations/AAI10826112
29. Salam A (2019) Design of subsurface phased array antennas for digital agriculture applications. In: Proceedings of the 2019 IEEE International Symposium on Phased Array Systems and Technology (IEEE Array 2019), Waltham, MA
30. Salam A (2019) A path loss model for through the soil wireless communications in digital agriculture. In: 2019 IEEE International Symposium on Antennas and Propagation. IEEE, New York, pp 1–2
31. Salam A (2019) Sensor-free underground soil sensing. In: ASA, CSSA and SSSA International Annual Meetings (2019), ASA-CSSA-SSSA
32. Salam A (2019) Subsurface MIMO: a beamforming design in internet of underground things for digital agriculture applications. J Sens Actuat Netw 8(3). https://doi.org/10.3390/jsan8030041. https://www.mdpi.com/2224-2708/8/3/41
33. Salam A (2019) Underground environment aware MIMO design using transmit and receive beamforming in internet of underground things. Springer International Publishing, Cham, pp 1–15
34. Salam A (2019) An underground radio wave propagation prediction model for digital agriculture. Information 10(4):147
35. Salam A (2020) Internet of Things for sustainability: perspectives in privacy, cybersecurity, and future trends. Springer International Publishing, Cham, pp 299–327. https://doi.org/10.1007/978-3-030-35291-2_10
36. Salam A (2020) Internet of Things for sustainable community development, 1st edn. Springer Nature, New York. https://doi.org/10.1007/978-3-030-35291-2
37. Salam A (2020) Internet of Things for sustainable community development: introduction and overview. Springer International Publishing, Cham, pp 1–31. https://doi.org/10.1007/978-3-030-35291-2_1
38. Salam A (2020) Internet of Things for sustainable forestry. Springer International Publishing, Cham, pp 147–181. https://doi.org/10.1007/978-3-030-35291-2_5
39. Salam A (2020) Internet of Things for sustainable human health. Springer International Publishing, Cham, pp 217–242. https://doi.org/10.1007/978-3-030-35291-2_7
40. Salam A (2020) Internet of Things for sustainable mining. Springer International Publishing, Cham, pp 243–271. https://doi.org/10.1007/978-3-030-35291-2_8

41. Salam A (2020) Internet of Things in sustainable energy systems. Springer International Publishing, Cham, pp 183–216. https://doi.org/10.1007/978-3-030-35291-2_6
42. Salam A (2020) Internet of Things in water management and treatment. Springer International Publishing, Cham, pp 273–298. https://doi.org/10.1007/978-3-030-35291-2_9
43. Salam A (2020) Wireless underground communications in sewer and stormwater overflow monitoring: radio waves through soil and asphalt medium. Information 11(2):98
44. Salam A, Karabiyik U (2019) A cooperative overlay approach at the physical layer of cognitive radio for digital agriculture. In: 3rd international Balkan conference on communications and networking (2019 BalkanCom), Skopje
45. Salam A, Shah S (2019) Internet of things in smart agriculture: enabling technologies. In: 2019 IEEE 5th World Forum on Internet of Things (WF-IoT). IEEE, New York, pp 692–695
46. Salam A, Vuran MC (2016) Impacts of soil type and moisture on the capacity of multi-carrier modulation in internet of underground things. In: Proceedings of the 25th ICCCN 2016, Waikoloa, HI
47. Salam A, Vuran MC (2017) EM-based wireless underground sensor networks, pp 247–285. https://doi.org/10.1016/B978-0-12-803139-1.00005-9
48. Salam A, Vuran MC (2017) Smart underground antenna arrays: a soil moisture adaptive beamforming approach. In: Proceedings of the 36th IEEE INFOCOM 2017, Atlanta
49. Salam A, Vuran MC (2017) Wireless underground channel diversity reception with multiple antennas for internet of underground things. In: Proceedings of the IEEE ICC 2017, Paris
50. Salam A, Vuran MC, Irmak S (2016) Pulses in the sand: impulse response analysis of wireless underground channel. In: Proceedings of the IEEE INFOCOM 2016, San Francisco
51. Salam A, Vuran MC, Irmak S (2016) Pulses in the sand: impulse response analysis of wireless underground channel. In: The 35th Annual IEEE International Conference on Computer Communications (INFOCOM 2016), San Francisco
52. Salam A, Vuran MC, Irmak S (2017) Towards internet of underground things in smart lighting: a statistical model of wireless underground channel. In: Proceedings of the 14th IEEE International Conference on Networking, Sensing and Control (IEEE ICNSC), Calabria
53. Salam A, Hoang AD, Meghna A, Martin DR, Guzman G, Yoon YH, Carlson J, Kramer J, Yansi K, Kelly M, et al (2019) The future of emerging IoT paradigms: architectures and technologies. https://doi.org/10.20944/preprints201912.0276.v1
54. Salam A, Vuran MC, Irmak S (2019) Di-sense: in situ real-time permittivity estimation and soil moisture sensing using wireless underground communications. Comput Netw 151:31–41. https://doi.org/10.1016/j.comnet.2019.01.001. http://www.sciencedirect.com/science/article/pii/S1389128618303141
55. Scott WR, Smith GS (1992) Measured electrical constitutive parameters of soil as functions of frequency and moisture content. IEEE Trans Geosci Remote Sens 30(3):621–623. https://doi.org/10.1109/36.142943
56. Small EE, Larson KM, Chew CC, Dong J, Ochsner TE (2016) Validation of GPS-IR soil moisture retrievals: comparison of different algorithms to remove vegetation effects. IEEE J Select Top Appl Earth Obs Remote Sens 9(10):4759–4770. https://doi.org/10.1109/JSTARS.2015.2504527
57. Smith G, Nordgard J (1985) Measurement of the electrical constitutive parameters of materials using antennas. IEEE Trans Anten Propag 33(7):783–792. https://doi.org/10.1109/TAP.1985.1143657
58. Smith GS, King RWP (1974) The resonant linear antenna as a probe for measuring the in situ electrical properties of geological media. J Geophys Res 79(17):2623–2628. http://dx.doi.org/10.1029/JB079i017p02623
59. Solimene R, D'Alterio A, Gennarelli G, Soldovieri F (2014) Estimation of soil permittivity in presence of antenna-soil interactions. IEEE J Select Top Appl Earth Obs Remote Sens 7(3):805–812. https://doi.org/10.1109/JSTARS.2013.2268576
60. Tan X, Sun Z, Akyildiz IF (2015) Wireless underground sensor networks: MI-based communication systems for underground applications. IEEE Anten Propag Mag 57(4). https://doi.org/10.1109/MAP.2015.2453917

61. Temel S, Vuran MC, Lunar MM, Zhao Z, Salam A, Faller RK, Stolle C (2018) Vehicle-to-barrier communication during real-world vehicle crash tests. Comput Commun 127:172–186
62. Tiusanen MJ (2013) Soil scouts: description and performance of single hop wireless underground sensor nodes. Ad Hoc Netw 11(5):1610–1618. http://dx.doi.org/10.1016/j.adhoc.2013.02.002
63. Topp GC, Davis JL, Annan AP (1980) Electromagnetic determination of soil water content: Measurements in coaxial transmission lines. Water Resour Res 16(3):574–582. https://doi.org/10.1029/WR016i003p00574
64. Toro-Vazquez J, Rodriguez-Solis RA, Padilla I (2012) Estimation of electromagnetic properties in soil testbeds using frequency and time domain modeling. IEEE J Select Top Appl Earth Obs Remote Sens 5(3):984–989. https://doi.org/10.1109/JSTARS.2012.2193610
65. Ulaby FT, Long DG (2014) Microwave radar and radiometric remote sensing. University of Michigan Press, Ann Arbor
66. van der Velde R, Salama MS, Eweys OA, Wen J, Wang Q (2015) Soil moisture mapping using combined active/passive microwave observations over the east of the Netherlands. IEEE J Select Top Appl Earth Obs Remote Sens 8(9):4355–4372. https://doi.org/10.1109/JSTARS.2014.2353692
67. Vereecken H, Schnepf A, Hopmans JW, Javaux M, Or D, Roose T, Vanderborght J, Young M, Amelung W, Aitkenhead M, et al (2016) Modeling soil processes: review, key challenges, and new perspectives. Vadose Zone J 15(5):57 pp.
68. Vuran MC, Dong X, Anthony D (2015) Antenna for wireless underground communication. http://www.google.com/patents/US20150181315. US Patent App. 14/415,455
69. Vuran MC, Salam A, Wong R, Irmak S (2018) Internet of underground things in precision agriculture: architecture and technology aspects. Ad Hoc Netw https://doi.org/10.1016/j.adhoc.2018.07.017. http://www.sciencedirect.com/science/article/pii/S1570870518305067
70. Vuran MC, Salam A, Wong R, Irmak S (2018) Internet of underground things: sensing and communications on the field for precision agriculture. In: 2018 IEEE 4th World Forum on Internet of Things (WF-IoT) (WF-IoT 2018), Singapore
71. Wang JR, Schmugge TJ (1980) An empirical model for the complex dielectric permittivity of soils as a function of water content. IEEE Trans Geosci Remote SensGE-18(4):288–295. https://doi.org/10.1109/TGRS.1980.350304
72. Weir WB (1974) Automatic measurement of complex dielectric constant and permeability at microwave frequencies. Proc IEEE 62(1):33–36. https://doi.org/10.1109/PROC.1974.9382
73. Welch PD (1967) The use of fast Fourier transform for the estimation of power spectra: a method based on time averaging over short, modified periodograms. IEEE Trans Audio Electroacoust 15(2):70–73. https://doi.org/10.1109/TAU.1967.1161901

Part IV
IOUT Applications in Decision Agriculture, Real-Time Soil Monitoring, and Irrigation Automation

Chapter 10
Current Advances in Internet of Underground Things

10.1 Introduction

Internet of Underground Things (IOUTs) is subset of IoT which consists of sensors and communication devices for real-time sensing and monitoring of soil. IOUT differs from IoT in that the sensors and communication devices are either partly or fully buried underground. IOUT extends Wireless Underground Sensor Networks (WUSNs) [6, 13, 18, 20, 35, 36, 41, 54, 71, 115, 117, 121, 137–139, 150, 156, 163, 164, 174] and includes autonomous devices that collects relevant earth information. These devices are connected via some communication and network systems which send information out from the field to farmers and growers for decision making. IOUT, through Internet, facilitates seamless access to agriculture information. IOUT operation includes: in-situ soil sensing such as soil's moisture, temperature, and salinity, etc., communication through soil and plants, and real-time environment information, e.g., rain, wind, and solar, etc. IOUT aims to improve and provide efficient food production mechanism. It achieves complete autonomy by connecting with the various farming equipment such as seeders, combines and irrigation systems [112]. The operating environment of IOUT is very unique, i.e., access of information from soil, communication through plants and soil, exposure to element and unexpected environmental conditions. Existing over-the-air (OTA) wireless solutions were not made for such environment, therefore, these OTA solutions face significant performance challenges in IOUT environment (Fig. 10.1). In attempt to deal with challenges of communicating in IOUT environment, a new type of wireless communication is proposed: wireless underground communications (UG) [6, 171]. Wireless UG communications is a communication method where radios are buried under earth and communicate through soil. IOUT along with the UG communication will be beneficial in conserving water and improving crop yields [156, 164]. Advancement in IOUT technologies can also improve operations of various applications (underground mining, pipeline assessment, and landslide monitoring) which utilizes earth underground resources [7, 20, 96, 115, 121, 171].

A. Salam, U. Raza, *Signals in the Soil*,
https://doi.org/10.1007/978-3-030-50861-6_10

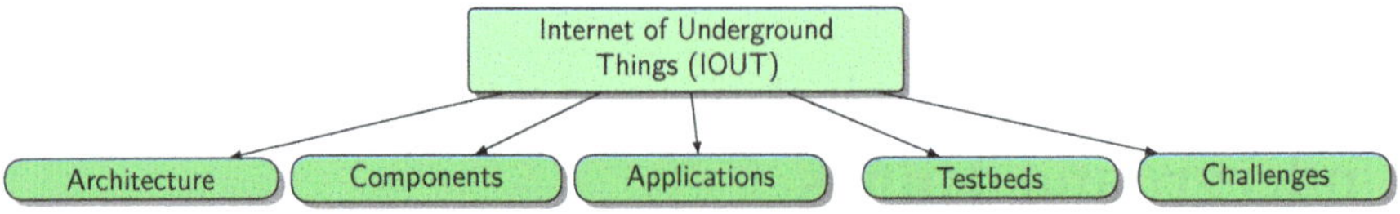

Fig. 10.1 Organization of the chapter

10.2 IOUT Architecture

As mentioned in the previous section, IOUT consists of interconnected heterogeneous nodes customized for field operations. IOUT is expected to provide the following functionalities:

10.2.1 Functionalities

- *In-situ Sensing*: In-situ sensing refers to a buried sensor collecting the information (soil moisture, salinity, and temperature) from the soil. In-situ sensing plays a very important role in gathering precise localized information of the soil. These sensors can be used in two ways: (1) they can be integrated on the chip with components in the architecture, (2) can be used as standalone separate sensors connected to the main components through wires [103, 125].
- *Wireless Communication in Challenging Environment*: Communication systems in IOUT are either deployed in the field or within the soil. As these systems are exposed to natural environment, they should sustain the unexpected rough and challenging environment. OTA solutions should be customized to changing environment because of irrigation and growth of the crop. UG solutions are shielded from the environment and should be able to communicate through the soil and adapt to dynamic changes in soil parameters.
- *Inter-Connection of Field Machinery, Sensors, Radios, and Cloud*: IOUT systems are mainly responsible for collecting data from the field and sending it to the cloud for further processing and decision making. It uses a multitude of diverse devices for successfully completing these operations. IOUT should be able to seamlessly integrate and link these large number of different devices. In addition to collection of information, IOUT must also be able to automate the field operations based on this information.

Figure 10.2 illustrates the IOUT architecture which aims to achieve these required functionalities. The components of this architecture are described below:

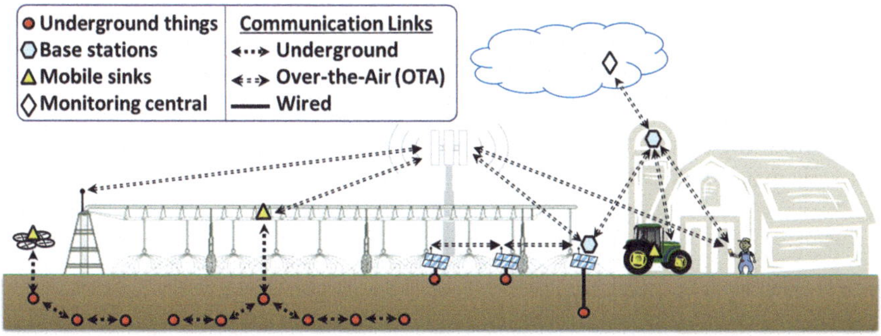

Fig. 10.2 IOUT paradigm in precision agriculture [172]

10.2.2 Elements

- *Underground Things (UTs)*: An UT is an embedded system with on board sensing and communication component. UTs can either partially or fully buried in the ground. UTs are protected from the environment with water proof enclosure and watertight containers. Buried UTs are protected from the farm's equipment. Soil temperature and moisture sensors are used most commonly, however, many other soil- and weather-related phenomena can also be monitored. Bluetooth, satellite, ZigBee, underground, and cellular are some of the existing communication schemes. Bluetooth [54, 173] and underground wireless [20, 116] provide short-range communication of about 100 meters. Commercial products operating in industrial, scientific, and medical (ISM) band covers three times larger distance of short-range communication systems. Cellular and satellite communication systems provide much longer range covering more distance. For large field size, a network can be configured to send the data to a collector sink and self-heal if nodes are unreachable. Nodes are powered by the combination of batteries and solar panels. UTs are cheaper as they are deployed in a very large number [36, 122].
- *Base Stations*: Base stations are used as gateways for transferring data from the field to the cloud. They are permanent structures and are installed in buildings, e.g., weather station. Due to high power of processing and communication capabilities, base stations are expensive and well protected [36].
- *Mobile Sinks*: Mobile sinks are attached with the moving farming equipment such as tractors or irrigation systems [20, 124]. However, sometimes turning the farming equipment just for data retrieval can sometimes be expensive. Hence, as an alternative, unmanned vehicle (ground robots or quadrotors) are also being used for such purposes.
- *Cloud*: The purpose of the cloud services in IOUT architecture is to provide permanent storage for information, real-time processing of information, decision making, and integration with other databases such as soil and weather.

Tables 1.1 and 1.2 provide the summary of existing academic and commercial architecture. Most of the commercial solutions uses OTA communication systems where UT connects to field tower via cellular or satellite communication. This diverse communication architecture makes it difficult to form a unified IOUT architecture fulfilling all the requirements. Lack of standard sensing and communication protocols makes this task more difficult.

10.3 IOUT Components

10.3.1 Communications

This section explores some communication systems which can be used in the implementation of IOUT systems. Here, all these systems are discussed at very basic level. There are different technologies that can be used for communication in IOUT. The wired IOUT solutions uses coaxial cables and fiber optics. Wired solutions provide high data rates and are being used in different applications [8, 40, 131]. Although, these wired solutions provide reliable and accurate communication, however, they increase the complexity and poses scalability issues. Given the disadvantages of wired systems, wireless solutions are termed as feasible solutions to provide relatively low complexity and scalable solutions.

Wireless solutions are classified on the basis of technology they are using to transmit information. These solutions includes: Electromagnetic waves (EM), magnetic induction (MI), acoustic waves, and VLC. Acoustic wave is used for detection purposes, e.g., detecting objects in underground environment [75] and detecting water content in the soil. The disadvantage of using acoustic-based system is the low data rate and high noise and attenuation [32].

Some of the technologies and their feasibility for IOUT systems are given below:

Magnetic Induction (MI) Magnetic Induction-based IOUT communication network consists of buried sensors (UTs) and aboveground (AG) devices. Aboveground devices can provide extended communication by using extra large and powerful dipole antenna. Therefore, downlink communication (AG2UG) is singlehop and upstream (UG2AG) is multi-hop due to limited transmission power [108].

MI transceiver is composed of induction coils producing magnetic fields. These magnetic fields can be sensed by nearby coil. Each induction coil is attached to a capacitor in a way that it operates at resonant frequency. In an attempt to achieve long transmission range (10–100 m) in IOUT moderate size coils are preferred. A time-varying magnetic field is produced by the coil of transmitter node which in-turn induces current at receiver antenna. This procedure is shown in Fig. 10.3c.

Decay rate of received signal strength (RSS) is the inverse cube factor in MI. Hence, high data rates and long-range are not an option in MI-based systems and both are primal requirements of IOUT. In MI, transmitter and receiver antenna are perpendicular to each other which prevent establishing communication in MI.

Table 10.1 Various IOUT systems developed by academic institutions [172]

Architecture	Sensors	Comm. tech.	Node density
Automated irrigation system [36]	DS1822 (temperature) VH400 (soil moisture)	OTA, ZigBee (ISM)	One node per indoor bed
Soil scout [164]	TMP122 (temperature) EC-5 (soil moisture)	UG, Custom (ISM)	Eleven scouts on field
Remote sensing and irrigation sys. [54]	TMP107 (temperature) CS616 (soil moisture) CR10 data logger	OTA, Bluetooth (ISM)	Five field stations
Autonomous precision agriculture [20]	Watermark 200SS-15 (soil moisture) Data logger	UG, Custom (ISM)	Up to 20 nodes per field
SoilNet [13]	ECHO TE (soil moisture) EC20 TE (soil conductivity)	OTA, ZigBee (ISM)	150 nodes covering 27 ha
MOLES [156]	Magnetic induction communications	Magnetic induction	Indoor testbed
Irrigation nodes in vineyards [169]	Yield NDVI	VRI	140 irrigation nodes
Sensor network for irrigation scheduling [17, 134]	Capacitance (soil moisture) Irromesh	OTA	6 nodes per acre
Cornell's digital agriculture [14]	E-synch, touch-sensitive soft robots Vineyard mapping technology, RTK	OTA	Field dependant
Plant water status network [83]	Crop water stress index (CWSI) Modified water stress index (MCWSI)	OTA	Two management zone
Real-time leaf temperature monitor system [63]	Leaf temperature Relative humidity and incident solar radiation	OTA	Soil and plant monitors
Thoreau [183]	Temperature, soil moisture Electric conductivity and water potential	OTA	Based on Sigfox

(continued)

Table 10.1 (continued)

Architecture	Sensors	Comm. tech.	Node density
FarmBeats [167]	Temperature, soil moisture orthomosaic and pH	OTA	Field size of 100 acres
Video-surveillance and data-monitoring WUSN [31]	Agriculture data-monitoring motion detection, camera sensor	OTA	In the order of several km
Purdue's digital agriculture initiative [81]	Adaptive weather tower PhenoRover sensor vehicle	OTA	Field dependant
Pervasive wireless sensor network [176]	Soil moisture, camera	OTA	Field dependant
Pilot sensor network [61]	Sensirion SHT75	OTA	100 nodes in a field
SoilBED [26]	Contamination detection	UG	Cross-well radar

Therefore, MI-based IOUT architecture are not scalable. Due to these factors and inability to communicate with above-ground devices, MI-based systems are not a reliable option for IOUT systems and their performance is dependent on using relay coil [60, 76, 95, 152].

Acoustic Acoustic waves have been used as communication and detection techniques in underground measurements. Geologist have been using to explore natural resources (oil and gas) by measuring the reflected acoustic waves from the ground. It has been used in drilling for communication with underground drilling equipment. Some of the major applications of acoustic waves include buried pipeline monitoring, earthquake monitoring, seismic exploration, smart drilling for oil and gas reservoirs. There exist many studies in the literature for acoustic which can also support other applications [107].

Acoustic methods can be active or passive depending on how signal is generated. Passive methods generates an acoustic signal, from the underground, after some major events like earthquake or volcanic explosion, or sudden underground changes such as structural transformation, pipeline leakage or rock crack formation. Sensors are placed in the vicinity of these events which detect infrasonic wave and help in predicting such events. Active method's signal is self-generated by an artificial explosion/vibration which is sent underground to understand the earth's property (see Fig. 10.3a). Reflection-based seismology is one popular application of such method though IOUT UG communication and underwater communication [149] have some similarities. However, they are not suitable choice for communication in IOUT because of low propagation speed and vibration limitations. Therefore, they are mostly used for the detection purposes.

Table 10.2 Various commercially available IOUT systems developed by industry stakeholders [172]

Architecture	Sensors	Comm. tech.	Node density
IRROmesh [49]	200TS (temperature)	OTA, custom (ISM)	Up to 20 nodes network mesh
	Watermark 200SS-15 (soil moisture)	OTA, cellular	
Field connect [50]	Leaf wetness		
	Temperature probe	OTA, Proprietary	
	Pyranometer	OTA, cellular	Up to eight nodes per gateway
	Rain gauge	OTA, Satellite	
	Weather station		
SapIP wireless mesh network [22]	Plant water use	OTA	25 SapIP nodes.
	Measure plant stress		
	Soil moisture profile		
	Weather and ET		
Automated irrigation advisor [165]	Tule Actual ET sensor	OTA	Field dependant
Internet of agriculture-BioSense [11]	Machinery auto-steering and automation	OTA	Field dependant
	EC probe & XRF scanner		
	Electrical conductivity map		
	NDVI map		
	Yield map		
	Remote sensing		
	Nano and micro-electronic sensors		
	Big data, and Internet of things		
EZ-Farm [43]	Water usage	OTA	IBM Bluemix & IoT foundation
	Big data, and internet of things		
	Terrain, soil, weather		
	Genetics		
	Satellite info		
	Sales		

(continued)

Table 10.2 (continued)

Architecture	Sensors	Comm. tech.	Node density
Internet of food and farm (IoF2020) [46]	Soil moisture Soil temperature Electrical conductivity and Leaf wetness	OTA	Field dependant
Cropx soil monitoring system [15]	Soil moisture Soil temperature and EC	OTA	Filed dependant
Plug & sense smart agriculture [78]	Temperature and humidity sensing Rainfall, wind speed and direction Atmospheric pressure Soil water content, and leaf wetness	OTA	Field dependant
Grain monitor-TempuTech [158]	Grain temperature and humidity	OTA	Multiple depths in grain elevator
365FarmNet [1]	Mobile device visualization tool for IOUT data	OTA	Field dependant
SeNet [133]	Sensing and control architecture	OTA	Field dependant
PrecisionHawk [79]	Drones for sensing Field map generation	OTA	Field dependant
HereLab [39]	Soil moisture Drip line psi and rain	OTA	Field dependant
IntelliFarms [45]	YieldFax Biological BinManager	OTA	Field dependant
IoT sensor platform [47]	IoT/M2M sensors	OTA	Field dependant
Symphony link [154]	Long-range communications	OTA	Field dependant

Unmanned Aerial Vehicle (UAV) Unmanned Aerial Vehicle (UAV) has recently seen an advancement in IOUT precision agriculture for sensing the communication of the field conditions [66, 87], imaging surveillance [73], and decision support [86]. Before the popularity of UAVs, the satellite imaging was only used for monitoring purposes, however, now it is also being used to create soil moisture maps for the field in timely and cost-efficient methods. Other applications include: crop growth monitoring, seed planting, and pesticide applications. When integrated with IOUT systems, UAVs require efficient communication protocols to communicate with the sensors, radio antennas, and make real-time decision. However, there are some challenges faced by the UG to UAVs communication, e.g., limitation on payloads antennas used by UAVs, limitations on time of the flight, need of special skills for operating UAVs, operational license for UAV, and shorter communication range. Advancement in technology and regulatory restrictions can lead to improved integration of UAVs in IOUT Ecosystem [104].

Electromagnetic Waves for IOUT Electromagnetic (EM) waves have been extensively used for communications in various applications of IOUT in agriculture [21, 181], seismic exploration [129], oil and gas (Fig. 10.3b), and drilling [5, 29, 52].

Low Power Wide Area Network (LPWAN) IOUTs are designed to operate for a longer period of time which make energy conservation a very important issue. Low Power Wide Area Network (LPWAN) not only conserves energy but also gives long-range connectivity [77, 109]. LPWANs are used in those IOUT applications which have primary requirement of energy conservation and range, therefore, high data rate operations are not required and latency is acceptable. As per one of the LPWAN technical workgroups, LPWAN can operate for several years and customized to the

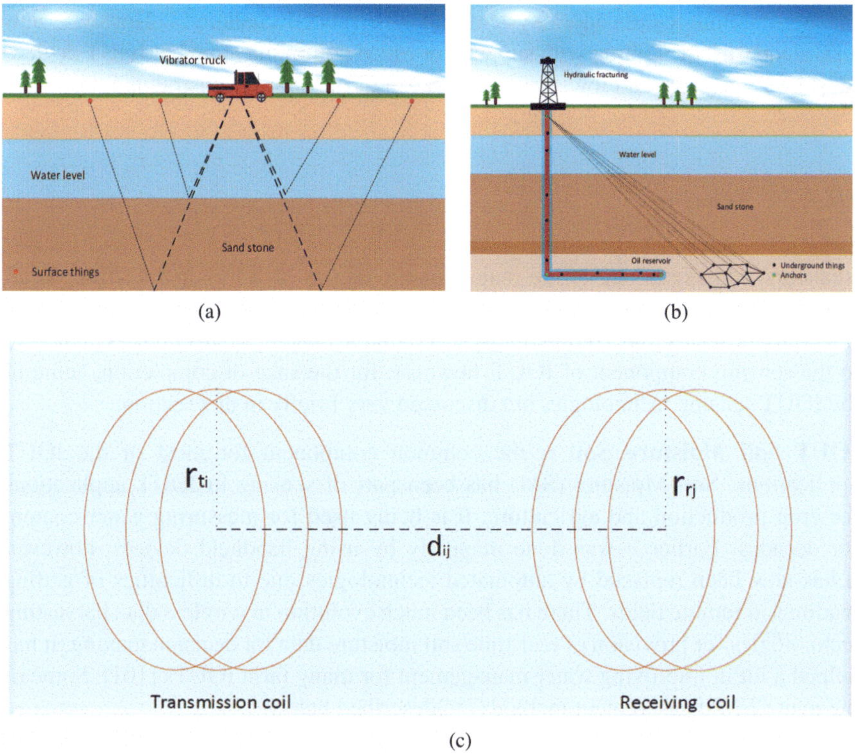

Fig. 10.3 Communication technologies in IOUT: (**a**) acoustic-based IoUT, (**b**) EM-based IoUT, (**c**) schematic of MI communication link [90]

applications which transmits small data packets intermittently. Some of the famous LPWAN technologies are: Long Range Wide Area Network (LoRaWAN) [16], Sigfox [136], NB-IoT [175], and Extended coverage GSM IoT (EC-GSM-IoT) [24].

Wireless PAN/LAN One of the shortcoming of LPWAN is its low data rate. This shortcoming can be overcome using wireless PAN/LAN. It can facilitate communication between machinery and equipment, base stations, and field workers. Wireless PAN/LAN includes technologies like Bluetooth [12] provides bandwidth up to 25 MHz and range of 100m, ZigBee [184]] provides bandwidth up to 1 MHz and range of 20–30m, Thread [162] provides secure support for up to 250 devices, and Wi-Fi provides single channel bandwidth up to 160 MHz [44, 105].

Cellular Technologies With more advancement in IOUT applications, there is an increasing demand of cellular and broadband connectivity in IOUT solutions. There is a scarcity of cellular broadband in rural area is the major hurdle in accessing the big data being generated from the IOUT field. One of the reasons of no or slower cellular communication speed, in rural area, is the cost of infrastructure. Currently, data is manually collected and transmitted to base stations. Cellular communications were designed for the human communications, therefore, Machine-to-Machine (M2M) communication faces system and cost related challenges. However, new LTE standards are coming with support for M2M communication but IOUT devices must be compatible and low-powered because of the energy constraints [106].

10.3.2 Sensing

Real-time sensing is one of the major functionalities of IOUT architecture. Real-time sensing is the cause of widespread adoption of IOUT. It also gives efficiency in IOUT applications [84, 100]. Although a complete chapter in this book is dedicated to the sensing component of IOUT, however, for the sake of completion, some of the IOUT sensing technologies are discussed very briefly in this section.

IOUT Soil Moisture Soil is the common component for most of the IOUT applications. Soil Moisture (SM) has been part of sensing in IOUT applications for crop production and agriculture. It is being used for measuring water content for decades. Earlier it was done manually by using handheld devices, however, it has now been replaced by automated technologies due to difficulties of getting readings in remote fields. There has been much evolution in wireless data harvesting technologies for provision of real-time soil moisture data for decision making. It has helped a lot in improving water management for many farm IOUTs [101]. Some of the major SM measurement methods are described below:

- Gravimetric Sampling is a method to measure the volumetric water content of the soil by using ration of dry and wet soil mass with pore spaces. The sampling is done manually and soil samples, taken from the field, are oven dried [28, 103].

- Resistive sensors [145] work on the electrical conductivity of water. It measures change in resistance due to water in soil. An accurate SM reading is highly dependent on calibration of sensors.
- Capacitive sensors works by detecting changes in capacitance caused by water presence in soil. An accurate SM reading is highly dependent on calibration of sensors. These sensors are more accurate then resistive sensors, however, they are every expensive and are being used by commercial UTs.
- Ground Penetrating Radar (GPR) [2, 42], Salam Chapter takes readings on the basis of absorption and reflection of electromagnetic (EM) waves. It uses frequency sweep, impulse and frequency modulated technologies for SM sensing. GPRs are used to measure near surface (up to 10cm) SM readings.
- Neutron scattering probes [23, 30] and gauges measure estimating change in neutron flux density with respect to water in the soil. It uses radiation scattering techniques and are most accurate probes used for taking SM readings in the field. However, they require specific licenses to be used.
- Other famous SM measuring techniques are: Gamma ray attenuation [58], frequency domain reflectometry (FDR) [140], and time-domain reflectometry (TDR) [85].

The burying depth of sensors ranges from 5 cm to 75 cm depending upon the root depth and type of the crop. It produces SM data which is used as input to create soil moisture maps for real-time decision making. The number of SM sensors deployed in the field are increasing at very fast rate. For example, Nebraska Agricultural Water Management Network [27, 48, 122] was initially built with only 20 grower in 2005. The number has increased to 1400 growers for adoption of SM sensors-based energy conservation and water management practices. Apart from in-situ sensing of soil moisture, other data sources for SM data are: NASA North American Land Data Assimilation System [64], NASA Soil Moisture Active Passive [141], US Climate Reference Network [166], TAMU North American Soil Moisture Database [118, 155], Soil Climate Analysis Network [144], and Soil Moisture and Ocean Salinity [143]. Web Soil Survey (WSS) [178] collects the US soil information and classifies them on the basis of region.

10.3.3 System Integration

IOUT generates glut of data from the field and it is not possible to locally process it because of limited processing power and energy resources. The data can be store privately, publicly or can be shared among different users depending upon the user requirement [127, 180]. Many online applications and market places use the big data sets to analyze region for better and maximized crop yield [179]. There are national databases for keeping the soil moisture data. In-situ SM sensors can be lined these databases to achieve accurate and detailed information on SM [4, 11, 15, 64, 141, 155]. Cloud services can be used to support real-time decision making and visualization. Therefore, cloud can be used as centralized data storage and processing system in IOUT. Integration of Cloud and IOUT adds scalability from field to big geographical area where unified network can be formed within various IOUT applications.

Moreover, after overcoming the storage and processing constraints, base stations can pull meteorological (from weather service) or soil data (from national service), combine this information with local UTs data for extended control on equipment. IOUT and Cloud integration opens new doors for more robust stakeholders in IOUT applications such as users, industry, trading companies. It would result in efficient and sustainable IOUT ecosystem. For example, in precision agriculture, in addition to integration of field data with different soil and weather databases, linking of UAVs and robotics can also be done in precision agriculture paradigm [102].

Regardless of whether the processing is being done locally or in the cloud, the major challenge is to the integrate so many heterogeneous devices in a system. Reliable delivery of data from field to cloud and vice versa is an important functionality IOUT cloud architecture. This will not only provide a unified cloud connection of fields to spread over vast geographical area but also enable to use data for assessment and improvement. There is also lack of standard interfaces which can provide seamless connectivity between various components of an IOUT system [91].

IOUT senses and communicates even the minor changes in the field including change in medium properties. IOUT applications generate big data and it is very important to correctly analyze this data to extract meaningful information and make real-time decisions to get better rate on investment. Therefore, developing a big data analytic is very important. For example, big data analytic in precision agriculture are factors that may affect crop yield, dividing the field into multiple zones based on particular applications such as nutrient, soil moisture, harvesting, and productivity. It is important to analyze the water and energy consumption, and its effect on labor cost resulting from adoption of IOUT solution. Big data analytic are very important in showing the productivity and efficiency of the system. It attracts stakeholder and helps in widespread adoption of IOUT systems [98, 112].

10.4 IOUT Applications

10.4.1 Smart Lighting

One of the application areas of IOUT is smart lighting [43, 44, 162]. In smart lighting, the cables are buried underground to provide intelligent lighting system. Figure 10.4 shows the architecture of IOUT implementation in lighting. The IOUT-based smart lighting architecture consists of the following basic components:

- Sensors—Attached to areas where lightning is required such as lamp posts, garage walls, and roadside poles, etc.

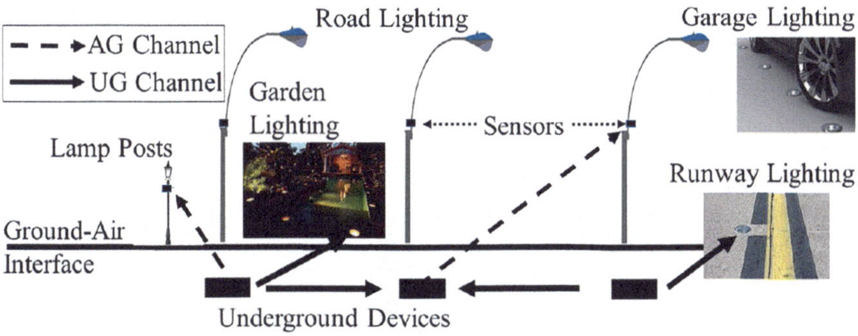

Fig. 10.4 IOUT-based smart lighting architecture [124]

- Aboveground (AG) Channel—Used for downlink communication from AG devices to buried UG devices.
- Underground (UG) Channel—Used for uplink communication from UG devices to AG devices.
- Lighting Infrastructure—IOUT lighting infrastructure includes: road lighting, airport runway lighting, household driveway, garage illumination, lamp posts, and household driveway, etc.

This architecture has many advantage over traditional over-the-air (OTA) lighting system. It eliminates wired networks completely. Moreover, installation of underground buried communication devices reduces the cabling complexity significantly which, otherwise, is needed to power AG nodes. Finally, it also reduces the interference and spectral congestion. Communication in IOUT smart lightning is carried out by UG2AG communication channels.

One of the important tasks is in-depth analysis of UG channel while designing the smart lighting. A 2-wave model is presented in [171] which do not consider the lateral waves. [151] models the UG communication in tunnels and mines, however, it cannot be applied to smart lighting because of different propagation environment.

References [119, 124] develops a statistical impulse response model for UG channel using the modeling approach presented in [37, 82, 128]. It modifies the model to suit the unique nature of underground channel. It assumes the correlation between the multipath wave components to be negligible. Furthermore, it also assumes the phases at receiver side to be uniformly random and distributed over $[0, 2\pi)$.

It does so by performing extensive experiments on indoor (see Fig. 10.5) and field testbed to analyze the UG channel in smart lighting. The statistical model is developed by analyzing the power delay profile (PDP). It can generate wireless UG channel impulse response for different types of soil and soil moisture level, delay spread, and coherence bandwidth statistics. The model gives the important feedback for designing the IOUT for smart lighting systems.

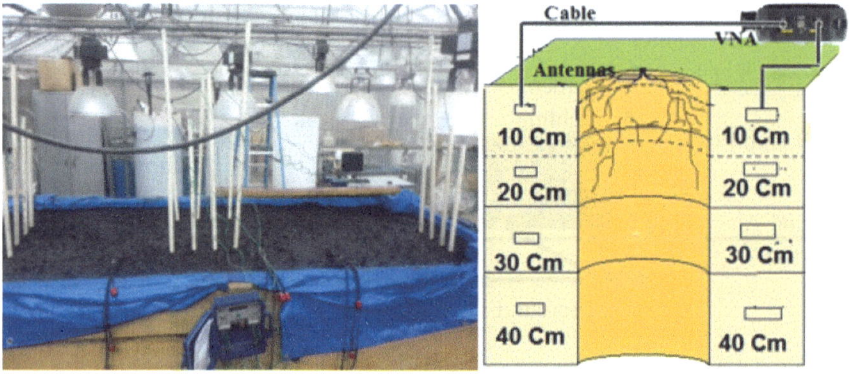

Fig. 10.5 Indoor testbed [124]

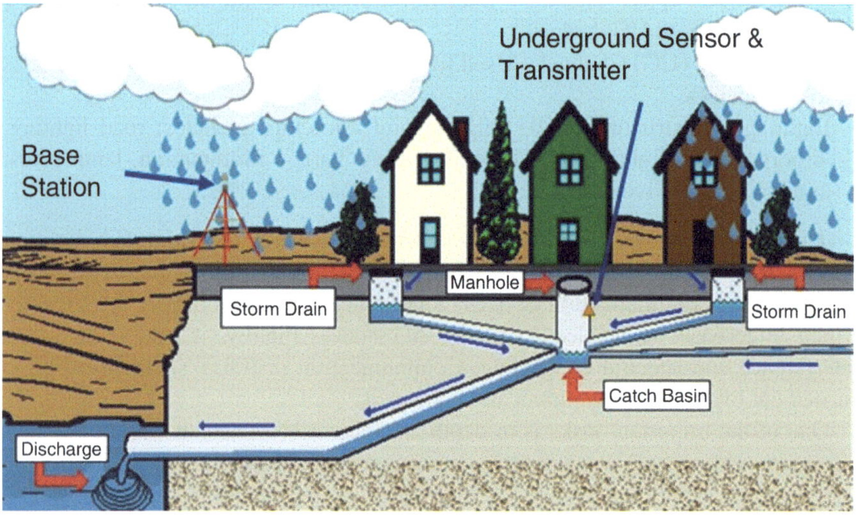

Fig. 10.6 IOUT-based urban underground infrastructure: Smart wastewater system [111]

10.4.2 Urban UG Infrastructure

10.4.2.1 Overview

An important application for implementation of IOUT is urban underground infrastructure. Storm drains and sewers can overflow due to large amount of wastewater entering the pipes. In urban areas, city management collects millions of gallons of waste water and treat them in their waste facilities. This activity needs to be monitored very carefully because extra waste water in the pipes can cause accumulation of water in pipes and ultimately leads to sewer overflows. Therefore, a smart monitoring of urban underground infrastructure is very important. IOUT can be used to develop smart applications for real-time monitoring of waste water or storm water over flow and timely warnings can be issued.

There are very limited solutions for this problem because of connectivity issues and extensive cabling requirement for implementation of such solutions. Figure 10.6 shows the architecture of IOUT implementation in urban underground infrastructure. The IOUT-based urban underground architecture consists of the following basic components:

- Sensors—Attached to underground pipes for sensing the incoming water and communicating the overflow, if happens.
- Wireless underground communication technology [170]—Communication infrastructure connects the sensors and base stations to communicate data to the control systems.
- Base Stations—Base stations collect data from the underground sensor nodes and send it to the control center/cloud for decision making.
- Urban Infrastructure—Urban infrastructure includes roadside traffic poles which are used to implement aboveground communication technology with the underground sensors.

10.4.2.2 Path Loss Model for Urban UG Infrastructure

As discussed in the previous section, underground sensors have to communicate with roadside pole for communicating the status of water in the pipes. The medium of communication in this case would be road. Road is a complex medium which is made up of multiple layers: asphalt, air, and soil (see Fig. 10.7). Therefore, it

Fig. 10.7 Layers of stratified underground medium [111]

is important to understand the effect of each layer of the medium to achieve long range of communication. To that end, authors in [95, 111, 114], authors have done a path loss analysis for improving wireless UG communication in urban underground IOUT applications of waster water monitoring. The path loss has been empirically evaluated in different UG communication media and with thickness of layers in striated medium. They used Friis equation [33] to calculate the path loss in each layer. The received signal power in a layered medium can be written as [111, 171]:

$$P_r = -L_m + G_r + P_t + G_t \ , \tag{10.1}$$

where L_m is the attenuation due to layered medium. L_m is calculated as:

$$L_m = L_{fs} + L_l, \tag{10.2}$$

where L_{fs} is the over-the-air path loss and L_l extra attenuation due to EM wave propagation in the layered medium. It is given as:

$$L_{fs} = 33.2 + 20log(d) + 20log(f), \tag{10.3}$$

$$L_l = \sum_{n=0}^{N-1} L_n, \tag{10.4}$$

where L_n gives the attenuation in the *n-th* layer. L_n depends upon the number of factors such as dielectric permittivity of the layer and the wavenumber of the medium. The wavenumber of the medium is given as $j\beta + \alpha = \gamma$ where

$$\alpha = \omega\sqrt{\frac{\mu\epsilon'}{2}\left[\sqrt{1 + (\frac{\epsilon''}{\epsilon'})^2} - 1\right]}\ , \tag{10.5}$$

$$\beta = \omega\sqrt{\frac{\mu\epsilon'}{2}\left[\sqrt{1 + (\frac{\epsilon''}{\epsilon'})^2} + 1\right]}\ , \tag{10.6}$$

where $\omega = 2\pi f$= angular frequency, μ is the magnetic permeability, ϵ' is the real, and $\epsilon^{''}$ is the real and imaginary part of permittivity. Propagation loss, L_n for n-th layer is given as:

$$L_n[dB] = 20.\gamma.d.\log 10(e), \tag{10.7}$$

where $e = 2.71828$, and d represents the thickness of the n-th layer. [111] also determined the dispersion in the medium layers, i.e., soil, asphalt, and base gravel aggravate layer. The experiments showed that, generally, path loss increased with the increase in the distance, however, it was less than 100 dB till 4km. Similarly,

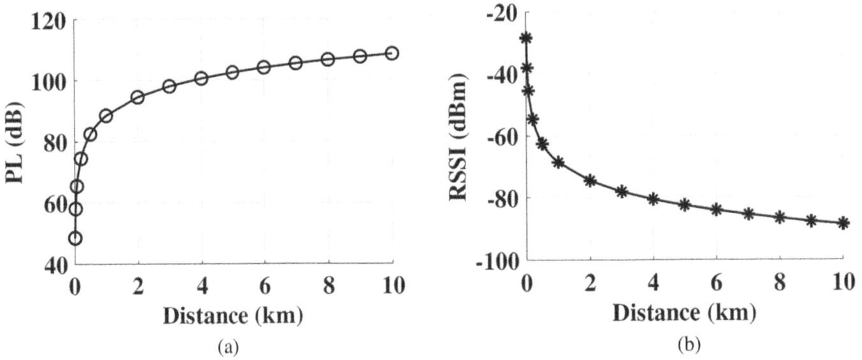

Fig. 10.8 (**a**) Effect of distance on path loss [111], (**b**) effect of distance on RSS [111]

received signal strength decreased rapidly till 2 km, however, after 2 km, i.e., greater than 2km, the smooth decrease is observed. Both results are shown in Fig. 10.8. Therefore, up to 4km of communication range can be achieved if propagation loss of a stratified medium is properly modeled.

10.4.3 Oil and Gas Reservoirs

10.4.3.1 Overview

As per international energy agency (IEA), the energy demand of the world is going to increase significantly by 2030 [38]. The major portion of this increase originates from oil and gas industries, however, the challenging environment of Oil and Gas sector is failing to meet such a huge energy requirements. An important aspect of oil and gas industry reservoir is to obtain real-time information. IOUT is an enabling technology for optimized operation of oil and gas industry. These operations include production, flow monitoring, and reservoir monitoring [55].

The challenging conditions of Oil and Gas sector have limited the application of traditional wireless networks. Therefore, magnetic induction (MI) based technology is being proposed as an enabling technology to develop a sensing system [6, 9, 146]. MI uses magnetic fields as major source of transferring communication. Its counterpart, EM waves, are highly effected by the properties of underground environment and requires large antennas for the implementation [29]. On contrary, MI is not affected by these properties. As large antennas are not feasible for IOUTs, MI uses tine coils as antennas, hence, are more practical choice for the implementation in IOUT.

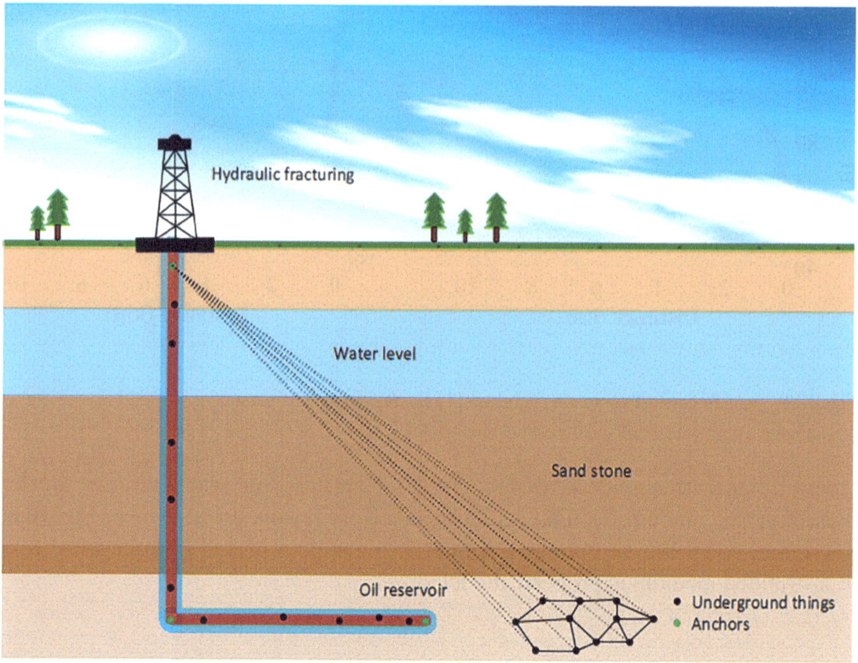

Fig. 10.9 Architecture of MI-based IOUT reservoirs [90]

10.4.3.2 MI-Based IOUT Setup

An MI-based IOUT network is shown in Fig. 10.9 which consists of randomly distributed N sensor nodes (underground things (UTs)) and M aboveground networking equipment (anchors). The UTs are injected to subsurface area by hydraulic fracturing [35]. UTs are uniformly distributed with position represented as $S = {s_i}_{i=1}^{N}$ where $s_i = x_i, y_i, z_i$ are 3D position coordinates of i-th UT. Similarly, position of the j-th anchor is given as ${s_j}_{j=1}^{M}$ where $s_j = x_j, y_j, z_j$. Moreover, position of the anchor is well known and is attached to the fracturing well.

Large dipole antennas are used by the anchors to communicate with UTs. Downlink channel (anchor $\rightarrow$ UT) is single hop because of extended transmission capabilities provided by the large dipole antenna. Downlink channel is represented by dotted black lines in Fig. 10.9. On contrary, uplink (UT $\rightarrow$ anchor) communication is multi-hop because of low power and transmission of UTs [153]. Uplink channel is represented by solid red lines in Fig. 10.9. High communication range for anchors is assumed because of the possibility of attaching them to external power sources. Waveguide structure of the coil is used to extend the communication range of MI coils [152].

UTs communicate with the other UTs and anchors using magnetic induction. A successful communication requires a proper coupling between the coils. Therefore, to receiver strong signal, a tri-directional coil receiver structure is proposed in [156]. The advantage of this coil structure is omni-directional coverage provided by it. From the perspective of magnetic induction, received and transmitted power are related as [86]:

$$P_{r_j} = \frac{\omega\mu P_{t_i} N_{r_j} N_{t_i} r_{t_i}^3 r_{r_j}^3 \sin^2 \alpha_{ij}}{16 R_0 d_{ij}^6}, \tag{10.8}$$

where P_{t_i} is the transmitted power, ω and μ denote the angular frequency and soil permeability. Diameter of receiver and sending coil is given by r_{r_i} and r_{t_i}, respectively. N_{r_j} denotes the total turns in receiver coil, α_{ij} denotes the angle between both transmitting and receiving coil, R_0 give $d_{ij} = \| \mathbf{s}_i - \mathbf{s}_j \|$ represents the distance between the sending and receiving destination coils. Equation (10.8) is also validated empirically in [156]. Equation (10.8) does not consider the skin depth effect because of low frequencies, however, high frequencies are used for the soil, therefore, it is necessary to consider skin depth effect for such cases. For skin depth effect, distance between two UTs can be modified as:

$$\hat{d}_{ij} = f(P_{r_j}) = \arg\left\{d_{ij}|\Theta\right\}, \tag{10.9}$$

where θ is given in [67] as:

$$\Theta = 10^{\frac{(P_{t_i} - P_{r_j})}{10}} - 1 = \frac{16 R_{0_t} R_{0_r} d_{ij}^3}{\omega^2 \mu^2 N_{t_i} N_{r_i} r_{t_i}^3 r_{r_j}^3 G^2(\sigma, \omega, d_{ij})}. \tag{10.10}$$

In Eq. (10.10), $G^2(\sigma, \omega, d_{ij})$ is the additional loss due to skin depth effect and σ denotes the soil electrical conductivity.

10.4.3.3 IOUT Application in Oil and Gas Reservoir

In [89], a MI-based localization technique for Oil and Gas reservoir is proposed and evaluated using Cramer Rao Lower Bound (CRLB). The distance between any two UTs i and j in MI-based IOUT is given by Eq. (10.9). As the transmission range of MI nodes is limited, so each node has to estimate only the distance between itself and nearby nodes. These distances are communicated to central control room, where missing distances are computed using matrix completion strategy as:

$$\rho_{ij} = \begin{cases} \hat{d}_{ij} & \text{if } d_{ij} \le d_m, \\ \hat{d}_{ih(1)} + \sum_{k=1}^{L-1} \hat{d}_{h(k),h(k+1)} + \hat{d}_{h(L)j} & \text{otherwise}, \end{cases}$$

where d_m is the maximum transmission distance, L is the total hops between nodes i and j, and $h(1), \ldots, h(L)$. All ρ_{ij} values will result in diagonal squared geodesic distance matrix (SGDM) as given below:

$$\boldsymbol{\Psi} = \begin{bmatrix} 0 & \cdots & \rho^2_{i,(N+M)} \\ \vdots & \ddots & \vdots \\ \rho^2_{(N+M),1} & \cdots & 0 \end{bmatrix}, \tag{10.11}$$

where Ψ is the square symmetric matrix where $\rho_{ij} = \rho_{ji}$ and $\rho_{ii} = 0$. After creating Ψ, the author employs dimensionality reduction technique to convert and visualize high dimensional distances into low-dimensional coordinates [88]. For this case three dimensional architecture, Isomap technique considered more suitable [159]. Isomap reduces the distances from high dimension to lower 3 dimensional coordinates. It does so by minimizing the following functions:

$$\Phi(\mathbf{S}) = \sum_{ij} (\rho^2_{ij} - \parallel \mathbf{s}_i - \mathbf{s}_j \parallel^2), \tag{10.12}$$

where $\parallel \mathbf{s}_i - \mathbf{s}_j \parallel^2$ is the euclidean distance between nodes i and j. Another approach by Kruskal [58] applies a centering operator to SDGM, i.e., $(-\frac{1}{2} = \mathbf{G\Psi G})$, and gives a double centered matrix $(\mathbf{H} = -\mathbf{G\Psi G}^T/2)$, where $\mathbf{H}$ is given as follows:

$$h_{ij} = -0.5 \left(\rho^2_{ij} - \frac{1}{T} \sum_{i=1}^{T} \rho^2_{ij} - \frac{1}{T} \sum_{j=1}^{T} \rho^2_{ij} - \frac{1}{T^2} \sum_{i=1}^{T} \sum_{j=1}^{T} \rho^2_{ij} \right),$$

In above equation $T = N + M$. Eigenvalue decomposition of $\mathbf{H}$ gives

$$\widetilde{\mathbf{S}} = \mathbf{V}\sqrt{\mathbf{U}}, \tag{10.13}$$

where $\mathbf{V}$ is the eigenvector of $\mathbf{H}$ and $\mathbf{U}$ is the eigenvalues of $\mathbf{H}$. The 3D coordinates are relative to each other not with respect to actual coordinate system. These local coordinates are converted to global geographical coordinates using Helmert transformation [177] or Procrustes analysis [142]. The location of the nodes is calculated as follows:

$$\hat{\mathbf{S}} = \varpi \varsigma (\widetilde{\mathbf{S}}) +, \tau \tag{10.14}$$

where ϖ, ς, and τ are the different factors for rotation, scaling an translation, respectively. These factors are dependent upon the number of anchors being used. The cost function of Procrustes analysis is given as:

$$f(\boldsymbol{\varpi}, \boldsymbol{\tau}, \varsigma) = \sum_{i=1}^{M} (\widetilde{\mathbf{s}}_i - \varsigma \boldsymbol{\varpi}^T \mathbf{s}_i - \boldsymbol{\tau})^T \times (\widetilde{\mathbf{s}}_i - \varsigma \boldsymbol{\varpi}^T \mathbf{s}_i - \boldsymbol{\tau}). \tag{10.15}$$

The optimal values of ϖ, ς, and τ are calculated by minimizing Eq. (10.15). To that end, it is assumed that centroid of the real and estimated anchor location is $\mathbf{c}_a = \frac{1}{M}\sum_{i=1}^{M} \mathbf{s}_i$ and ,$\mathbf{c}_e = \frac{1}{M}\sum_{i=1}^{M} \hat{\mathbf{s}}_i$, respectively. Rewriting Eq. (10.15) by putting values of c_a and c_e as:

$$f(\boldsymbol{\varpi}, \boldsymbol{\tau}, \varsigma) = \sum_{i=1}^{M} \Big((\tilde{\mathbf{s}}_i - \mathbf{c}_e) - \varsigma \boldsymbol{\varpi}^T (\mathbf{s}_i - \mathbf{c}_a) + \tilde{\mathbf{s}}_i - \varsigma \boldsymbol{\varpi}^T \mathbf{s}_i - \boldsymbol{\tau} \Big)^T \times \Big((\tilde{\mathbf{s}}_i - \mathbf{c}_e) - \varsigma \boldsymbol{\varpi}^T (\mathbf{s}_i - \mathbf{c}_a) + \tilde{\mathbf{s}}_i - \varsigma \boldsymbol{\varpi}^T \mathbf{s}_i - \boldsymbol{\tau} \Big). \tag{10.16}$$

The optimal values of τ are obtained by solving Eq. (10.16):

$$\boldsymbol{\tau} = \mathbf{c}_e - \varsigma \boldsymbol{\varpi}^T \mathbf{c}_a. \tag{10.17}$$

Furthermore, assuming that $\mathbf{c_e} = \mathbf{c_a} = 0$ and putting value of τ in Eq. (10.16) gives

$$f(\boldsymbol{\varpi}, \boldsymbol{\tau}, \varsigma) = \sum_{i=1}^{M} (\tilde{\mathbf{s}}_i - \varsigma \boldsymbol{\varpi}^T \mathbf{s}_i)^T (\tilde{\mathbf{s}}_i - \varsigma \boldsymbol{\varpi}^T \mathbf{s}_i). \tag{13}$$

It is important to note that Eq. (10.4.3.3) is a convex function. Hence, this equation is differentiated with respect to ς to obtain the optimal value of ς:

$$\varsigma = \frac{\mathrm{Tr}\left(\mathbf{S}_a \boldsymbol{\varpi} \widetilde{\mathbf{S}}_e^T\right)}{\mathrm{Tr}\left(\mathbf{S}_a \widetilde{\mathbf{S}}_e^T\right)}, \tag{14}$$

where the function Tr(·) is the trace operator. Lastly, the eigenvalue decomposition of $\mathbf{S}_a \widetilde{\widetilde{\mathbf{S}}}_e^T$ is done to obtain the optimal value of ϖ.

A proposed method is then analyzed by calculating CRLB as follows:

$$\mathrm{CRLB} = \mathbf{I}_{x,x}^{-1} + \mathbf{I}_{y,y}^{-1} + \mathbf{I}_{z,z}^{-1}. \tag{10.18}$$

The proposed technique is evaluated for effect of coil size, number of coil turns, and transmit power on localization accuracy of the technique and shown that it surpasses the performance of other localization technique in both aspects.

10.5 IOUT Testbeds

Academic IOUT Testbed IOUT can be used to estimate the water and fertilizer quantity to be applied using irrigation control system. A testbed in South Central Agricultural Lab (SCAL) in Clay Center, Nebraska [21, 100] covers an area of 41 acres with advanced center pivot irrigation system installed in it. The purpose of this testbed is to study long-term effect of crop water and nutrient consumption, variable rate irrigation and fertigation, relation between the crop water stress and yield, development of crop production function, and other related topics under the settings of full and limited irrigation [21, 107, 112]. The testbed consists of a solar panel to provide sustainable energy, transmitting and receiving antennas, UTs with capabilities of measuring temperature and soil moisture. It is a fully functional testbed and IOUT sensing and communication can be investigated on it.

Another testbed developed for dynamically controlling soil moisture for IOUT wireless communications experiments inside greenhouse [106, 123]. It is enclosed in 100 in×36 in×48 wooden box. It has capacity of holding 90 cubic feet of packed soil and also has a drainage system. Controlled wireless communication experiments are carried out but using antennas buried at different depths and distances.

Another MI-based testbed is developed in [156]. This testbed consists of coil buried in lab settings. It is used to study the effect of MI wave guide effect with different soil configurations. SoilBED [26, 96] is used for the cross-well radar experiments. They are used for detection of contaminated materials in the soil and studying EM wave propagation. Thoreau [140] is another university level underground testbed which collects and curates the time related data on the cloud. It works on Sigfox design and operates in unlicensed band of 900 MHz. It measures soil moisture, water potential, temperature, and electrical conductivity with a very low data rate. There are many other IOUT testbed which are being used for the academic purpose. Table 10.1 provides a summary of all such testbeds.

Commercial IOUT Solutions Most of the commercial IOUT solutions use OTA wireless communication and UTs with high-end sensors for measuring various properties of interest, hence, measurement is centralized. UTs can be connected to each other to form a mesh, however, mostly UTs deployed in the field connects with any base stations in the field. This base station has some sort of cellular or satellite communication capabilities. Figure 10.10 shows the classification of various commercial IOUT systems. A highly desirable feature of IOUT system is modularity. Modularity allows for change and customization of application. Customized solutions can be made for some specific application which will work out-of-the-box. Table 10.2 provides the summary of IOUT commercial solutions. These commercial solutions can be classified into the following major classes:

- *Agricultural solutions*: Field Connect, by John Deere, uses eight sensor probes to transmits data for measuring temperature, wind speed, wind direction Sm at different depth, and leaf wetness. These probes are located at 1 mile, and

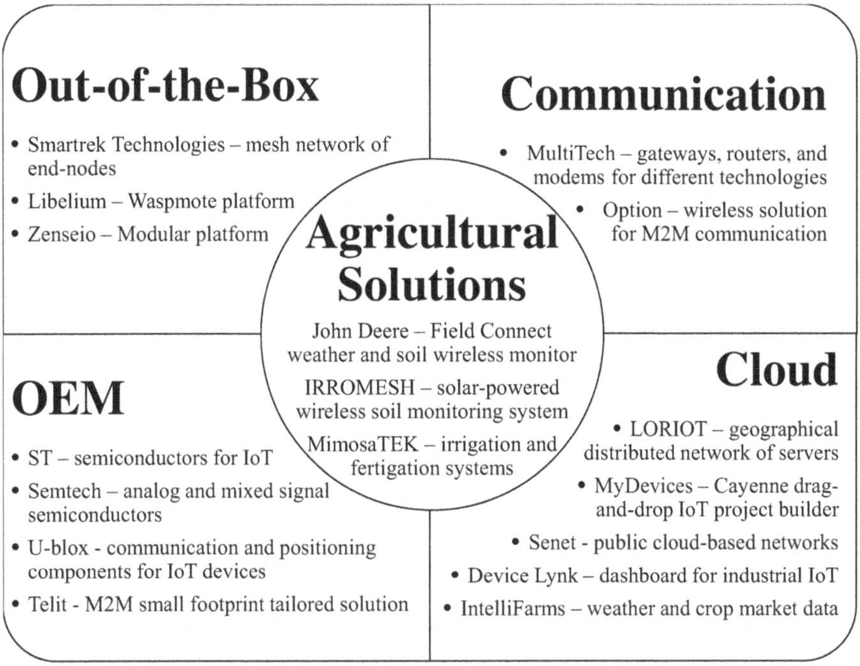

Fig. 10.10 Different types of commercially available IOUT systems [173]

in case of satellite communication, it is located at three miles. Mi-mosaTEK gives services of irrigation and fertilization solutions from small to large-scale farms [72, 108]. FarmBeats, developed by Microsoft, incorporate AI & IoT for agricultural solution.

- *Out-of-the-box packages*: Smartrek Technologies provide support for various sensors and gateways by developing weatherproof wireless nodes for outdoor settings and can easily be integrated into network mesh [93, 142]. Libelium provide solution, named as Plug & Sense Smart Agriculture solution [78], for sensing various farm parameters (SM, temperature, etc.). It develops platforms and end-user devices with support for various communication standards such as LoRaWAN, Sigfox, ZigBee, Wi-Fi, Bluetooth, RFID, and LoRa, etc. Their Waspmote platform has ability to attach 120 different sensors with connected sensor boards. Cropx's [15] IOUT comes with the hardware and software components for measuring soil properties (moisture, temperature, EC) for real-time decision making and irrigation. Precision Hawk's IOUT platform employs drones for sensing and generating field map using thermal, visual and multispectral imaging.
- *OEM components*: OEM components are mainly used for manufacturing nodes at large scale, however, OEM devices are also required for prototyping and small scale production of specific UT. IoT internal components, e.g., accelerometers,

MEMS microphones and gyroscope are developed by ST [147]. Various high-performance semiconductors and advance algorithms are supplied by Semtech [132].
- *Cloud-based services*: Cloud services can provide worldwide access to the information without having an technical knowledge of web programming. Stakeholders are not required to hire third party for configuring servers and make sense of collected data for decision making. LORIoT is one such cloud service which connect multiple distributed low latency networks through LoRa gateway. Some web service includes device management, cloud data storage, safe keeping of keys used for the encryption and translation of LoRaWAN to IP/IPv6 [68]. MyDevices helps IoT developer by providing their drag-and-drop IoT project builder called Cayenne. User can create accounts and use Cayenne (both web and mobile) for their IT devices registration and visualize the sensed data in customized dashboard.

10.5.1 Challenges

The combination of soil and communication components for wireless UG communications is very unique. This combination requires us to study the fundamental concept of communications from a completely different perspective. The factors which directly affect the soil may also influence the performance of UG communications. The network topology should be robust enough to support and cope with the rapidly changing channel conditions. One of the most important soil properties to consider while designing IOUT is Volumetric Water Content (VWC). Therefore, it is very important to study the spatial and temporal variation of VWC in the region of IOUT deployment. Soil composition of a field location plays an important role in tailoring the topology design to meet the criteria of underground channel of that location, hence, it should be thoroughly investigated. For example, if the IOUT is being deployed in a region where soil composition has significant spatial heterogeneity, it will be beneficial to study the different node densities and inter-node distances. In addition to the soil type, VWC variations due to seasonal changes also significantly affect the communication performance [119]. Some of the environmental parameters were studied by performing experiments. These experiments show that VWC has an adverse effect on the UG communications. Therefore, while designing protocol for IOUT, these environment parameters must be considered. IOUT protocols must allow the dynamic adjustment of operational parameters to adapt to changes in surroundings. Lastly, IOUT feasibility is dependent on the investigation of multiple other factors that were not considered for traditional WSNs, e.g., soil composition and VWC. To this end, detailed channel characterization of wireless UG communication is required. There is some positive aspect of UG environment as well, e.g., its temporal stability. These positive aspects need to be exploited and studied further to achieve reliable and energy-efficient communication [120].

Characteristics of UG wireless channel is the key factor which determines the data rate of communication in IOUT. If it is not well modeled, communication performance of IOUT suffers. Therefore, experimentation is needed for its characterization. Moreover, soil with communication components (antenna and wireless UG channel) gives unique IOUT performance characteristics. Figure 1.3 shows the empirical measurements [115, 121] for the soil effect on coherence bandwidth of UG channel and antenna bandwidth.

Soil characteristics such as soil type, soil moisture, soil depth, and burial distance have an effect on communication performance [171]. It can cause dynamic changes in root mean square (RMS) delay spread, antenna return loss, and impulse response of the channel. Figure 10.11a shows the empirical values for antenna return loss (at depth of 40cm in sandy soil) in response to change in soil moisture. Soil moisture is represented by soil matric potential (CB) and both have an inverse relation, i.e., large values of matric potential indicate low soil moisture. Similarly, zero matric potential represents a near situation condition. It can be seen that resonant antenna frequency jumps from 244 MHz to 289 MHz when matric

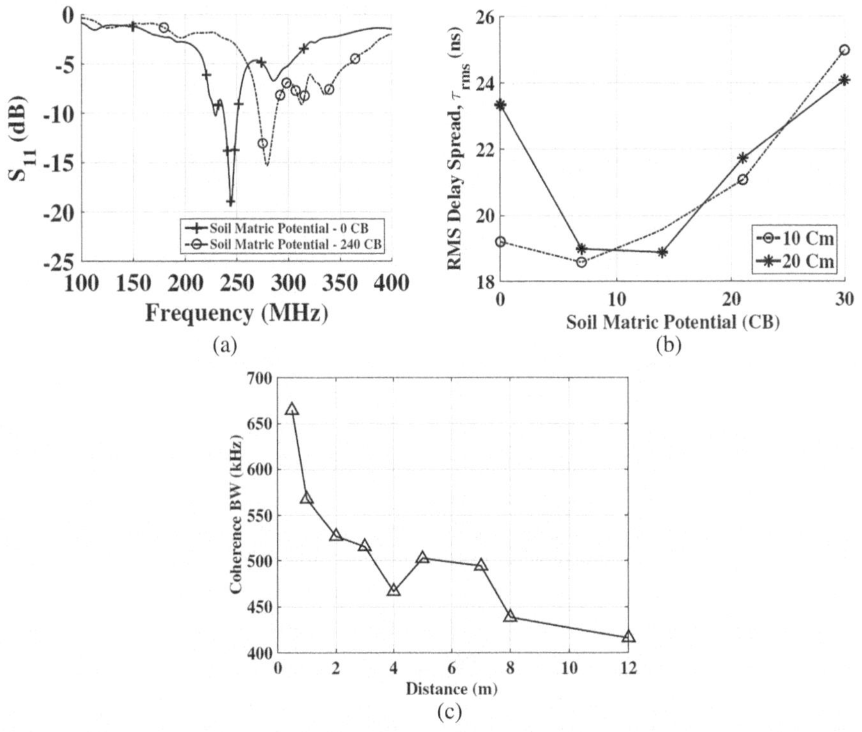

Fig. 10.11 (**a**) Effect of soil moisture on antenna return loss at varying frequencies and fixed depth of 40cm in sandy soil [115], (**b**) effect of soil moisture on RMS delay spread at T-R separation distance of 50cm and two depths of 10cm and 20cm [115], (**c**) effect of T-R separation (distance) on coherence bandwidth at fixed depth of 20cm in silty clay loam soil [115]

potential value increases from 0 CB to 240 CB. This significant increase requires a dynamic change in operational frequency to achieve maximum bandwidth otherwise operation frequency will exceed the range of resonant frequency and antenna bandwidth causing the degradation in performance [19, 174]. Similarly, decreasing the soil moisture causes the antenna bandwidth to increase from 14 MHz to 20 MHz. Accordingly, the soil moisture will also have a significant impact on system bandwidth [56, 92].

Soil texture is the measurement of percentage of sand, clay, and silt in the soil. Table 1.3 lists the classification of soil on the basis of texture and corresponding particle size distribution. Figure 10.11b plots change in RMS delay spread with soil moisture at a distance of 50cm and depths of 10cm and 20cm in silt loam. It can be seen that RMS delay spread, initially, decreases when soil moisture is decreased (0 CB to 8 CB). Afterwards, RMS delay spread is increased consistently. The occurrence of these variations with short period of time due to external impact, e.g., rain, may cause wireless UG channel to be frequency selective.

Figure 10.11c shows the statistics of coherence bandwidth as a function of distance. For distances up to 12m, coherence bandwidth lies in the range of 411 kHz and 678 kHz. Use of traditional communication techniques with this small coherence bandwidth may limit the achievable data rate in IOUT communications.

These sources provide extensive data on soil moisture and temperature for vast geographical areas and extend the Web Soil Survey (WSS) [110, 178].

IOUT Other Soil Medium Properties Soil properties, other than the soil moisture, which can be measured using sensing technologies are: acidity (pH) [130, 157], organic matter in the soil, sand percentage, nutrients such as P, Mg, Ca, OM, base saturation K, base saturation Mg, base saturation Ca, K/Mg, Ca/Mg ratios, and CEC [53, 62, 65], and clay and silt particles [113, 135]. These properties can be used to develop soil map. However, due to cost, size, and technology limitations, real-time and in-situ measurement of these properties is still a challenge.

IOUT Yield Monitoring Yield monitoring is the application of IOUT in agriculture. It is used to give spatial distribution of crop yield when the growing season is ending. It is used to make long-term decision in IOUT agriculture [59, 70, 99, 126]. The yield data is collected automatically by deploying yield monitors on the IOUT farm moving equipment. These equipment collect data during the harvesting season. Grain containers are equipped with mass flow sensors (e.g., Force Sensor by Ag Leader) which records grain inflow with location. Different geographic information systems (GIS), e.g., Mapinfo, ArchInfo, and Environment System Research International tools can be used to analyze the data.

IOUT Electrical Conductivity and Topography Surveys Electrical conductivity (EC) of the soil can be defined as the soil ability to conduct electricity [69]. EC data, combined with the field topography (slope and elevation), can be used to get insights on crop yield. EC measures the nitorgen usage, drainage, water holding and cation-exchange capacity, and rooting depth. EC classifies whole field into multiple zones then various precision agriculture technologies (e.g., VRT) are applied based on zoning. There are various methods to perform EC mapping such as visible-near infrared reflectance spectroscopy (VNIR) [10, 127], apparent electrical conductivity (ECa) [25], and electromagnetic Induction (EMI) [112, 148]. Commercial tools for EC mappings include: EC400 sensors combined with GPS systems [57, 94] and Veris 3100 [96, 168].

IOUT Weather and Environmental Sensing The performance of IOUT systems is highly dependent upon weather and environmental conditions such as wind speed, wind direction, temperature of soil and air, humidity, rainfall, and solar emissions. Weather and environmental sensors are used to measure all the mentioned effects. Such information is very useful in realization of real-time, informed and timely decisions in IOUT systems. Some of the examples of these sensors are: Field Connect solution [51] and Mesoscale Network (MesoNet) [161]. MesoNet is a large-scale network of weather and environment sensing nodes covering large geographical areas. It detects major weather patterns and can be used to provide real-time information when combined with IOUT systems

IOUT Macro-Nutrients Sensing Underground environment provides various important natural resources also including macro-nutrients. Macro-Nutrients (e.g., potassium, phosphorous and nitrogen) are very important for some IOUT applications. In agricultural IOUT, for example, calculating and assessing these nutrients help in determining the fertilizer application and impact in the future. In [74, 97], a sensing method is presented for detection of sulfate and nitrate concentration in natural water resources. This method uses planar electromagnetic sensors and senses nitrate and sulfate levels using correlation of their concentration and sensor array impedance. The study concludes that the sensor impedance is inversely proportional to the concentration of the chemicals. Other macro-nutrient sensing approaches that can be used in IOUT system include: ATR spectroscopy, VIS-NIRS spectroscopy, and Electrochemical approach. However, these approaches only sense on desired ion because their membrane respond to only one ion [65, 97]. Major challenge would be to develop a detector array of macro-nutrients sensing for accomplishing multi-ion sensing [53, 93].

IOUT Precision Agriculture Technologies As discussed earlier, precision agriculture is one of the major areas and precision agriculture is major application of IOUT system. There are tons of technologies playing a very important role in adoption of precision agriculture practices which will be presented here as useful IOUT tools. These technologies include: auto-steering and VRT, precision planting, GIS systems, geolocation, soil sampling, field analysis map generation and drones. Precision planting [80] involves seeding based on fine predetermined inter-plant distance and laser robot with an ability of automated weed zapping. Farm devices are aligned using robovator technology. Multiple zones can be created on the basis for field conditions using GPS [102, 111, 182]. For the improvement of crop yield, variable rate fertilizer application [3, 125] is also very important. Another major component of IOUT system is the drone with wireless communication capabilities. Deere & Co. has developed a GreenStar Lightbar [34] to measure width and location in the row crops. Another device, TK-GPS [160], is capable of performing real-time soil mapping.

These sensing technologies can become the stepping stone for the development of IOUT systems. Inexpensive sensors and wireless communication make their integration with IOUT control system possible. One of the major components, for realization of real-time decision making in IOUT systems, is the wireless communication between the heterogeneous sensor equipment. Furthermore, adoption rate of sensor technologies can be increased by a connected, secured, and reliable IOUT systems which in-turn will be very helpful in the development of improved sensing technologies in IOUT [91].

Some of the challenges in design and implementation of IOUT systems are given below:

1. When being deployed in large areas, cost-efficient and simple IOUT devices are desired which can sustain all harsh environmental effects.
2. Improving UTs will consequently increase the energy demand leading to less battery life. Therefore, energy harvesting, sustainable energy resources, and energy-efficient operations are the major challenges in IOUT systems [125].
3. Due to heterogeneity in UTs, seamless integration of UTs with communication systems is required.
4. For farm IOUT, multi-modal and inexpensive sensors are required which can sense physical parameters of soil in addition to moisture. Although, SM provides important information for irrigation system, however, in-situ sensing of soil chemicals is required for variable rate fertigation.
5. Secure mechanisms are required to store and transfer information from the fields. Moreover, solutions are required to merge data from all the fields for improved decision making in private and secure manner.
6. Seasonal changes can alter the working of IOUT equipment, e.g., freezing temperature can increase power consumption. Equipment can be set to sleep mode when monitoring is not needed [98].
7. UTs must be able to dynamically change their operational parameters such as frequency, modulation schemes, and error encoding schemes to adapt to changes in communication medium.
8. Impact of medium properties, e.g., soil in agriculture, must be modeled. A detailed analysis for these properties can support in building scalable and reliable IOUT architecture.
9. There is a dire need for specialized link and network layer protocols for UG communications which can lead to robust data transfer in IOUT.

References

1. 365 FarmNet (2017) https://www.365farmnet.com/en/
2. Adamchuk V, Hummel J, Morgan M, Upadhyaya S (2004) On-the-go soil sensors for precision agriculture. Comput Electron Agric 44(1):71–91

3. Adapt-N (2020) http://www.adapt-n.com/
4. Agriculture of UD (2014) Census of agriculture. http://www.agcensus.usda.gov/
5. Akkaş MA, Sokullu R, Balcı A (2015) Wireless sensor networks in oil pipeline systems using electromagnetic waves. In: 2015 9th international conference on electrical and electronics engineering (ELECO), pp 143–147. IEEE, Houston, TX, USA
6. Akyildiz IF, Stuntebeck EP (2006a) Wireless underground sensor networks: Research challenges. Ad Hoc Netw J 4(6):669–686
7. Akyildiz IF, Sun Z, Vuran MC (2009) Signal propagation techniques for wireless underground communication networks. Phys Commun J (Elsevier) 2(3):167–183
8. Algeroy J, Lovell J, Tirado G, Meyyappan R, Brown G, Greenaway R, Carney M, Meyer JH, Davies JE, Pinzon ID (2010) Permanent monitoring: taking it to the reservoir. Oilfield Review 22(1):34–41
9. Bansal R (2004) Near-field magnetic communication. IEEE Antennas Propag Mag 46(2):114–115
10. Bilgili AV, van Es H, Akbas F, Durak A, Hively W (2010) Visible-near infrared reflectance spectroscopy for assessment of soil properties in a semi-arid area of turkey. J Arid Environ 74(2):229–238. https://doi.org/10.1016/j.jaridenv.2009.08.011, http://www.sciencedirect.com/science/article/pii/S0140196309002559
11. Biosense (2020) www.biosens.rs
12. Bluetooth (2020) https://www.bluetooth.com/bluetooth-technology
13. Bogena HR, Herbst M, Huisman JA, Rosenbaum U, Weuthen A, Vereecken H (2010) Potential of wireless sensor networks for measuring soil water content variability. Vadose Zone J 9(4):1002–1013
14. Cornell digital agriculture (2020) https://cuaes.cals.cornell.edu/digital-agriculture
15. Cropx soil monitoring system (2020) https://www.cropx.com/
16. de Carvalho Silva J, Rodrigues JJ, Alberti AM, Solic P, Aquino AL (2017) Lorawan–a low power wan protocol for internet of things: A review and opportunities. In: 2nd International multidisciplinary conference on computer and energy science (SpliTech), 2017, pp 1–6. IEEE, Houston, TX, USA
17. DeJonge KC, Taghvaeian S, Trout TJ, Comas LH (2015) Comparison of canopy temperature-based water stress indices for maize. Agric Water Manag 156:51–62. https://doi.org/10.1016/j.agwat.2015.03.023, http://www.sciencedirect.com/science/article/pii/S0378377415001067
18. Dong X, Vuran MC (2011) A channel model for wireless underground sensor networks using lateral waves. In: Proc. of IEEE Globecom '11, Houston, TX
19. Dong X, Vuran MC (2013) Impacts of soil moisture on cognitive radio underground networks. In: Proc. IEEE BlackSeaCom, Georgia
20. Dong X, Vuran MC, Irmak S (2012) Autonomous precision agriculture through integration of wireless underground sensor networks with center pivot irrigation systems. Ad Hoc Netw (Elsevier) http://dx.doi.org/10.1016/j.adhoc.2012.06.012
21. Dong X, Vuran MC, Irmak S (2013) Autonomous precision agriculture through integration of wireless underground sensor networks with center pivot irrigation systems. Ad Hoc Netw 11(7):1975–1987
22. Dynamax SapIP wireless mesh network (2020) http://dynamax.com/products/data-loggers/sapip-wireless-mesh-network
23. Evett S, Steiner J (1995) Precision of neutron scattering and capacitance type soil water content gauges from field calibration. Soil Sci Soc Am J 59(4):961–968
24. Extended coverage - GSM – I (EC-GSM-IoT) (2020) https://www.gsma.com/iot/extended-coverage-gsm-internet-of-things-ec-gsm-iot/
25. Farahani HJ, Khosla R, Buchleiter G (2020) Field EC mapping: a new tool to make better decisions. Crop series Soil; no 0568
26. Farid A, Alshawabkeh A, Rappaport C (2006) Validation and calibration of a laboratory experimental setup for cross-well radar in sand. Am Soc Test Mater (ASTM) Geotech Test J 29(2):158–167

27. Ferguson H, Gardner W (1962) Water content measurement in soil columns by gamma ray absorption. Soil Sci Soc Am J 26(1):11–14
28. Foth HD (1978) Fundamentals of soil science. Soil Sci 125(4):272
29. Franconi NG, Bunger AP, Sejdić E, Mickle MH (2014) Wireless communication in oil and gas wells. Energy Technol 2(12):996–1005
30. Franz TE, Wahbi A, Vreugdenhil M, Weltin G, Heng L, Oismueller M, Strauss P, Dercon G, Desilets D (2016) Using cosmic-ray neutron probes to monitor landscape scale soil water content in mixed land use agricultural systems. Appl Environ Soil Sci https://doi.org/10.1155/2016/4323742
31. Garcia-Sanchez AJ, Garcia-Sanchez F, Garcia-Haro J (2011) Wireless sensor network deployment for integrating video-surveillance and data-monitoring in precision agriculture over distributed crops. Comput Electron Agric 75(2):288–303
32. Gardner WR, Hyden RE, Linyaev EJ, Gao L, Robbins C, Moore J, et al. (2006) Acoustic telemetry delivers more real-time downhole data in underbalanced drilling operations. In: IADC/SPE drilling conference, society of petroleum engineers
33. Goldsmith A (2005) Wireless communications. Cambridge University Press, New York, NY, USA
34. Greenstar lightbar (2020) https://www.deere.com/en_INT/products/equipment/agricultural_management_solutions/guidance_and_machine_control/greenstar_lightbar/greenstar_lightbar.page
35. Guo H, Sun Z (2014a) Channel and energy modeling for self-contained wireless sensor networks in oil reservoirs. IEEE Trans Wirel Commun 13(4):2258–2269. https://doi.org/10.1109/TWC.2013.031314.130835
36. Gutiarrez J, Villa-Medina JF, Nieto-Garibay A, Porta-Gandara MA (2014) Automated irrigation system using a wireless sensor network and GPRS module. IEEE Trans Instrum Meas 63(1):166–176. https://doi.org/10.1109/TIM.2013.227648
37. Hashemi H (1993) Impulse response modeling of indoor radio propagation channels. IEEE J Sel Areas Commun 11(7):967–978
38. Helsinki F (2013) Dynamic communications for oil and gas. Technical Report, Alcatel Lucent
39. Herelab (2020) www.Herelab.io
40. Hernandez M, MacNeill DW, Reeves M, Kirkwood AD, Ruszka JP, Zaeper R, Lemke SR, et al. (2008) High-speed wired drill string telemetry network delivers increased safety, efficiency, reliability, and productivity to the drilling industry. In: SPE Indian oil and gas technical conference and exhibition. Society of Petroleum Engineers
41. Hopkins J (2016) USDA ERS - ARMS farm financial and crop production practices: Tailored reports: Crop production practicesl. ersusdagov, Available: http://www.ers.usda.gov/data-products/arms-farm-financial-and-crop-production-practices/tailored-reports-crop-production-practices.aspx
42. Huisman J, Hubbard S, Redman J, Annan A (2003) Measuring soil water content with ground penetrating radar. Vadose Zone J 2(4):476–491
43. IBM EZ-Farm (2020) https://www.ibm.com/developerworks/community/blogs/dfa2dc54-5a14-4cf8-91e0-978bfd59d0d4/entry/IBM_Research_Africa_Enhancing_the_way_we_farm?lang=en
44. IEEE 802.11 (2020) The working group setting the standards for wireless LANs. http://www.ieee802.org/11/
45. Intellifarms (2020) www.intellifarms.com
46. IoF2020 (2020) https://www.iof2020.eu/
47. Iot sensor platform (2020) http://zenseio.com/zenseio-platform
48. Irmak S (2010) Nebraska water and energy flux measurement, modeling, and research network (NEBFLUX). Trans ASABE 53(4):1097–1115
49. Irromesh wireless mesh system (2020) http://www.irrometer.com/loggers.html#975
50. John Deere field connect (2020) https://www.deere.com/
51. John Deere field connect (2020) https://www.deere.com/en/technology-products/precision-ag-technology/field-and-water-management/field-connect/

52. Karli R, Bouchalkha A, Alhammadi K (2017) Investigation of electromagnetic (EM) wave attenuation in oil pipeline. In: 2017 International conference on electrical and computing technologies and applications (ICECTA), pp 1–4. IEEE, Houston, TX, USA
53. Kim HJ, Sudduth KA, Hummel JW (2009) Soil macronutrient sensing for precision agriculture. J Environ Monit 11(10):1810–1824
54. Kim Y, Evans RG, Iversen WM (2008) Remote sensing and control of an irrigation system using a distributed wireless sensor network. IEEE Trans Instrum Meas 57(7):1379–1387. https://doi.org/10.1109/TIM.2008.917198
55. Kisseleff S, Akyildiz IF, Gerstacker WH (2018) Survey on advances in magnetic induction-based wireless underground sensor networks. IEEE Internet Things J 5(6):4843–4856
56. Konda A, Rau A, Stoller MA, Taylor JM, Salam A, Pribil GA, Argyropoulos C, Morin SA (2018) Soft microreactors for the deposition of conductive metallic traces on planar, embossed, and curved surfaces. Adv Funct Mater 28(40):1803020. https://doi.org/10.1002/adfm.201803020
57. Krishna K (2016) Push button agriculture: robotics, drones, satellite-guided soil and crop management. Apple Academic Press https://books.google.com/books?id=NlmzDAAAQBAJ
58. Kruskal JB (1964) Multidimensional scaling by optimizing goodness of fit to a nonmetric hypothesis. Psychometrika 29(1):1–27
59. Kukal MS, Irmak S (2017) Spatial and temporal changes in maize and soybean grain yield, precipitation use efficiency, and crop water productivity in the US great plains. Trans ASABE 60(4):1189–1208
60. Kulkarni A, Kumar V, Dhok SB (2018) Enabling technologies for range enhancement of mi based wireless non-conventional media communication. In: 2018 9th international conference on computing, communication and networking technologies (ICCCNT), pp 1–7. IEEE, Houston, TX, USA
61. Langendoen K, Baggio A, Visser O (2006) Murphy loves potatoes: Experiences from a pilot sensor network deployment in precision agriculture. In: 20th International parallel and distributed processing symposium, 2006, IPDPS 2006. pp 8–pp. IEEE, Houston, TX, USA
62. Laskar S, Mukherjee S (2016) Optical sensing methods for assessment of soil macronutrients and other properties for application in precision agriculture: a review. ADBU J Eng Technol 4:206–210
63. Leaf monitor system (2020) http://www.westernfarmpress.com/tree-nuts/uc-research-explores-continuous-leaf-monitor-system-maximize-irrigation-almonds
64. LDAS land data assimilation systems (2020) https://ldas.gsfc.nasa.gov/nldas/
65. Lin J, Wang M, Zhang M, Zhang Y, Chen L (2007) Electrochemical sensors for soil nutrient detection: Opportunity and challenge. In: International conference on computer and computing technologies in agriculture, pp 1349–1353. Springer, Boston, MA
66. Lin SC, Akyildiz IF, Wang P, Sun Z (2014) Optimal energy-throughput efficiency for magneto-inductive underground sensor networks. In: 2014 IEEE international black sea conference on communications and networking (BlackSeaCom), pp 22–27. IEEE, Houston, TX, USA
67. Lin SC, Alshehri AA, Wang P, Akyildiz IF (2017) Magnetic induction-based localization in randomly deployed wireless underground sensor networks. IEEE Internet Things J 4(5):1454–1465
68. LOROT | LoRaWAN services and software (2020) www.loriot.io
69. Lund E, Christy C, Drummond P (1999) Practical applications of soil electrical conductivity mapping. Precis Agric 99:771–779
70. Magri A, Van Es HM, Glos MA, Cox WJ (2005) Soil test, aerial image and yield data as inputs for site-specific fertility and hybrid management under maize. Precis Agric 6(1):87–110
71. Markham A, Trigoni N (2012) Magneto-inductive networked rescue system (miners): Taking sensor networks underground. In: Proceedings of the 11th ICPS, ACM, IPSN '12, pp 317–328. https://doi.org/10.1145/2185677.2185746
72. Mimosatek (2017) www.mimosatek.com [Accessed: 2017-09-30]

73. Mwachaka SM, Wu A, Fu Q (2019) A review of mud pulse telemetry signal impairments modeling and suppression methods. J Pet Explor Prod Technol 9(1):779–792
74. Nor ASM, Faramarzi M, Yunus MAM, Ibrahim S (2015) Nitrate and sulfate estimations in water sources using a planar electromagnetic sensor array and artificial neural network method. IEEE Sensors J 15(1):497–504. https://doi.org/10.1109/JSEN.2014.2347996
75. Oelze ML, O'Brien WD, Darmody RG (2002) Measurement of attenuation and speed of sound in soils. Soil Sci Soc Am J 66(3):788–796
76. Pathak V, Kumar V, Barik RK (2018) Magnetic induction communication based transceiver coil and waveguide structure modeling for non-conventional wsns. In: 2018 9th international conference on computing, communication and networking technologies (ICCCNT), pp 1–7. IEEE, Houston, TX, USA
77. Petäjäjärvi J, Mikhaylov K, Hämäläinen M, Iinatti J (2016) Evaluation of LoRa LPWAN technology for remote health and wellbeing monitoring. In: 10th international symposium on medical information and communication technology (ISMICT), pp 1–5. IEEE, Houston, TX, USA
78. Plug and sense smart agriculture (2020) http://www.libelium.com/products/plug-sense/models/#smart-agriculture
79. Precision hawk drone data platform (2017) http://www.precisionhawk.com/agriculture
80. Precision planting (2020) http://www.precisionplanting.com
81. Purdue's digital agriculture initiative (2020) http://mailimages.purdue.edu/vo/?FileID=ac13556e-8057-487f-a01b-6914f848cc4c
82. Rappaport TS, Seidel SY, Takamizawa K (1991) Statistical channel impulse response models for factory and open plan building radio communicate system design. IEEE Trans Commun 39(5):794–807
83. Rojo F, Kizer E, Upadhyaya S, Ozmen S, Ko-Madden C, Zhang Q (2016) A leaf monitoring system for continuous measurement of plant water status to assist in precision irrigation in grape and almond crops. IFAC-PapersOnLine 49(16):209–215. https://doi.org/10.1016/j.ifacol.2016.10.039, http://www.sciencedirect.com/science/article/pii/S2405896316316020, 5th IFAC conference on sensing, control and automation technologies for agriculture AGRICONTROL 2016
84. Rossel RAV, Bouma J (2016) Soil sensing: A new paradigm for agriculture. Agric Syst 148:71–74
85. Roth K, Schulin R, Flühler H, Attinger W (1990) Calibration of time domain reflectometry for water content meaurement using a composite dielectric approach. Water Resour Res 26(10):2267–2273
86. Saeed N, Nam H (2016) Cluster based multidimensional scaling for irregular cognitive radio networks localization. IEEE Trans Signal Process 64(10):2649–2659
87. Saeed N, Celik A, Alouini MS, Al-Naffouri TY (2018a) Performance analysis of connectivity and localization in multi-hop underwater optical wireless sensor networks. IEEE Trans Mobile Comput 18(11):2604–2615
88. Saeed N, Nam H, Haq MIU, Muhammad Saqib DB (2018b) A survey on multidimensional scaling. ACM Comput Surv (CSUR) 51(3):1–25
89. Saeed N, Alouini M, Al-Naffouri TY (2019) 3d localization for internet of underground things in oil and gas reservoirs. IEEE Access 7:121769–121780. https://doi.org/10.1109/ACCESS.2019.2937915
90. Saeed N, Alouini MS, Al-Naffouri TY (2019) Toward the internet of underground things: A systematic survey. IEEE Commun Surv Tutorials 21(4):3443–3466
91. Salam A (2018) Pulses in the sand: Long range and high data rate communication techniques for next generation wireless underground networks. ETD collection for University of Nebraska - Lincoln (AAI10826112), http://digitalcommons.unl.edu/dissertations/AAI10826112
92. Salam A (2019a) A comparison of path loss variations in soil using planar and dipole antennas. In: 2019 IEEE international symposium on antennas and propagation. IEEE, Houston, TX, USA

93. Salam A (2019b) Design of subsurface phased array antennas for digital agriculture applications. In: Proc. 2019 IEEE international symposium on phased array systems and technology (IEEE Array 2019), Waltham, MA, USA
94. Salam A (2019c) A path loss model for through the soil wireless communications in digital agriculture. In: 2019 IEEE international symposium on antennas and propagation, pp 1–2. IEEE, Houston, TX, USA
95. Salam A (2019d) Sensor-free underground soil sensing. In: ASA, CSSA and SSSA international annual meetings (2019), ASA-CSSA-SSSA
96. Salam A (2019e) Subsurface MIMO: A beamforming design in internet of underground things for digital agriculture applications. J Sens Actuator Netw 8(3). https://doi.org/10.3390/jsan8030041 https://www.mdpi.com/2224-2708/8/3/41
97. Salam A (2019f) Underground environment aware MIMO design using transmit and receive beamforming in internet of underground things, pp 1–15. Springer International Publishing, Cham
98. Salam A (2019g) An underground radio wave propagation prediction model for digital agriculture. Information 10(4):147
99. Salam A (2019h) Underground soil sensing using subsurface radio wave propagation. In: 5th global workshop on proximal soil sensing, Columbia, MO
100. Salam A (2020a) Internet of things for environmental sustainability and climate change, pp 33–69. Springer International Publishing, Cham. https://doi.org/10.1007/978-3-030-35291-2_2
101. Salam A (2020b) Internet of things for sustainability: perspectives in privacy, cybersecurity, and future trends, pp 299–327. Springer International Publishing, Cham. https://doi.org/10.1007/978-3-030-35291-2_10
102. Salam A (2020c) Internet of Things for sustainable community development, 1st edn. Springer Nature, Basel. https://doi.org/10.1007/978-3-030-35291-2
103. Salam A (2020d) Internet of things for sustainable community development: introduction and overview, pp 1–31. Springer International Publishing, Cham. https://doi.org/10.1007/978-3-030-35291-2_1
104. Salam A (2020e) Internet of things for sustainable forestry, pp 147–181. Springer International Publishing, Cham. https://doi.org/10.1007/978-3-030-35291-2_5
105. Salam A (2020f) Internet of things for sustainable human health, pp 217–242. Springer International Publishing, Cham. https://doi.org/10.1007/978-3-030-35291-2_7
106. Salam A (2020g) Internet of things for sustainable mining, pp 243–271. Springer International Publishing, Cham. https://doi.org/10.1007/978-3-030-35291-2_8
107. Salam A (2020h) Internet of things for water sustainability, pp 113–145. Springer International Publishing, Cham. https://doi.org/10.1007/978-3-030-35291-2_4
108. Salam A (2020i) Internet of things in agricultural innovation and security, pp 71–112. Springer International Publishing, Cham. https://doi.org/10.1007/978-3-030-35291-2_3
109. Salam A (2020j) Internet of things in sustainable energy systems, pp 183–216. Springer International Publishing, Cham. https://doi.org/10.1007/978-3-030-35291-2_6
110. Salam A (2020k) Internet of things in water management and treatment, pp 273–298. Springer International Publishing, Cham. https://doi.org/10.1007/978-3-030-35291-2_9
111. Salam A (2020l) Wireless underground communications in sewer and stormwater overflow monitoring: Radio waves through soil and asphalt medium. Information 11(2):98
112. Salam A, Karabiyik U (2019) A cooperative overlay approach at the physical layer of cognitive radio for digital agriculture. In: Proc. 3rd international balkan conference on communications and networking (IEEE BalkanCom 2019), Skopje, North Macedonia
113. Salam A, Shah S (2019a) Internet of things in smart agriculture: Enabling technologies. In: 2019 IEEE 5th world forum on internet of things (WF-IoT), pp 692–695. IEEE, Houston, TX, USA
114. Salam A, Shah S (2019b) Urban underground infrastructure monitoring IoT: the path loss analysis. In: 2019 IEEE 5th world forum on internet of things (WF-IoT) (WF-IoT 2019), Limerick, Ireland

115. Salam A, Vuran MC (2016a) Impacts of soil type and moisture on the capacity of multi-carrier modulation in internet of underground things. In: Proc. ICCCN 2016, Waikoloa, Hawaii, USA
116. Salam A, Vuran MC (2016b) Impacts of soil type and moisture on the capacity of multi-carrier modulation in internet of underground things. In: Proc. of the 25th ICCCN 2016, Waikoloa, Hawaii, USA
117. Salam A, Vuran MC (2017a) Em-based wireless underground sensor networks. In: Pamukcu S, Cheng L (eds) Underground sensing: Monitoring and hazard detection for environment and infrastructure, chap 5, 1st edn. Elsevier, Cambridge
118. Salam A, Vuran MC (2017b) EM-based wireless underground sensor networks, pp 247–285. https://doi.org/10.1016/B978-0-12-803139-1.00005-9
119. Salam A, Vuran MC (2017c) Smart underground antenna arrays: A soil moisture adaptive beamforming approach. In: Proc. IEEE INFOCOM 2017, Atlanta, USA
120. Salam A, Vuran MC (2017d) Wireless underground channel diversity reception with multiple antennas for internet of underground things. In: Proc. IEEE ICC 2017, Paris, France
121. Salam A, Vuran MC, Irmak S (2016a) Pulses in the sand: Impulse response analysis of wireless underground channel. In: Proc. IEEE INFOCOM 2016, San Francisco, USA
122. Salam A, Vuran MC, Irmak S (2016b) Pulses in the sand: Impulse response analysis of wireless underground channel. In: The 35th annual IEEE international conference on computer communications (INFOCOM 2016), San Francisco, USA
123. Salam A, Vuran MC, Irmak S (2016c) Pulses in the sand: Impulse response analysis of wireless underground channel. In: IEEE INFOCOM 2016-The 35th annual IEEE international conference on computer communications, pp 1–9. IEEE, Houston, TX, USA
124. Salam A, Vuran MC, Irmak S (2017) Towards internet of underground things in smart lighting: A statistical model of wireless underground channel. In: Proc. 14th IEEE international conference on networking, sensing and control (IEEE ICNSC), Calabria, Italy
125. Salam A, Hoang AD, Meghna A, Martin DR, Guzman G, Yoon YH, Carlson J, Kramer J, Yansi K, Kelly M, et al. (2019a) The future of emerging IoT paradigms: Architectures and technologies. https://doi.org/10.20944/preprints201912.0276.v1
126. Salam A, Vuran MC, Dong X, Argyropoulos C, Irmak S (2019b) A theoretical model of underground dipole antennas for communications in internet of underground things. IEEE Trans Antennas Propag 67(6):3996–4009
127. Salam A, Vuran MC, Irmak S (2019c) Di-sense: In situ real-time permittivity estimation and soil moisture sensing using wireless underground communications. Comput Netw 151:31–41. https://doi.org/10.1016/j.comnet.2019.01.001, http://www.sciencedirect.com/science/article/pii/S1389128618303141
128. Saleh AA, Valenzuela R (1987) A statistical model for indoor multipath propagation. IEEE J Sel Areas Commun 5(2):128–137
129. Savazzi S, Spagnolini U, Goratti L, Molteni D, Latva-aho M, Nicoli M (2013) Ultra-wide band sensor networks in oil and gas explorations. IEEE Commun Mag 51(4):150–160
130. Schirrmann M, Gebbers R, Kramer E, Seidel J (2011) Soil pH mapping with an on-the-go sensor. Sensors 11(1):573–598
131. Schroeder RJ (2002) The present and future of fiber optic sensors for the oilfield service industry: where is there a role? In: 2002 15th optical fiber sensors conference technical digest. OFS 2002 (Cat. No. 02EX533), pp 39–42. IEEE, Houston, TX, USA
132. Semtech (2020) www.semtech.com
133. Senet (2020) www.senetco.com
134. Sensor network for irrigation scheduling (2020) http://soilphysics.okstate.edu/research/moisst/2017-moisst-workshop/Taghvaeian%20MOISST%202017.pdf/at_download/file
135. Shamma'a A, Tanner R, Shaw A, Lucas J (2003) On line EM wave sand monitoring sensor for oil industry. In: 33rd European microwave conference, vol 2, pp 535–538. IEEE, Houston, TX, USA
136. Sigfox (2020) https://www.sigfox.com/en

137. Silva AR, Vuran MC (2009) Empirical evaluation of wireless underground-to-underground communication in wireless underground sensor networks. In: Proc. of IEEE DCOSS '09, Marina del Rey, CA, pp 231–244
138. Silva AR, Vuran MC (2010a) Communication with aboveground devices in wireless underground sensor networks: An empirical study. In: Proc. of IEEE ICC'10, Cape Town, South Africa, pp 1–6. https://doi.org/10.1109/ICC.2010.5502315
139. Silva AR, Vuran MC (2010b) $(CPS)^2$: integration of center pivot systems with wireless underground sensor networks for autonomous precision agriculture. In: Proc. of ACM/IEEE international conf. on cyber-physical systems, pp 79–88, Stockholm, Sweden. http://doi.acm.org/10.1145/1795194.1795206
140. Skierucha W, Wilczek A (2010) A FDR sensor for measuring complex soil dielectric permittivity in the 10–500 MHz frequency range. Sensors 10(4):3314–3329
141. SMAP (2020) https://smap.jpl.nasa.gov/
142. Smartrek technologies (2020) www.smartrektechnologies.ca
143. SMOS (2020) www.esa.int/Our_Activities/Observing_the_Earth/SMOS
144. Soil climate analysis network (scan) data & products (2020) www.wcc.nrcs.usda.gov/scan/
145. Soil moisture sensor (2020) http://www.irrometer.com/sensors.html
146. Sojdehei JJ, Wrathall PN, Dinn DF (2001) Magneto-inductive (mi) communications. In: MTS/IEEE Oceans 2001. An Ocean Odyssey. Conference Proceedings (IEEE Cat. No. 01CH37295), vol 1, pp 513–519. IEEE, Houston, TX, USA
147. St (2020) www.st.com
148. Sudduth KA, Kitchen N, Bollero G, Bullock D, Wiebold W (2003) Comparison of electromagnetic induction and direct sensing of soil electrical conductivity. Agron J 95(3):472–482
149. Sun L, Li Y (2010) Acoustic emission sound source localization for crack in the pipeline. In: 2010 Chinese control and decision conference, pp 4298–4301. IEEE, Houston, TX, USA
150. Sun Z, Akyildiz I (2010a) Channel modeling and analysis for wireless networks in underground mines and road tunnels. IEEE Trans Commun. http://doi.acm.org/10.1109/TCOMM.2010.06.080353
151. Sun Z, Akyildiz IF (2010b) Channel modeling and analysis for wireless networks in underground mines and road tunnels. IEEE Trans Commun 58(6):1758–1768
152. Sun Z, Akyildiz IF (2013) Optimal deployment for magnetic induction-based wireless networks in challenged environments. IEEE Trans Wirel Commun 12(3):996–1005
153. Sun Z, Zhu B (2013) Channel and energy analysis on magnetic induction-based wireless sensor networks in oil reservoirs. In: 2013 IEEE international conference on communications (ICC), pp 1748–1752. IEEE, Houston, TX, USA
154. Symphony link (2020) https://www.link-labs.com/symphony
155. Tamu North American Soil Moisture Database (2020) www.soilmoisture.tamu.edu
156. Tan X, Sun Z, Akyildiz IF (2015a) Wireless underground sensor networks: Mi-based communication systems for underground applications. IEEE Antennas Propag Mag 57(4):74–87. https://doi.org/10.1109/MAP.2015.2453917
157. Temel S, Vuran MC, Lunar MM, Zhao Z, Salam A, Faller RK, Stolle C (2018) Vehicle-to-barrier communication during real-world vehicle crash tests. Comput Commun 127:172–186
158. Temputech wireless sensor monitoring (2020) http://www.temputech.com/
159. Tenenbaum JB, De Silva V, Langford JC (2000) A global geometric framework for nonlinear dimensionality reduction. Science 290(5500):2319–2323
160. TK-GPS (2020) http://www.tk-star.com/
161. The Oklahoma Mesonet (2020) http://www.mesonet.org/
162. Thread (2020) https://www.threadgroup.org/technology/ourtechnology
163. Tiusanen MJ (2006) Wideband antenna for underground soil scout transmission. IEEE Antenna Wirel Propag Lett 5:517–519
164. Tiusanen MJ (2013) Soil scouts: Description and performance of single hop wireless underground sensor nodes. Ad Hoc Netw 11(5):1610–1618. http://dx.doi.org/10.1016/j.adhoc.2013.02.002
165. Tule (2020) https://www.tuletechnologies.com/

166. U.S. climate reference network (2020) www.ncdc.noaa.gov/crn/
167. Vasisht D, Kapetanovic Z, Won J, Jin X, Sudarshan M, Stratman S (2020) Farmbeats: An iot platform for data-driven agriculture. In: NSDI'17: Proceedings of the 14th USENIX conference on networked systems design and implementation, March 2017, pp 515–528
168. Veris technology (2020) https://www.veristech.com/the-sensors/v3100
169. VRI study (2017) http://proceedings.esri.com/library/userconf/proc15/papers/185_435.pdf
170. Vuran M, Dong X, Anthony D (2016) Antenna for wireless underground communication. US Patent 9,532,118
171. Vuran MC, Akyildiz IF (2010a) Channel model and analysis for wireless underground sensor networks in soil medium. Phys Commun 3(4):245–254. https://doi.org/10.1016/j.phycom.2010.07.001
172. Vuran MC, Salam A, Wong R, Irmak S (2018a) Internet of underground things in precision agriculture: Architecture and technology aspects. Ad Hoc Netw 81:160–173
173. Vuran MC, Salam A, Wong R, Irmak S (2018b) Internet of underground things in precision agriculture: Architecture and technology aspects. Ad Hoc Netw. https://doi.org/10.1016/j.adhoc.2018.07.017, http://www.sciencedirect.com/science/article/pii/S1570870518305067
174. Vuran MC, Salam A, Wong R, Irmak S (2018c) Internet of underground things: Sensing and communications on the field for precision agriculture. In: 2018 IEEE 4th world forum on internet of things (WF-IoT) (WF-IoT 2018), Singapore
175. Wang YPE, Lin X, Adhikary A, Grovlen A, Sui Y, Blankenship Y, Bergman J, Razaghi HS (2017) A primer on 3GPP narrowband internet of things. IEEE Commun Mag 55(3):117–123
176. Wark T, Corke P, Sikka P, Klingbeil L, Guo Y, Crossman C, Valencia P, Swain D, Bishop-Hurley G (2007) Transforming agriculture through pervasive wireless sensor networks. IEEE Pervasive Comput 6(2):50–57
177. Watson G (2006) Computing Helmert transformations. J Comput Appl Math 197(2):387–394
178. Web soil survey home (2017) https://websoilsurvey.sc.egov.usda.gov/
179. Wolfert S, Ge L, Verdouw C, Bogaardt MJ (2017) Big data in smart farming – a review. Agric Syst 153:69. https://doi.org/10.1016/j.agsy.2017.01.023, http://www.sciencedirect.com/science/article/pii/S0308521X16303754
180. Yan Q, Yang H, Vuran MC, Irmak S (2017) Spride: Scalable and private continual geo-distance evaluation for precision agriculture. In: IEEE conference on communications and network security (IEEE CNS), Las Vegas, NV, USA
181. Yu X, Han W, Zhang Z (2017) Path loss estimation for wireless underground sensor network in agricultural application. Agric Res 6(1):97–102
182. Zhang C, Kovacs JM (2012) The application of small unmanned aerial systems for precision agriculture: a review. Precis Agric 13(6):693–712. https://doi.org/10.1007/s11119-012-9274-5
183. Zhang X, Andreyev A, Zumpf C, Negri MC, Guha S, Ghosh M (2017) Thoreau: A subterranean wireless sensing network for agriculture and the environment. In: 2017 IEEE conference on computer communications workshops (INFOCOM WKSHPS), pp 78–84. https://doi.org/10.1109/INFCOMW.2017.8116356
184. Zigbee alliance (2017) www.zigbee.org

Chapter 11
Decision Agriculture

11.1 Introduction

Decision agriculture (also referred to as precision agriculture (PA)) can be termed as the process of adequate amount of farming techniques, i.e., only applying sufficient and required amount of nutrients, seeds, water, and other farm resources. It helps in avoiding waste of resources while increasing the crop production [101]. This is achieved using various technologies such as creating soil maps, image analysis, variable rate application, measuring soil moisture and soil chemical content, and automation of farming equipment [14, 116, 122].

PA's objective is not just to increase the sustainable crop production. It also help towards improving workers condition by automating the repetitive task based on soil moisture and weather readings [35]. It also impact socially by providing more control over environment. It can be used locally or can be expanded over large fields spanning over several kilometers. For example, it is being used to monitor individual testbed in a greenhouse [30] as well as used for controlling pollination of crops [50] (Fig. 11.1).

Soil moisture provides information on vegetation and climate state. It also effect the underground communication; therefore, it is important to monitor soil. Scale of soil monitoring can vary from application to application. For a large geographical area, satellites or unmanned aerial vehicles (UAV) are used to perform spectral image analysis. However, this approach is limited by bad weather conditions such as heavy clouds which can block the view [55, 71]. These methods rely on number of factors such as quality and post-processing of image, quality of captured videos, instrument calibration, cloud screen, atmospheric correction, normalization of off-nadir effects [54]. The analysis of global position system (GPS) microwaves signals in the L-band and cosmic-ray neutron probe can be counted as medium level soil monitoring approaches [57]. In-situ measurement techniques are invasive, e.g., gravimetric sampling and can be very time consuming if samples are collected manually. In-situ measurements can be used as a calibration step for large-scale and

A. Salam, U. Raza, *Signals in the Soil*,
https://doi.org/10.1007/978-3-030-50861-6_11

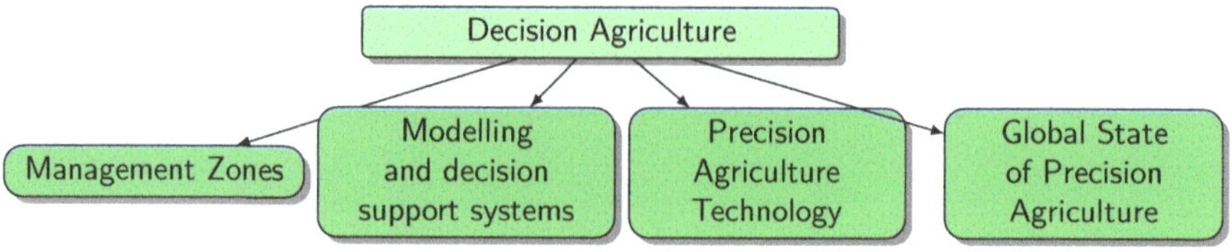

Fig. 11.1 Organization of the chapter

medium-scale soil monitoring mechanisms. The wireless technology saves the cost for installation and maintenance and provides the facility of real-time information access [69, 121].

The probability of mismatch between the spatial scale of the soil moisture measurement and the one required by the application is high [66, 118]. For example, UG node is deployed several meters way from the sensor which it is assigned to monitor. Upscaling the soil moisture readings to the area where UG node is deployed can help in achieving better communication. The soil moisture readings by WUSN are used as an input to control the CP irrigation system rate. The overall status of the complete field can be obtained by upscaling the data [13, 77, 80]. Just as in-situ soil moisture can calibrate the medium- and large-scale soil moisture sensing products, similarly, these remote solutions can be used for upscaling data in WUSN or self-calibration [38, 86]. In the following sections, various solutions for sensing soil moisture are described.

Cyber-physical system (CPS) is an integration of computational components into engineering systems which are used to detect, react, and communicate certain environmental conditions [16, 69, 102, 117]. The CPS aims to create a smart physical and cyber domain by improving machine-to-machine (M2M), human-to-human, and human-to-machine [120]. In precision agriculture (PA), a network of farm machinery can be transformed to a CPS by making them environment-aware and proactive to environmental changes and conditions. To that end, WUSNs extend the PA by transmitting data to an open access media, providing high spatio-temporal resolution of environment properties [17], and not being effected by the farm activities due to buried devices [4, 77]. Moreover, these activities can be automated and controlled by using cloud services. To summarize, a WUSN can be converted to a CPS system by adding actuators which reacts to physical world phenomenons and cloud control applications [79, 100, 117, 120].

A successful agricultural development impacts the local community, the nation, and the world in a very positive way. It also plays an important role in driving country's economy because everyone is either a producer or a consumer of food, hence it cannot be ignored. There is also a social responsibility of making better use of limited resources while fulfilling the food needs of the people. The most important resources used in farming are soil and water which makes need of developing underground detection and communication methods more inevitable. To that end, WUSN aims to detect and communicate the farming conditions and parameters in soil, however, face many challenges.

The United Nation resolves to eliminate malnutrition and poverty by 2030. However, the current food production rate of the world is much bigger than the UN's agenda [23, 84]. By 2050, the World's population is expected to be 9 billion. The increase in population will eventually increase the demand of human's as well as livestock food by 25–70% [37, 93]. Current farming practices are not climate-friendly affecting the food production methods, hence developing environmental- and climate-friendly farming practices is an urgent need of the hour. Meadows and forests can be converted to arable field to increase the food production [28, 90], however, this can disrupt the ecosystem by increasing the greenhouse effect due to deforestation. Extreme weather conditions, e.g., drought or prolonged hot weather, can harm crop fields wasting the farmer's hard work of a complete season.

Instead of creating new fields, a possible solution can be to optimize the usage of current available resources by use of information and technology. Technology can predict exact dose of fertilizers and pesticides to be applied in the crop and produce genetically modified crops. However, consumers are more attracted towards organic food; therefore, these practices are not being adopted in excess [25, 88]. Farms in underdeveloped countries lack means to adopt these improvements. Conversely, optimized water usage has significant advantages to implement agriculture in areas where water is scarce.

Three-fourth ($\frac{3}{4}$) area of the Earth is covered by water, of which only 2.5% of the water is fresh water. From that 2% of fresh water, 70% is frozen (glaciers and ice caps) and rest (residual 0.75% of freshwater) is in swamps, underground, lakes, rivers, living things, and the atmosphere. 70% of the residual freshwater is used for irrigation [61, 91]. Therefore, water is a very valuable and scarce natural resource and must be optimally used by adopting efficient food production technologies [95, 119].

The practical application of wireless underground communication is limited because of high signal attenuation by soil moisture [4, 6, 7, 17, 18, 103, 108, 112, 115]. There have been extensive studies which provide guidance on how communication can be made better in an underground channel. For example, channel model for soil–air communication is presented in [19, 89], channel model considering impacts of various deployment and environmental parameters is presented in [45, 87, 112], and a three-wave channel model is given in [18, 113]. These models can be validated by testing them empirically. An outdoor testbed is created to perform this research. This testbed simulate real-life scenarios such as crop rotation, fertigation events which are faced by sensors and transceivers, dynamic weather conditions, and irrigation. Soil moisture level can vary because of irrigation and storm, therefore, it would be interesting to study the effect of such variation on communication performance. The performance of communication systems must be measured and analyzed for a long operational period of time to get better understanding of the communication in underground environments.

Internet of Things (IoT) provides many sensors (commercial and proprietary) for sensing soil condition and solution for reacting on the basis of sensed information, e.g., applying right amount of water [3, 20, 22, 24, 26, 52, 55–58, 63, 73, 92, 122].

However, most solutions are application specific such as related to data storage, information acquisition, or communication [1]. Therefore, a complete general framework is required with capability of sensing, modeling, decision making, and performing actuation. A cloud-based middleware platform accomplishes information exchange, provides global data access and management facilities, and gives on-demand processing services [46, 96]. This work develops an autonomous cloud-based application for controlling irrigation system. This model can be used as a guideline for development of PA middleware projects.

CPS models physical processes using data abstraction. Therefore, it is important to give theoretical reasoning for each CPS component and decision while explaining this system [67, 97]. Introduction of engineering systems into the CPS necessitates the development of a framework containing modeling and abstraction of CPS. Abstract modeling of CPS architecture makes it easier to integrate the components developed by other disciplines [42, 65]. Nevertheless, it has helped many fields to make quick substantial improvements, e.g., wireless sensor networks (WSNs).

During the past decade, developments in WSNs have led to discovering long-range and/or power-efficient communication protocols [5]. To prolong WSNs lifetime, novel methods of harvesting energy from natural resources (light, vibration, heat) and electromagnetic sources are being investigated. Micro-electro-mechanical systems (MEMS) have contributed to the success of WSNs by providing precise, small, and affordable sensors. CPS abstraction enables this alliance of these independent and interdisciplinary studies, e.g., a middleware issued command or transfer data to various devices without worrying about the data types or underlying network [36, 70].

Wireless underground sensor networks (WUSNs) are a subset area of WSNs which enables the communication between buried underground (UG) nodes and aboveground (AG) devices [5]. Water retention in soil affects the wireless communication. Path loss modeling is also an active research area in UG communication. WUSNs application includes border surveillance, natural disaster detection (e.g., earthquakes and landslides), mine safety, and precision agriculture [4, 5].

WUSNs are well-suited for measuring the amount of water and nutrients in the soil. UG devices can be deployed for taking measurement without being visible to farm equipment and machinery. AG nodes can also be deployed in such location where they can communicate wirelessly without being obstructive [4], e.g., nodes can be attached to moving equipment to collect data from all over the farm. Typically, a central data storage and processing is also implemented.

Precision agriculture (PA) can be termed as the process of adequate amount of farming techniques, i.e., only applying sufficient and required amount of nutrients, seeds, water, and other farm resources. It helps in avoiding waste of resources while increasing the crop production [101]. This is achieved using various technologies such as creating soil maps, image analysis, variable rate application, measuring soil moisture and soil chemical content, and automation of farming equipment [14, 70, 116, 122].

PA's objective is not just to increase the sustainable crop production. It also help towards improving workers condition by automating the repetitive task based on soil moisture and weather readings [35]. It also impact socially by providing more control over environment. It can be used locally or can be expanded over large fields spanning over several kilometers. For example, it is being used to monitor individual testbed in a greenhouse [30] as well as used for controlling pollination of crops [50].

Precision irrigation applications require high spatio-temporal resolution for proper working which is provided by the sensor networks. Wireless communication helps in providing remote information access. This information is provided in real-time so that manual manipulation can be avoided to get an idea about the field conditions. Dong and Vuran [17] used small amount of sensors for measuring soil moisture because of precipitation duration and rainfall cell radius magnitude.

Vereecken et al. [111] discusses various techniques for estimating exact location of the sensor nodes. In general, examining different soil properties with varying soil moisture level can give an average soil moisture value for the field. Delgado and Martínez [15] calculates location-specific solar radiation intensity values to estimate the evaporation rate corresponding to that certain location. A mobile application is used for this approach. The calculations matched the reading from agriculture station which helped in generalizing the result to all those locations which have same radiation conditions. Sensor cluster is then placed in chosen area with following the recommendations given in [17, 71].

The primary task in the WUSN is to determine number of sample readings. It is important because sensing and communicating data consumes large amount of power [17, 41]. Shallow roots and high porosity cause speedy water infiltration and evaporation in soil. Therefore, large sampling rate is needed to overcome this highly fluctuating effect.

11.2 Management Zones in Digital Agriculture

The regions in the agricultural fields with similar characteristic are known as management zones (MZs). Identifying MZs is an ultimate goal of precision agriculture. MZs differ from each other in major factors such as topography, soil type and water, and nutrition availability [9]. MZ must be large enough for variable rate application (VRT) and small enough for the precise application. MZ can help in increasing output of farms in terms of cost, yield, and quality of production [40]. MZs can be identified on the basis of large variety of attributes, e.g., soil sampling, sensing, or plant-based. Combination of these attributes can also be used. Any information obtained for these variables can open new possibilities for crop management.

Soil-based measurement can be used to provide more temporally stable zones. For example, ECa has been used extensively for MZ delineation [39, 53] along with elevation mapping using real-time kinematic-GPS [60, 98, 113, 115]. Soil ECa is suitable for explaining spatial variability for static soil properties, however, it does not explain the spatial variability of yield or quality which strengthens the argument of using the combination of different parameters in the process of MZ delineation for meaningful results. References [98, 115] recommends using the combination of crop-based and soil-based MZ for the analyses for temporal variability.

NDVI is also a useful parameter for delineating MZs [21, 71]. It is related to various crop properties such as chlorophyll content, biomass, leaf area [32], crop yield [105] and quality. MZ delineation can be classified on the basis of application to inputs, e.g., nutrient MZs, irrigation MZs [110], and MZs for herbs and pesticides [68, 107, 109].

Nutrition MZs have been found in citrus orchard on the basis of soil variation [99], in wheat crop on the basis of ECa, soil depth and elevation, and in variable rate nitrogen application for increased nitrogen efficiency [60]. Nitrogen management in spring wheat has been optimized by using a spatio-temporal Bayesian network [44, 90]. References [31, 68] used soil available water, yield data, and ECa satellite images to delineate an irrigation MZ and found that soil ECa is extremely useful in forming irrigation MZ for variable rate application of irrigation in pivot systems.

11.2.1 Methodology for Delineating MZs

There is no specific methodology to analyze the multi-layer information obtained from the field and delineate MZ [29, 91], hence evaluating different algorithms and methods for MZ delineation is a challenging task. Several methods have been proposed to delineate MZ based on nature of data sources, agronomic knowledge of the field, and inter-dependencies of data variables. The most common methods for delineating MZ in digital agriculture framework are described below.

Empirical This simplest and easiest method requires farmers to visually inspect the attribute map or aerial images and delineate the MZ based on their personal experiences from the field. It is an accurate method since the field observations are not scientifically validated and may also incur the personal bias in judgment of the farmer.

Geostatistical Geostatistical methods have been used to analyze and combine different parameters for producing MZ maps. Due to values spatial dependence, sampling points can be interpolated using geostatistical methods. Maps for the whole field can be generated to study the variability of the properties across the field. All layers of information can be overlaid over each other using interpolation. [2, 65] did MZ delineation using multivariate analysis. In [43, 95], MZ delineation for soil moisture is done for management of water under stressed condition using multivariate geostatistics.

Clustering Cluster analysis classifies the data into discrete classes or clusters. K-means (also known as c-means), a non-hierarchical clustering technique, divides a multi-dimensional data into k-clusters. Centroid in each cluster is Euclidean distance away from the data points. Ping et al. [62] uses k-means to evaluate procedure for MZ delineation in cotton. Fuzzy k-means, an extension to k-means, accounts for the uncertainties related to class boundaries and membership. Molin and Castro [51] show that principal component analysis (PCA) and fuzzy logic application on soil data and ECa can reliably delineate soil MZ. Many universities have also developed software that perform fuzzy k-means. For example, Management Zone Analyst (MZA) was developed by University of Missouri [27] and FUZME developed by University of Sydney, Australia, have been extensively used by researcher all over the world. ZoneMAP automates delineation of MZ from satellite imagery and field data as an input from the user and use the fuzzy k-means as a processing algorithm.

Degree of Agreement References [39, 82] delineate soil productivity zone (SPZ) and yield productivity zone (YPZ) using unsupervised fuzzy k-means clustering. They calculated degree of agreement by matching outcomes of YPZs and SPZs with overall accuracy statistics (matched cells divided by total cells in data sets). Same concept is applied for delineation of MZ in vineyards [85, 104]. Field data is collected and analyzed by using crop canopy and soil sensors. Mono-parametric zone maps are generated using fuzzy clustering. This data analysis method can be extended by including comparison between more parameters, and enhanced information management during digital agriculture experiments.

Other studies have used combination of statistical and geostatistical methods along with fuzzy clustering method to create management zone. Chlorosis MZs were created in soybean and maize by examining data for pH, yield, NDVI, and ECa. They selected NDVI and ECa, using data interpolation and regression analysis, to delineate MZs using MZA [27, 84].

11.3 Modeling and Decision Support Systems in Digital Agriculture

Decision support systems (DSSs) can be defined as a computerized system which are used for making decisions using the models and databases (see Fig. 11.2). Motivation behind using DSSs is to choose the best alternative from social, economic, and environmental point of view. An overview and role of DSSs in agriculture, farm planning, and management is presented by [47, 75]. However, due to difference in the context of farmer and scientist, adoption of DSSs in digital agriculture is very low.

For a long period of time, the only objective of agricultural production was to secure profits and no considerations were given to the environmental impact caused by it. Farmers are now facing lot of pressure from environmental protection agencies, public, consumers, and bank. Farming practices now need to be adjusted for the environment protection. However, all production regulations and specifications still cannot guarantee a clean agriculture [49, 78]. As per [47, 81]: The planning and development process in agriculture is a complex problem which, if not face thoroughly, cannot be solved easily. Due to diverse agricultural conditions, same

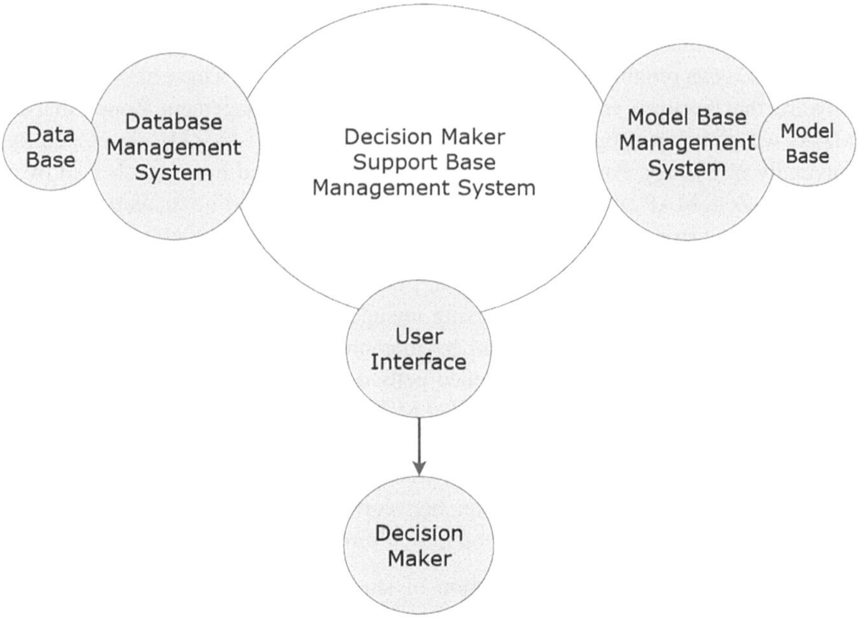

Fig. 11.2 Structure of a basic decision support system (DSS)

practice may produce different results in different fields. Therefore, decision making in agriculture is not based on set of well-defined rules and regulations instead it is based on experience, knowledge, and skills of producers.

In linear model, several controlled experiments are performed with each experiment observing some variables while others being controlled and standard recommendations are made as an outcome [11, 83]. When linear model is applied in agriculture, research data is analyzed and interpreted as recommendations to farmer. However, in reality, few farmers apply these recommendations, therefore, the use of linear model in agriculture is very rare. Cook et al. [12] summarized that scientific results from experiments in agriculture are not well translated by DSS and adopted by farmers because scientist and farmers do not see the problem in same way. For example, scientists gave more importance to the information related to the quantifiable effect of factors while ignoring the farming environment in which they are occurring.

Scientific publications have to satisfy certain strict rules regarding experiments, technical requirements, and statistical procedure to get published. Most of the research is published in peer-reviewed journals with requiring scientists to demonstrate that finding applies to any particular condition of end user. This difference between scientific and a more complex farming world makes it difficult for the farmers to apply the insights from the formal experiments in the real world. This is the area where modeling and validated DSSs can help to bridge the gap between the research and adoption of the research by farmers. Recent advancement in communication technologies have given access of key information to the farmers via cell phones or sophisticated agricultural machinery.

Agriculture sector is disorganized due to uncontrolled variations. These uncontrolled variations (due to site- and time-specific effect) led to the development of statistical methods by Fisher and co. (Rothamsted Experimental Station, Harpenden, UK) for clarification of experimental effects in agriculture. Fisher's statistical methods allow for the simple evaluation method for publication of scientific work at the expense of adaption and recognition from the end user.

11.3.1 Issues in DSSs

There is a lack of proper DSSs in digital agriculture [48, 76]. Although, DSS is more user-friendly today, however, does not completely addresses the needs of the user. Historically, digital agriculture has more focused on spatial variability, however, temporal variability for the crop and animal production is also being considered now. In addition to spatial and temporal variabilities, digital agriculture also has to deal with uncertainties, human and social issues. Some of the issues are discussed below.

Soil (Spatial) Majority of research in DA is mainly focused on spatial variability. Soil information must be brought down to useful and properly quantified parameters for integration to DSS. Soil sampling is done to assess the physio-chemical properties of the soil and aligns with the goals of digital agriculture if done with adequate planning and spatial statistics. However, it is not possible to always use interpolation for the mapped points due to sampling resolution limitations [33, 74]. Moreover, applications map created by the interpolation may be misleading and detrimental. With the development of proximal and remote methods, sampling can be replaced with sensing at much higher spatial density. The spatial density is so high that interpolation and MZ production can be eliminated. The rapid and high turnover of technology has produced a gap between scientific understanding and technological capability to measure and apply input and outputs of the crop [8, 79].

References [59, 76, 94] studied soil properties for site-specific protection and found that heterogeneous soil properties effect the weeds, pests, and pesticides behavior. They concluded that information on soil variable properties combined with precision crop protection (PCP) can benefit both economically and ecologically. Another study by Christensen et. al. [10] proposed a computerized management system which assists in making decision on choosing herbicides and doses of herbicides. The system uses a huge database of herbicides performance on various crops for ranges of weeds at different growth stage. It allowed to rank and make recommendations on efficient herbicides or number of doses to use against weeds. References [64, 73] reported localized mechanical stress (e.g., compaction during traffic) as one of the possible causes for the high variability in the soil structure within the field.

11.4 Precision Agriculture Technology

A total of 32% increase is expected in world's population by 2050. This increase in population will also result in doubling of food demand. Food production is the cause of 70% of water withdrawal in the world. Therefore, there is a dire need of novel technologies which can produce more crop for drop. USDA Agricultural Resource Management Survey (ARMS) publishes information on food production practices, financial conditions, resource utilization, and economic well-being of America's farm household and business. ARMS data shows that precision agriculture is being widely adopted by American farmers. In Fig. 11.3, adoption rate of major precision agriculture techniques (represented by bars) is plotted along with the adoption of precision agriculture for corn (represented by line) based on the ARMS data published yearly (USDA ARMS 2015). It can be clearly seen that, for corn, adoption rate for precision agriculture has significantly increased from 17.29% to 72.74% between the years 1997 and 2010. Similar trends have been noticed for other crops such as peanuts and soybean. Apart from high adoption rate of precision agriculture in corn production, farmers are adopting new technologies as they emerge [88, 95].

Crop yield monitoring is one of the most adopted (61.4%) among all precision agriculture techniques. For guidance and auto-steering, the adoption rate increased from 5.34% to 45.16% in 9 years. Auto-steering information allows precise control by using equipment and spatial information of the crop while reducing maintenance cost and extra work for farmers. Although other precision agriculture techniques have seen drastic increase in adoption rates, adoption of variable rate technology (VRT) has been steady relatively. In 8 years (1998–2005), VRT's adoption rate has increased from 8.04% to 11.54% only. Crop production has increased significantly by adaptive application of agricultural resources (e.g., fertilizers, pesticides, and water, etc.) but gathering correct and timely information from the field is also key factor for improving crop growth. It is evident from the fact that after the adoption of crop moisture sensing technology, the adoption of VRT doubled (from 11.54% to 22.44%) in the period of 2005–2010. The same period sees the increase in adoption of crop moisture sensing from 36.21% to 51.68% [65, 91]. This shows that VRT

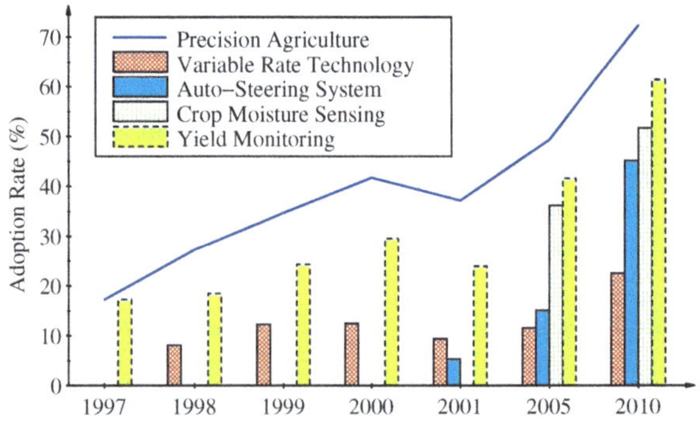

Fig. 11.3 Data by USDA ARMS on PA adoption rate in corn fields [114]

adoption is highly dependent upon the advancement in soil monitoring technologies. Crop moisture sensing has become the most adopted and popular technique in a very short period of time. Yet the techniques are being used with manual data collection or very limited coverage of the field [67].

11.5 Global State of Precision Agriculture

There is not much information available on precision agriculture (PA) adoption all over the world. The most desirable form of information would be published statistics, by known and credible organization, about the farming sector, however, it is very rare. Other useful information may include data obtained from the surveys and interviews of the farmers, opinions from the academics doing research in concerned field, and industry reports, etc. Though this data is rare and very low, it is still better than guessing [78]. This section aims to provide information on the state of PA adoption all over the world [34].

11.5.1 Bulgaria

Although, there is no official data available on PA in Bulgaria [34], however, the first implementation of PA was using lightbar displays for fertilizer application somewhere around 2002–2003, and were extended to be used in sowing and weed control after the invention of automatic steering systems. After 2009, these technologies became famous and were commonly used in Bulgaria. For example, farm machines come with the advanced equipment such as steering systems, global navigation satellite systems (GNSS) displays, guidance systems for slopes, yield monitoring systems for creating yield maps during harvesting season, and spray control systems for sections. However, these technologies are limited to big farms for oilseeds and grain production located in North Bulgaria only. As per 2016 agricultural census, 2350 large farms which cover 68% of agricultural area have crop and livestock output greater than 250,000 EUR. These farms produce extensively and represent around 45% of the farm output of the country. These big farms are using variable rate application, soil sampling, Geoscan, and weather monitoring technology and some of them use precision planting and precision irrigation. Due to lack of access to technology, small scale farms do not implement PA technologies and livestock sector is even far behind in adoption than crop production [72, 115].

PA systems (precision farming systems and different software) are being offered by few companies but they are being implemented by only large farms. Moreover, PA implementation relies majorly on economic profitability and affordability of farmer. To that end, detailed investment information on PA technologies is lacking. Although, the advantages of PA are being debated over many forums, however, a farmer's perspective is lacking. To that end, a survey study results were published in 2018 which shows that, out of 258 randomly sampled set, 49% do not have knowledge about the technologies, 38% have no plan to implement PA, and only 4% plan to implement PA [93].

PA technologies are discussed in various universities and scientific institutes of Bulgaria. For example, remote control methods, such as satellite imagery, are being used to monitor the crop and land conditions at the Space Research and Technology Institute since its inception from 1970s. The Institute is working on many satellite systems projects funded under the 6th and 7th Framework Program, Horizon 2020, and others. Other Institutes, e.g., *Kliment Ohridski*, a Geological-Geographical Faculty from The Sofia University has developed remote methods involving satellite images and drones. The Agricultural University-Plovdiv is offering PA specific master courses under Erasmus projects. It is conducting experiments near Plovdiv city for the development of crops and varieties of maize, sunflower, and wheat. Agricultural Hubs also play an important role in PA adoption. For example, SmartAgriHubs coordinated by Wageningen University under Horizon 2020 Program and Bulgarian AgroHub comes under the European SmartAgriHubs project.

11.5.2 Denmark

In 2019, PA technologies were being used in 28% of the Danish farms which cover 66% of farming land. Denmark is one of the very few countries which officially collects data on PA via government offices using stratified sampling and sample size of 6005 respondents. The survey includes the PA technologies such as: drone images and crop sensors, global navigation satellite systems (GNSS), software for planning nitrogen applications, sprayer section control. It was found that GNSS guidance, sprayer section control, drone or satellite images, and crop sensors were used 24% (covering 59% of the farmland), 40% (covering 40% of the farmland), 5% (covering 15% of the farmland), and 2% (covering 8% of the farmland) of the Danish farms, respectively. Average farm size in Denmark was 83 ha and the average farm size adopting GNSS guidance and crop sensor was 202 ha and 342 ha. It shows how, like other countries, large farms have more PA adoption rate than smaller farms [34].

11.5.3 France

France has a largest agricultural area (292,800 km^2 (53.2% of the surface area of France), about 1/2 ha per inhabitant) and employs 3.4% of the population generating 4.5% of GDP (over 72 billion euros) [34]. It is the largest agricultural country in European Union covering 18% of agricultural land in Europe. Some of the main productions include wheat (5th in the world), corn (8th in the world), sugar (7th in the world), wine and milk, livestock and meat products (5th in the world for beef).

The Observatoire des Usages de L'Agriculture Numérique (Digital Agriculture Adoption Observatory—http://agrotic.org/observatoire/chaire-agrotic/) is a partnership of 8 research institutes, 2 universities, and 28 agricultural related companies and collects PA data in France. Some of the observations from the data are given below:

- Software-based variable rate input application is used by 4% French farmers.
- By 2017, remote sensing is used by 1 million ha of farm land with distribution of 85% satellite, 15% using drones/aircraft. Moreover, 10% and 1% of crop and viticulture area, respectively, were being managed by remote sensing.
- Electrical conductivity or resistance sensors are used to map approx. 135,000 ha (>1%) of French farmland in the last decade.
- 20% of the French farms uses soil maps developed by the professionals.
- In 2018, milking robots were being used by 10% of the dairies and other 70% were thinking to acquire it; 10 robots were being used for weeding vineyards and 100 for vegetable crops.

11.5.4 Malaysia

Prof. Siva Kumar Balasundram from Universiti Putra Malaysia (UPM), Department of Agriculture Technology and a Malaysian representative in ISPA says that, *PA is limited to research purposes and community in Malaysia and there is no adoption statistics reported yet.* It is also confirmed by Dr. Hari Krishna of Sime Darby stating the cost as a reason of low adoption. Dr. Redmond Shamshiri, formerly with UPM and now with Bioeconomy Institute (ATB) Potsdam, Germany, also says that PA in Malaysia is being studied by universities, government research institutes (Malaysian Agricultural Research and Development Institute, the Malaysian Cocoa Board, and Malaysian Palm Oil Board), and some startups. The efforts to promote PA in Malaysia include study tour of Malaysian farmers to New Zealand in 2017 organized by Space Exploration Asia in partnership with Adaptive AgroTech Research Group International, the New Zealand Center for Precision Agriculture, and Massey University of New Zealand. Malaysian Remote Sensing Agency (MRSA) conducted a PA project in rice, however, adoption by farms is very low [34].

11.5.5 Pakistan

In January, 2019, a delegation headed by Dr. Yubin Lane from South China University visited University of Sargodha in Pakistan to introduce and promote PA to researchers and farmers of Punjab province of Pakistan. They, for the first time, practically demonstrated the application and effectiveness of drones technology for precise application of spray in the agricultural campus of the University of Sargodha. The technology efficiently uses the available resources as per requirement and also enables secure crop production [34].

11.5.6 United Kingdom

In 2018, Harper Adams University (HAU) conducted an electronic survey from thousands of UK farmers on farmer mailing lists. The survey shows that 48% of a total of 186,000 respondents use map-based variable rate technology (VRT) fertilizer, 63% use GNSS autosteer, 53% use GNSS sprayer boom control, and 15% use optical nitrogen sensors. It shows them as a leading users of VRT fertilizers as its adoption hardly exceeds 20% in any other country/region of the world. UK farmers prefers personal training but open to use electronic tutorial materials and willing to pay average 278 euro (339 USD) for annual subscription with web-chat support, if available. 40% were willing to share data for technical help [34].

11.5.7 Ukraine

PA is readily available in Ukraine with many startup companies providing PA services (drones, soil mapping, and testing, etc.). Big farming companies like Kernel (www.kernel.ua) are providing services such as: yield mapping, satellite images, variable rate technology for fertilizer and seed, and intensive soil sampling. There is no official data on PA adoption in Ukraine, however, all indications show that PA is very limited in Ukraine except for GNSS. Iaroslav Beiko, ISPA member and co-founder of AgriLabs, says that, *"Ukraine is at beginner stage of adopting PA components with 20% usage of automatic section control. The conscience and systematic use of PA complex (field mapping to VRT fertilizers and seeds) is less than 5% of the area."* Some other industrial sources state higher adoption rate with same underlying story, e.g., SmartFarming website shows adoption rate of 30% but with limitation to single technique—an autopilot/guidance system for preventing overlaps and gaps in the fields [34].

11.5.8 Poland

By 2017, there was 14.6 million ha of agricultural land and 10.8 million ha sown land (with 70.7% of Cereal) in Poland [34]. There were 1.4 million farms; 2.5% of which are more than 50 ha covering 31% of agricultural land area. The agricultural land area depends on region of the country, e.g., in southern part it ranges from 4.1

to 30.8 ha and even bigger farms in northern parts. This difference of range explains the difference in adoption of PA in Poland farms. Glaciation increases the potential adoption of PA [73, 77]. PA is also being taught in few universities for research purposes. The current condition of PA in Poland is as follows:

- The PA tools being used include: GNSS autosteers and lightbars, and auto section control of spreaders, sprayers, and planters. These technologies are saving 5–15% for the farmers. Software-based GNSS technology is being used by very large farms.
- Yield mapping is being used for grains, however, is quite limited due to lack of proper calibration system.
- Soil sampling and soil fertility map is being created by companies and consultants; variable rate application of the fertilizers, using active optical sensors, is being applied on very big farms. Few dozen farms are also using variable rate application of pesticides and variable rate seeding.
- A very few companies are offering soil mapping by management zones using soil electro-conductivity and facility of satellite images for estimating yield potential and biomass production.

11.5.9 Sweden

The ISPA newsletter (May, 2013) presents the report on development in last 6 years and the current condition of PA adoption in Sweden. PA is regarded as one of the important means to increase sustainable food production which also stressed by the National Food Strategy for Sweden by 2017 (Government bill 2016/17:104). The access to useful digital data and digital transformation is beneficial for development of digital equipment in an effort to increase PA adoption [94, 106]. Approximately all farmers in Sweden has access to the digital decision support system which has been proved to be key aspect in spreading of PA [34].

CropSAT, a satellite-based imaging system was launched in 2014 and since its inception is being used in variable rate application (VRA) of grain crops and nitrogen. This system has paved the way for the development of other imagery systems. It was the first system developed in partnership with university researchers, authorities, private enterprises, and advisory organizations. It is available in multiple languages and can be used globally.

Digital Soil Map of Sweden (DSMS) is an open access soil database. It predicts values for soil texture every 50 m and has been used in different decision support systems, e.g., a web-based application Markdata.se. Mark.se can be sued by farmers to produce VRA seeding and prescription files for liming. It is an interactive application which can upload users soil data to become more accurate locally. Drone is still not being used in practical PA in Sweden, however, availability of Solvi.nu, a decision support system for drone images, can make the adoption process faster [41, 113].

References

1. Abbasi AZ, Islam N, Shaikh ZA, et al. (2014) A review of wireless sensors and networks' applications in agriculture. Comput Stand Inter 36(2):263–270
2. Aggelopooulou K, Castrignanò A, Gemtos T, De Benedetto D (2013) Delineation of management zones in an apple orchard in Greece using a multivariate approach. Comput Electron Agric 90:119–130
3. Ajwa H, Trout T, Mueller J, Wilhelm S, Nelson S, Soppe R, Shatley D (2002) Application of alternative fumigants through drip irrigation systems. Phytopathology 92(12):1349–1355. https://doi.org/10.1094/phyto.2002.92.12.1349
4. Akyildiz IF, Stuntebeck EP (2006) Wireless underground sensor networks: research challenges. Ad Hoc Netw 4(6):669–686
5. Akyildiz IF, Vuran MC (2011) Wireless sensor networks (Advanced texts in communications and networking). Wiley, Hoboken
6. Bogena H, Huisman J, Meier H, Rosenbaum U, Weuthen A (2009) Hybrid wireless underground sensor networks: quantification of signal attenuation in soil. Vadose Zone J 8(3):755–761
7. Bogena H, Herbst M, Huisman J, Rosenbaum U, Weuthen A, Vereecken H (2010) Potential of wireless sensor networks for measuring soil water content variability. Vadose Zone J 9(4):1002–1013
8. Brisco B, Brown R, Hirose T, McNairn H, Staenz K (1998) Precision agriculture and the role of remote sensing: a review. Can J Remote Sens 24(3):315–327
9. Bullock DS, Ruffo ML, Bullock DG, Bollero GA (2009) The value of variable rate technology: an information-theoretic approach. Am J Agric Econ 91(1):209–223
10. Christensen MG, Teicher HB, Streibig JC (2003) Linking fluorescence induction curve and biomass in herbicide screening. Pest Manag Sci (Formerly Pestic Sci) 59(12):1303–1310
11. Cock J, Oberthür T, Isaacs C, Läderach PR, Palma A, Carbonell J, Victoria J, Watts G, Amaya A, Collet L, et al. (2011) Crop management based on field observations: case studies in sugarcane and coffee. Agric Syst 104(9):755–769
12. Cook S, Cock J, Oberthür T, Fisher M, et al. (2013) On-farm experimentation. Better Crops Plant Food 97:17–20
13. Cosh M (2017) Ongoing research at the Marena, Oklahoma, in situ sensor testbed (MOISST). http://soilphysics.okstate.edu/research/moisst/2017-moisst-workshop/Cosh_MOISST_2017.pdf. 2017 MOISST Workshop
14. Das J, Cross G, Qu C, Makineni A, Tokekar P, Mulgaonkar Y, Kumar V (2015) Devices, systems, and methods for automated monitoring enabling precision agriculture. In: Automation Science and Engineering (CASE), International Conference on 2015 IEEE. IEEE, New York, pp 462–469
15. Delgado B, Martínez M (2015) Software application for calculating solar radiation and equivalent evaporation in mobile devices. Agric Water Manage 151:30–36

16. Ding W, Engel W, Goode A, Santostasi G (2016) Declarative modeling cases of cyber physical systems. In: 2016 International conference on logistics, informatics and service sciences (LISS). IEEE, New York, pp 1–6
17. Dong X, Vuran MC (2010) Spatio-temporal soil moisture measurement with wireless underground sensor networks. In: 2010 the 9th IFIP Annual Mediterranean Ad Hoc Networking Workshop (Med-Hoc-Net). IEEE, New York, pp 1–8
18. Dong X, Vuran MC (2011) A channel model for wireless underground sensor networks using lateral waves. In: 2011 IEEE Global Telecommunications Conference-GLOBECOM 2011. IEEE, New York, pp 1–6
19. Dong X, Vuran MC, Irmak S (2013) Autonomous precision agriculture through integration of wireless underground sensor networks with center pivot irrigation systems. Ad Hoc Netw 11(7):1975–1987
20. Dukes M, Simonne E, Davis W, Studstill D, Hochmuth R (2003) Length of irrigation and soil humidity as basis for delivering fumigants through drip lines in Florida spodosols. In: Proceedings of the Florida State Horticultural Society, vol 116, pp 85–87
21. El Nahry A, Ali R, El Baroudy A (2011) An approach for precision farming under pivot irrigation system using remote sensing and GIS techniques. Agric Water Manage 98(4):517–531
22. Entekhabi D, Njoku EG, O'Neill PE, Kellogg KH, Crow WT, Edelstein WN, Entin JK, Goodman SD, Jackson TJ, Johnson J, et al (2010) The soil moisture active passive (SMAP) mission. Proc IEEE 98(5):704–716. https://doi.org/10.1109/jproc.2010.2043918
23. FAO of the UN (2016) The State of Food and Agriculture 2016 (SOFA): climate change, agriculture and food security. FAO of the UN, available at: http://www.fao.org/3/a-i6030e.pdf
24. Ferrández-Pastor FJ, García-Chamizo JM, Nieto-Hidalgo M, Mora-Pascual J, Mora-Martínez J (2016) Developing ubiquitous sensor network platform using internet of things: application in precision agriculture. Sensors 16(7):1141
25. Foley J (2014) Where will we find enough food for 9 billions. Nat Geograph 225(5):48–77
26. Franz TE, Wahbi A, Vreugdenhil M, Weltin G, Heng L, Oismueller M, Strauss P, Dercon G, Desilets D (2016) Using cosmic-ray neutron probes to monitor landscape scale soil water content in mixed land use agricultural systems. Applied and Environmental Soil Science 2016(5):1–11
27. Fridgen JJ, Kitchen NR, Sudduth KA, Drummond ST, Wiebold WJ, Fraisse CW (2004) Manage Zone Analyst (MZA). Agron J 96(1):100–108
28. Godfray HCJ, Beddington JR, Crute IR, Haddad L, Lawrence D, Muir JF, Pretty J, Robinson S, Thomas SM, Toulmin C (2010) Food security: the challenge of feeding 9 billion people. Science 327(5967):812–818
29. Guastaferro F, Castrignanò A, De Benedetto D, Sollitto D, Troccoli A, Cafarelli B (2010) A comparison of different algorithms for the delineation of management zones. Prec Agric 11(6):600–620
30. Gutiérrez J, Villa-Medina JF, Nieto-Garibay A, Porta-Gándara MÁ (2014) Automated irrigation system using a wireless sensor network and GPRS module. IEEE Trans Instrum Meas 63(1):166–176. 10.1109/TIM.2013.2276487
31. Haghverdi A, Leib BG, Washington-Allen RA, Ayers PD, Buschermohle MJ (2015) Perspectives on delineating management zones for variable rate irrigation. Comput Electron Agric 117:154–167
32. Hansen P, Schjoerring J (2003) Reflectance measurement of canopy biomass and nitrogen status in wheat crops using normalized difference vegetation indices and partial least squares regression. Remote Sens Environ 86(4):542–553
33. Heege HJ (2013) Precision in Crop Farming: Site Specific Concepts and Sensing Methods: Applications and Results. Springer Science & Business Media, Dordrecht
34. Home | International Society of Precision Agriculture (2020). https://www.ispag.org/
35. Hoornweg D, Munro-Faure P (2008) Urban agriculture for sustainable poverty alleviation and food security. Position paper, FAO Africa

36. Huang HM, Tidwell T, Gill C, Lu C, Gao X, Dyke S (2010) Cyber-physical systems for real-time hybrid structural testing: a case study. In: Proceedings of the 1st ACM/IEEE international conference on cyber-physical systems. ACM, pp 69–78
37. Hunter MC, Smith RG, Schipanski ME, Atwood LW, Mortensen DA (2017) Agriculture in 2050: recalibrating targets for sustainable intensification. BioScience 67(4):386–391. https://doi.org/10.1093/biosci/bix010
38. Kim KD, Kumar PR (2012) Cyber–physical systems: a perspective at the centennial. In: Proceedings of the IEEE 100(Special Centennial Issue), pp 1287–1308
39. Kitchen N, Sudduth K, Myers D, Drummond S, Hong S (2005) Delineating productivity zones on claypan soil fields using apparent soil electrical conductivity. Comput Electron Agric 46(1–3):285–308
40. Koch B, Khosla R, Frasier W, Westfall D, Inman D (2004) Economic feasibility of variable-rate nitrogen application utilizing site-specific management zones. Agron J 96(6):1572–1580
41. Konda A, Rau A, Stoller MA, Taylor JM, Salam A, Pribil GA, Argyropoulos C, Morin SA (2018) Soft microreactors for the deposition of conductive metallic traces on planar, embossed, and curved surfaces. Adv Funct Mater 28(40):1803020. https://doi.org/10.1002/adfm.201803020
42. Kopetz H (2011) Real-Time Systems: Design Principles for Distributed Embedded Applications. Real-Time Systems Series. Springer, New York. https://link.springer.com/content/pdf/10.1007%2F978-1-4419-8237-7.pdf
43. Landrum C, Castrignanò A, Mueller T, Zourarakis D, Zhu J, De Benedetto D (2015) An approach for delineating homogeneous within-field zones using proximal sensing and multivariate geostatistics. Agric Water Manage 147:144–153
44. Lawrence PG, Rew LJ, Maxwell BD (2015) A probabilistic Bayesian framework for progressively updating site-specific recommendations. Prec Agric 16(3):275–296
45. Li L, Vuran MC, Akyildiz IF (2007) Characteristics of underground channel for wireless underground sensor networks. In: Proceedings of the Med-Hoc-Net, vol 7, pp 13–15
46. López-Riquelme J, Pavón-Pulido N, Navarro-Hellín H, Soto-Valles F, Torres-Sánchez R (2017) A software architecture based on FIWARE cloud for precision agriculture. Agric Water Manage 183:123–135. 10.1016/j.agwat.2016.10.020. https://doi.org/10.1016/j.agwat.2016.10.020
47. Manos BD, Ciani A, Bournaris T, Vassiliadou I, Papathanasiou J (2004) A taxonomy survey of decision support systems in agriculture. Agric Econ Rev 5(389-2016-23416):80–94
48. McBratney A, Whelan B, Ancev T, Bouma J (2005) Future directions of precision agriculture. Prec Agric 6(1):7–23
49. Meynard JM, Cerf M, Guichard L, Jeuffroy MH, Makowski D (2002) Which decision support tools for the environmental management of nitrogen? Agronomie 22:817–829
50. Miller-Struttmann NE, Heise D, Schul J, Geib JC, Galen C (2017) Flight of the bumble bee: buzzes predict pollination services. PloS One 12(6):e0179273
51. Molin JP, Castro CND (2008) Establishing management zones using soil electrical conductivity and other soil properties by the fuzzy clustering technique. Scientia Agricola 65(6):567–573
52. Monaghan JM, Daccache A, Vickers LH, Hess TM, Weatherhead EK, Grove IG, Knox JW (2013) More 'crop per drop': constraints and opportunities for precision irrigation in European agriculture. J Sci Food Agric 93(5):977–980
53. Moral F, Terrón J, Da Silva JM (2010) Delineation of management zones using mobile measurements of soil apparent electrical conductivity and multivariate geostatistical techniques. Soil Tillage Res 106(2):335–343
54. Moran MS, Inoue Y, Barnes E (1997) Opportunities and limitations for image-based remote sensing in precision crop management. Remote Sens Environ 61(3):319–346
55. Mulla DJ (2013) Twenty five years of remote sensing in precision agriculture: key advances and remaining knowledge gaps. Biosyst Eng 114(4):358–371
56. Muñoz-Huerta RF, Guevara-Gonzalez RG, Contreras-Medina LM, Torres-Pacheco I, Prado-Olivarez J, Ocampo-Velazquez RV (2013) A review of methods for sensing the nitrogen status in plants: advantages, disadvantages and recent advances. Sensors 13(8):10823–10843

57. Ochsner TE, Cosh MH, Cuenca RH, Dorigo WA, Draper CS, Hagimoto Y, Kerr YH, Njoku EG, Small EE, Zreda M (2013) State of the art in large-scale soil moisture monitoring. Soil Sci Soc Am J 77(6):1888–1919
58. Panciera R, Walker JP, Jackson TJ, Gray DA, Tanase MA, Ryu D, Monerris A, Yardley H, Rudiger C, Wu X, et al (2014) The soil moisture active passive experiments (SMAPEx): toward soil moisture retrieval from the smap mission. IEEE Trans Geosci Remote Sens 52(1):490–507
59. Patzold S, Mertens FM, Bornemann L, Koleczek B, Franke J, Feilhauer H, Welp G (2008) Soil heterogeneity at the field scale: a challenge for precision crop protection. Prec Agric 9(6):367–390
60. Peralta NR, Costa JL, Balzarini M, Franco MC, Córdoba M, Bullock D (2015) Delineation of management zones to improve nitrogen management of wheat. Comput Electron Agric 110:103–113
61. Perlman UH (2020) The world's water. https://water.usgs.gov/edu/earthwherewater.html
62. Ping JL, Green CJ, Bronson KF, Zartman RE, Dobermann A (2005) Delineating potential management zones for cotton based on yields and soil properties. Soil Sci 170(5):371–385
63. Risinger M, Carver K (2020) Neutron moisture meters. http://sanangelo.tamu.edu/extension/agronomy/agronomy-publications/grain-sorghum-production-in-west-central-texas/how-to-estimate-soil-moisture-by-feel/soil-moisture-measuring-devices/neutron-moisture-meters/
64. Roger-Estrade J, Richard G, Dexter A, Boizard H, De Tourdonnet S, Bertrand M, Caneill J (2009) Integration of soil structure variations with time and space into models for crop management. a review. Agron Sustain Develop 29(1):135–142
65. Salam A (2019) A comparison of path loss variations in soil using planar and dipole antennas. In: 2019 IEEE International Symposium on Antennas and Propagation. IEEE, New York
66. Salam A (2019) Design of subsurface phased array antennas for digital agriculture applications. In: Proceedings of the 2019 IEEE international symposium on phased array systems and technology (IEEE Array 2019), Waltham, MA
67. Salam A (2019) A path loss model for through the soil wireless communications in digital agriculture. In: 2019 IEEE international symposium on antennas and propagation. IEEE, New York, pp 1–2
68. Salam A (2019) Sensor-free underground soil sensing. In: ASA, CSSA and SSSA international annual meetings (2019). ASA-CSSA-SSSA
69. Salam A (2019) Subsurface MIMO: a beamforming design in internet of underground things for digital agriculture applications. J Sens Actuat Netw 8(3). https://doi.org/10.3390/jsan8030041. https://www.mdpi.com/2224-2708/8/3/41
70. Salam A (2019) Underground environment aware MIMO design using transmit and receive beamforming in internet of underground things. Springer International Publishing, Cham, pp 1–15
71. Salam A (2019) An underground radio wave propagation prediction model for digital agriculture. Information 10(4). https://doi.org/10.3390/info10040147. http://www.mdpi.com/2078-2489/10/4/147
72. Salam A (2019) An underground radio wave propagation prediction model for digital agriculture. Information 10(4):147
73. Salam A (2019) Underground soil sensing using subsurface radio wave propagation. In: 5th global workshop on proximal soil sensing, Columbia, MO
74. Salam A (2020a) Internet of Things for environmental sustainability and climate change. Springer International Publishing, Cham, pp 33–69. https://doi.org/10.1007/978-3-030-35291-2_2
75. Salam A (2020b) Internet of Things for sustainability: perspectives in privacy, cybersecurity, and future trends. Springer International Publishing, Cham, pp 299–327. https://doi.org/10.1007/978-3-030-35291-2_10
76. Salam A (2020c) Internet of Things for sustainable community development, 1st edn. Springer, New York. https://doi.org/10.1007/978-3-030-35291-2

77. Salam A (2020d) Internet of Things for sustainable community development: introduction and overview. Springer International Publishing, Cham, pp 1–31. https://doi.org/10.1007/978-3-030-35291-2_1
78. Salam A (2020e) Internet of Things for sustainable forestry. Springer International Publishing, Cham, pp 147–181. https://doi.org/10.1007/978-3-030-35291-2_5
79. Salam A (2020f) Internet of Things for sustainable human health. Springer International Publishing, Cham, pp 217–242. https://doi.org/10.1007/978-3-030-35291-2_7
80. Salam A (2020g) Internet of Things for sustainable mining. Springer International Publishing, Cham, pp 243–271. https://doi.org/10.1007/978-3-030-35291-2_8
81. Salam A (2020h) Internet of Things for water sustainability. Springer International Publishing, Cham, pp 113–145. https://doi.org/10.1007/978-3-030-35291-2_4
82. Salam A (2020i) Internet of Things in agricultural innovation and security. Springer International Publishing, Cham, pp 71–112. https://doi.org/10.1007/978-3-030-35291-2_3
83. Salam A (2020j) Internet of Things in sustainable energy systems. Springer International Publishing, Cham, pp 183–216. https://doi.org/10.1007/978-3-030-35291-2_6
84. Salam A (2020k) Internet of Things in water management and treatment. Springer International Publishing, Cham, pp 273–298. https://doi.org/10.1007/978-3-030-35291-2_9
85. Salam A (2020l) Wireless underground communications in sewer and stormwater overflow monitoring: radio waves through soil and asphalt medium. Information 11(2):98
86. Salam A, Karabiyik U (2019) A cooperative overlay approach at the physical layer of cognitive radio for digital agriculture. In: Proceedings of the 3rd international Balkan conference on communications and networking (2019 BalkanCom), Skopje
87. Salam A, Shah S (2019) Internet of things in smart agriculture: enabling technologies. In: 2019 IEEE 5th world forum on Internet of Things (WF-IoT). IEEE, New York, pp 692–695
88. Salam A, Vuran MC (2016) Impacts of soil type and moisture on the capacity of multi-carrier modulation in internet of underground things. In: Proceedings of the 25th ICCCN 2016, Waikoloa, HI
89. Salam A, Vuran MC (2017a) Em-based wireless underground sensor networks, pp 247–285. https://doi.org/10.1016/B978-0-12-803139-1.00005-9
90. Salam A, Vuran MC (2017b) Smart underground antenna arrays: a soil moisture adaptive beamforming approach. In: Proceedings of the IEEE INFOCOM 2017, Atlanta
91. Salam A, Vuran MC (2017c) Wireless underground channel diversity reception with multiple antennas for internet of underground things. In: Proceedings of the IEEE ICC 2017, Paris
92. Salam A, Vuran MC, Irmak S (2016) Pulses in the sand: impulse response analysis of wireless underground channel. In: The 35th Annual IEEE International Conference on Computer Communications (INFOCOM 2016), San Francisco
93. Salam A, Vuran MC, Irmak S (2017) Towards internet of underground things in smart lighting: a statistical model of wireless underground channel. In: Proceedings of the 14th IEEE International Conference on Networking, Sensing and Control (IEEE ICNSC), Calabria
94. Salam A, Hoang AD, Meghna A, Martin DR, Guzman G, Yoon YH, Carlson J, Kramer J, Yansi K, Kelly M, et al (2019) The future of emerging IoT paradigms: architectures and technologies. https://doi.org/10.20944/preprints201912.0276.v1
95. Salam A, Vuran MC, Dong X, Argyropoulos C, Irmak S (2019) A theoretical model of underground dipole antennas for communications in internet of underground things. IEEE Trans Anten Propag 67(6):3996–4009
96. Salam A, Vuran MC, Irmak S (2019) Di-sense: in situ real-time permittivity estimation and soil moisture sensing using wireless underground communications. Comput Netw 151:31–41. https://doi.org/10.1016/j.comnet.2019.01.001. http://www.sciencedirect.com/science/article/pii/S1389128618303141
97. Sanislav T, Miclea L (2012) Cyber-physical systems-concept, challenges and research areas. J Control Eng Appl Inf 14(2):28–33
98. Schepers AR, Shanahan JF, Liebig MA, Schepers JS, Johnson SH, Luchiari A (2004) Appropriateness of management zones for characterizing spatial variability of soil properties and irrigated corn yields across years. Agron J 96(1):195–203

99. Schumann AW, et al (2006) Nutrient management zones for citrus based on variation in soil properties and tree performance. Prec Agric 7(1):45–63
100. Silva AR, Vuran MC (2010) $(CPS)^2$: integration of center pivot systems with wireless underground sensor networks for autonomous precision agriculture. In: Proc. of ACM/IEEE international conference on cyber-physical systems, Stockholm, pp 79–88. http://doi.acm.org/10.1145/1795194.1795206
101. Srbinovska M, Gavrovski C, Dimcev V, Krkoleva A, Borozan V (2015) Environmental parameters monitoring in precision agriculture using wireless sensor networks. J Cleaner Prod 88:297–307
102. Stojmenovic I (2014) Machine-to-machine communications with in-network data aggregation, processing, and actuation for large-scale cyber-physical systems. IEEE Internet of Things J 1(2):122–128
103. Stuntebeck EP, Pompili D, Melodia T (2006) Wireless underground sensor networks using commodity terrestrial motes. In: 2nd IEEE Workshop on Wireless mesh networks, 2006, WiMesh 2006. IEEE, New York, pp 112–114
104. Tagarakis A, Liakos V, Fountas S, Koundouras S, Gemtos T (2013) Management zones delineation using fuzzy clustering techniques in grapevines. Prec Agric 14(1):18–39
105. Tagarakis AC, Ketterings QM, Lyons S, Godwin G (2017) Proximal sensing to estimate yield of brown midrib forage sorghum. Agron J 109(1):107–114
106. Temel S, Vuran MC, Lunar MM, Zhao Z, Salam A, Faller RK, Stolle C (2018) Vehicle-to-barrier communication during real-world vehicle crash tests. Comput Commun 127:172–186
107. Timmermann C, Gerhards R, Kühbauch W (2003) The economic impact of site-specific weed control. Prec Agric 4(3):249–260
108. Tiusanen MJ (2013) Soil scouts: Description and performance of single hop wireless underground sensor nodes. Ad Hoc Netw 11(5):1610–1618. http://dx.doi.org/10.1016/j.adhoc.2013.02.002
109. Vasileiadis V, Moonen A, Sattin M, Otto S, Pons X, Kudsk P, Veres A, Dorner Z, Van der Weide R, Marraccini E, et al (2013) Sustainability of European maize-based cropping systems: economic, environmental and social assessment of current and proposed innovative IPM-based systems. Eur J Agron 48:1–11
110. Vellidis G, Tucker M, Perry C, Kvien C, Bednarz C (2008) A real-time wireless smart sensor array for scheduling irrigation. Comput Electron Agric 61(1):44–50
111. Vereecken H, Huisman J, Pachepsky Y, Montzka C, Van Der Kruk J, Bogena H, Weihermüller L, Herbst M, Martinez G, Vanderborght J (2014) On the spatio-temporal dynamics of soil moisture at the field scale. J Hydrol 516:76–96
112. Vuran MC, Akyildiz IF (2010) Channel model and analysis for wireless underground sensor networks in soil medium. Phys Commun 3(4):245–254
113. Vuran MC, Salam A, Wong R, Irmak S (2018a) Internet of underground things in precision agriculture: architecture and technology aspects. Ad Hoc Netw. https://doi.org/10.1016/j.adhoc.2018.07.017. http://www.sciencedirect.com/science/article/pii/S1570870518305067
114. Vuran MC, Salam A, Wong R, Irmak S (2018b) Internet of underground things in precision agriculture: architecture and technology aspects. Ad Hoc Netw 81:160–173
115. Vuran MC, Salam A, Wong R, Irmak S (2018c) Internet of underground things: sensing and communications on the field for precision agriculture. In: 2018 IEEE 4th world forum on Internet of Things (WF-IoT) (WF-IoT 2018), Singapore
116. Wang N, Zhang N, Wang M (2006) Wireless sensors in agriculture and food industry–recent development and future perspective. Comput Electron Agric 50(1):1–14
117. Wan J, Chen M, Xia F, Di L, Zhou K (2013) From machine-to-machine communications towards cyber-physical systems. Comput Sci Inf Syst 10(3):1105–1128
118. Western AW, Blöschl G (1999) On the spatial scaling of soil moisture. J Hydrol 217(3):203–224
119. Wong R (2017) Towards cloud-based center pivot irrigation automation based on in-situ soil information from wireless underground sensor networks. Master's thesis, University of Nebraska-Lincoln

120. Wu FJ, Kao YF, Tseng YC (2011) From wireless sensor networks towards cyber physical systems. Perv Mob Comput 7(4):397–413
121. Wu X, Wang Q, Liu M (2015) In-situ soil moisture sensing: measurement scheduling and estimation using sparse sampling. ACM Trans Sensor Networks 11(2):26
122. Zhang N, Wang M, Wang N (2002) Precision agriculture–a worldwide overview. Comput Electron Agric 36(2):113–132

Chapter 12
Autonomous Irrigation Management in Decision Agriculture

12.1 Introduction

The application of uniform irrigation leads to under- or over-watering, if there is no information available on soil and crop conditions [4]. The variable-rate systems can be utilized to efficiently use water. Although saving water is the biggest driving factor behind switching, but other advantages include: erosion prevention, reduced maintenance cost, avoiding fines for not using enough water, and limiting nutrient runoff [25]. Using the current irrigation systems, Chemigation, without involving any human, uniformly applies chemicals that too without danger of exposure. Smart irrigation system is accepted as beneficial technology; however, its adoption can be delayed because of cost involved in replacing or modifying the current equipment. A balanced approach would be to use cheaper valves in place of expensive pressure control nozzles [4, 46].

There have been a lot of investigations done towards the application of smart irrigation for delivering cheap decision-making solution in an effort to improve its adoption rate [25, 28]. A challenge could be the lack of interdisciplinary approach to fill the gap between machinery and agricultural requirements. A hardware-based system might be cumbersome and expensive to adopt for different crop and soil conditions, whereas a software-based solution can flexibly address this challenge.

12.2 Types of Sensor-Guided Irrigation Systems

Many different types of techniques are used for autonomous irrigation systems (e.g., drip, surface, and sprinkler). The percentage of land irrigated by sprinkler, gravity, and drip/trickle systems in the Great Plains is shown in Fig. 12.1. In this section, the different types of irrigation systems are discussed (Fig. 12.2).

A. Salam, U. Raza, *Signals in the Soil*,
https://doi.org/10.1007/978-3-030-50861-6_12

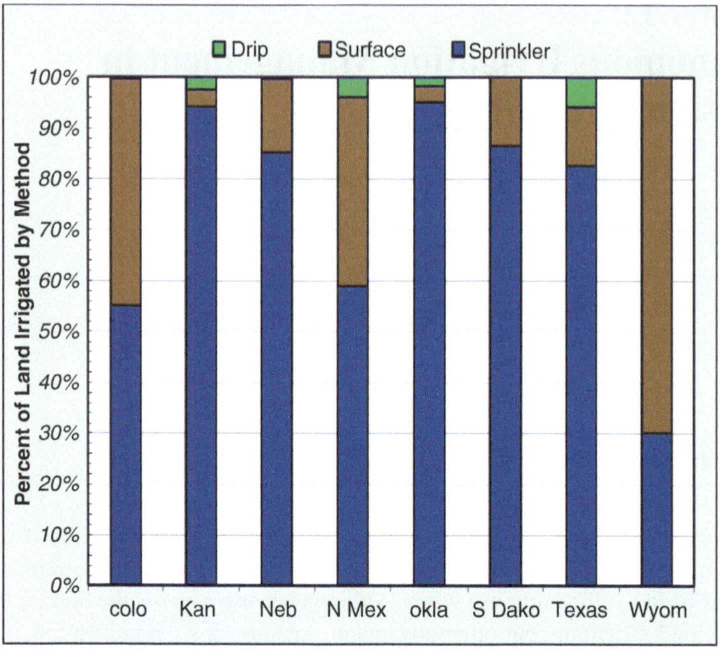

Fig. 12.1 Ratio of different sources (gravity, sprinkler, drip/trickle) used for land irrigation in great plains [15]

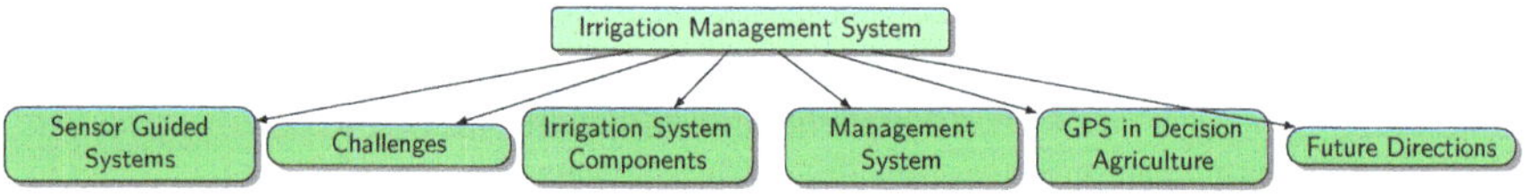

Fig. 12.2 Organization of the chapter

12.2.1 Center Pivot System

Infrastructure is required for field management practices such as center pivot irrigation system. This system needs more advanced communication capabilities to be able to communicate and control the center pivot using IOUT soil moisture measurements. A list of required features in a center pivot system for autonomous irrigation is given below [65]:

- The center pivot location/angle can be graphically viewed online.
- The pivot can be remotely controlled online for speed, irrigation application rate, etc.
- Operational information and data can be viewed and stored.
- Pivot point operating water pressure can be monitored.
- Operating direction of the pivot can be changed.

- AS/RS status can be monitored.
- End gun operations, including the water application rates, can be controlled.
- The system can be run "dry" or "wet."
- Text alerts can be sent if any operational or safety issues are encountered.

In addition to the remote control and operation hardware and software, the center pivot nozzle sprinkler nozzles should support variable water application capabilities as well as support to enhance the uniformity of the water application. The variable rate center pivot and all of the hardware and software components as well as all agronomic and soil management practices lead to an efficient design of autonomous irrigation system [36, 44].

12.2.2 Drip Irrigation System

Drip irrigation system (DIS) is a water system which uses low pressure to water small lawns and gardens using different methods. It differs from other irrigation methods in that it uses less amount of water to keep the roots moist instead of soaking it. The drip system can be hidden under the mulch with water spraying part above the mulch or kept above the soil or mulch with plants hiding it as their size increases with the growth. As discussed above it gives an advantage of using less water as compared to the other counterpart options such as UG sprinkler systems and lawn sprinklers. Moreover, customized water can be applied using drip irrigation system by restricting watering in some areas and allowing in others.

Following are some advantages of drip irrigation system:

1. Conserves water by preventing over-spraying, hence, reducing evaporation and environmental effect.
2. It can directly connect to the hose spigot with having to do cuts in home water supplies.
3. It can be placed underground eliminating the need of trenches.
4. It provides customization for water spray control and container in which it can be placed.
5. It prevents over moistening which, otherwise, can cause fungal diseases.

Components of Drip Irrigation System

While creating a DIS yourself, make sure to buy all components from the same manufacturer. *Soaker hoses* is a simplest DIS with small holes allowing water to come out. However, this cost-efficient solution uses more water than a normal DIS. Another alternative to create a DIS is using an entire system *kit* to create a customized DIS. KIT integrates the components to create DIS for specific applications, e.g., for flowerbeds, vegetable gardens, landscape plants (shrubs and trees), and container plants. These kits can also be extended as per the need of the user. Various components of DIS are given below:

- *Backflow preventers*—Also known as anti-siphon devices, do not allow the DIS water to go back to water supply so that drinking water is not contaminated while system is switched off.
- *Pressure regulator*—Balances the pressure between the home water supply and DIS; otherwise, the water pressure of home supply is much greater than the water pressure at DIS.
- *Filters*—are used to prevent clogging in the tube caused by the debris.
- *Fittings*—are used to connect all the system components together.
- *Stakes*—are used to secure all the system components together.
- *Risers*—are used to bring the water emitters to the top, i.e., at the level of plants.
- *Timers*—are used to prevent overwatering by setting the time to automatically switch DIS on or off. It can also be controlled via smartphone or computers by connecting it with the home automation system.
- *Hole punches*— are used to create the holes in the tube which is connecting emitters. *Plugs* can be used to stop holes punched by mistake.
  Emitters—Emitters are connected with the tubing system and it emits water through tubes. Its flow rate is given by gallons-per-hour (GPH) and it depends on the type of soil and plant being watered. Every emitter has maximum water pressure mentioned in pounds per square inch (PSI).

12.3 Development Challenges

The challenges which were addressed while developing the system are [65]:

- **Deployment of a field network capable of measuring the soil moisture levels, temperature, and radio communication metrics.** The desired network must be low-maintenance, be durable in all-weather condition, and not affect other farm activities.
- **Characterization of the underground/aboveground wireless channel to determine communication reliability.** Communication and channel characteristics are quantified by using different metrics, e.g., coverage distance and received signal strength indicator [31, 60].
- **Back-end software development for data management and visualization.** WUSN networks produce glut of important data. All this data must be dynamically received and stored on a cloud application online (via GPRS/3G/4G link) and offline mode (locally). A cloud application can be used for monitoring, e.g., sensor nodes communication, and visualizing, e.g., real-time the power status, purposes. Furthermore, visualization can also be used to locate disconnected

sensor nodes, for visualizing historical readings such as soil moisture readings, radio packets location, last reported pivot angle, timestamp, and signal strength of packet received from each underground node [12, 63].

- **Usage of test software to interact with commercial irrigation solutions.** Some studies are done on irrigation system controllers; however, most commercial deployments use industrial solutions. These solutions come with a proprietary cloud-based control system with no facility of customized programs development using APIs. The control systems are programmed using a test software [26, 50, 52].

12.4 Irrigation System Components

An advanced CP irrigation system can be used to study the long-term effects of water stress and yield relationships, variable rate irrigation and fertigation, crop water and nutrient uptake, develop crop production functions, and other related issues. These studies can be done under full and limited rainfed and irrigation settings [10, 11]. Figure 12.3 shows the overview of irrigation control system [20]. The GPS unit at the end tower in the center pivot can be used to control various irrigation and other operational aspects in addition to these state-of-the-art features in control units of the center pivot. The characteristics of a typical center pivot, where 45% of land is under the outer two spans while 2% is under the first span are shown in Fig. 12.4. The components of irrigation system are described in the upcoming section:

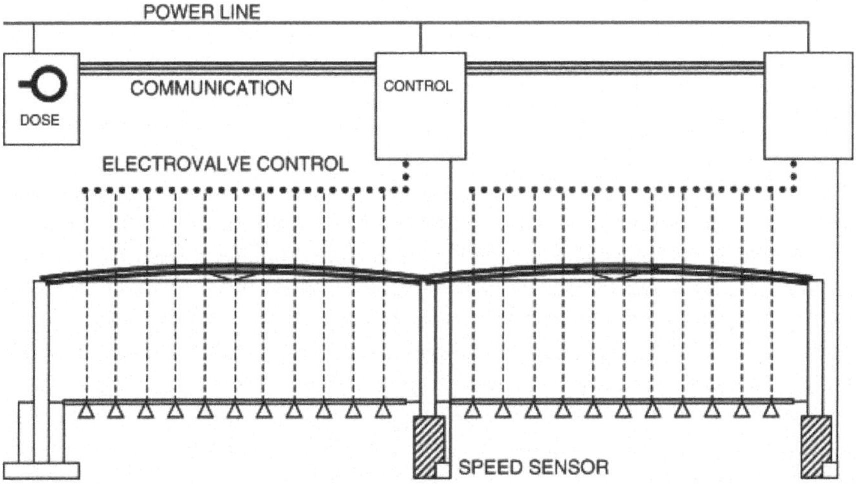

Fig. 12.3 The control system diagram

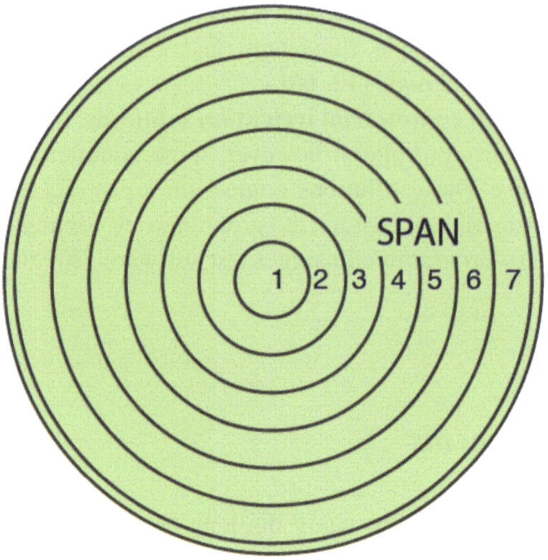

Span	End of Span, ft	Acres Within Span	Discharge from Span, gpm
1	180	2.3	14
2	360	7.0	42
3	540	11.7	71
4	720	16.4	99
5	900	21.0	127
6	1080	25.7	156
7	1260	30.4	184
OH	1310	9.3	56
Total		124	750

Fig. 12.4 A typical center pivot [15]

12.4.1 Sensing and Communication Nodes

Sensors are used to get the real information from the field. This information can be taken using in-situ sensors, i.e., inside the field, or using remote sensors. The real information of the field can be lost, if inaccurate equipment are used. Most sensing systems are used for sensing soil parameters; however, spectral imaging can also be used to detect plant stress [17, 54]. The sensed data, if not sent immediately to the cloud, can be stored locally for later retrieval. In the absence of direct link with the controller, an optional gateway is used.

12.4.1.1 Subsurface Nodes

The subsurface node is an embedded cyber-physical hardware for sensing and communication between different low-powered low-frequencies sensors and radios. The subsurface nodes use underground antennas for communications [29, 62]. Soil sensors along with dataloggers are also connected to these modes. The information from datalogger act as the basis for calibration of soil sensors. The subsurface nodes can be buried at different depth in soil and should also support connections with soil sensors at different depths generally up to 4 feet [32, 46]. Accordingly, these observations can be used for better deployment of advanced irrigation systems. The subsurface node deployment should be done considering the impact of crop growth and density should be taken into account in the field terrain. Some of the nodes should be used in the field as a reference point for measurement calibrations and for comparison of sensor readings.

For irrigation applications, different types of subsurface nodes can be selected (e.g., off-the-shelf and customized) to speed up the development process and achieve communication between devices, respectively. These nodes and platform should include efficient and high-speed processors, I/O ports, on-board flash storage, and serial ports. These are some of the important parameters to be considered while subsurface node development because they are used to connect with the microcontrollers and GPS modules which communicate with underground nodes [27, 34, 38, 43].

To provide soil moisture calibration data to the IOUT, the soil moisture sensors, temperature sensors, and electrical conductivity sensors can be installed in the center location of different grids (see Fig. 12.4) that can be created in the center pivot field in spatial and temporal domains that measure soil moisture, soil temperature, and electrical conductivity on an hourly basis throughout the growing year. External batteries are the power source of subsurface node that can be complemented by intelligent uninterrupted power supply (UPS) [19] to control the power management and also to report the status of power. A step-down voltage converter can be utilized to meet the 5 V power requirement of nodes [31, 54]. Furthermore, during daytime, power can be provided by solar panels that can also recharge lithium-ion polymer batteries for night usage when solar energy is not available [65].

In addition to system upgrades, a prototype enclosure (Fig. 12.5) is required to make the nodes fully underground, where connectors can be used for software updates along with battery connections for easy aboveground access. The enclosure can host these connectors underground, with easy access through a structure close to the surface. Accordingly, this new enclosure should be tested during the harvesting period for accuracy and durability.

Due to high signal attenuation in the soil, the low operation frequency and low-power transceivers are preferred in underground communication such as WizziMote developed by WizziLab [64]. WizziMote combines a Texas Instruments CC430F513 16-bit ultra-low-power microcontroller unit with four analog to digital input/output

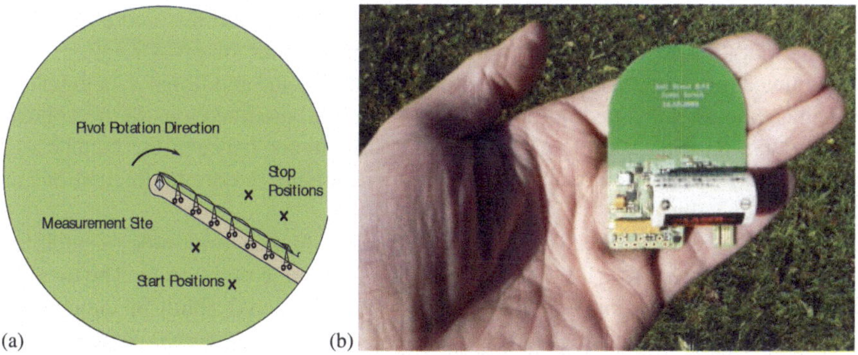

(a) (b)

Fig. 12.5 (**a**) Underground node placement layout [15]. (**b**) A prototype of an underground node, [61] pins, a CC1101 low-power sub-1 GHz RF transceiver, eight configurable GPIOs, and female SMA antenna connector. The microcontroller unit comes with a flash memory of 32 KB and a 12-bit analog to digital converter (ADC). It consumes, at its maximum capacity, 3.2 mA; however, it has various power saving feature which, if enabled, can tailor down the consumption to 1.0 micro Ampere. WizzMote works with 433 MHz applications and reports the transceiver sensitivity of −116 dBm at 0.6 kBaud with a 1% packet error rate, at this frequency [28, 37, 65].

The protection of subsurface nodes is an important issue. The waterproof subsurface nodes are very important for efficient underground operation [23, 42] and can be achieved by using waterproof enclosures. These enclosures come with an automatic equalization valve that helps to balance the internal pressure and avoid water condensation [65]. It should also provide ease of access through a soil structure close to the surface. This enclosure should also be able to handle the weight of the farm machinery that is used in harvesting season for collecting crop [33, 35]. The electronic equipment can be protected by the PVC enclosures but it has the risks of being damaged by the leaks in connectors.

12.4.1.2 OTA Nodes

The OTA nodes are designed to study and validate the impact of irrigation system in the real-time. Therefore, data is collected using opposite-facing directional antennas before and after the completion of irrigation. The improved communications in CP help in studying changes in soil moisture. Similarly, customized solution is developed by updating on-board models for irrigation systems, in real-time, to reach soil moisture level [41, 45]. A modem [21] is used to send data from OTA node to server using a wireless mobile communication network. A USB dongle can operate at different cellular bands depending upon the availability of services from telecom service providers. A voltage datalogger can be used to register the battery voltage over time in order to record data for battery replacement needs. The duty cycling in underground nodes is important to conserve energy. This can be achieved by disabling ports. Moreover, the power can be disconnected and restored after some optimal duration to duty cycle the device in the absence of pivot rotation. The communication between devices can allow transmitting the additional data together, which can be implemented in the control. In addition, modifications to the voltage regulation circuit for sampling allow placing a regulator between the battery and embedded system, which can solve the problems arising due to the fluctuations in the reference voltage.

Many GPS modules are available with the ability to track multitude of satellites on different channels [2]. These are used for localization in this system. The module selected should be equipped with high sensitivity receiver, built-in datalogger, radiofrequency connector for an external active antenna, and internal patch antenna for efficient power performance. Moreover, the high performance antenna with thread mount and waterproof capabilities is desirable [9].

The aboveground nodes are also used in the system for retrieving sensing data, monitoring the power consumption and battery capacity. These nodes enable the power management thread to gather power consumption information from the battery charging board. This component also enhances the system performance through design of a watchdog system and provides better control of the power system. Moreover, the antennas can be deployed with a better polarization angle in the deployments to enhance connectivity [49, 55]. Figure 12.6 shows the aboveground components. The multi-element directional antennas [58] can be employed for OTA

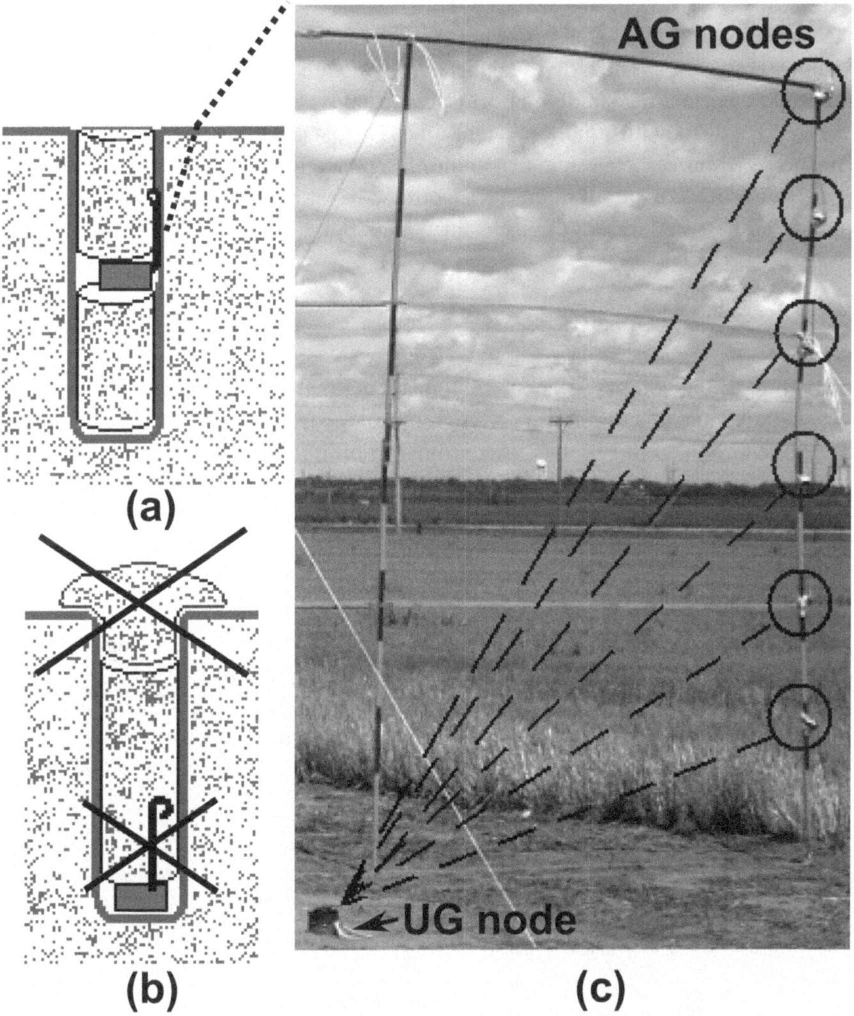

Fig. 12.6 Experiments for AG2UG and UG2AG communication: (**a**) Antenna placed in non-obstructed direction of AG device, (**b**) UG2UG experimental aspects which cannot be used for AG experiments, and (**c**) AG nodes grid [57]

nodes with support for a wide range of operational frequency in MHz and GHz bands. The antennas should support high power and gain with a front to back ratio greater than 20 dB. Generally, lightweight antennas are preferable for these applications.

12.4.2 Software, Flow Rate, and System Capacity

12.4.2.1 Software System

The software are needed in terms of both functionality and to control state-of-the-art system architectures. The individual pieces of software that needed to be maintained separately can run on a single platform. Accordingly, a visualization system can be developed to display information in real-time. The database can be hosted in a server with more resources ensuring that multiple connection requests can be handled properly. As part of the system, an adaptive control system can be developed to support wireless underground sensor-aided irrigation control.

12.4.2.2 System Capacity

In addition to the software system components, a real-time capacity and flow rate system based on this framework can be developed. A definition of the system capacity is provided in Fig. 12.7. Accordingly, soil moisture, location of the center pivot, and wireless communication quality information can be displayed in real-time

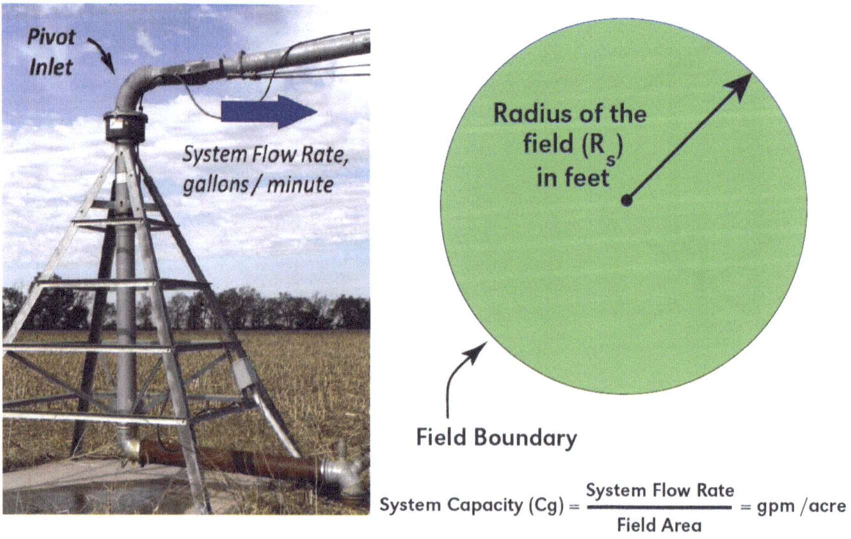

Fig. 12.7 Definition of system capacity for a field [15]

as data is received. A group of underground nodes can provide individual sensor status that is very useful to detect shorted or disconnected sensors. A gauge can be utilized to display the OTA battery voltage. Similarly, a map of the underground nodes can show the location of the irrigation pivot and the trail of the recent radio packages received by each subsurface node.

12.4.3 Data Collection

The data collection from the field is important in all seasons particularly in the growing season and winter season. Its significance becomes even higher during farming operations that include cultivation, spraying, planting, determining the amount of irrigation and fertilizer applications, irrigation and fertigation management, harvesting, herbicide, insecticide, pesticide, and fungicide applications, and other soil and plant management practices [30, 42, 51, 60]. This section discusses the message exchange among UG nodes, the AG nodes, and the server.

To enable an out-of-the-box operation, WizziLab provides a Software Development Kit (SDK). This SDK comes with a sample code and environmental installation procedures. A quick 433 MHz solution can be developed using the wireless communication module of SDK; however, no modification can be made to it as it comes as a library. This limits the development of advanced programs for WizziMote because of a bug which, upon hardware interrupt, disables the radio [39, 48]. SDK handles the radio operation and development is done by application, which is the payload of the packets transmitted. The system mainly can use different packets for communications. The AG node sends a packet to request soil moisture from UG nodes. For soil moisture data collection, different sampling periods of soil moisture can be used. UG nodes store these readings with timestamp in device's flash memory. Accordingly, the AG node uses a cellular link to connect and query the status irrigation system from the cloud [40].

12.5 Management System

Table 12.1 shows the monitoring, control, communication, and data reporting capabilities of center pivot control panels for irrigation and water management research infrastructures. The loud-based and mobile applications are used in many commercial irrigation system for control and monitoring of the system [59]. The programming of irrigation system is done by using a very simple panel having couple of buttons. These buttons are enough to implement all options such as increasing/decreasing of value and accept/cancel an action. The cloud-based applications add more flexibility for the manipulation of irrigation system, e.g., scheduling events to occur at certain time. One such event may include start/stop the CP at certain time to synchronize the real-time clock. The cloud system provides interactive graphical interface with all command options appearing on single screen.

Table 12.1 Different panels of center pivot control [13]

	Reinke	T-L	Valmont	Zimmatic
Monitors				
Position in field and travel direction	Y	Y	Y	Y
Speed of travel	Y	Y	Y	Y
Wet or dry operation	Y	Y	Y	Y
Pipeline pressure	Y	Y	Y	Y
Pump status	Y	Y	Y	Y
Auxiliary components[a]	Y (7)	Y (2)	Y (6)	Y (3)
Stop-in-slot and auto restart	Y	Y	Y	Y
Wind speed	Y	N	Y	Y
Controls				
Start and stop	Y	Y	Y	Y
Speed of travel	Y	Y	Y	Y
Auto restart and auto reverse	Y	Y	Y	Y
End gun	Y	Y	Y	Y
High and low pressure shutdown	Y	Y	Y	Y
High and low voltage shutdown	N/Y	N/Y	Y/Y	N/Y
System stall shutdown	Y	Y	Y	Y
Auxiliary components	Y(7)	Y(2)	Y(6)	Y(3)
System guidance	Y	Y	Y	Y
Maximum control points per circle	3600	180	72	180
Sprinkler application zones	2	3	30	NL
Remote communications				
Cell phone	Y	Y	Y	Y
Radio	Y	Y	Y	Y
Computer	Y	Y	Y	Y
Subscription required	Y	Y	Y	Y
Data collection and reports				
Soil water content	Y	Y	Y	N
Precipitation per season	Y	Y	Y	Y
Application date and depth	Y	Y	Y	Y
Irrigation events per season	Y	Y	Y	N
Chemical application rate	N	N	N	Y
Chemical application per season	N	N	N	Y
System position by date	Y	Y	Y	Y

[a] Y(x) means that panel can control up to x number of auxiliary components

Moreover, the IOUT can collect the soil information from the system using a mobile sink. The sink can be attached on the top of the controller tower where two opposite-facing antennas for receiving data from the nodes at different distances can be employed using solar panel as an energy source. Accordingly, the subsurface devices installed in the field can measure soil moisture and temperature from different types of soil sensors buried at different depths. The nodes can employ batteries as a power source. The sink nodes then collect the spatio-temporal information from the subsurface nodes and send it over to cloud using OTA communication which further interacts with the CP controller to automate irrigation control [26, 47, 49, 55]. A data analysis flow chart is shown in Fig. 12.8.

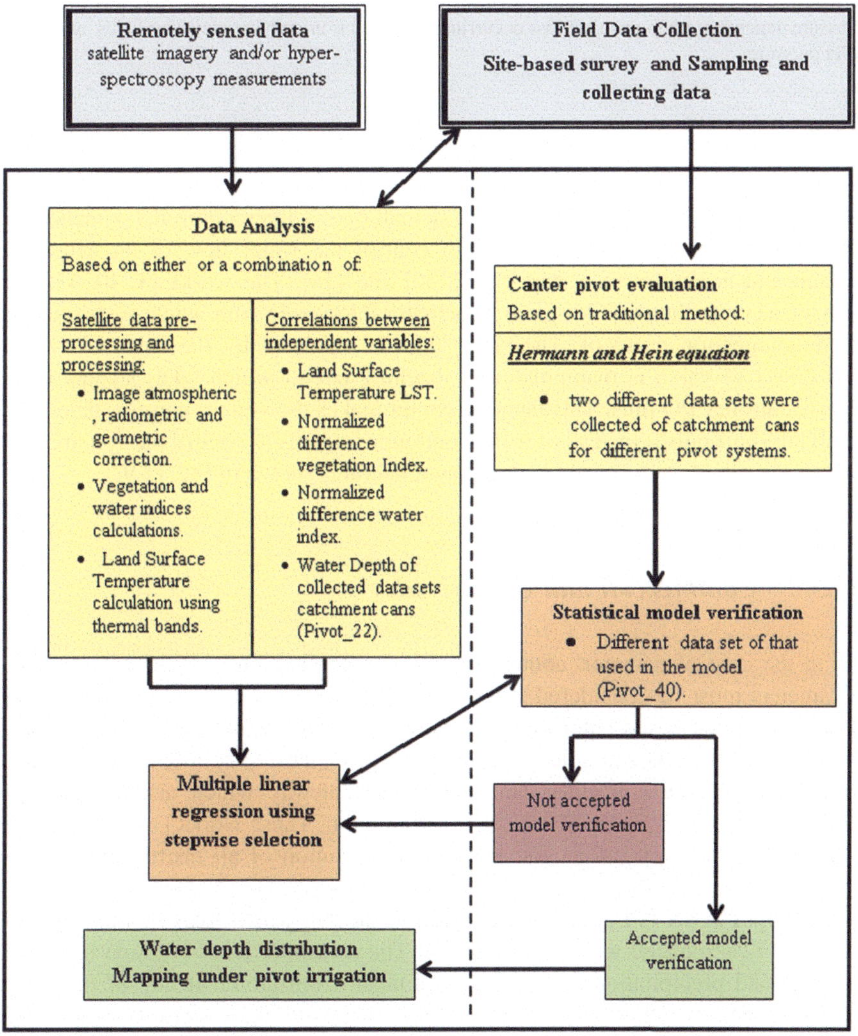

Fig. 12.8 Data analysis flow chart [6]

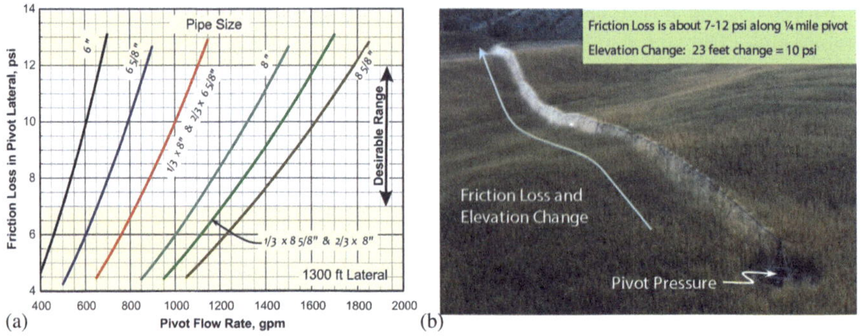

Fig. 12.9 (**a**) Pressure loss because of pipe friction due to flow of water [15]. (**b**) Sprinkler pressure dependent upon pressure loss occurring due to friction, varying elevation fields, and pivot inlet pressure

12.5.1 Control and Actuation Nodes

Studies such as [3, 7, 14, 18, 30] use actuator as a microcontroller attached to an on/off valve. This valve is used to control the water flow of an irrigation system and is similar to a drip [5, 22, 56] and [14]. The works [3, 40, 43, 53] provide a comprehensive design of an actuator interacting with whole system using a communication protocol. The water flow is regulated by the solenoid valves and power to valve is manipulated by latching circuit which takes signals from microcontroller I/O pins. This latching circuit makes the system power efficient by applying short pulses instead of using continuous signals to control the system. The pressure available to a sprinkler along pivot lateral is shown in Fig. 12.9.

12.5.2 Commercial and Cloud Nodes

As in the case of actuator, controllers are also application specific and multiple parameters must be considered while designing controllers. There can be in-situ, remote, or cloud-based controller; standalone or shared with other task; hardware could be a cost-efficient microcontroller or an expensive full-fledged computer system. An in-situ controller can also act as gateway which makes irrigation decision after receiving and processing information. These works [14, 35] suggest a use of real-time operating systems so that execution of all instructions can be ensured.

In [56], a WSN is implemented for application of peach trees monitoring and achieves 72% reduced water consumption. The controller takes decision based on weather and precipitation probability data taken from cloud. If any of the three deployed sensors values goes below the lower threshold, the system starts watering the tree. Due to deep deployment, the upper threshold overshot which, otherwise, would have added significant value to savings. In [14], the authors have used hydrological model to determine exact amount of water to apply. The value for average correlation coefficient of the model estimation and field measurements is 0.9331. For both studies, a valve was used as an actuator to control the water flow from irrigation system. Gutiérrez et al. [8] and Salam [41] recommend to consult farmers for marking threshold values for the irrigation system. In addition to soil moisture value, they also measured soil temperature and started the irrigation every time the threshold of soil moisture and temperature was crossed.

There are no commercial controllers with specific features. This fact motivated [7, 38] towards designing a Wi-Fi re-programmable precision irrigation control system. It is powered by solar radiations and uses the threshold approach as given in [36, 56]. The commercial irrigation system mostly comes with a proprietary control and modification can end up in losing warranty or service term; however, they come with a cloud-based or graphical interface controls and can be extended by writing macro-like scripts for inclusion of data from other sources. Major applications of microcontroller agriculture are to sense crop and soil properties. It reduces the role of humans by remotely collecting the data using wireless transceivers. The system can be customized to one's need by using different communication technologies such as Bluetooth, ZigBee, and Wi-Fi and too without losing long range and power-efficiency. However, it would be difficult to combine solutions from different vendors because there is no standard protocol [1, 39].

The difference between the prototype and commercial irrigation control system is that the former can be customized for water flow control [22, 24, 36]. There is very limited space of customizing control in commercial systems and due to warranty restrictions internal circuitry cannot be manipulated. The commercial systems come with the support of cloud-based interface and/or mobile applications for sending operational commands and visualizing status of the field. Figure 12.10 shows the system architecture.

The cloud-based controllers have their own limitations which must be addressed in order to implement a reliable system. A reliable network is the key to reliability of the system because it provides access to Internet. Loading time of the web-page depends upon the load on the network. A re-designed website may warrant rendering of replay commands and redesigning of control interface. It is also important to mention here that using web-page also brings its own challenges, e.g., time out between the commands can add extra time in executing an action. A modular approach can be followed while designing the adaptive control system to effectively address these limitations with major changes to the system.

12.6 Global Positioning System in Decision Agriculture

The decision irrigation relies on accurate location information to control the irrigation system. However relying solely on GPS information can lead to significant errors in precision irrigation applications. While higher-end GPS devices may

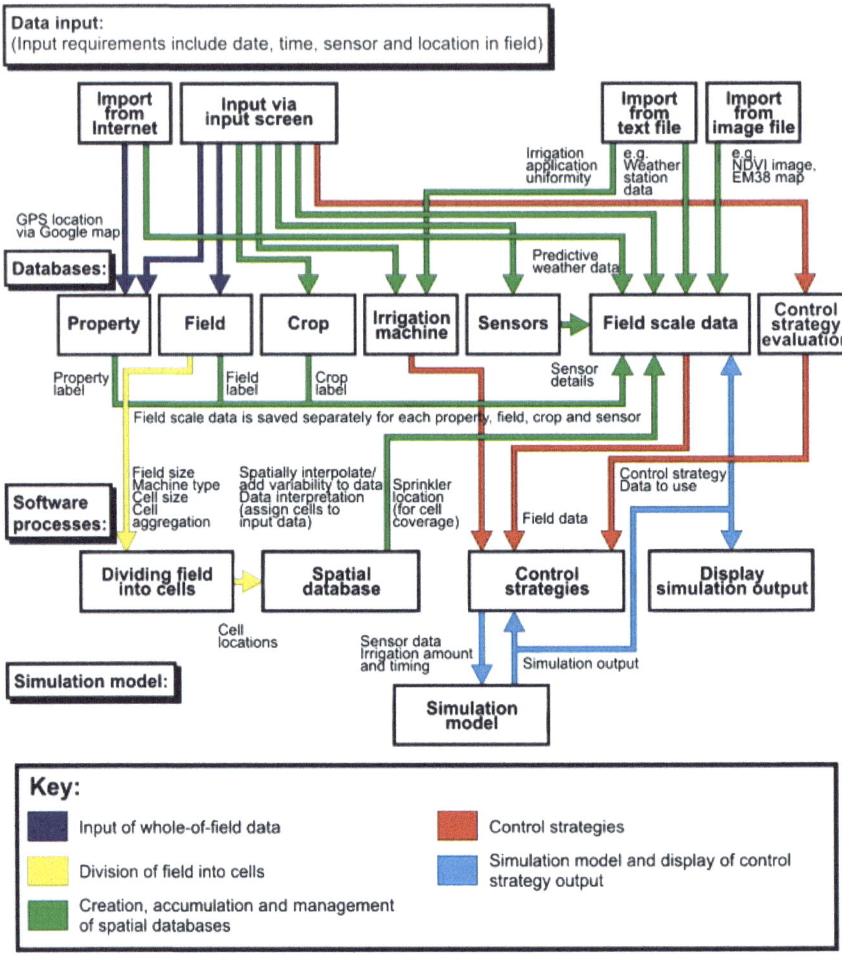

Fig. 12.10 The variable rate irrigation [16]

decrease the errors to some extent, GPS errors should be considered in system design due to the wireless nature of the GPS system and cost considerations. Due to the inherent errors and delays in pivot angle and GPS measurements, a multi-modal design with other sources for system localization should be developed to solve this problem. Alternatives include using image detection to determine the field location.

12.7 Future Research Directions

The freshwater is important resource and it should be preserved in huge amount for a center pivot irrigation system to automatically irrigate based on soil conditions. However, decision from such system is likely to have an adverse effect on crop yield which requires a long-term commitment in order to study in detail. The controller advances should be used to receive information from UG node, buried in the field, and also to accordingly set the irrigation rate. However, there is still need of improvements.

Accordingly, such systems should be capable of self-calibrating. To that end, remote soil sensing technologies should be used for the sensors adjustment close to the surface. GPS systems are very important part of control system and GPS interferometric reflectometry is a perfectly viable method. The center pivot movements can be used to calibrate the UG nodes in the vicinity of irrigation system path on the go. The current systems are incapable of employing more sensors and adding software components that reduces the efficiency of embedded system in terms of low-power consumption. Therefore, there are potential areas in which underground nodes can be improved.

References

1. Abbasi AZ, Islam N, Shaikh ZA, et al (2014) A review of wireless sensors and networks' applications in agriculture. Comput Stand Inter 36(2):263–270
2. Adafruit Ultimate GPS (2020) Adafruit ultimate GPS breakout - 66 channel w/10 Hz updates. https://www.adafruit.com/product/746
3. Coates RW, Delwiche MJ, Broad A, Holler M (2013) Wireless sensor network with irrigation valve control. Comput Electron Agric 96:13–22
4. Daccache A, Knox JW, Weatherhead E, Daneshkhah A, Hess T (2015) Implementing precision irrigation in a humid climate–recent experiences and on-going challenges. Agric Water Manag 147:135–143
5. Dukes M, Simonne E, Davis W, Studstill D, Hochmuth R (2003) Length of irrigation and soil humidity as basis for delivering fumigants through drip lines in florida spodosols. In: Proceedings of the florida state horticultural society, vol 116, pp 85–87
6. Farg E, Arafat S, El-Wahed MA, El-Gindy A (2017) Evaluation of water distribution under pivot irrigation systems using remote sensing imagery in eastern Nile delta. Egypt J Remote Sens Space Sci 20:S13–S19. Water scarcity and harvesting. https://doi.org/10.1016/j.ejrs.2016.12.001. http://www.sciencedirect.com/science/article/pii/S1110982316301387
7. Fue K, Schueller J, Schumann A, Tumbo S (2014) A solar-powered, Wi-Fi Re-programmable precision irrigation controller. In: 2014 Pan African conference on science, computing and telecommunications (PACT). IEEE, Piscataway, pp 171–176
8. Gutiérrez J, Villa-Medina JF, Nieto-Garibay A, Porta-Gándara MÁ (2014) Automated irrigation system using a wireless sensor network and GPRS module. IEEE Trans Instrum Meas 63(1):166–176. https://doi.org/10.1109/TIM.2013.2276487
9. Hercules GPS Antenna (2020) Hercules A.40 GPS/GLONASS/GALILEO permanent mount, 3M RG-174 - taoglas. http://www.taoglas.com/product/hercules-a40-permanent-mount-gps-glonass/

10. Irmak S (2014) Interannual variation in long-term center pivot–irrigated maize evapotranspiration and various water productivity response indices. I: Grain yield, actual and basal evapotranspiration, irrigation-yield production functions, evapotranspiration-yield production functions, and yield response factors. J Irrig Drain Eng 141(5):04014068
11. Irmak S (2014) Interannual variation in long-term center pivot–irrigated maize evapotranspiration and various water productivity response indices. II: Irrigation water use efficiency, crop WUE, evapotranspiration WUE, irrigation-evapotranspiration use efficiency, and precipitation use efficiency. J Irrig Drain Eng 141(5):04014069
12. Konda A, Rau A, Stoller MA, Taylor JM, Salam A, Pribil GA, Argyropoulos C, Morin SA (2018) Soft microreactors for the deposition of conductive metallic traces on planar, embossed, and curved surfaces. Adv Funct Mat 28(40):1803020. https://doi.org/10.1002/adfm.201803020
13. Kukal MS, Irmak S, Sharma K, et al (2019) Development and application of a performance and operational feasibility guide to facilitate adoption of soil moisture sensors. Sustainability 12(1):1–19
14. Lozoya C, Mendoza C, Aguilar A, Román A, Castelló R (2016) Sensor-based model driven control strategy for precision irrigation. J Sens 2016:1–12
15. Martin D, Kranz W, Smith T, Irmak S, Burr C, Yoder R (2017) Center pivot irrigation handbook (ec3017). Nebraska Extension Publications, Lincoln
16. McCarthy AC, Hancock NH, Raine SR (2010) VARIwise: a general-purpose adaptive control simulation framework for spatially and temporally varied irrigation at sub-field scale. Comput Electron Agric 70(1):117–128.https://doi.org/10.1016/j.compag.2009.09.011. http://www.sciencedirect.com/science/article/pii/S0168169909002026
17. Muñoz-Huerta RF, Guevara-Gonzalez RG, Contreras-Medina LM, Torres-Pacheco I, Prado-Olivarez J, Ocampo-Velazquez RV (2013) A review of methods for sensing the nitrogen status in plants: advantages, disadvantages and recent advances. Sensors 13(8):10823–10843
18. Nemali KS, van Iersel MW (2006) An automated system for controlling drought stress and irrigation in potted plants. Sci Horticult 110(3):292–297
19. OpenUPS (2020) Openups. http://www.mini-box.com/OpenUPS
20. Palacín J, Salse J, Clua X, Arnó J, Blanco R, Zanuy C (2005) Center-pivot automatization for agrochemical use. Comput Electron Agricul 49(3):419–430. Modelling and control in agricultural processes. https://doi.org/10.1016/j.compag.2005.08.006. http://www.sciencedirect.com/science/article/pii/S0168169905001353
21. Pantech UML290 (2020) Pantech 4G LTE global usb modem UML290. https://www.verizonwireless.com/support/4g-lte-global-usb-modem-uml290/
22. Parameswaran G, Sivaprasath K (2016) Arduino based smart drip irrigation system using internet of things. Int J Eng Sci 6:5518
23. Pelican (2020) Pelican - consumers. http://www.pelican.com/us/en/consumer/
24. Rasin Z, Hamzah H, Aras MSM (2009) Application and evaluation of high power zigbee based wireless sensor network in water irrigation control monitoring system. In: ISIEA 2009. IEEE symposium on industrial electronics & applications, 2009, vol 2. IEEE, Piscataway, pp 548–551
25. Sadler E, Evans R, Stone K, Camp C (2005) Opportunities for conservation with precision irrigation. J Soil Water Conserv 60(6):371–378
26. Salam A (2018) Pulses in the sand: long range and high data rate communication techniques for next generation wireless underground networks. ETD collection for University of Nebraska - Lincoln (AAI10826112). http://digitalcommons.unl.edu/dissertations/AAI10826112
27. Salam A (2019) A comparison of path loss variations in soil using planar and dipole antennas. In: 2019 IEEE international symposium on antennas and propagation. IEEE, Piscataway
28. Salam A (2019) Design of subsurface phased array antennas for digital agriculture applications. In: Proceeding of the 2019 IEEE international symposium on phased array systems and technology (IEEE Array 2019), Waltham
29. Salam A (2019) A path loss model for through the soil wireless communications in digital agriculture. In: 2019 IEEE international symposium on antennas and propagation. IEEE, Piscataway, pp 1–2

30. Salam A (2019) Sensor-free underground soil sensing. In: ASA, CSSA and SSSA international annual meetings (2019), ASA-CSSA-SSSA
31. Salam A (2019) Subsurface MIMO: a beamforming design in internet of underground things for digital agriculture applications. J Sens Actuator Netw 8(3):41. https://doi.org/10.3390/jsan8030041. https://www.mdpi.com/2224-2708/8/3/41
32. Salam A (2019) Underground environment aware MIMO design using transmit and receive beamforming in internet of underground things. Springer International Publishing, Cham, pp 1–15
33. Salam A (2019) An underground radio wave propagation prediction model for digital agriculture. Information 10(4), 147. https://doi.org/10.3390/info10040147. http://www.mdpi.com/2078-2489/10/4/147
34. Salam A (2019) Underground soil sensing using subsurface radio wave propagation. In: 5th global workshop on proximal soil sensing, Columbia
35. Salam A (2020) Internet of things for environmental sustainability and climate change. Springer International Publishing, Cham, pp 33–69. https://doi.org/10.1007/978-3-030-35291-2_2. https://doi.org/10.1007/978-3-030-35291-2_2
36. Salam A (2020) Internet of things for sustainability: perspectives in privacy, cybersecurity, and future trends. Springer International Publishing, Cham, pp 299–327. https://doi.org/10.1007/978-3-030-35291-2_10. https://doi.org/10.1007/978-3-030-35291-2_10
37. Salam A (2020) Internet of things for sustainable community development, 1st edn. Springer Nature, Berlin. https://doi.org/10.1007/978-3-030-35291-2
38. Salam A (2020) Internet of things for sustainable forestry. Springer International Publishing, Cham, pp 147–181. https://doi.org/10.1007/978-3-030-35291-2_5. https://doi.org/10.1007/978-3-030-35291-2_5
39. Salam A (2020) Internet of things for sustainable human health. Springer International Publishing, Cham, pp 217–242. https://doi.org/10.1007/978-3-030-35291-2_7. https://doi.org/10.1007/978-3-030-35291-2_7
40. Salam A (2020) Internet of things for sustainable mining. Springer International Publishing, Cham, pp 243–271. https://doi.org/10.1007/978-3-030-35291-2_8. https://doi.org/10.1007/978-3-030-35291-2_8
41. Salam A (2020) Internet of things for water sustainability. Springer International Publishing, Cham, pp 113–145. https://doi.org/10.1007/978-3-030-35291-2_4. https://doi.org/10.1007/978-3-030-35291-2_4
42. Salam A (2020) Internet of things in agricultural innovation and security. Springer International Publishing, Cham, pp 71–112. https://doi.org/10.1007/978-3-030-35291-2_3. https://doi.org/10.1007/978-3-030-35291-2_3
43. Salam A (2020) Internet of things in sustainable energy systems. Springer International Publishing, Cham, pp 183–216. https://doi.org/10.1007/978-3-030-35291-2_6. https://doi.org/10.1007/978-3-030-35291-2_6
44. Salam A (2020) Internet of things in water management and treatment. Springer International Publishing, Cham, pp 273–298. https://doi.org/10.1007/978-3-030-35291-2_9. https://doi.org/10.1007/978-3-030-35291-2_9
45. Salam A (2020) Wireless underground communications in sewer and stormwater overflow monitoring: radio waves through soil and asphalt medium. Information 11(2):96
46. Salam A, Karabiyik U (2019) A cooperative overlay approach at the physical layer of cognitive radio for digital agriculture. In: Third international Balkan conference on communications and networking 2019 (BalkanCom'19). Republic of Skopje, Macedonia, the former Yugoslav
47. Salam A, Shah S (2019) Internet of things in smart agriculture: Enabling technologies. In: 2019 IEEE 5th world forum on internet of things (WF-IoT). IEEE, Piscataway, pp 692–695
48. Salam A, Vuran MC (2016) Impacts of soil type and moisture on the capacity of multi-carrier modulation in internet of underground things. In: Proceedings of the 25th ICCCN 2016, Waikoloa

49. Salam A, Vuran MC (2017) Em-based wireless underground sensor networks. In: Underground sensing. Academic, Cambridge, pp 247–285. https://doi.org/10.1016/B978-0-12-803139-1.00005-9
50. Salam A, Vuran MC (2017) Smart underground antenna arrays: a soil moisture adaptive beam-forming approach. In: IEEE INFOCOM 2017-IEEE conference on computer communications, Atlanta
51. Salam A, Vuran MC (2017) Wireless underground channel diversity reception with multiple antennas for internet of underground things. In: 2017 IEEE international conference on communications (ICC), Paris
52. Salam A, Vuran MC, Irmak S (2016) Pulses in the sand: impulse response analysis of wireless underground channel. In: The 35th annual IEEE international conference on computer communications (INFOCOM 2016), San Francisco
53. Salam A, Vuran MC, Irmak S (2017) Towards internet of underground things in smart lighting: a statistical model of wireless underground channel. In: Proceedings of the 14th IEEE international conference on networking, sensing and control (IEEE ICNSC), Calabria
54. Salam A, Hoang AD, Meghna A, Martin DR, Guzman G, Yoon YH, Carlson J, Kramer J, Yansi K, Kelly M, et al (2019) The future of emerging IoT paradigms: architectures and technologies. https://doi.org/10.20944/preprints201912.0276.v1
55. Salam A, Vuran MC, Irmak S (2019) Di-sense: in situ real-time permittivity estimation and soil moisture sensing using wireless underground communications. Comput Netw 151:31–41. https://doi.org/10.1016/j.comnet.2019.01.001. http://www.sciencedirect.com/science/article/pii/S1389128618303141
56. Sales N, Remédios O, Arsenio A (2015) Wireless sensor and actuator system for smart irrigation on the cloud. In: 2015 IEEE 2nd World Forum on Internet of Things (WF-IoT). IEEE, Piscataway, pp 693–698
57. Silva AR, Vuran MC (2010) Development of a testbed for wireless underground sensor networks. EURASIP J Wirel Comm Netw 2010:1–14
58. Sirio (2020) Sirio WY400-10N UHF 400–470 MHz base station 10 element Yagi Antenna. http://www.sirioantenna.com/index.php?main_page=product_info&productsid=214
59. T-L Irrigation (2020) Pivot management and control- T-L pivot irrigation system management. http://tlirr.com/products/pivot_management_options/
60. Temel S, Vuran MC, Lunar MM, Zhao Z, Salam A, Faller RK, Stolle C (2018) Vehicle-to-barrier communication during real-world vehicle crash tests. Comput Comm 127:172–186
61. Tiusanen MJ (2013) Soil scouts: description and performance of single hop wireless underground sensor nodes. Ad Hoc Netw 11(5):1610–1618. https://doi.org/10.1016/j.adhoc.2013.02.002. http://www.sciencedirect.com/science/article/pii/S157087051300022X
62. Vuran M, Dong X, Anthony D (2016) Antenna for wireless underground communication. US Patent 9,532,118
63. Vuran MC, Salam A, Wong R, Irmak S (2018) Internet of underground things in precision agriculture: architecture and technology aspects. Ad Hoc Netw 81:160–173. https://doi.org/10.1016/j.adhoc.2018.07.017. http://www.sciencedirect.com/science/article/pii/S1570870518305067
64. Wizzimote (2020) Wizzilab - connecting things. http://old.wizzilab.com/shop/wizzimote/
65. Wong R (2017) Towards cloud-based center pivot irrigation automation based on in-situ soil information from wireless underground sensor networks. Master's thesis, University of Nebraska-Lincoln

Chapter 13
Variable Rate Applications in Decision Agriculture

13.1 Introduction

Precision agriculture can be considered as a menu of numerous management techniques from which farmer can select one or multiple technologies, e.g., one farmer can choose variable rate application (VRA) for nitrogen application and other can use VRA for application of all fertilizers. Irrespective of the chosen technology, PA can be viewed as a cyclic approach (see Fig. 13.1) which involves collecting data, developing plan to manage the farms, implementing those plans, and finally evaluating those plans [36, 40]. VRA is one of the most popular PA method for adjustable application of fertilizer and chemicals for crop protection. Early VRA systems consist of adjusting flow rate through adjusting the speed of applicator [11] and provided a starting point for VRA systems (Fig. 13.2).

13.2 Properties of VRA Control Systems

Figure 13.3 shows a complex schematic of a VRA system. A typical VRA system includes data interpenetration, devising management plans, determining application rate, and vehicle related task for application. It is important to note that not all VRA system consist of all the elements shown in Fig. 13.3. Similarly, some VRA system may have functionalities of multiple elements combined into one.

Control decision for VRA system can be applied using map-based (also known as offline) or sensor-based approach (also known as online). Sensor-based approach senses the data and uses it immediately in real-time for automatic control. Map-based operation can be divided into two phases: Phase one includes gathering and storing the data and phase 2 includes the usage of information by controlled equipment in a separate field operation [23, 52]. Although map-based systems are popular nowadays; however, real-time sensing systems are becoming mature and

A. Salam, U. Raza, *Signals in the Soil*,
https://doi.org/10.1007/978-3-030-50861-6_13

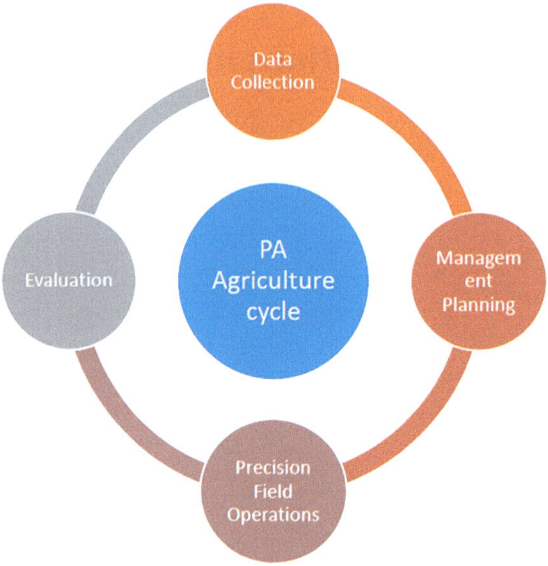

Fig. 13.1 The precision agriculture cycle

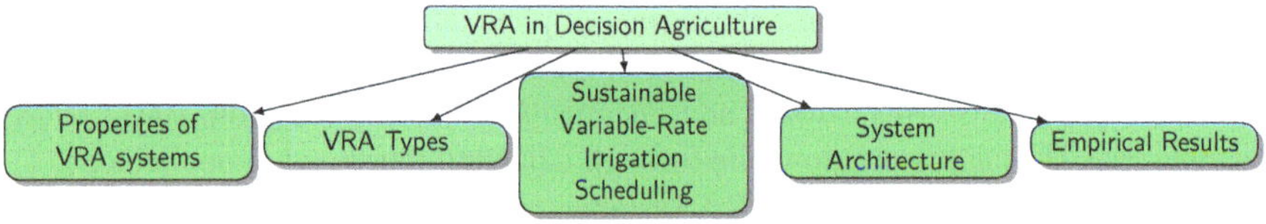

Fig. 13.2 Organization of the chapter

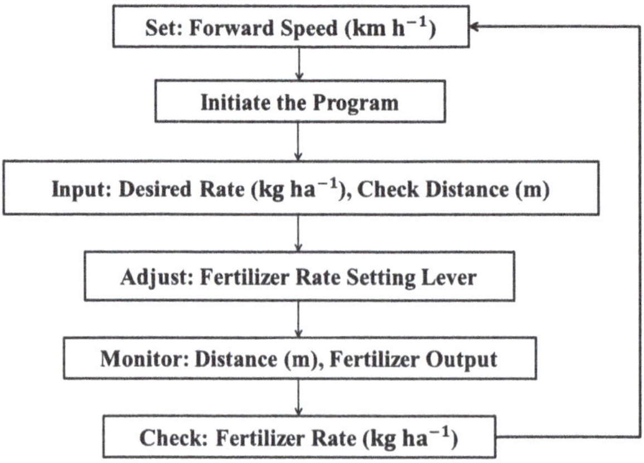

Fig. 13.3 A generalized schematic of data flow in VRA system [2]

Table 13.1 Comparison of sensor-based and map-based systems [61]

Sensor-based systems	Map-based systems
Human intervention is eliminated for data collection	Non-real data can be used
Reduced spatial interpolation errors with dense sensor data	Required amount is known before calculation
Suitable for applications with high temporal variability (soil nitrate level)	Gives more time between application and data collection allowing for intensive processing and analysis

it is possible that sensor-based VRA may takeover in the future. Hybrid systems consisting both of the technologies may also become popular. Benefits of both map-based and sensor-based systems are given in Table 13.1.

Map-based technologies can use the historical agricultural and soil data collected from different states or region. However, this technique is often questioned because it is widely established that crop response may vary from site to site and sometimes it even changes within the field [17]. Therefore, integrated decision support systems (DSS) can be used with the combination of expert knowledge and data from different sources.

Sensing based technologies use sensors on the applicator vehicle and determine the application rate. Most widely used sensing based technology is crop canopy reflectance sensing for the assessment and application of nitrogen. Various sensing technologies have been used for detection, identification, and quantification of weeds as an input to VRA systems. Other online sensors, e.g., soil electrical conductivity (EC_a) are also used to develop offline control maps for VRA nematicide [37].

13.2.1 Control Hardware

Following hardware enables the operation of a VRA system:

- *Application rate processor.* Application rate processor and the associated software are the core part of a VRA system. It combines the speed, position, and sensed data with the application rate map, and issues the rate command to application rate controller.
- *Application controller.* It receives the instruction from the application rate processor and controls the actuators on the applicator based on those commands. It compares the desired application rate with actual application rate and adjusts the control signal based on the error rate to minimize the rate. This is a closed-loop control operation [25].

- *Operator interface.* It is a very important part of VRA system. An effective operator interface must quickly communicate the complex and continuously changing information to the driver. This requirement comes important especially when the VRA system is being used in multiple area simultaneously. It must allow the operator to dynamically change the operating parameters, fix any detected fault, and override the default parameters. The information can be provided in the form of readable, audible audio alarms, and graphical interface. Similarly, data can be entered using touchscreen [34].
  Some systems also combine the all or multiple components (application rate processor, operator interface, or application controller) into one box and connect them with sensor and actuator [31].
- *Equipment actuators and sensors.* Most of the equipment works on the hydraulic and electric motors. Liquid fertilizers are pumped through hose and dry fertilizers from a holding tank. Application controller controls the delivery rate by taking the input signal from the actuator motor speed. In some cases, delivery rate can also be controlled by adjusting the size of flow passage.
- *Documentation of application: as-applied maps.* The application rates maps generated by the VRA systems can be logged with time, distance intervals, GPS position, and current rates reported by the controller. This documentation can be used to keep record what and how much of it was applied, check compliance with environmental regularities for chemical and fertilizer application [18, 28].

13.3 Types of VRA Systems

In this section, two different types of VRA systems are explained.

13.3.1 Liquid VRA Systems

This section discusses the following liquid-based VRA systems.

- *Flow Control Methods.* There are two ways of controlling flow for VRA. One includes varying the concentration of an active ingredient (a.i) and is known as *Variable Concentration Method.* Second method includes varying the whole solution, i.e., a.i + carrier solution and is known as *Total Output Control Method.* A comparative study of both method is given in [14, 32].
  Historically, total output method is implemented by controlling the pressure at the nozzle which then transformed to electronic spray providing closed-loop flow control using pressure and flow sensors and sometime ground speed sensor to compensate for variations in travel speed.
  Variable concentration method was first reported in mid-70s [14, 44, 48] and [68] tested a laboratory system for injecting concentrated pesticides. However,

disadvantage of such is system includes a non-uniform application of pesticides and frequent transient error due to operating speed.

- *Section Control.* Section control method involves division of applicator into multiple sections and maintaining the input for each individual section through application rate processor. This gives much more independence and control over regions within the field. The primary goal of the method is to avoid (1) overlapping application in irregular shape of the area and (2) spraying on the non-target area, e.g., grassed waterway. This method provides good spatial resolution of control, therefore, resulted in 15–17% reduction in spray [19, 64]. However, it may result in large variations from desired application rate while turning nozzle on and off [42, 56].

13.3.2 Dry VRA Systems

This section discusses the following liquid-based VRA systems.

- *Flow Control Methods.* There are two ways of controlling flow for VRA. One includes varying the concentration of an active ingredient (a.i) and is known as *Variable Concentration Method.* Second method includes varying the whole solution, i.e., a.i + carrier solution and is known as *Total Output Control Method.* A comparative study of both method is given in [14, 70].
  Historically, total output method is implemented by controlling the pressure at the nozzle which then transformed to electronic spray providing closed-loop flow control using pressure and flow sensors and sometime ground speed sensor to compensate for variations in travel speed.
  Variable concentration method was first reported in mid-70s [14] and [33, 68] tested a laboratory system for injecting concentrated pesticides. However, disadvantage of such is system includes a non-uniform application of pesticides and frequent transient error due to operating speed.
- *Spinner Spreaders.* They use spinning disc to spread the dry granular fertilizer (dropped on the disc) in the wide area. They are not used for banded operation and are mostly used for the broadcast application. They relatively cover large per unit area as compared to other VRA systems and are suitable for the application where large volume of application is required. These VRA systems are commonly used in the application of fertilizers and pH balancing products during harvesting and planting season [26, 43, 51, 71].

13.3.3 A Case Study

A conventional spray application uses a huge amount of pesticides in horticultural crop production system to effectively control the pesticides. This problem was

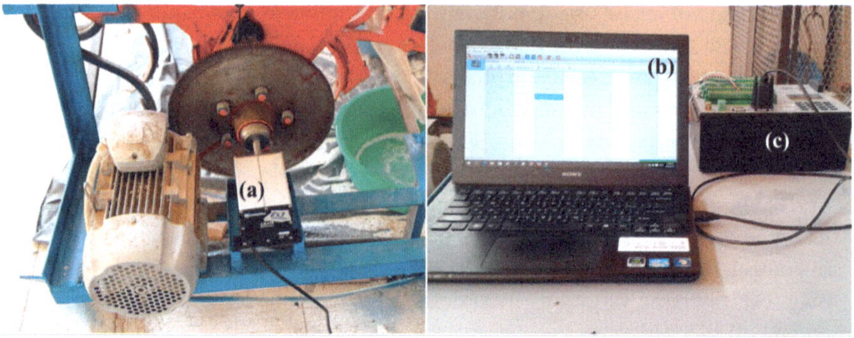

Fig. 13.4 VRA case study of a variable rate granular fertilizer [2]: (**a**) the shaft encoder, (**b**) software program, and (**c**) data logger

solved by an automated sprayer shown in Fig. 13.4a. It uses sensor technology to apply pesticides as per the crop needs and requirements. The sprayer uses the size and shape, presence and foliage density of target area to apply the optimum amount of pesticide that too with minimum human involvement. It consists of the following components: a laser scanning sensor, a travel speed and nozzle flow control sensor, a touch screen, an embedded computer, 40-variable rate nozzles, and a switch box. It significantly reduced the over-spray of pesticides while benefiting the environmental ecosystem and saving time by fewer refill [46, 53].

A system is developed by modifying a commercial self-propelled sprayer (Spra-Coupe 3640, Melroe Co., Bismarck, ND, USA). It has the following design requirement: (1) Fast rate change, integrated sensing and associated algorithm for online N-rate determination and application and (2) ability of using offline data. Moreover, system uses a flow feedback system which changes the application rate based on pressure of the nozzle to achieve high turn-down ratio. System is controlled by a Visual Basic program. This system has hugely evolved by including advanced sensors and efficient algorithms for in-season VRA application. it is being used by many research projects and is involved in many on-farm research projects [45, 47].

13.4 Sustainable Variable-Rate Irrigation Scheduling

Increasing food demand has given rise to popularity of new generation farming technique: Precision Agriculture (PA). The main focus of PA techniques is the variable amount of natural components (nutrients, water content, drainage, runoff, chemical leaching, and soil components) present in the field [10, 12]. The main purpose of PA is to accurately analyze the field variations using modern day technologies such as GPS, aerial and remote sensing, satellites, and sensors. As a result the farming methods (irrigation and fertilizer management, sowing) can be scheduled and applied autonomously [29, 43].

The success of PA-based farming systems depends upon the effective real-time evaluation of field which is then used to make timely decisions. For example, it is important to have complete knowledge of *quantity* and *time* of applying water in irrigation scheduling. A successful irrigation management demands accurate monitoring of water status for the field under observation. In order to obtain optimum crop yield, water in the crop root-zone must be maintained at a certain desirable level. A good and accurate irrigation management system can help in avoiding financial losses which otherwise may occur because of over- or under-irrigation, pesticides, nutrients movement, and other water bodies, etc. Hand-feel method is the qualitative method and soil sensor is the quantitative method for estimating soil moisture levels. If there is no cost-effective soil moisture sensor available, then hand-feel method is used for irrigation management purposes. However, hand-feel method relies on the person ability of feeling and perceiving the soil. This method is prone to human-error and can sometimes be less accurate causing financial loss. Therefore, a quantitative method is needed for reporting soil moisture status for accurate irrigation management [16, 22, 44].

A proper irrigation management is needed to decide the timing and quantity of irrigation to mitigate the farmers loss which may occur due to water stress. An accurate irrigation can result in increased yield response to other management technologies. The farmers profitability is increased by adopting these practices. Other advantages of proper irrigation management include: reduced runoff, soil erosion, and pesticides in surface and ground water. To summarize, combination of an irrigation management system and in-situ water monitoring through sensors has great advantage over unmanaged system in that it can prevent water wastage while increasing crop yield [15].

Traditional soil sensing techniques involve installation of sensor at the start of season and are needed to be removed before harvesting begins. These techniques are not considered effective for real-time in-situ soil sensing in PA. To deal with this situation, Wireless Underground Sensor Networks (WUSNs) have become popular recently for such unattended soil monitoring [1, 4, 26, 51, 57, 58, 65]. WUSN can be considered as the wireless network of underground sensors nodes communicating through soil. A WUSN-based cyber-physical system (CPS) can be considered as an efficient solution for the PA which provides detailed soil information in timely manner [23, 30, 55]. WUSN proved to be better than satellites and aerial remote sensing because of their cost-efficient methodology of providing accurate information. Another advantage of WUSN is that they do not interfere with the farming operations (planting process and machinery operations) while deployment which is a significant improvement over the wired networks which are frequently installed/removed while planting process.

This chapter discuss a Wireless Underground Sensor-Aided Center Pivot (WUSA-CP) irrigation system. This system is an application of cyber-physical system and presented here as a proof-of-concept. This application uses a center pivot system [20, 24, 33]. Center pivot system is a physical system that moves through the field and collects soil moisture data from the sensors buried in the ground. To that end, some of the challenges faced in implementation of such irrigation system are presented. An important challenge is the difference between underground (UG) channel and aboveground (AG) channel as communication in UG channel is affected by the soil properties. A WUSN channel model is analyzed and two antenna designs are evaluated for verification of the model and WUSA-CP. Moreover, analysis for burstiness of the packet error rate and range of communication is also analyzed.

13.4.1 Central Pivot System

Cost-efficient and productive methods are constantly being developed in an effort to improve the agricultural crop yield. To that end, irrigation with center pivot (CP) system [10, 20] is being used to efficiently use and apply water to the fields. Figure 13.5 shows the different components of CPS system. It consists of segmented pipes with sprinklers mounted on the wheels [3, 10, 20, 35, 38]. The pipe is connected to a pivot which is placed at the center of the irrigation field and is known as pivot point. The water is sprayed through sprinklers as the machine moves in circular pattern.

A well-managed CP irrigation systems differ from traditional surface irrigation systems in that they reduces surface runoff and deep percolation, and requires less water application (e.g., up to 40%). Saving water makes more water available for crop transpiration while increasing the productivity of the crop. Using chemigation [10, 39, 40], CP irrigation system can also apply the nutrient to crop canopy. Chemigation, in contrast to ground application, enhances the absorption of nutrients, fertilizers, herbicide, pesticide, and insecticides by the crop leaves which consequently increases the productivity of applied chemical and nutrients.

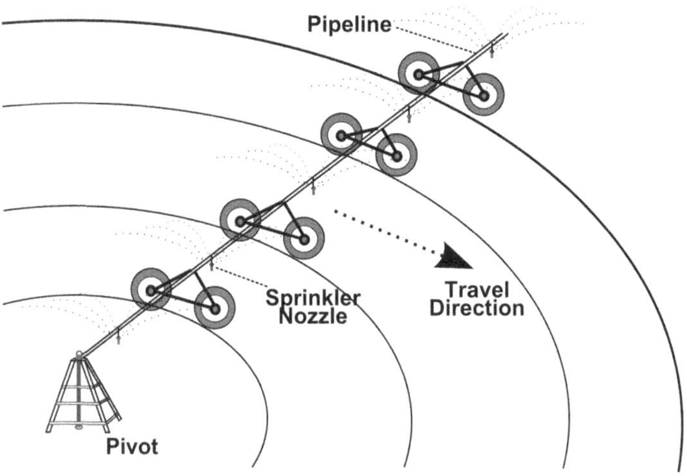

Fig. 13.5 Center pivot system (CPS) [8]

CP cost prevents it to be used in small irrigation areas and is used in larger areas of 3.5–65 ha [10, 20, 25, 28]. A testbed of 22 ha is developed at the South Central Agricultural Laboratory (SCAL) for experimentation to analyze CP system.

CP water application can be controlled by either (1) controlling its traveling speed or (2) electronically controlling the application rate of sprinklers. First method is preferred because of its simplicity, accuracy, and low cost as there is no requirement to change sprinklers. For a given flow rate, high speed of CP applies less water to the field. Furthermore, water applied through CP system can be adjusted by reading the real-time soil moisture data from the field through soil moisture.

WUSN can be used for failure detection in CP irrigation systems. Normally it takes time to detect the failure in the system which can cause sprinkler to stop irrigating. One method is to measure the sprinkler rate at discharge point to determine the rate at which water is being discharged. However, this requires very expensive and sophisticated small sized flow meters which are impractical for the real-life implementation. To this end, sensors can be used to generate alert if soil moisture level of a region does not rise to a particular level when CP passes through that region. Farmers can use these alerts to mitigate the effect of reduced irrigation. To that end, an efficient communication system is required for communicating real-time data from UG sensors. Next section discusses the one such option for implementing communication infrastructure [52].

13.5 System Architecture

Experiments are conducted at South Central Agricultural Laboratory (SCAL), Clay Center, Nebraska, using 433 MHz Mica2 [1] sensor nodes. The purpose of the experiment is to provide proof-of-concept of autonomous irrigation system and investigate the associated challenges. Table 13.2 shows the data (bulk density, particle density, and soil texture) of the site, gathered from laboratory analysis, where center pivot is located [27].

Figure 13.6 shows the experiment setup for central pivot irrigation system in the corn field. AG node is deployed on the system's arm at 2.5 m along with UG nodes buried at the depth of 35 cm in circular arrangement around the field. This

Table 13.2 Characteristics of soil used in the experiments [8]

Depth	Texture	Sand	Silt	Clay
0–20 cm	Silt loam	17	55	28
20–60 cm	Silt clay loam	16	46	38

Particle density	Bulk density
2.66 g/cm^3	1.3

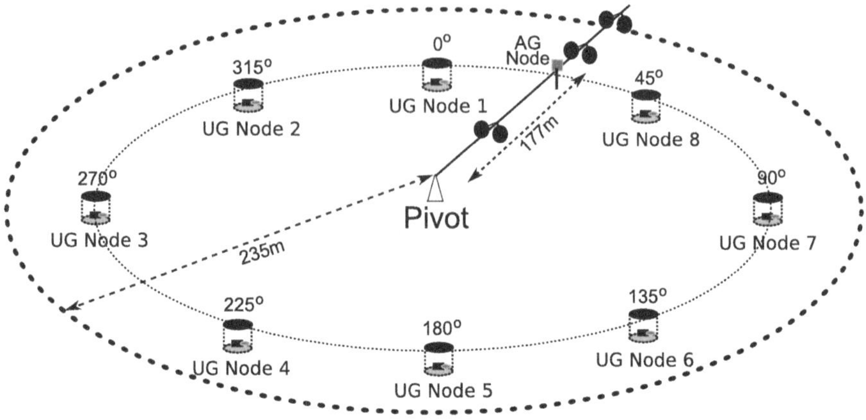

Fig. 13.6 CPS testbed with one central AG node and eight buried UG nodes [8]

depth keeps the UG nodes safe from farming machinery. For AG and UG nodes communication, CP has to be within communication range. CP operates in both clockwise and counter clockwise direction.

13.5.1 Hardware Architecture

Soil decreases the signal strength (attenuation) and wavelength of the signal [16, 22]. Phase shifting constant β in equation is related to wavelength λ as: $\lambda = 2\pi/\beta$. This relation between λ and soil properties requires an antenna designed specifically to underground communication. the effective soil permittivity ϵ defines phase shifting constant β and attenuation constant α. It is highly affected by the different soil properties such as soil type, soil structure and moisture, and salinity. Peplinski model has this property for the frequency range of 0.3–1.3 GHz [11]. The operating frequency of Mica2 nodes is 433 MHz. Maximum and minimum values for VWC of site are observed for the experimental site. Given this frequency and VWC values, UG antenna works between 30–69 cm wavelength. The frequency range for free space is 1–1.8 GHz. Therefore, operation frequency of an underground antenna must be in the range of 1–1.8 GHz in order to communicate with Mica2 nodes in air. AG node uses the antennas with high gain [31].

Two different schemes are used for both AG and UG node. In first scheme, AG node uses Full-Wave (FW) dipole antenna and UG node uses Ended Elliptical Antenna (SEA) [13, 16]. The gain for the dipole antenna is 3 dB. In second scheme, AG node uses Yagi antenna and UG node uses a circular planar antenna. The gain for the Yagi antenna is 3 dB. The operational range of SEA and circular planar antenna is customized as per the application requirement.

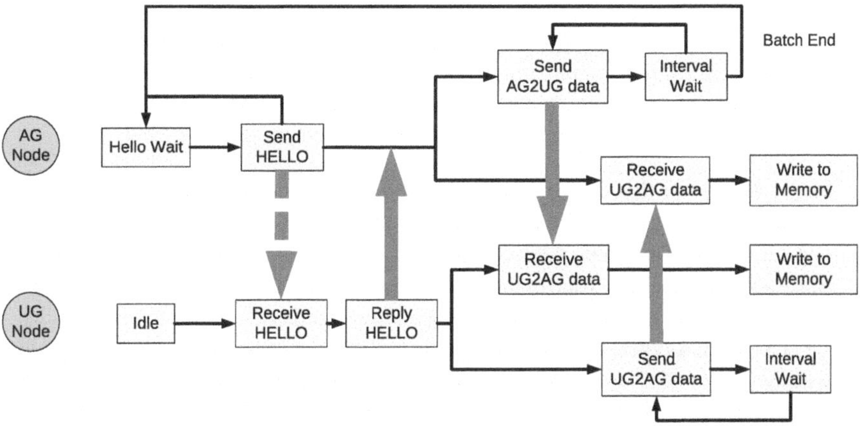

Fig. 13.7 The program structure for the experiments [8]

13.5.2 Software Architecture

In order to avoid the reprogramming of the sensor node for every experiment, an application, TinyOS was developed and all experiments were performed with the transmit power of 10 dBm. Each transaction involves transmission of 100 packets, with 100 ms interval between each packet, in both direction (UG–AG nodes and AG–UG nodes). Size of each test packet is 37 bytes.

Figure 13.7 shows the program structure. UG and AG nodes send and receive packet from each other. For each received packet, UG node extracts the timestamp, id of the AG node which sent the packet, and RSS of the signal, and stores it in the flash memory. Similarly, for each packet received by AG node, it extracts the timestamp using its own clock, id of the UG node which sent the packet, and RSS of the signal, and stores it in the flash memory. Experiment data is read from the flash memory of each node. An important thing to note is that timestamp of only AG node is used for the experiment. As the speed of the CP is very slow (0.704 m/min), hence, all 100 packets can be considered to be sent from same location. Therefore, the timestamp is used to determine the location of AG node [71].

13.6 Empirical Results

The difference between the antenna and transceiver of each Mica2 node is significant, therefore, a qualification test is performed before every experiment [61]. For this purpose, a *through-the-air* test with 200 packets of 30 bytes are sent. The test identifies the complaint nodes and check if the battery level of a node is above the safe threshold. A node will be considered compliant if it satisfies the two conditions:

(1) node's packet error rate (PER) is within the 10% of average PER calculated for all nodes and (2) the node's RSS varies ± 1 dB from average RSS of all nodes. The safety threshold for battery level is set as 2.5 V.

A total of five different experiments were performed. Each experiment considered different soil moisture conditions and vegetation canopy as given below:

- **Static-Dry:** For this experiment, CP was not used and corn field was used with VWC [5] of 16.6%. As the crops were recently harvested, therefore, impact of vegetation canopy can be neglected.
- **Static-Wet:** For this experiment, corn field was used with VWC of 22.7%. As field was wet, therefore, no vegetation canopy was there.
- **CP-Crop-SEA-Vert:** For this experiment, CP was used and corn field was used with VWC of 22.7%. In this case, the crops reached their maximum height of 2.85 m, hence, wireless communication had impact of vegetation canopy can be neglected. SEA and FW antennas were used for the experiment and vertical placement of SEA.
- **CP-Crop-SEA-Hori:** For this experiment, corn field was used with VWC of 32% and SEA antenna was placed horizontally.
- **CP-Circular-Yagi:** Second antenna scheme, with Yagi and circular planar antenna, was used for this experiment. Experiments were considered right after harvesting with horizontal placement of circular planar antenna and VWC of 32%.

First two experiments were used to analyze how soil moisture affects the communication. Last three experiments were used CP arrangement similar to shown in Fig. 13.6. In order to observe the worst case scenario, the maximum speed of CP was used, i.e., 43°/h.

This section discusses the experiment results to provide proof-of-concept for autonomous irrigation system. It also discusses the effect of distance between the sending and receiving on UG2AG and AG2UG communication link.

13.6.1 Communication Range

A communication window is defined as the time duration in which AG and UG nodes communicate with each other and is analyzed first. Figure 13.8 shows the results from CP-Crop-SEA-Hori experiment. With traveling speed of 43°, CPS travels the complete field in 8.37 h. It can be observed in Fig. 13.8 that nodes communicate for 1.33 h and which is 16% of total traveling time. Communication time also varies significantly with longest time being 29 min and shortest is 10 s only.

Figure 13.9 shows the communication windows for each of the AG and UG node in experiment. For UG node, communication range is shown as horizontal distance and time of AG node while moving. AG approaching UG is indicated by negative distance and positive distance represents AG moving away from UG node [48].

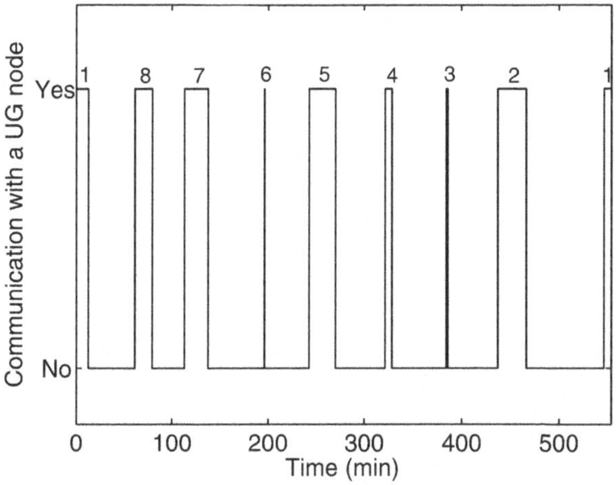

Fig. 13.8 Travel timeline for CP-Crop-SEA-Hori experiment in clockwise direction [8]

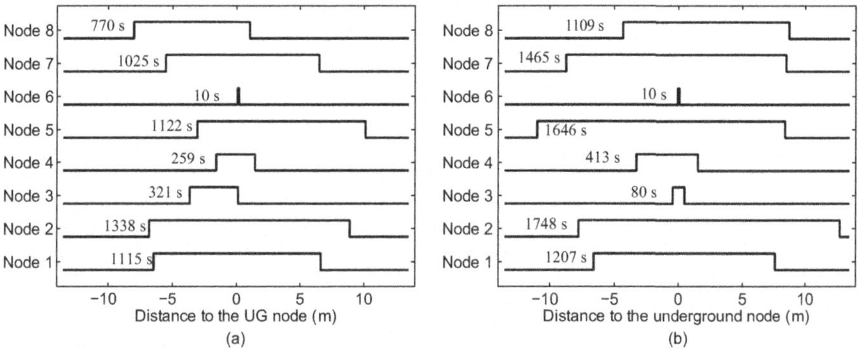

Fig. 13.9 AG and UG node communication in both clockwise and anti-clockwise direction for CP-Crop-SEA-Hori experiment [8]

It can be observed that communication window varies significantly. In Fig. 13.9b, best case scenario occurs when node 2 communicates for a total of 20.6 m by starting the communication when AG node is 7.8 m away from UG node and continues the communication 12.8 m after the AG node has passed the UG node. Similarly a worst case scenario is just 0.5% of best case and occurs when node 6 communicates for only 0.11 m. The average communication distance in Fig. 13.9a and b is 8.75 m and 11.27 m with standard deviation of $\delta_{CCW} = 5.73$ and $\delta_{CW} = 8.19$, respectively.

For both cases, CCW and CW, the communication distances are quite similar which shows that communication quality is independent of the CP movement and related to specific location. The variations could be because of irregular soil surface which may occur due to farming activities (e.g., plowing) on the ground affecting the EM waves dispersion. Even if the nodes are installed carefully, the soil surface above the UG nodes can change because of working of agricultural machinery. UG

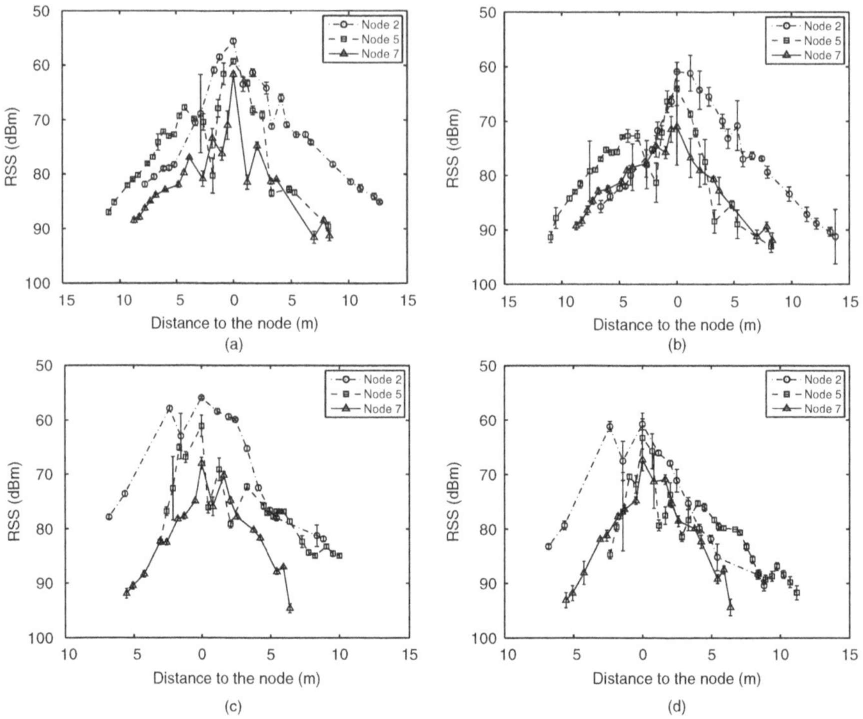

Fig. 13.10 RSS v/s Horizontal inter-node distance using SEA and FW antennas [8]: (**a**) AG2UG-Clockwise, (**b**) UG2AG-Clockwise, (**c**) AG2UG-Anticlockwise, (**d**) UG2AG-Anticlockwise

nodes are buried so that AG node can be right above them. This deployment may result in some nodes (e.g., node 2) being close to crop canopy and some being farther (e.g., node 7). The irregularity can also be seen in the results since crops hinder the EM waves propagation. The phenomenon is known as canopy effect and empirically analyzed in coming sections.

Figure 13.10 plots the RSS results from CP-Crop-SEA-Hori experiment for AG2UG and UG2AG link. Figure 13.10a and c shows results for AG2UG link in CW and CCW direction, respectively. Similarly, Fig. 13.10b and d shows results for UG2AG link in CW and CCW direction, respectively. Figure 13.11a and b plots the RSS results from CP-Circular-Yagi experiment for AG2UG and UG2AG link, respectively [41].

The Yagi and circular planar antenna pair achieves a maximum of 65 m communication range and that of FW and SEA antenna pair achieves maximum of 14 m communication range. Hence, the Yagi and circular planar relatively increases the communication range by 364–400%. In worst case scenario the communication range for Yagi-circular pair is 40 m and that of SEA-FW pair is 8 m. This high difference is because of two facts: (1) The return loss of planar antenna (−10 dB) is less than that of SEA antenna (−3 dB) in UG deployment and (2) the antenna gain

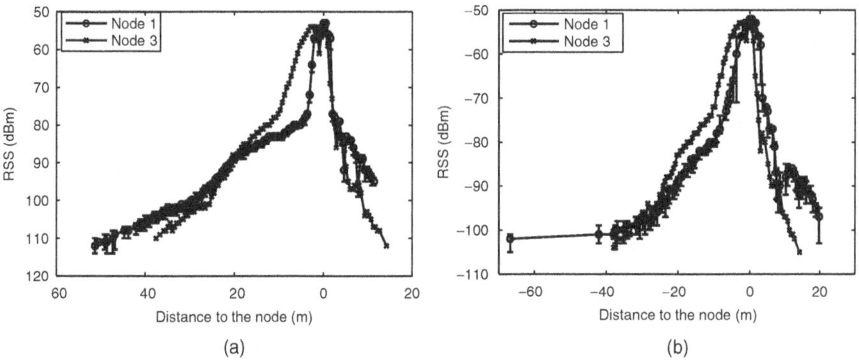

Fig. 13.11 RSS v/s horizontal inter-node distance using circular and Yagi antennas [8]: (**a**) AG2UG, (**b**) UG2AG

of Yagi is also higher. The communication distance for Yagi-circular pair is also not symmetrical because of directivity of Yagi antenna and asymmetry for circular planar are analyzed in Sect. 13.6.2

13.6.2 Numerical Analysis of the Channel Model

This section analyzes the model developed from the results. Minimum mean square (MMSE) is used empirically to calculate the air attenuation η^* and soil-dependent component $\mathcal{K}^*$.[1] $\mathcal{K}$ is compared with the model developed. The values for constant c are 13.57 dB and 3.57 dB for CP-Crop-SEA-Hori and CP-Circular-Yagi experiment, respectively. Tables 13.3 and 13.4 show the results for UG2AG and AG2UG links. Model comparison is shown in Table 13.4. $\mathcal{K}$ is calculated from Peplinski model.

It can be observed in Table 13.2 that attenuation model shows the UG2AG and AG2UG link with very low error with max MSE $= 5.94$. However, large variations can be seen based on location. Attenuation coefficient η varies because of plants and $\mathcal{K}$ varies because of variations in α, β, and $L_{R,\rightarrow}$ due to different locations. Effective soil permittivity ϵ mainly determines these values. These variation shows how soil characteristics even within the field. It can be observed in Table 13.4 that $\mathcal{K}$ predicted by model is similar to empirical results obtained in CP-Crop-SEA-Hori experiments but its accuracy is limited for CP-Circular-Yagi experiment. Yagi antenna gain G_r becomes dependent on distance with moving CP because of its high directivity. Moreover, the model results are focused for only one point in the field. The error in the values predicted by the model is due to spatial variance in

[1]For rest of the discussion, empirical estimates are represented by $*$ superscript.

Table 13.3 Channel model parameters [8]

	Node	η^*	$\mathcal{K}^*$	MSE
CP-Crop-SEA-Hori (UG2AG)	*Approaching*			
	2	4.67	34.27	2.32
	5	4.21	29.52	5.94
	7	2.68	48.11	1.33
	Departing			
	2	4.05	29.7	1.52
	5	5.25	34.19	3.70
	7	3.30	47.42	1.51
CP-Crop-SEA-Hori (AG2UG)	*Approaching*			
	2	5.11	23.87	1:54
	5	4.48	24.43	5.94
	7	3.15	44.38	3.10
	Departing			
	2	3.53	31.52	2.03
	5	5.58	27.35	3.57
	7	3.84	42.43	4.47
CP-Circular-Yagi (UG2AG)	1	5.10	72.06	4.69
	3	4.91	60.65	3.22
CP-Circular-Yagi (UG2AG)	1	5.62	73.02	4.60
	3	5.34	63.91	3.34

Table 13.4 Comparison theoretical model computations and measured results [6]

	Node	η^*	$\mathcal{K}^*$	MSE
CP-Crop-SEA-Hori (UG2AG)	3.16	40.78	5.41	48.59
CP-Crop-SEA-Hori (AG2UG)	3.29	38.46	5.84	55.50
CP-Circular (UG2AG)	5.01	66.36	7.24	48.59
CP-Circular (AG2UG)	5.48	68.47	6.33	55.50

soil. To summarize, the results from the channel model are accurate enough but also underscore the importance of semi-empirical models because of uncertainty due to soil properties [49, 50].

13.6.3 Asymmetry of the Communication Over Distance

With same horizontal distance, RSS value changes depending on the fact that AG node is approaching or departing from the UG node. For Yagi-circular pair results in Fig. 13.11a and b, it happens because of high directivity of Yagi antenna. However, for SEA-FW pair results in Fig. 13.10a and d has symmetric propagation pattern, hence, generating symmetric RSS results. Figure 13.10c and d further analyzes

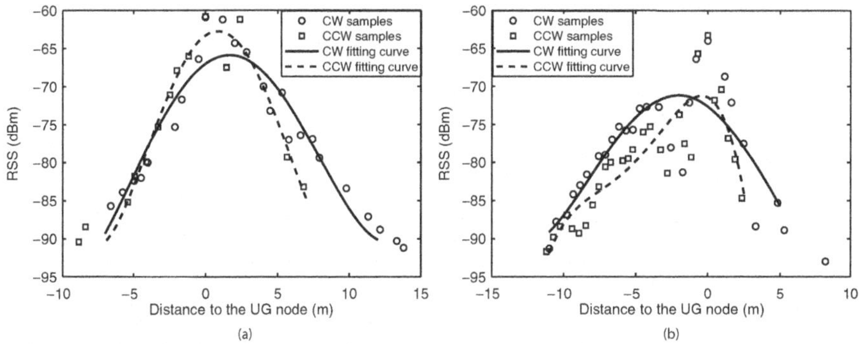

Fig. 13.12 Comparison between communication of two nodes with AG node (UG2AG link) in both directions [8]: (**a**) Node 2, (**b**) Node 5

this phenomenon of SEA-FW pair for UG2Ag and AG2US links, respectively, by experiments while CP is moving in opposite direction A asymmetric communication quality is observed for both AG2UG and UG2AG cases. For example, in Fig. 13.10c and d, communication between node 5 and AG nodes starts 3 m before the AG node approaches node 5, however, both keeps on communicating till 11 m after AG node departs from the UG node. Furthermore, RSS values can be different when absolute distance between AG and UG node is same but side is different. For example, for a distance of −5.65 m and 5.4 m, RSS is −73.6 dBm and −78 dBm, respectively, in Fig. 13.10c (CCW). Similarly, for a distance of −5.65 m and 5.4 m, RSS is −78.9 dBm and −72.7 dBm, respectively, in Fig. 13.10a. A comparison of CW and CCW results is performed to determine the reason for this asymmetry [29, 30].

Figure 13.12 shows RSS values for UG2AG links of node 2 and 5. it can be observed that node 5 communicates better in CCW direction when Ag is approaching it and, in CW direction, it is better when AG is departing it. Hence, it can be concluded that asymmetry is mainly because of the environment conditions (soil surface and crops), antenna placement, and asymmetric propagation patterns, rather than the AG node movement [24].

13.6.4 Burstiness of the Packet Error

In an attempt to understand the communication characteristics completely, a CP-Crop-SEA-Hori experiment is performed to understand the burstiness of packet error. This experiment sends a burst of 100 packets after the communication is established between the nodes. The time interval is kept 100 ms between each packet within the burst. The number of packet loss between the two successfully received packet is selected as the metric for burstiness of the packet error and average error is shown in Fig. 13.13.

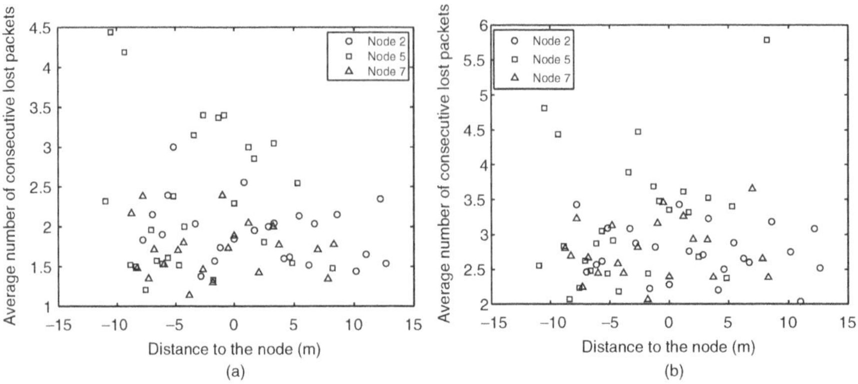

Fig. 13.13 Effect of distance on average burstiness of packet error in clockwise direction [8]. (**a**) AG2UG communication. (**b**) UG2AG communication

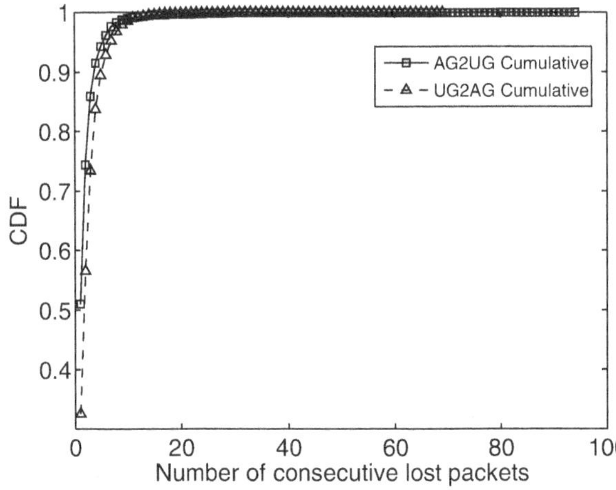

Fig. 13.14 Lost packet percentage v/s burstiness of packet error [8]

In the figure, average length of the consecutive packet errors is represented by a single point. It is observed that there is no relation between the burstiness of the packet error and the distance which confirms the findings of [16] about the transitional region of UG communication being very small from over-the-air communication. Hence, the error rate is stable while nodes are communicating. Figure 13.14 shows the distribution of the burstiness of the packet error. It can be observed that with the increase in continuous packet loss, a decrease is seen in probability of high burst error. Both UG2AG and AG2UG observe one packet loss of 33% and 50%, respectively, between two successfully received packets. The maximum size of burst error is 94 packets. Figure 13.14 concludes that AG2UG communication link performs better than the UG2AG communication with average number of consecutive packet loss of 2.25 which 2.89 in case of UG2AG [38, 39].

13.6.5 Effects of Canopy and Soil Moisture

Wireless communication is highly effected by the crop growth (increased vegetation canopy) [20] and soil moisture [1, 15]. This section discusses how communication is effected by these parameters by performing the three experiments explained in Sect. 13.5, i.e., Static-Dry, Static-Wet, and CP-Crop-SEA-Vert experiments. Only Static-Dry experiment uses dry soil, whereas other two use wet soil. Static-Wet and CP-Crop-SEA-Vert experiments are performed with and without canopy present, whereas Static-Dry experiment without canopy.

Figure 13.15 shows average RSS, RSS variance, and PER for AG2UG and UG2AG links in each experiments. The horizontal inter-node distance is kept at 3 m. It is observed that PER values are very small to make any meaningful comparison but RSS values confirm the expected attenuation differences for all experiments. Static-Dry soil suffers the least attenuation because of least amount of soil moisture (16.6%) and almost no canopy effect. Second least attenuation is seen in Static-Wet experiment with no canopy effect but high soil moisture (22.7%). CP-Crop-SEA-Vert is the worst case with most attenuation (6 dB decrease in the RSS) because of soil moisture and canopy effect contributing towards attenuation [27, 35].

To understand the effect of vegetation canopy on signal attenuation, the results from CP-Crop-SEA-Vert and Static-Wet are compared. It can be observed that there is an attenuation increase of 3 dB in both AG2UG and UG2AG communication links which confirms the results from [6, 20] and is important for developing

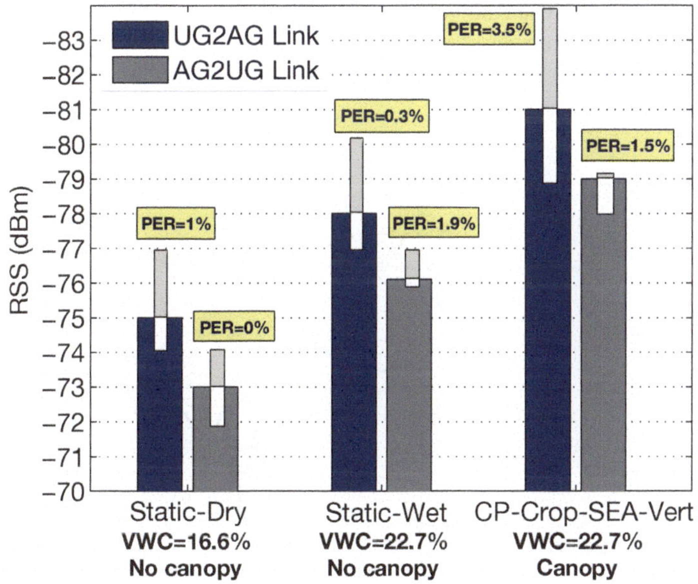

Fig. 13.15 The received signal strength for different vegetation canopy and VWC combination

environmental-aware networking solutions. Similarly, to understand the effect of soil moisture on signal attenuation, the results from Static-Dry (VWC = 16.6%) and Static-Wet (VWC = 22.7%) are compared. It can be observed that for an increase of 6.1% in VWC there is an attenuation increase of 3 dB in both AG2UG and UG2AG communication links.

Soil moisture is an important parameter to consider while designing UG communication as it can have negative impact on the signal depending on the soil path that signal may have to traverse. These results play an important role in designing WUSA-CP.

13.6.6 The Impact of Crop Growth on Received Signal Strength

To understand the impact of the crop growth on communication quality, the received signal strength between an UG and an AG node is recorded for a growing season. The RSS is measured when the center pivot is stopped at the south side of the field, 20 m away from the UG node buried at the south side.

In Fig. 13.16, the RSS values for both the UG2AG and AG2UG links for the period of June 23rd to October 5th are shown. It can be observed that the growth of the crops, in addition to the soil moisture level, has a strong impact on the communication quality, especially on the AG2UG link. When at the peak of the crop growth (2.7 m), the RSS value for the AG2UG link is deteriorated by 25 dB (−75 dB to −100 dB. The RSS values vary during July 21st and July 30th due to the change of soil moisture caused by precipitation and irrigation. From August 4th to October 5th, the RSS values increase gradually due to the drying of the crops at the end of the growing season.

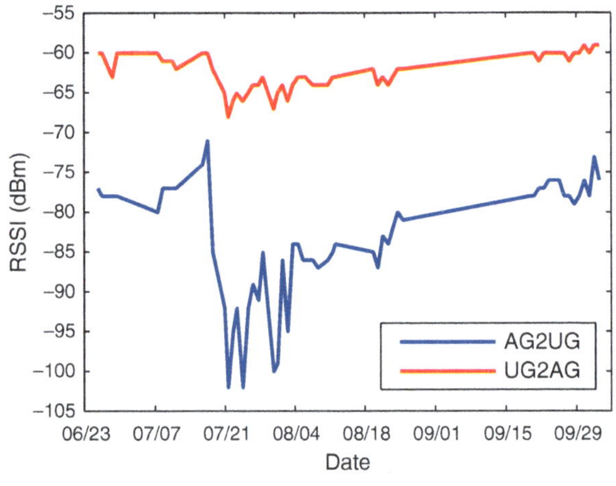

Fig. 13.16 The received signal strength during the growth of the crops

The result of the RSS over time shows that for the application of WUSNs in agriculture, the impact of the crops growth should be incorporated into communication decisions. As each crop may have different growth rates in each growing season, tailored decisions are needed to maintain communication quality high with low energy cost. This impact on RSS call for adaptive transmit power control, which adjusts the transmit power based on the environment factors, such as soil moisture level and the growth of the crops.

13.7 Recent Advances and Future Trends

VRA systems are widely adopted by many farmers [7, 21, 59, 60, 62, 66, 67, 69]. A survey conducted from retail crop input dealers of USA shows that 70% of the customers has acquired some sort of VRA fertilizers application and 27% acquired VRA pesticide application [9]. Many producers implemented the VRA without taking help from their dealers. Producers were more interested in area specific application of fertilizers in a controlled manner because 74% of the dealers reported average of 33% customers interested in GPS-based section control sprayer.

The same survey reported incompatibility (mechanical or electrical/electronic) of PA technology and equipment as a hindrance to adoption on much wider scale. The reason of this could be that many companies offering the PA services uses their own equipment and data formats for the data. A standard PA data format is needed so that interoperability can be achieved in PA system.

VRA has provided many environmental and economical benefits by minimizing the overall application of fertilizers and pesticides. Environmental benefits of VRA are reviewed by Bongiovanni and Lowenberg-DeBoer [5]. Several studies report the economical benefits of VRA application [5, 13, 54]. It is also shown that VRA system with real-time sensing was more popular than the ones based on soil variability. However, [63] reported that if all cost is accounted, VRA systems rarely give profit but he also reported that advancement in technologies (real-time sensing and DSS) can reduce the cost significantly making VRA a viable option.

Currently, economic and sustainability features of VRA system are challenging aspect for researcher and producer. Development of robust and accurate application algorithm can help in widespread deployment of VRA systems [22].

Future systems will see a significant improvement in terms of accuracy in spatial application of fertilizers and pesticides. Enhanced spatial accuracy can transform VRA systems from section control to independent nozzle control. Advanced equipment design and control systems can be used for accuracy in application rate by reducing error rate. An improved sensing and decision support system will optimize the future VRAs.

Future VRA systems will use the combination of real-time sensed data, weather forecast, and mapped soil information to predict the need of crop. However, issues like interoperability, security, connectivity, and privacy should be resolved to accomplish this.

References

1. Akyildiz IF, Stuntebeck EP (2006) Wireless underground sensor networks: research challenges. Ad Hoc Netw J 4(6):669–686
2. Alameen AA, Al-Gaadi KA, Tola E (2019) Development and performance evaluation of a control system for variable rate granular fertilizer application. Comput Electron Agric 160:31–39. https://doi.org/10.1016/j.compag.2019.03.011. http://www.sciencedirect.com/science/article/pii/S0168169918318088
3. Al-Ghobari HM (2006) Effect of maintenance on the performance of sprinkler irrigation systems and irrigation water conservation. Technical Report Res. Bult. No. 141, Food Science & Agriculture Research Center, King Saud University
4. Bogena HR, Huismana JA, Meierb H, Rosenbauma U, Weuthena A (2009) Hybrid wireless underground sensor networks: Quantification of signal attenuation in soil. Vadose Zone J 8(3):755–761
5. Bongiovanni R, Lowenberg-DeBoer J (2004) Precision agriculture and sustainability. Precis Agric 5(4):359–387
6. Do T, Gan L, Nguyen N, Tran T (2012) Fast and efficient compressive sensing using structurally random matrices. IEEE Trans Signal Proc 60(1):139–154
7. Dong X, Vuran MC (2011) A channel model for wireless underground sensor networks using lateral waves. In: 2011 IEEE global telecommunications conference-GLOBECOM 2011. IEEE, Piscataway, pp 1–6
8. Dong X, Vuran MC, Irmak S (2012) Autonomous precision agriculture through integration of wireless underground sensor networks with center pivot irrigation systems. Ad Hoc Netw 11:1975–1987
9. Erickson B, Widmar DA (2015) Precision agricultural services dealership survey results. Purdue University, West Lafayette, p 37
10. Foth HD (1978) Fundamentals of soil science. Soil Sci 125(4):272
11. Gebhardt M, Day C, Goering C, Bode L (1974) Automatic sprayer control system. Trans ASAE 17(6):1043–1047
12. Giacomin J, Vasconcelos F (2006) Wireless sensor network as a measurement tool in precision agriculture. In: In Proceedings of XVIII IMEKO world congress - metrology for a sustainable development, Rio de Janeiro
13. Godwin R, Richards T, Wood G, Welsh J, Knight S (2003) An economic analysis of the potential for precision farming in UK cereal production. Biosyst Eng 84(4):533–545
14. Hughes K, Frost A (1985) A review of agricultural spray metering. J Agric Eng Res 32(3):197–207
15. Irmak S, Rathje WR (2008) Plant growth and yield as affected by wet soil conditions due to flooding or over-irrigation. Technical Report G1904, University of Nebraska-Lincoln Extension NebGuide
16. Irmak S, Payero JO, Eisenhauer DE, Kranz WL, Martin D, Zoubek GL, Rees JM, VanDeWalle B, Christiansen AP, Leininger D (2006) Ec06-783 watermark granular matrix sensor to measure soil matric potential for irrigation management. Nebraska Extension Publications, Lincoln
17. Kitchen NR, Sudduth KA, Drummond ST, Scharf PC, Palm HL, Roberts DF, Vories ED (2010) Ground-based canopy reflectance sensing for variable-rate nitrogen corn fertilization. Agron J 102(1):71–84
18. Konda A, Rau A, Stoller MA, Taylor JM, Salam A, Pribil GA, Argyropoulos C, Morin SA (2018) Soft microreactors for the deposition of conductive metallic traces on planar, embossed, and curved surfaces. Adv Funct Mater 28(40):1803020. https://doi.org/10.1002/adfm.201803020
19. Luck J, Pitla S, Shearer S, Mueller T, Dillon C, Fulton J, Higgins S (2010) Potential for pesticide and nutrient savings via map-based automatic boom section control of spray nozzles. Comput Electron Agric 70(1):19–26

20. Phocaides A (2007) Handbook on pressurized irrigation techniques, 2nd edn. Food and Agriculture Organization of the United Nations, Rome
21. Powell J, Chandrakasan A (2004) Differential and single ended elliptical antennas for 3.1–10.6 GHz ultra wideband communication. In: Antennas and propagation society international symposium, Sendai, vol 2
22. Salam A (2018) Pulses in the sand: Long range and high data rate communication techniques for next generation wireless underground networks. ETD collection for University of Nebraska - Lincoln (AAI10826112). http://digitalcommons.unl.edu/dissertations/AAI10826112
23. Salam A (2019) A comparison of path loss variations in soil using planar and dipole antennas. In: 2019 IEEE international symposium on antennas and propagation, IEEE
24. Salam A (2019) Design of subsurface phased array antennas for digital agriculture applications. In: Proceedings of the 2019 IEEE international symposium on phased array systems and technology (IEEE Array 2019), Waltham
25. Salam A (2019) A path loss model for through the soil wireless communications in digital agriculture. In: 2019 IEEE international symposium on antennas and propagation. IEEE, Piscataway, pp 1–2
26. Salam A (2019) Sensor-free underground soil sensing. In: ASA, CSSA and SSSA international annual meetings (2019), ASA-CSSA-SSSA
27. Salam A (2019) Subsurface MIMO: a beamforming design in internet of underground things for digital agriculture applications. J Sens Actuator Netw 8(3):41. https://doi.org/10.3390/jsan8030041. https://www.mdpi.com/2224-2708/8/3/41
28. Salam A (2019) Underground environment aware MIMO design using transmit and receive beamforming in internet of underground things. Springer International Publishing, Cham, pp 1–15
29. Salam A (2019) An underground radio wave propagation prediction model for digital agriculture. Information 10(4):147
30. Salam A (2019) Underground soil sensing using subsurface radio wave propagation. In: 5th global workshop on proximal soil sensing, Columbia
31. Salam A (2020) Internet of things for environmental sustainability and climate change. Springer International Publishing, Cham, pp 33–69. https://doi.org/10.1007/978-3-030-35291-2_2. https://doi.org/10.1007/978-3-030-35291-2_2
32. Salam A (2020) Internet of things for sustainability: perspectives in privacy, cybersecurity, and future trends. Springer International Publishing, Cham, pp 299–327. https://doi.org/10.1007/978-3-030-35291-2_10. https://doi.org/10.1007/978-3-030-35291-2_10
33. Salam A (2020) Internet of things for sustainable community development, 1st edn. Springer Nature, Berlin. https://doi.org/10.1007/978-3-030-35291-2
34. Salam A (2020) Internet of things for sustainable community development: introduction and overview. Springer International Publishing, Cham, pp 1–31. https://doi.org/10.1007/978-3-030-35291-2_1. https://doi.org/10.1007/978-3-030-35291-2_1
35. Salam A (2020) Internet of things for sustainable forestry. Springer International Publishing, Cham, pp 147–181. https://doi.org/10.1007/978-3-030-35291-2_5. https://doi.org/10.1007/978-3-030-35291-2_5
36. Salam A (2020) Internet of things for sustainable human health. Springer International Publishing, Cham, pp 217–242. https://doi.org/10.1007/978-3-030-35291-2_7. https://doi.org/10.1007/978-3-030-35291-2_7
37. Salam A (2020) Internet of things for sustainable mining. Springer International Publishing, Cham, pp 243–271. https://doi.org/10.1007/978-3-030-35291-2_8. https://doi.org/10.1007/978-3-030-35291-2_8
38. Salam A (2020) Internet of things for water sustainability. Springer International Publishing, Cham, pp 113–145. https://doi.org/10.1007/978-3-030-35291-2_4. https://doi.org/10.1007/978-3-030-35291-2_4
39. Salam A (2020) Internet of things in agricultural innovation and security. Springer International Publishing, Cham, pp 71–112. https://doi.org/10.1007/978-3-030-35291-2_3. https://doi.org/10.1007/978-3-030-35291-2_3

40. Salam A (2020) Internet of things in sustainable energy systems. Springer International Publishing, Cham, pp 183–216. https://doi.org/10.1007/978-3-030-35291-2_6. https://doi.org/10.1007/978-3-030-35291-2_6
41. Salam A (2020) Internet of things in water management and treatment. Springer International Publishing, Cham, pp 273–298. https://doi.org/10.1007/978-3-030-35291-2_9. https://doi.org/10.1007/978-3-030-35291-2_9
42. Salam A (2020) Wireless underground communications in sewer and stormwater overflow monitoring: radio waves through soil and asphalt medium. Information 11(2):98
43. Salam A, Karabiyik U (2019) A cooperative overlay approach at the physical layer of cognitive radio for digital agriculture. In: Third international Balkan conference on communications and networking 2019 (BalkanCom'19)
44. Salam A, Shah S (2019) Internet of things in smart agriculture: enabling technologies. In: 2019 IEEE 5th world forum on internet of things (WF-IoT). IEEE, Piscataway, pp 692–695
45. Salam A, Vuran MC (2016) Impacts of soil type and moisture on the capacity of multi-carrier modulation in internet of underground things. In: Proceedings of the 25th ICCCN 2016, Waikoloa
46. Salam A, Vuran MC (2017) Em-based wireless underground sensor networks. In: Underground sensing. Academic, Cambridge, pp 247–285. https://doi.org/10.1016/B978-0-12-803139-1.00005-9
47. Salam A, Vuran MC (2017) Smart underground antenna arrays: a soil moisture adaptive beamforming approach. In: Proceedings of the IEEE INFOCOM 2017, Atlanta
48. Salam A, Vuran MC (2017) Wireless underground channel diversity reception with multiple antennas for internet of underground things. In: Proceedings of the IEEE ICC 2017, Paris
49. Salam A, Vuran MC, Irmak S (2016) Pulses in the sand: impulse response analysis of wireless underground channel. In: The 35th annual IEEE international conference on computer communications (INFOCOM 2016), San Francisco
50. Salam A, Vuran MC, Irmak S (2017) Towards internet of underground things in smart lighting: a statistical model of wireless underground channel. In: Proceedings of the 14th IEEE international conference on networking, sensing and control (IEEE ICNSC), Calabria
51. Salam A, Hoang AD, Meghna A, Martin DR, Guzman G, Yoon YH, Carlson J, Kramer J, Yansi K, Kelly M, et al (2019) The future of emerging IoT paradigms: architectures and technologies. https://doi.org/10.20944/preprints201912.0276.v1
52. Salam A, Vuran MC, Dong X, Argyropoulos C, Irmak S (2019) A theoretical model of underground dipole antennas for communications in internet of underground things. IEEE Trans Antenn Propag 67(6):3996-4009
53. Salam A, Vuran MC, Irmak S (2019) Di-sense: in situ real-time permittivity estimation and soil moisture sensing using wireless underground communications. Comput Netw 151:31–41. https://doi.org/10.1016/j.comnet.2019.01.001. http://www.sciencedirect.com/science/article/pii/S1389128618303141
54. Scharf PC, Shannon DK, Palm HL, Sudduth KA, Drummond ST, Kitchen NR, Mueller LJ, Hubbard VC, Oliveira LF (2011) Sensor-based nitrogen applications out-performed producer-chosen rates for corn in on-farm demonstrations. Agron J 103(6):1683–1691
55. Sha L, Gopalakrishnan S, Liu X, Wang Q (2008) Cyber-physical systems: a new frontier. In: Proceedings of the IEEE international conference on sensor networks, ubiquitous, and trustworthy computing (SUTC 2008), Taichung
56. Sharda A, Fulton JP, McDonald TP, Zech WC, Darr MJ, Brodbeck CJ (2010) Real-time pressure and flow dynamics due to boom section and individual nozzle control on agricultural sprayers. Trans ASABE 53(5):1363–1371
57. Silva AR, Vuran MC (2009) Empirical evaluation of wireless underground-to-underground communication in wireless underground sensor networks. In: Proceedings of the IEEE DCOSS '09, Marina Del Rey
58. Silva AR, Vuran MC (2010) Communication with aboveground devices in wireless underground sensor networks: an empirical study. In: 2010 IEEE international conference on communications, Cape Town

59. Silva AR, Vuran MC (2010) $(CPS)^2$: integration of center pivot systems with wireless underground sensor networks for autonomous precision agriculture. In: Proceedings of the ACM/IEEE international conference cyber-physical systems, Stockholm, pp 79–88. http://doi.acm.org/10.1145/1795194.1795206
60. Silva AR, Vuran MC (2010) Development of a testbed for wireless underground sensor networks. EURASIP J Wireless Comm Netw 2010:1–14 https://doi.org/10.1155/2010/620307
61. Sudduth HZSTD KA, Franzen AJ (2019) Variable-rate application technologies in precision agriculture, chap 7. Burleigh Dodds Science, Cambridge, pp 171–193
62. Sun Z, Akyildiz I (2011) Dynamic connectivity in wireless underground sensor networks. IEEE Trans Wireless Comm 10(12):4334–4344
63. Swinton SM (2005) Economics of site-specific weed management. Weed Sci 53(2):259–263
64. Temel S, Vuran MC, Lunar MM, Zhao Z, Salam A, Faller RK, Stolle C (2018) Vehicle-to-barrier communication during real-world vehicle crash tests. Comput Comm 127:172–186
65. Tiusanen MJ (2005) Attenuation of a soil scout radio signal. Biosyst Eng 90(2):127–133
66. Tiusanen MJ (2006) Wideband antenna for underground soil scout transmission. IEEE Anten Wireless Prop Lett 5:517–519
67. Tiusanen MJ (2008) Wireless soil scout prototype radio signal reception compared to the attenuation model. Precis Agric 10(5):372–381
68. Vidrine CG, Goering C, Day C, Gebhardt MR, Smith DB (1975) A constant pesticide application rate sprayer model. Trans ASAE 18(3):439–0443
69. Vuran MC, Akyildiz IF (2010) Channel model and analysis for wireless underground sensor networks in soil medium. Phys Comm 3(4):245–254
70. Vuran MC, Salam A, Wong R, Irmak S (2018) Internet of underground things in precision agriculture: architecture and technology aspects. Ad Hoc Netw 81:160–173. https://doi.org/10.1016/j.adhoc.2018.07.017. http://www.sciencedirect.com/science/article/pii/S1570870518305067
71. Vuran MC, Salam A, Wong R, Irmak S (2018) Internet of underground things: sensing and communications on the field for precision agriculture. In: 2018 IEEE 4th world forum on internet of things (WF-IoT) (WF-IoT 2018)

Index

A. Salam, U. Raza, *Signals in the Soil*,
https://doi.org/10.1007/978-3-030-50861-6

The manufacturer's authorised representative in the EU is Springer Nature Customer Service Centre GmbH, Europaplatz 3, 69115 Heidelberg, Germany. If you have any concerns regarding our products, please contact ProductSafety@springernature.com

Printed and bound by CPI Group (UK) Ltd, Croydon, CR0 4YY
30/09/2026
02225651-0014